The Traveling Salesman Problem

A Computational Study

David L. Applegate
Robert E. Bixby
Vašek Chvátal
William J. Cook

PRINCETON UNIVERSITY PRESS
PRINCETON AND OXFORD

Published by Princeton University Press, 41 William Street,
Princeton, New Jersey 08540
In the United Kingdom: Princeton University Press, 3 Market Place,
Woodstock, Oxfordshire OX20 1SY

Library of Congress Control Number: 2006931528

ISBN 978-0-691-12993-8

The publisher would like to acknowledge the authors of this volume for
providing the camera-ready from which this book was printed.

British Library Cataloging-in-Publication Data is available

Printed on acid-free paper.

press.princeton.edu

Printed in the United States of America

10 9 8 7 6 5 4 3 2 1

Bash on regardless.

J. P. Donleavy, *The Destinies of Darcy Dancer, Gentleman*

Contents

Preface

The traveling salesman problem has a long history of capturing researchers; we fell under its spell in January of 1988 and have yet to break free. The goal of this book is to set down the techniques that have led to the solution of a number of large instances of the problem, including the full set of examples from the TSPLIB challenge collection.

The early chapters of the book cover history and applications and should be accessible to a wide audience. This is followed by a detailed treatment of our solution approach.

In the summer of 1988, Bernhard Korte took three of us into his Institut für Diskrete Mathematik in Bonn, where we could get our project off the ground under ideal working conditions; his moral and material support has continued through the subsequent years. We want to thank him for his unswerving faith in us and for his hospitality.

We would also like to thank AT&T Labs, Bellcore, Concordia University, Georgia Tech, Princeton University, Rice University, and Rutgers University for providing homes for our research and for their generous computer support. Our work has been funded by grants from the National Science Foundation, the Natural Sciences and Engineering Research Council, and the Office of Naval Research.

Chapter One

The Problem

Given a set of cities along with the cost of travel between each pair of them, the *traveling salesman problem*, or *TSP* for short, is to find the cheapest way of visiting all the cities and returning to the starting point. The "way of visiting all the cities" is simply the order in which the cities are visited; the ordering is called a *tour* or *circuit* through the cities.

This modest-sounding exercise is in fact one of the most intensely investigated problems in computational mathematics. It has inspired studies by mathematicians, computer scientists, chemists, physicists, psychologists, and a host of non-professional researchers. Educators use the TSP to introduce discrete mathematics in elementary, middle, and high schools, as well as in universities and professional schools. The TSP has seen applications in the areas of logistics, genetics, manufacturing, telecommunications, and neuroscience, to name just a few.

The appeal of the TSP has lifted it to one of the few contemporary problems in mathematics to become part of the popular culture. Its snappy name has surely played a role, but the primary reason for the wide interest is the fact that this easily understood model still eludes a general solution. The simplicity of the TSP, coupled with its apparent intractability, makes it an ideal platform for developing ideas and techniques to attack computational problems in general.

Our primary concern in this book is to describe a method and computer code that have succeeded in solving a wide range of large-scale instances of the TSP. Along the way we cover the interplay of applied mathematics and increasingly more powerful computing platforms, using the solution of the TSP as a general model in computational science.

A companion to the book is the computer code itself, called *Concorde*. The theory and algorithms behind Concorde will be described in detail in the book, along with computational tests of the code. The software is freely available at

www.tsp.gatech.edu

together with supporting documentation. This is jumping ahead in our presentation, however. Before studying Concorde we take a look at the history of the TSP and discuss some of the factors driving the continued interest in solution methods for the problem.

1.1 TRAVELING SALESMAN

The origin of the name "traveling salesman problem" is a bit of a mystery. There does not appear to be any authoritative documentation pointing out the creator of the name, and we have no good guesses as to when it first came into use. One of the most influential early TSP researchers was Merrill Flood of Princeton University and the RAND Corporation. In an interview covering the Princeton mathematics community, Flood [183] made the following comment.

> Developments that started in the 1930s at Princeton have interesting consequences later. For example, Koopmans first became interested in the "48 States Problem" of Hassler Whitney when he was with me in the Princeton Surveys, as I tried to solve the problem in connection with the work of Bob Singleton and me on school bus routing for the State of West Virginia. I don't know who coined the peppier name "Traveling Salesman Problem" for Whitney's problem, but that name certainly caught on, and the problem has turned out to be of very fundamental importance.

This interview of Flood took place in 1984 with Albert Tucker posing the questions. Tucker himself was on the scene of the early TSP work at Princeton, and he made the following comment in a 1983 letter to David Shmoys [527].

> The name of the TSP is clearly colloquial American. It may have been invented by Whitney. I have no alternative suggestion.

Except for small variations in spelling and punctuation, "traveling" versus "travelling," "salesman" versus "salesman's," etc., by the mid-1950s the TSP name was in wide use. The first reference containing the term appears to be the 1949 report by Julia Robinson, "On the Hamiltonian game (a traveling salesman problem)" [483], but it seems clear from the writing that she was not introducing the name. All we can conclude is that sometime during the 1930s or 1940s, most likely at Princeton, the TSP took on its name, and mathematicians began to study the problem in earnest.

Although we cannot identify the originator of the TSP name, it is easy to make an argument that it is a fitting identifier for the problem of finding the shortest route through cities in a given region. The traveling salesman has long captured our imagination, being a leading figure in stories, books, plays, and songs. A beautiful historical account of the growth and influence of traveling salesmen can be found in Timothy Spears' book *100 Years on the Road: The Traveling Salesman in American Culture* [506]. Spears cites an 1883 estimate by *Commercial Travelers Magazine* of 200,000 traveling salesmen working in the United States and a further estimate of 350,000 by the turn of the century. This number continued to grow through the early 1900s, and at the time of the Princeton research the salesman was a familiar site in most American towns and villages.

THE 1832 HANDBOOK BY THE ALTEN COMMIS-VOYAGEUR

The numerous salesmen on the road were indeed interested in the planning of economical routes through their customer areas. An important reference in this context is the 1832 German handbook *Der Handlungsreisende—wie er sein soll und was er zu thun hat, um Aufträge zu erhalten und eines glücklichen Erfolgs in seinen*

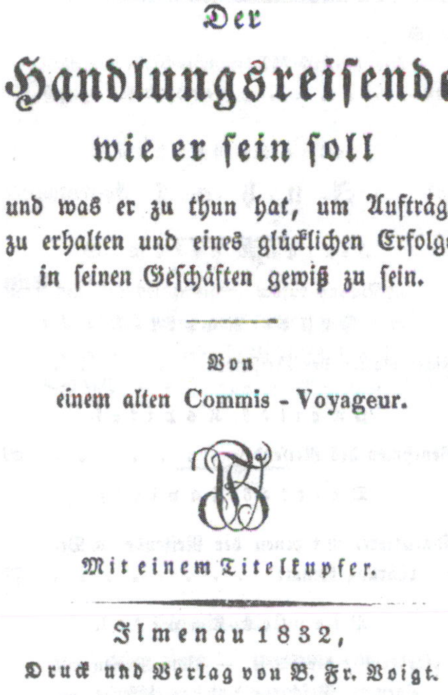

Der

Handlungsreifende

wie er fein foll

und was er zu thun hat, um Aufträge
zu erhalten und eines glücklichen Erfolgs
in feinen Geschäften gewiß zu fein.

Von
einem alten Commis - Voyageur.

Mit einem Titelkupfer.

Ilmenau 1832,
Druck und Verlag von B. Fr. Voigt.

Figure 1.1 1832 German traveling salesman handbook.

Geschäften gewiss zu sein—Von einem alten Commis-Voyageur, first brought to the
attention of the TSP research community in 1983 by Heiner Müller-Merbach [410].
The title page of this small book is shown in Figure 1.1.

The Commis-Voyageur [132] explicitly described the need for good tours in the
following passage, translated from the German original by Linda Cook.

> Business leads the traveling salesman here and there, and there is not a
> good tour for all occurring cases; but through an expedient choice and
> division of the tour so much time can be won that we feel compelled to
> give guidelines about this. Everyone should use as much of the advice
> as he thinks useful for his application. We believe we can ensure as
> much that it will not be possible to plan the tours through Germany
> in consideration of the distances and the traveling back and fourth,
> which deserves the traveler's special attention, with more economy.
> The main thing to remember is always to visit as many localities as
> possible without having to touch them twice.

This is an explicit description of the TSP, made by a traveling salesman himself!

The book includes five routes through regions of Germany and Switzerland. Four
of these routes include return visits to an earlier city that serves as a base for that
part of the trip. The fifth route, however, is indeed a traveling salesman tour, as
described in Alexander Schrijver's [495] book on the field of combinatorial opti-

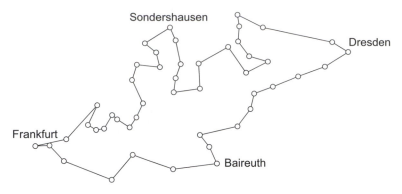

Figure 1.2 The Commis-Voyageur tour.

mization. An illustration of the tour is given in Figure 1.2. The cities, in tour order, are listed in Table 1.1, and a picture locating the tour within Germany is given in Figure 1.3.   One can see from the drawings that the tour is of very good quality,

Table 1.1 A 47-city tour from the Commis-Voyageur.

1. Frankfurt	17. Meißen	33. Mühlhausen
2. Hanau	18. Leipzig	34. Langensalza
3. Aschaffenburg	19. Halle	35. Gotha
4. Würzburg	20. Merseburg	36. Eisenach
5. Schweinfurt	21. Weißenfels	37. Salzungen
6. Bamberg	22. Zeitz	38. Meiningen
7. Baireuth	23. Altenburg	39. Möllrichstadt
8. Kulmbach	24. Gera	40. Neustadt
9. Kronach	25. Naumburg	41. Bischofsheim
10. Hof	26. Weimar	42. Gersfeld
11. Plauen	27. Rudolstadt	43. Brückenau
12. Greiz	28. Ilmenau	44. Zunderbach
13. Zwickau	29. Arnstadt	45. Schlichtern
14. Chemnitz	30. Erfurt	46. Fulda
15. Freiberg	31. Greußen	47. Gelnhausen
16. Dresden	32. Sondershausen	

and Schrijver [495] comments that it may in fact be optimal, given the local travel conditions at that time.

    The Commis-Voyageur was not alone in considering carefully planned tours. Spears [506] and Friedman [196] describe how salesmen in the late 1800s used guidebooks, such as L. P. Brockett's [95] *Commercial Traveller's Guide Book*, to

Figure 1.3 The Commis-Voyageur tour in Germany.

map out routes through their regions. The board game *Commercial Traveller* created by McLoughlin Brothers in 1890 emphasized this point, asking players to build their own tours through an indicated rail system. Historian Pamela Walker Laird kindly provided the photograph of the McLoughlin Brothers' game that is displayed in Figure 1.4.

The mode of travel used by salesmen varied over the years, from horseback and stagecoach to trains and automobiles. In each of these cases, the planning of routes would often take into consideration factors other than simply the distance between the cities, but devising good TSP tours was a regular practice for the salesman on the road.

1.2 OTHER TRAVELERS

Although traveling salesmen are no longer a common sight, the many flavors of the TSP have a good chance of catching some aspect of the everyday experience of most people. The usual errand run around town is a TSP on a small scale, and longer trips taken by bus drivers, delivery vans, and traveling tourists often involve a TSP through modest numbers of locations. For the many non-salesmen of the world, these natural connections to tour finding add to the interest of the TSP as a subject of study.

Spears [506] makes a strong case for the prominence of the traveling salesman in

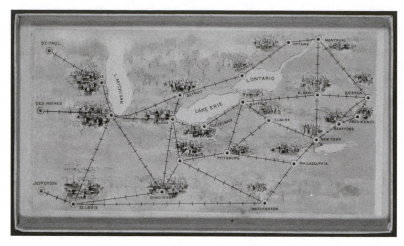

Figure 1.4 The game of *Commercial Traveller*. Image courtesy of Pamela Walker Laird.

recent history, but a number of other tour finders could rightly lay claim to the TSP moniker, and we discuss below some of these alternative salesmen. The goal here is to establish a basis to argue that the TSP is a naturally occurring mathematical problem by showing a wide range of originating examples.

CIRCUIT RIDERS

In its coverage of the historical usage of the word "circuit," the *Oxford English Dictionary* [443] cites examples as far back as the fifteenth century, concerning the formation of judicial districts in the United Kingdom. During this time traveling judges and lawyers served the districts by riding a circuit of the principal population centers, where court was held during specified times of the year. This practice was later adopted in the United States, where regional courts are still referred to as circuit courts, even though traveling is no longer part of their mission.

The best-known circuit-riding lawyer in the history of the United States is the young Abraham Lincoln, who practiced law before becoming the country's sixteenth president. Lincoln worked in the Eighth Judicial Circuit in the state of Illinois, covering 14 county courthouses. His travel is described by Guy Fraker [194] in the following passage.

> Each spring and fall, court was held in consecutive weeks in each of the 14 counties, a week or less in each. The exception was Springfield, the state capital and the seat of Sangamon County. The fall term opened there for a period of two weeks. Then the lawyers traveled the fifty-five miles to Pekin, which replaced Tremont as the Tazewell County seat in 1850. After a week, they traveled the thirty-five miles to Metamora, where they spent three days. The next stop, thirty miles to the southeast, was Bloomington, the second-largest town in the circuit. Because of its size, it would generate more business, so they

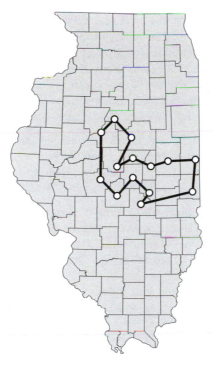

Figure 1.5 Eighth Judicial Circuit traveled by Lincoln in 1850.

would probably stay there several days longer. From there they would travel to Mt. Pulaski, seat of Logan County, a distance of thirty-five miles; it had replaced Postville as county seat in 1848 and would soon lose out to the new city of Lincoln, to be named for one of the men in this entourage. The travelers would then continue to another county and then another and another until they had completed the entire circuit, taking a total of eleven weeks and traveling a distance of more than four hundred miles.

Fraker writes that Lincoln was one of the few court officials who regularly rode the entire circuit. A drawing of the route used by Lincoln and company in 1850 is given in Figure 1.5. Although the tour is not a shortest possible one (at least as the crow flies), it is clear that it was constructed with an eye toward minimizing the travel of the court personnel. The quality of Lincoln's tour was pointed out several years ago in a TSP article by Jon Bentley [57].

Lincoln has drawn much attention to circuit-riding judges and lawyers, but as a group they are rivaled in fame by the circuit-riding Christian preachers of the eighteenth and nineteenth centuries. John Hampson [248] wrote the following passage in his 1791 biography of John Wesley, the founder of the Methodist church.

> Every part of Britain and America is divided into regular portions, called circuits; and each circuit, containing twenty or thirty places, is supplied by a certain number of travelling preachers, from two to three or four, who go around it in a month or six weeks.

The difficult conditions under which these men traveled is part of the folklore in Britain, Canada, and the United States. An illustration by Alfred R. Waud [550] of a traveling preacher is given in Figure 1.6. This drawing appeared on the cover of *Harper's Weekly* in 1867; it depicts a scene that appears in many other pictures and sketches from that period.

If mathematicians had begun their study of the TSP some hundred years earlier, it may well have been that circuit-riding lawyers or preachers would be the users that gave the problem its name.

KNIGHT'S TOUR

One of the first appearances of tours and circuits in the mathematical literature is in a 1757 paper by the great Leonhard Euler. The Euler Archive [169] cites an estimate by historian Clifford Truesdell that "in a listing of all of the mathematics, physics, mechanics, astronomy, and navigation work produced during the 18th Century, a full 25% would have been written by Leonhard Euler." The particular paper we mention concerns a solution of the *knight's tour* problem in chess, that is, the problem of finding a sequence of knight's moves that will take the piece from a starting square on a chessboard, through every other square exactly once and returning to the start. Euler's solution is depicted in Figure 1.7, where the order of moves is indicated by the numbers on the squares.

The chess historian Harold J. Murray [413] reports that variants of the knight's tour problem were considered as far back as the ninth century in the Arabic liter-

Figure 1.6 Traveling preacher from *Harper's Weekly*, October 12, 1867.

42	57	44	9	40	21	46	7
55	10	41	58	45	8	39	20
12	43	56	61	22	59	6	47
63	54	11	30	25	28	19	38
32	13	62	27	60	23	48	5
53	64	31	24	29	26	37	18
14	33	2	51	16	35	4	49
1	52	15	34	3	50	17	36

Figure 1.7 Knight's tour found by Euler.

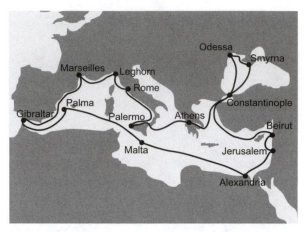

Figure 1.8 Mark Twain's tour in *Innocents Abroad*.

ature. Despite the 1,200 years of work, the problem continues to attract attention today. Computer scientist Donald Knuth [319] is one of the many people who have recently considered touring knights, including the study of generalized knights that can leap x squares up and y squares over, and the design of a font for typesetting tours. An excellent survey of the numerous current attacks on the problem can be found on George Jellis' web page Knight's Tour Notes [285].

The knight's problem can be formulated as a TSP by specifying the cost of travel between squares that can be reached via a legal knight's move as 0 and the cost of travel between any two other squares as 1. The challenge is then to find a tour of cost 0 through the 64 squares. Through this view the knight's problem can be seen as a precursor to the TSP.

THE GRAND TOUR

Tourists could easily argue for top billing in the TSP; traveling through a region in a limited amount of time has been their specialty for centuries. In the 1700s taking a Grand Tour of Europe was a rite of passage for the British upper class [262], and Thomas Cook brought touring to the masses with his low-cost excursions in the mid-1800s [464]. The *Oxford English Dictionary* [443] defines *Cook's tour* as "a tour, esp. one in which many places are viewed," and Cook is often credited with the founding of the modern tourism industry.

Mark Twain's *The Innocents Abroad* [528] gives an account of the author's passage on a Grand Tour organized by a steamship firm; this collection of stories was Twain's best-selling book during his lifetime. His tour included stops in Paris, Venice, Florence, Athens, Odessa, Smyrna, Jerusalem, and Malta, in a route that appears to minimize the travel time. A rough sketch of Twain's tour is given in Figure 1.8.

MESSENGERS

In the mathematics literature, it appears that the first mention of the TSP was made by Karl Menger, who described a variant of the problem in notes from a mathematics colloquium held in Vienna on February 5, 1930 [389]. A rough translation of Menger's problem from the German original is the following.

> We use the term *Botenproblem* (because this question is faced in practice by every postman, and by the way also by many travelers) for the task, given a finite number of points with known pairwise distances, to find the shortest path connecting the points.

So the problem is to find only a path through the points, without a return trip to the starting point. This version is easily converted to a TSP by adding an additional point having travel distance 0 to each of the original points.

Bote is the German word for messenger, so with Menger's early proposal a case could be made for the use of the name *messenger problem* in place of TSP.

It is interesting that Menger mentions postmen in connection with the TSP, since in current mathematics terminology "postman problems" refers to another class of routing tasks, where the target is to traverse each of a specified set of roads rather than the cities joined by the roads. The motivation in this setting is that mail must be brought to houses that lie along the roads, so a full route should bring a postman along each road in his or her region. Early work on this topic was carried out by the Chinese mathematician Mei Gu Guan [240], who considered the version where the roads to be traversed are connected in the sense that it is possible to complete the travel without ever leaving the specified collection of roads. This special case was later dubbed the *Chinese postman problem* by Jack Edmonds in reference to Guan's work. Remarkably, Edmonds [165] showed that the Chinese postman problem can always be solved efficiently (in a technical sense we discuss later in the chapter), whereas no such method appears on the horizon for the TSP itself.

In many settings, of course, a messenger or postman need not visit every house in a region, and in these cases the problem comes back to the TSP. In a recent publication, for example, Graham-Rowe [226] cites the use of TSP software to reduce the travel time of postmen in Denmark.

FARMLAND SURVEYS

Two of the earliest papers containing mathematical results concerning the TSP are by Eli S. Marks [375] and M. N. Ghosh [204], appearing in the late 1940s. The title of each of their papers includes the word "travel," but their research was inspired by work concerning a traveling farmer rather than a traveling salesman. The statistician P. C. Mahalanobis described the original application in a 1940 research paper [368]. The work of Mahalanobis focused on the collection of data to make accurate forecasts of the jute crop in Bengal. A major source of revenue in India during the 1930s was derived from the export of raw and manufactured jute, accounting for roughly 1/4 of total exports during 1936–37. Furthermore, nearly 85% of India's jute was grown in the Bengal region.

A complete survey of the land in Bengal used in jute production was impractical, owing to the fact that it was grown on roughly 6 million small farms, scattered among some 80 million lots of all kinds. Mahalanobis proposed instead to make a random sample survey by dividing the country into zones comprising land of similar characteristics and within each zone select a random number of points to inspect for jute cultivation. One of the major components in the cost of making the survey was described by Mahalanobis as follows.

> *Journey*, which includes the time spent in going from camp to camp (camp being defined as a place where the night is spent); from camp to a sample unit; and from sample unit to camp. This therefore covers all journeys undertaken for the purpose of carrying out the field enumeration.

This is the TSP aspect of the application, to find efficient routes for the workers and equipment between the selected sites in the field.

It is interesting that, perhaps owing to his research interests as a statistician, Mahalanobis did not discuss the problem of finding tours for specific data but rather the problem of making statistical estimates of the expected length of an optimal tour for a random selection of locations. These estimates were included in projected costs of carrying out the survey, and thus were an important consideration in the decision to implement Mahalanobis' plan in a small test in 1937 and in a large survey in 1938.

A similar "traveling farmer" problem was considered in the United States by Raymond J. Jessen, who wrote a report in 1942 on sampling farming regions in Iowa. Jessen [286] cites Mahalanobis' results, but notes an important difference.

> This relationship is based upon the assumption that points are connected by direct routes. In Iowa the road system is a quite regular network of mile square mesh. There are very few diagonal roads, therefore, routes between points resemble those taken on a checkerboard.

The paper includes an empirical study of particular point sets under this checkerboard distance function. We will see this type of travel cost again in Section 2.3 when we discuss applications of the TSP in the design of computer chips.

SCHOOL BUS ROUTING

We mentioned earlier a quote from Merrill Flood's 1984 interview with Albert Tucker, citing a school bus routing problem as a source of his early interest in the TSP. Flood also made the following statement in a 1956 paper [182].

> I am indebted to A. W. Tucker for calling these connections to my attention, in 1937, when I was struggling with the problem in connection with a school-bus routing study in New Jersey.

Note that Flood mentions New Jersey here and West Virginia in the earlier quote; it may well have been that he was involved in applications in several districts. Given the great importance of Flood's contributions to the development of the TSP in the

1940s and 1950s, the "school bus problem" would certainly have been a suitable alternative to the traveling salesman.

A van routing application did inspire one early research group to at least temporarily adopt the term "laundry van problem" in place of TSP, as described in a 1955 paper by G. Morton and A. H. Land [403].

> In the United States this problem is known as the Travelling-salesman problem; the salesman wishes to visit one city in each of the 48 States and Washington, D. C., in such a sequence as to minimize total road distance travelled. Thinking of the problem independently and on a smaller geographical scale, we used to call it the laundry van problem, where the conditions were a daily service by a one-van laundry. Since the American term was used by Robinson (1949) we propose to adopt it here.

The Morton-Land name is clearly less "peppy" than the traveling salesman, but van routing is indeed an important application of the TSP.

WATCHMEN

In the early 1960s, plotting a tour through the United States was in the public eye owing to an advertising campaign featuring a 33-city TSP. The original ad from Procter & Gamble is reproduced in Figure 1.9. The $10,000 prize for the shortest tour was huge at the time, enough to purchase a new house in many parts of the country. The proposed drivers of the tour were Toody and Muldoon, the police heroes of the *Car 54, Where Are You?* television series. Although the duty of these two officers did not actually go beyond the boundaries of New York's Bronx borough, police and night watchmen must often walk or drive tours in their daily work.

The large prize money in the Procter & Gamble contest caught the attention of applied mathematicians; several papers on finding TSP tours cite this particular example. In fact, one of the winners of the contest was the TSP researcher Gerald Thompson of Carnegie Mellon University. Thompson coauthored a paper [304] with R. L. Karg in 1964 that reports the optimal tour displayed in Figure 1.10. It should be noted that Thompson did not know for sure that he had the shortest tour; there is a difference between finding a good tour and producing a convincing argument that there is no tour of lesser cost. We discuss this point further in Chapters 3 and 4.

In a recent conversation Professor Thompson pointed out that a number of contestants, including himself, had submitted the identical tour. The tiebreaker for the contest asked for a short essay on one of Procter & Gamble's products, and we are happy to report that Thompson came through as a prize winner.

BOOK TOURS

Authors have a long history of being sent on far-reaching tours in promotion of their most recent works. Manil Suri, the author of the novel *The Death of Vishnu* [515], made the following remark in an interview in *SIAM News* in 2001 [516].

Figure 1.9 The 33-city contest. Image courtesy of Procter & Gamble.

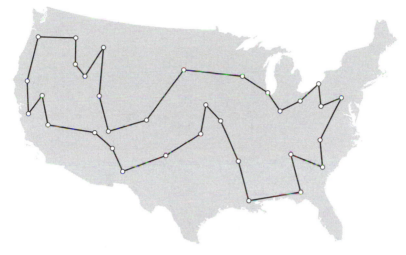

Figure 1.10 Optimal 33-city tour.

The initial U.S. book tour, which starts January 24, 2001, will cover 13 cities in three weeks. When my publisher gave me the list of cities, I realized something amazing. I was actually going to live the Traveling Salesman Problem! I tried conveying my excitement to the publicity department, tried explaining to them the mathematical significance of all this, and how we could perhaps come up with an optimal solution, etc., etc. They were quite uneasy about my enthusiasm and assured me that they had lots of experience in planning itineraries, and would get back to me if they required mathematical assistance. So far, they haven't.

Despite the reluctance of Suri's publishers, book touring is another natural setting for the TSP and one that we see in action quite often.

To get a feeling for the scope of tours made by authors, even as far back as the 1800s, consider the itinerary for a lecture tour made by Mark Twain in the winter of 1884–85, shown in Figure 1.11. The list in the figure describes only a portion of the travels of Twain and fellow writer George Washington Cable, who visited over 60 cities in their four-month tour of the United States and Canada [35].

1.3 GEOMETRY

When investigating a problem for solution ideas, most mathematicians will get out a pencil and draw a few figures to get things rolling. Geometric instances of the TSP, where cities are locations and the travel costs are distances between pairs, are tailor-made for such probing. In a single glance at a drawing of a tour we get a good measure of its quality.

For example, in Figure 1.12 we give a closer view of the tour traveled by Abra-

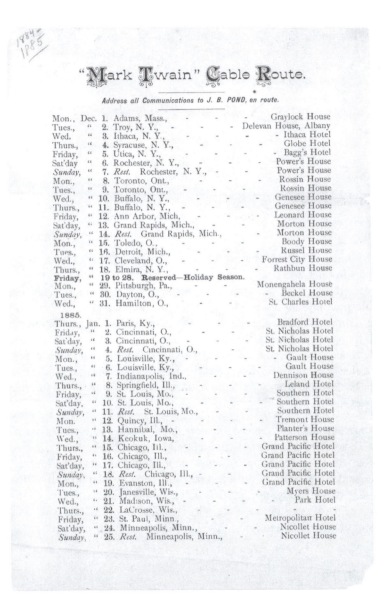

Figure 1.11 Mark Twain's lecture tour in 1884–85. Image courtesy of the Mark Twain
Project, Bancroft Library, University of California, Berkeley.

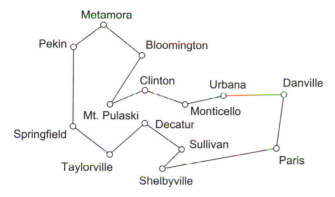

Figure 1.12 Lincoln tour.

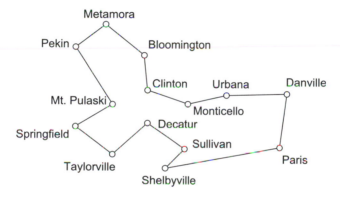

Figure 1.13 Improved Lincoln tour.

ham Lincoln. Although it is clear that the route is quite good, the geometry indicates some likely points of improvement where the tour forms acute angles around Mt. Pulaski and Shelbyville. The first angle suggests the obvious improvement of picking up Mt. Pulaski on the way from Springfield to Pekin. This gives the adjusted tour indicated in Figure 1.13. The remaining acute angle around Shelbyville suggests the following two-step improvement: insert Decatur between Clinton and Monticello, then visit Sullivan before Shelbyville, after leaving Paris. This change gives the tour in Figure 1.14, which is in fact the optimal tour.

   This ability to visualize tours and to easily manipulate them by hand has certainly contributed to the widespread appeal of the problem, making the study of the TSP accessible to anyone with a pencil and a clean sheet of paper. In our discussion below we cover a number of other ways in which geometry has helped to shape the growth and interest in the TSP.

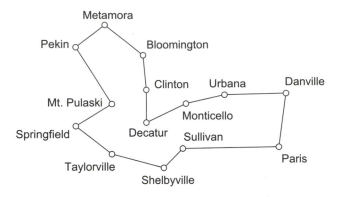

Figure 1.14 Optimal Lincoln tour.

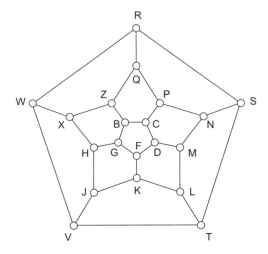

Figure 1.15 The Icosian.

HAMILTON'S ICOSIAN

In mathematical circles a TSP tour is often referred to as a *Hamiltonian circuit* in honor of Sir William Rowan Hamilton, one of the greatest mathematicians of the nineteenth century. A comprehensive study of Hamilton's life and works can be found on David Wilken's web site [554].

Hamilton's connection with the history of the TSP has roots in classic geometry, arising in a study of the dodecahedron, the 12-sided Platonic solid. Inspired by a mix of geometry and algebra, Hamilton considered the structure of tours through the 20 corner points of this geometric object.

In his tours, Hamilton permitted travel only along the dodecahedron's geometric edges. As an aid, he made use of an abstract drawing of the edges and corner points given in Figure 1.15. The lines in the drawing represent the edges and the circles represent the corners. This drawing was dubbed the *Icosian* owing to a connection with the 20 faces of the icosahedron. Hamilton showed that tours in the Icosian are both plentiful and flexible in that no matter what subpath of five points is chosen to start the path, it is always possible to complete the tour through the remaining 15 points and end up adjacent to the originating point. Fascinated with the elegance of this structure, Hamilton described his work as a game in a letter from 1856 [246].

> I have found that some young persons have been much amused by trying a new mathematical game which the Icosian furnishes, one person sticking five pins in any five consecutive points, such as *abcde*, or *abcde'*, and the other player then aiming to insert, which by the theory in this letter can always be done, fifteen other pins, in cyclincal succession, so as to cover all the other points, and to end in immediate proximity to the pin wherewith his antagonist had begun.

In the following year, Hamilton presented his work on the Icosian at a scientific meeting held in Dublin. The report from the meeting [94] included the following passage.

> The author stated that this calculus was entirely distinct from that of quaternions, and in it none of the roots concerned were imaginary. He then explained the leading features of the new calculus, and exemplified its use by an amusing game, which he called the Icosian, and which he had been led to invent by it, — a lithograph of which he distributed through the Section, and examples of what the game proposed to be accomplished were lithographed in the margin, the solutions being shown to be exemplifications of the calculus. The figure was the projection on a plane of the regular pentagonal dodecahedron, and at each of the angles were holes for receiving the ivory pins with which the game was played.

Two versions of the Icosian game were actually marketed several years later by a British manufacturer of toys. One variant was a wooden board where pegs could be inserted into holes to mark the visited points; an image of this version of the game appeared on the cover of the TSP book by Lawler et al. [338]. The second variant was a handheld device shaped as a partially flattened dodecahedron, with pegs for the points and a string to trace out the tour. This handheld version is called *The Traveller's Dodecahedron: A Voyage Round the World*, with the 20 points labeled by letters representing interesting cities around the globe. A photograph of the Traveller's Dodecahedron is given in Figure 1.16 and an image of a document that was included in the packaging of the game is given in Figure 1.17. These two figures are courtesy of James Dalgety, whose Puzzle Museum obtained an example of the game in a 2002 auction.

Although the Traveller's Dodecahedron does not include the notion of travel costs, its general description of finding a route through a set of cities is certainly very near to the TSP itself.

THE GEOMETRIC TSP CONSTANT β

Randomly generated geometric TSP instances have received special attention from researchers in mathematics and statistics. In this class of problems, each city corre-

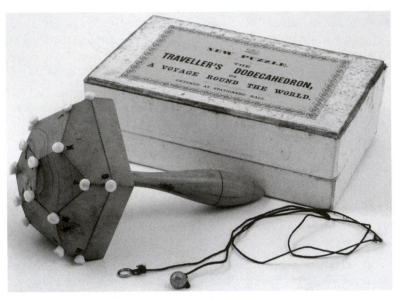

Figure 1.16 The Traveller's Dodecahedron handheld game. Copyright ©2006 Hordern-Dalgety Collection. http://puzzlemuseum.com.

sponds to a point chosen independently and uniformly from the unit square, that is, each point (x, y) with both x and y between 0 and 1 is equally likely to be selected as a city. The cost of travel between cities is measured by the usual Euclidean distance. These instances were targeted for study by Mahalanobis [368] in connection with his work on farmland surveys.

Although it may seem surprising, instances from this class of random problems are actually quite similar to one another. For example, consider the pair of 1,000-city tours shown in Figures 1.18 and 1.19. The displayed tours are optimal for the problems obtained by scaling the travel costs by 1,000,000 and rounding to the nearest integer, to avoid the difficulty of working with square roots when comparing distances. Now, at first glance, it is not so easy to see the difference between the two pictures.

The tours in the figures have lengths approximately 23.269 and 23.041 respectively. The proximity of these values is not a coincidence. In Table 1.2 we list the lengths for 10 random 1,000-city tours; a quick inspection shows that they are all relatively close in value. The problem that intrigued Mahalanobis [368] and many others is to determine what can be said in general about the distribution of tour lengths for random geometric problems on n cities.

Clearly, the length of tours we expect to obtain will increase as n gets larger. Mahalanobis [368] makes an argument that the lengths should grow roughly in proportion to $\sqrt{n}$, and a formal proof of this was given by Marks [375] and Ghosh [204]. This pair of researchers attacked the problem from two sides, Ghosh showing that the expected length of an optimal tour is no more than $1.27\sqrt{n}$ and Marks showing that the expected length is at least $(\sqrt{n} - 1/\sqrt{n})/\sqrt{2}$.

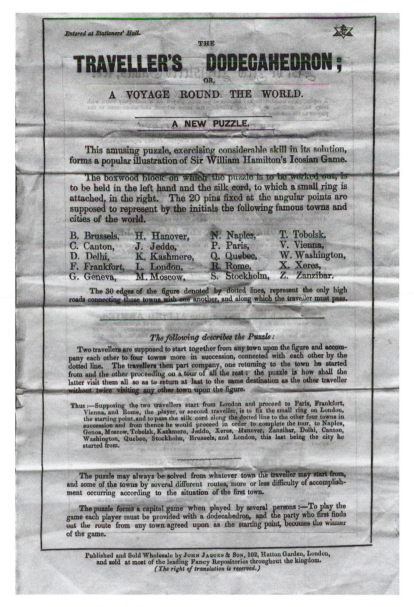

Figure 1.17 The Traveller's Dodecahedron. Copyright ©2004 Hordern-Dalgety Collection. http://puzzlemuseum.com.

Table 1.2 Tour lengths for 10 random 1,000-city instances (seeds $= 90, 91, \ldots, 99$).

23.215	23.297	23.249	23.277	23.367
23.059	23.100	22.999	23.269	23.041

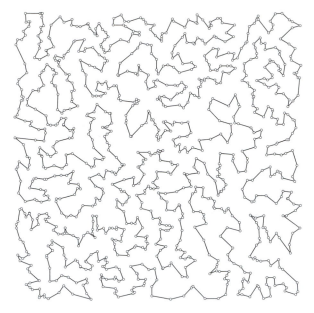

Figure 1.18 Optimal tour through 1,000 random points (seed = 98).

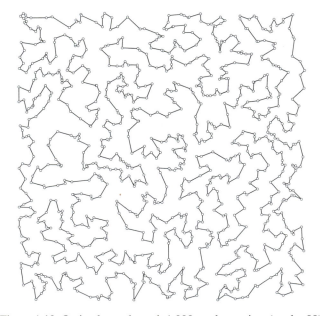

Figure 1.19 Optimal tour through 1,000 random points (seed = 99).

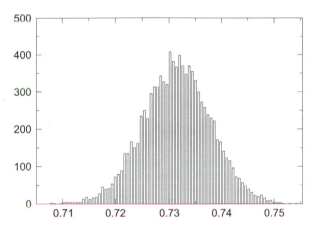

Figure 1.20 Distribution of tour lengths for 10,000 random geometric 1,000-city instances.

The work of Marks and Ghosh, carried out in the late 1940s, led to a famous result of Beardwood, Halton, and Hammersley [45] published in 1959. This result states that with probability 1, as n approaches infinity the optimal tour length divided by $\sqrt{n}$ will approach a constant value β. Some feeling for this result can be obtained by examining the histogram given in Figure 1.20, displaying the tour lengths divided by $\sqrt{1,000}$ for 10,000 random geometric instances, each with 1,000 cities. The tours, computed with Concorde, are again optimal for the problems obtained by scaling and rounding to the nearest integer. With only 1,000 cities there is still some variance in the tour values, but the results form a nice bell curve around the mean 0.7313. The Beardwood-Halton-Hammersley result implies that as n gets larger, the distribution of tour lengths will spike around the constant β. Estimates for β have been made through both analytical and experimental studies, but the actual value is not yet known. We will come back to this point in our computational study of Concorde later in the book.

The remarkable theorem of Beardwood-Halton-Hammersley has received considerable attention in the research community. In probability theory, the collection of techniques used to prove the result and various strengthenings have grown into an important subfield, as described in the survey works by Karp and Steele [311] and Steele [509]. In physics, the study of random geometric TSP instances is used as an approach to understanding more complicated stochastic physical models, as described in Percus and Martin [459], Cerf et al. [112], and Jacobsen et al. [280]. In operations research and computer science, the Beardwood-Halton-Hammersley constant has been a target of empirical studies, leading to general work on estimating the quality of heuristic algorithms.

JULIAN LETHBRIDGE PAINTING

An easy argument shows that in a geometric TSP, any tour that crosses itself can be shortened by replacing a pair of crossing edges, where an "edge" is a tour segment going directly from one city to another. Indeed, suppose the edges between cities

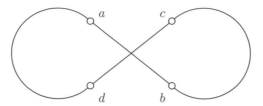

Figure 1.21 A crossing in a geometric tour.

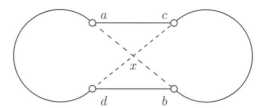

Figure 1.22 Uncrossing the tour segments.

a and b and between cities c and d cross as in Figure 1.21. Whenever we remove two edges from a tour we are left with two paths through the entire set of cities, and these paths can be rejoined by adding a new pair of edges. In the tour indicated in Figure 1.21, after deleting the crossing pair (a, b) and (c, d), we can reform the tour by adding the new pair (a, c) and (b, d), as shown in Figure 1.22. To argue that this operation shortens the tour, let x denote the point where the edges (a, b) and (c, d) cross. The edge (a, c) is shorter than the sum of the lengths of (a, x) and (x, c), while (b, d) is shorter than the sum of (b, x) and (x, d). Since the sum of the lengths of edges (a, b) and (c, d) is the same as the sum of the lengths of the four segments (a, x), (x, b), (c, x), and (x, d), we have in fact improved the tour by uncrossing the edges.

Flood [182] pointed out this uncrossing operation and used it as motivation for a general approach to improving tours by replacing pairs of edges whenever it is of direct advantage to do so, whether the pair actually crosses or not. We will have much more to say about this general algorithm later in the book.

For geometric TSP instances, the noncrossing rule means that a good tour gives a closed curve that divides the plane into "inside" and "outside" regions. The artist Julian Lethbridge commented that he was struck by the economy of thought and space given by these curves for good-quality tours for the TSP. He illustrated this in a beautiful painting titled *Traveling Salesman*, displayed in Figure 1.23. In the painting Lethbridge uses different textures for the two regions of space created by the tour.

CONTROL ZONES AND MOATS

Michael Jünger and William Pulleyblank [295] use geometry to make valid statements of the form "No tour through this set of points can have length less than 100" where 100 is replaced by a suitable length B for a given TSP instance. Such

Figure 1.23 *Traveling Salesman*, by Julian Lethbridge. Image courtesy of Julian Lethbridge and United Limited Art Editions.

a number B is called a *lower bound* for the TSP since it bounds from below the length of all tours.

The importance of lower bounds is that they can be used to certify that a tour we may have found is good, or even best, when compared to all of the other tours that we have not considered. A more direct approach would of course be to simply consider all possible tours, but this number grows so quickly that checking all of them for a modest-size instance, say 50 cities, is well beyond the capabilities of even the fastest of today's supercomputers. Carefully constructed lower bounds provide a more subtle way to make conclusions about the lengths of all tours through a set of points.

There are a number of different strategies for finding good lower bounds for the TSP. In Concorde and in other modern solvers, the bounding technique of choice goes back to George Dantzig, Ray Fulkerson, and Selmer Johnson [151], who in 1954 had the first major success in the solution of the TSP. For geometric problems, Jünger and Pulleyblank have shown that some of the basic ideas employed by this research team have a simple geometric interpretation, as we now describe.

Suppose our TSP consists of a set of points and that the cost of traveling between two cities is the Euclidean distance between the corresponding points. The first step in the geometric lower-bound procedure is to draw a disk of radius r centered at city 1 in such a way that the disk does not touch any of the other cities. Jünger and Pulleyblank call such a disk a *control zone*; see Figure 1.24.

Figure 1.24 A control zone.

The salesman must at some point in his or her tour visit city 1, and to do so they will need to travel at least distance r to arrive at the city and at least distance r to leave the city, since each other city is at least distance r from city 1. We can conclude that every tour has length at least $2r$, that is, we can set $B = 2r$ and have a lower bound for this TSP instance.

When judging a lower bound, bigger is better. Unfortunately, for almost any TSP you can think of, the bound $2r$ will be rather small when compared to the length of any tour through the points. To counter this objection, we can draw a separate disk for each of the cities, as long as the disks do not overlap. In this way we get twice the sum of the radii of the disks as a lower bound for the TSP.

For each city i let us denote by r_i the radius of its disk. If we have five cities, as in Figure 1.25, then by specifying values for each of the five radii $r_1, \ldots, r_5$ we can obtain a lower bound. Since we want this bound to be as large as possible, we would like to choose the five radii so as to maximize twice their sum, subject to the condition that the disks do not overlap.

The nonoverlapping condition can be expressed succinctly as follows. For a pair of cities i and j, let $dist(i, j)$ denote the distance from i to j. To ensure that the

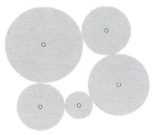

Figure 1.25 Five control zones.

disks for i and j do not overlap, we must choose the radii so that their sum is at most the distance between the two cities, that is, r_i and r_j must satisfy

$$r_i + r_j \leq dist(i, j).$$

Putting everything together, the problem of getting the best TSP bound from our collection of disks can be written as follows.

$$\text{maximize } 2r_1 + 2r_2 + 2r_3 + 2r_4 + 2r_5$$

subject to

$$r_1 + r_2 \leq dist(1, 2)$$
$$r_1 + r_3 \leq dist(1, 3)$$
$$r_1 + r_4 \leq dist(1, 4)$$
$$r_1 + r_5 \leq dist(1, 5)$$
$$r_2 + r_3 \leq dist(2, 3)$$
$$r_2 + r_4 \leq dist(2, 4)$$
$$r_2 + r_5 \leq dist(2, 5)$$
$$r_3 + r_4 \leq dist(3, 4)$$
$$r_3 + r_5 \leq dist(3, 5)$$
$$r_4 + r_5 \leq dist(4, 5)$$
$$r_1 \geq 0, r_2 \geq 0, r_3 \geq 0, r_4 \geq 0, r_5 \geq 0.$$

Such a model is known as a *linear programming problem*, that is, it is a problem of maximizing or minimizing a linear function subject to linear equality and inequality constraints. We postpone until Chapter 3 an introduction to linear programming, when we discuss in detail the work of Dantzig, Fulkerson, and Johnson. Here the important point is that large models such as these can be solved very efficiently, for example, via the *simplex algorithm* for linear programming developed by Dantzig [149]. Thus we have techniques for easily computing the best collection of control zones.

A problem with this approach is that the quality of the bound we obtain is often not very good. The nice packing of disks in the our five-city example is a bit misleading; usually the bound we get from the best zone collection is significantly below the length of the best tour. An example of what can go wrong is given

in Figure 1.26, where the disks bump into each other and cannot cross the gap
between the two clusters of three cities. To improve the lower bound, Jünger and

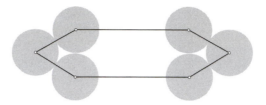

Figure 1.26 Bad example of control zones.

Pulleyblank add in a structure to take advantage of the fact that any tour must cross
such a gap at least twice.

The idea of Jünger and Pulleyblank is the following. Given a subset S of cities,
we can draw a region separating the points corresponding to S from those points
not in S. Now since any tour must at some time visit the cities in S, we can add
twice the width of the geometric region to our lower bound, as in Figure 1.27.
These geometric regions are called *moats*.

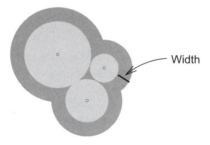

Figure 1.27 A moat.

The use of moats allows us to close gaps that control zones alone cannot touch.
This is illustrated in Figure 1.28, where we use two moats (one large one would
have been enough) to close the gap in the example given earlier. In this case,

Figure 1.28 Use of moats to fill a gap.

adding twice the widths of the two moats together with twice the sum of the radii
of the disks gives a lower bound equal to the length of the indicated tour.

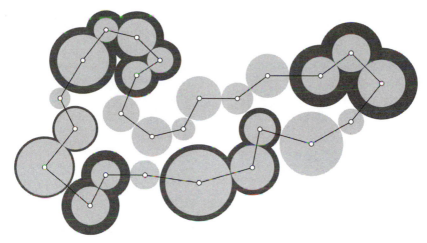

Figure 1.29 A packing of control zones and moats.

A larger example of a packing of TSP control zones and moats is given in Figure 1.29. Note that again the lower bound provided by the zones and moats proves that the indicated tour is in fact optimal. Although this is not typical, it is often true that the zone and moat bound is fairly close to the value of the best tour.

Computing the best possible collection of zones and moats is more involved than with zones alone, due to the many possibilities for choosing sets of cities for the moats. As you can see in the figure, however, it is sometimes the case that we do not need very many different moats to get a good lower bound. It is this observation that motivates Dantzig, Fulkerson, and Johnson's *cutting-plane method* for efficiently computing lower bounds for the TSP, including an algebraic method to generate moats on the fly as they are needed. We discuss this further in Chapter 3.

The technique of Jünger and Pulleyblank that we have described is not only a nice way to introduce lower bounds for the TSP, it has also been adopted in important work by Goemans and Williamson [209] and others to develop approximation algorithms for problems in other domains.

CONTINUOUS LINE DRAWINGS

Robert Bosch and Adrianne Herman [82] had the interesting idea to use geometric TSPs as a means for creating continuous line drawings of images. In such a drawing the pencil or pen should never leave the paper as the line is being created. Tracing out a tour will of course give a drawing of this form, but the trick is to find a layout of cities so that the tour gives a good reproduction. To accomplish this, Bosch and Herman make a digital version of the image they would like to draw, using a gray-scale scanning device. The cities are then placed with density corresponding to the various shades of gray in the digital document. More precisely, they divide the image into a grid and place between 0 and k cities at random locations in each grid cell, depending on the average level of gray in the cell, where 0 is nearly white and k is nearly black. The value of k and the size of the grid control the number of

Figure 1.30 Continuous line drawings via the TSP. Images courtesy of Robert Bosch and
          Craig Kaplan.

cities in the resulting TSP.

The technique for placing the cities was refined by Kaplan and Bosch [303],
using an algorithm developed by Secord [496]. In this approach the tone of the
image is used as a density fuction to build a weighted Voronoi diagram, that is, a
partition of the space into regions such that all points in region i have city i as their
nearest neighbor. In computing the nearest neighbor, the density function is used to
weight the geometric distance, so that dark Voronoi regions will tend to be smaller
than light regions. In an iteration of Secord's placement algorithm, the cities are
replaced by the centers of the regions and a new diagram is computed.

Bosch and Kaplan used the above methodology to create the drawings of the
Mona Lisa and the Colosseum given in Figure 1.30. The *Mona Lisa* TSP has 10,000
cities and the Colosseum TSP has 11,999 cities. In each case the drawn tour was
computed with Concorde, using a heuristic search module (the tours are most likely
not optimal).

CHALLENGE PROBLEMS AND THE TSPLIB

Geometric examples have long been the primary source of challenge problems for
TSP researchers. The example solved in 1954 by Dantzig, Fulkerson, and Johnson
[151] consists of finding a tour through 49 cities in the United States. This began a
tradition of creating instances by selecting sets of actual cities and defining the cost
of travel as the distance between the city locations. Over the years, increasingly
larger examples were considered by taking 120 cities in Germany, 532 cities in the
USA, 666 cities all over the world, and so on. Test instances of this type, together
with geometric instances from industrial sources and a number of other examples,
were gathered together by Gerhard Reinelt [472] in the early 1990s. His library is
called *TSPLIB* and contains over 100 challenge instances, with sizes ranging up to
85,900 cities. This collection has been of great help in TSP research, providing a
common test bed for both new and old solution approaches.

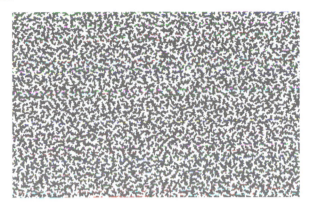

Figure 1.31 Portion of a million-city tour.

Random Euclidean problems, with the locations of the cities selected at random from a square, form an alternative class of geometric test instances. The ease with which such examples can be created made them a common target in computational studies, particularly before the availability of the TSPLIB. More recently, the search for computational evidence for the value of the Beardwood, Halton, and Hammersley [45] TSP constant has driven the interest in this class of problems. Using Lethbridge's style, a portion of a tour through one million uniformly distributed points is displayed in Figure 1.31. This particular tour is known to be at most 0.04% greater in length than an optimal tour for the given point set.

1.4 HUMAN SOLUTION OF THE TSP

An interesting example of the range of research interest in the TSP is its use by teams of psychologists. In this work the TSP is adopted in experiments aimed at understanding the native problem-solving abilities of humans. The easy-to-grasp goal of the TSP permits large-scale studies among subjects unaided by computer algorithms or mathematical techniques.

An initial experiment by N. I. Polivanova [463] in 1974 compared human performance on geometrically represented problems versus performance on problems where the travel costs are given explicitly for each pair of cities. The small examples used in this test (having at most 10 cities) allowed for easy look-up in the explicit lists, but the participants performed distinctly better on the geometric instances. This result is not surprising, given the geometric appeal of the TSP discussed in the previous section. It does, however, indicate that humans may rely on perceptual skills in approximately solving the TSP, rather than purely cognitive skills.

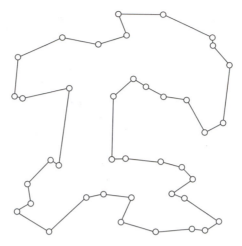

Figure 1.32 Tour found by member of the Gestalt group. Vickers et al. [539].

TOUR AESTHETICS

Polivanova's research was pursued in a number of intriguing studies over the past decade. One of the implications of this body of work is that high-quality geometric tours "feel right" when compared with tours of lesser quality. Indeed, the economy of space that attracted Julian Lethbridge may be indicative of a human desire for minimal structures.

This theme was emphasized in an experiment carried out by Vickers et al. [539]. In their study two groups were presented with identical 10-city, 25-city, and 40-city instances of the TSP (two of each size), but with different instructions on how to proceed. The *Optimization* group was asked to find the shortest tour in each example, while the *Gestalt* group was asked to find a tour such that "the overall pathway looked most natural, attractive, or aesthetically pleasing." The results showed a striking similarity in the quality of tours obtained by the two groups. In fact, the 40-city tour in Figure 1.32 found by a member of the Gestalt group was shorter than all tours found by the Optimization group for this example. Vickers et al. [539] remark that this 40-city tour was found by a fashion designer, who also produced the shortest tour for four of the other five examples.

A preference for high-quality tours was also observed in a study by Ormerod and Chronicle [440]. In one of their experiments, participants were shown a set of tours for a collection of 10-city TSP instances. In each case the participants were asked to rate the tour on a numerical scale "where 1 represented *a good figure* and 5 a *poor figure.*" The results indicated a direct relation between the quality of a tour and its perceived goodness of figure.

A further study was made by Vickers et al. [542] using a set of 25-city problems. This study aimed at learning the geometric properties of tours that make them more attractive or less attractive to humans; the figures in the experiment were not described as potential solutions to the TSP. The participants were requested to give a "rating of the aesthetic appeal of the figure, of how attractive they found it as an

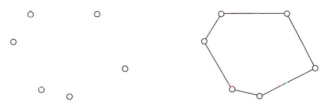

Figure 1.33 Convex hull of a set of points.

abstract configuration." The authors found that a measure of the compactness of the tours had high correlation with their perceived attractiveness. An interesting point is that the 40 participants in the study could be split into a group of 30 that consistently preferred tours with a compact figure and a group of 10 that preferred those having the least compact figure.

Tour-Finding Strategies

The studies of the perceived attractiveness of tours is closely tied with research by psychologists into the general strategies adopted by humans when faced with the TSP. In the many reported experiments on this topic, human subjects consistently produce good-quality tours for modest-sized geometric instances. Although there are many mathematical methods that can easily construct solutions superior to those found by humans, it seems clear that humans use a smaller number of explicit calculations in arriving at their good results.

An early study of human performance on the TSP was carried out by MacGregor and Ormerod [363]. In this work the researchers focused on the degree to which the global shape of a set of points serves as a guide in tour finding. A measure of this shape can be obtained by considering how a rubber band will enclose the set of points X, as in Figure 1.33. The figure traced by the rubber band is the border of the *convex hull* of X, that is, the smallest set S containing X and having the property that any straight line joining two points in S is contained entirely within the set.

The starting point of the MacGregor-Ormerod study is the observation that if all points lie on the border of the convex hull, then the border itself is an optimal tour. Since the convex hull is a natural structure that can readily be constructed by humans, TSPs such as the small example in Figure 1.33 are particularly easy to solve.

Of course, not all geometric TSP instances have this nice property, as illustrated in Figure 1.34 with the convex hull of the city locations from the TSP faced by Lincoln, described earlier in the chapter. Even in this case, however, the border of the convex hull has a connection with optimal tours. Indeed, it is easy to check that the noncrossing rule implies that an optimal tour must trace the border cities in the order in which they appear as we walk around the border. This convex-hull rule is indicated in Figure 1.35 with the optimal tour for Lincoln's TSP. The seven points on the border of this example appear in the same order as in the optimal tour. In this context, the cities not on the border are called *interior points*.

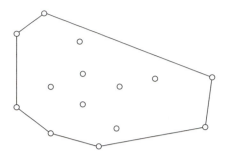

Figure 1.34 Convex hull of points in Lincoln's TSP.

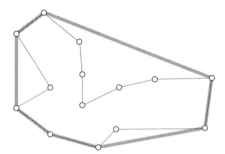

Figure 1.35 Optimal tour following the convex-hull ordering.

Following a detailed analysis of experimental results on human solutions to specially constructed 10-city and 20-city instances, MacGregor and Ormerod conclude that for the examples in their study, the complexity of finding an approximate solution to the TSP is in direct proportion to the number of interior points. Moreover, MacGregor and Ormerod [381] write the following.

> The evidence presented here indicates that human subjects reach solutions based on the perception of global spatial properties of a TSP, and in particular, of the boundary of the convex hull.

The degree to which this convex-hull hypothesis carries over to more general distributions of points is the subject of a lively debate in the human-problem-solving community. Research papers on this theme include Lee and Vickers [343], MacGregor and Ormerod [364], Graham et al. [225], van Rooij et al. [486], Vickers et al. [541], and MacGregor et al. [365]

Much of the technical debate on the convex-hull hypothesis centers on the types of data sets that are used in the experimental studies, but there is also a general discussion on whether humans make use of global-to-local strategies or rather local-to-global strategies. In the former we perceive an overall structure and then make local choices to fit the cities into this structure, while in the latter we carry out local analysis (such as clustering points with nearest-neighbor computations) and then try to best put the local information into an overall tour.

The TSP is clearly playing a useful role in motivating studies in this community.

Vickers et al. [539] summarize the general goals of the research in the following statement.

> Meanwhile, the link with intelligence, and the occurrence of optimal structure in the natural world, suggest that the perception of optimal structure may have some adaptive utility. This suggests that the type of task studied here may not only be interesting from the perspective of perception, and cognition, but may help to provide a conceptual framework of optimization, within which to study intelligence in general.

TOURS FOUND BY CHILDREN

A study by van Rooij et al. [485] examined how children perform in tour-finding experiments, compared with adults facing the same examples. This approach gave the researchers a new means to consider perceptual versus cognitive skills, since young children would be expected to rely primarily on their perception of good structure when searching for tours.

The participants in the study were classes of 7-year-old and 12-year-old elementary school students, and a group of university students; the elementary school participants received a sticker as a reward for their work. The TSP test set consisted of randomly generated 5-city, 10-city, and 15-city instances, including five of each size. A summary of the results for each age group can be found in Table 1.3, expressed as the average percentage that the tours exceeded the optimal values. These

Table 1.3 Average tour quality found by participants. van Rooij et al. [485].

Number of Cities	7-Year-Olds	12-Year-Olds	Adult
5	3.8%	2.5%	1.7%
10	5.8%	3.4%	1.7%
15	9.4%	5.0%	2.7%

averages show a clear improvement in performance as we move from the children to the adults, but even the young children obtained reasonably short tours. This combination of results supports the idea that both perception and cognition are used in approximately solving the TSP.

Some of the themes studied by van Rooij et al. can be seen in the tours illustrated in Figures 1.36 and 1.37. The example is a 1,173-city problem from the TSPLIB. The tours were drawn by the same child, first as a 7-year-old and then six years later as a teenager.

THE TSP IN NEUROSCIENCE

Examples of the TSP that are either very small or constructed with a large convex-hull border are routinely solved by humans, with little variation among study participants. Individual differences in performance quickly arise, however, when general problem instances have 20 or more cities, as the fashion designer in the study

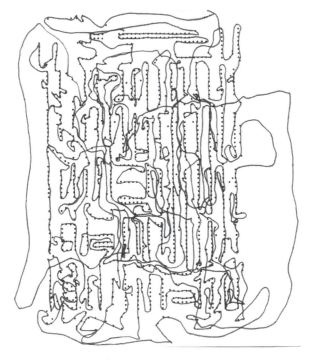

Figure 1.36 Tour for fl1173 by Benjamin Cook (7 years old).

of Vickers et al. [539] demonstrates. On even larger examples, having 50 cities, Vickers et al. [543] report consistent differences in the tour quality produced by individuals in their experiments. The researchers also note a modest correlation between TSP performance and the scores of individuals on a standard nonverbal intelligence test.

Performance differences in aspects of TSP-like problems have long been a resource for clinical tests in neuropsychology. Indeed, a great achievement of TSP tours comes from this realm, namely the Trail Making test from the Halstead-Reitan Battery, described in Reitan and Wolfson [477]. The first part of Trail Making consists of the 25 labeled cities displayed in Figure 1.38. The test is administered by asking the subject to draw a path connecting the cities in consecutive order, requesting that the subject complete the drawing as quickly as possible and pointing out any errors as they occur. The correct path, displayed in Figure 1.39, is clearly not an optimal way to visit the cities, but it is a relatively short path that follows the noncrossing rule. A second part of the Trail Making test consists of a similar task, but where the cities are labeled $1, A, 2, B, \ldots, 12, L, 13$. The two-part test was developed by US Army psychologists in the 1940s; the commonly used scoring system introduced by Reitan is based entirely on the times taken by the subject to complete the tasks.

Trail Making is an exceptionally important tool in clinical neuropsychology. A 1990 survey by Butler et al. [101] singled out Trail Making as the most widely

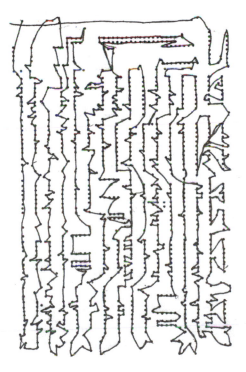

Figure 1.37 Tour for fl1173 by Benjamin Cook (13 years old).

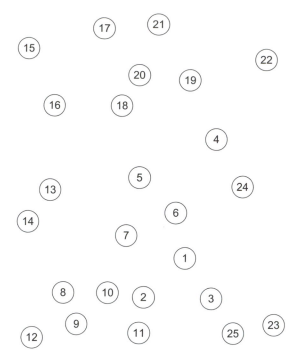

Figure 1.38 Trail Making (Part A).

used test adopted by members of the International Neuropsychological Society. Numerous experiments point out the sensitivity of the test in distinguishing patients suffering from brain damage, as described in the text by Lezak [355].

It is interesting that research work on Trail Making makes use of the specific distributions of city locations given in the two parts of the test; there does not appear to be a reliable method to generate alternative distributions with well-established clinical properties. This point is taken up from a TSP perspective in Vickers and Lee [540], where a tour-finding algorithm is presented as a means to generate noncrossing paths with features common to those used in Trail Making.

Basso et al. [43] looked more generally at how the TSP can be used in neurological testing. In a clinical experiment they measured the delay incurred by participants in deciding how to proceed to the next city while constructing TSP tours. They found a large difference in the variation of the delay when measuring closed-head-injury patients versus the variation measured in normal participants.

The remarkable success of Trail Making, as well as recent findings such as those by Basso et al., adds considerable weight to the studies of human performance on the TSP.

ANIMALS SOLVING THE TSP

Animals other than humans have also been the subject of TSP experiments. Of course, a new challenge here is to convince the animal participants to actually seek

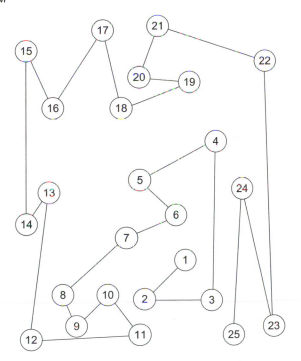

Figure 1.39 Path for Trail Making (Part A).

short tours.

A central paper in this field is the 1973 study of Emil Menzel [392], working with a team of chimpanzees. Menzel describes the purpose of his experiment as follows.

> If a chimpanzee has in the past seen the locations of several hidden objects in the field, how does he manage to get to them again, and how does he organize his travel route? What does his itinerary tell us about the nature of his "cognitive mapping," his strategy, and his criteria of "efficiency"?

In this work Menzel devised a beautiful scheme to coax the subjects into traveling along an efficient tour. To begin a trial, his team of six chimpanzees was kept in a cage on the edge of a field. A trainer would take a selected animal from the cage and carry it around the field while an assistant hid 18 pieces of fruit at random locations. The animal was returned to the cage, and after a two-minute waiting period all six chimpanzees were released. The selected animal would then make use of its memory of the food locations to quickly gather the treats before the other animals located them by simple foraging.

The route taken by Bido, one of the chimpanzees in the study, is depicted in Figure 1.40. Bido started at the point marked "S" on the boundary of the field and finished at the food location marked "F"; the arrows on some of the links indicate

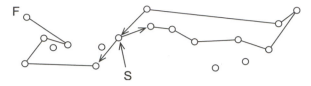

F

S

Figure 1.40 Chimpanzee tour (Bido).

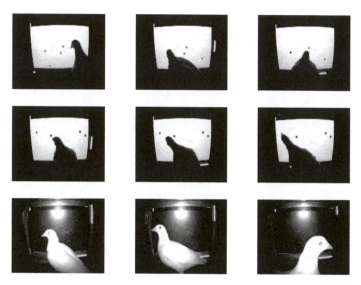

Figure 1.41 Pigeon solving a TSP. Images courtesy of Brett Gibson.

the direction of travel. The chimpanzee missed four pieces of fruit, but altogether Bido made a remarkably good tour working from its memory of the food locations.

Other studies of animals solving the TSP include vervet monkeys (Cramer and Galliestel [143]), marmosets (MacDonald et al. [362]), rats (Bures et al. [98]), and pigeons (Gibson [205]). In Figure 1.41 we display pictures from Gibson's experiment with pigeons. Gibson writes that the birds were trained to find tours through five cities by striking the locations on a touchscreen. To encourage the birds to find short tours, a reward was given only if the constructed solution was in the top 25% of all possible tours. The TSP task provided a good means for testing the spatial cognitive skill of pigeons, and Gibson plans further studies on Clark's nutcrackers, a bird species known for its geometric abilities (Kamil and Jones [302]).

1.5 ENGINE OF DISCOVERY

Looking back on the first 50 years of work on the TSP, it is easy to argue that the most important reason for continued effort on the problem has been its great success as an engine of discovery for general-purpose techniques in applied mathematics. We discuss below a few of the many areas to which TSP research has made fundamental contributions.

MIXED-INTEGER PROGRAMMING

One of the major accomplishments of the study of the TSP has been the aid it has given to the development of the flourishing field of *mixed-integer programming* or *MIP*. An MIP model is a linear programming problem with the additional constraint that some of the variables are required to take on integer values. This extension allows MIP to capture problems where discrete choices are involved, thus greatly increasing its reach over linear programming alone. In the 50 years since its introduction, MIP has become one of the most important models in applied mathematics, with applications spanning nearly every industry. An illustration of the widespread use of the model is the fact that licenses for commercial MIP solvers currently number in the tens of thousands.

The subject of mixed-integer programming has its roots in the Dantzig, Fulkerson, and Johnson [151] paper, and nearly every successful solution method for MIP was introduced and studied first in the context of the TSP. Much of the book deals with aspects of this field, and in several chapters we discuss direct contributions to MIP solution methods made through the development of the Concorde TSP code. General treatments of mixed-integer programming can be found in the texts by Schrijver [494], Nemhauser and Wolsey [428], and Wolsey [559].

BRANCH-AND-BOUND METHOD

The well-known *branch-and-bound* search method has its origins in work on the TSP. The name was introduced in a 1963 TSP paper by Little, Murty, Sweeney, and Karel [360] and the concept was introduced in TSP papers from the 1950s by Bock [69], Croes [145], Eastman [159], and Rossman and Twery [488].

Branch-and-bound is an organized way to make an exhaustive search for the best solution in a specified set. Each branching step splits the search space into two or more subsets in an attempt to create subproblems that may be easier than the original. For example, suppose we have a TSP through a collection of cities in the United States. In this case, if it is not clear whether or not we should travel directly between Philadelphia and New York, then we could split the set of all tours into those that use this edge and and those that do not. By repeatedly making such branching steps we create a collection of subproblems that need to be solved, each defined by a subset of tours that include certain edges and exclude certain others.

Before searching a subproblem and possibly splitting it further, a bound is computed on the cost of tours in its subset. In our example, a simple bound is the sum of the travel costs for all edges that we have insisted be in the tours; much better bounds are available, but we do not want to go into the details here. The purpose of the bounding step is to attempt to avoid a fruitless search of a subproblem that contains no solution better than those we have already discovered. The idea is that if the bound is greater than or equal to the cost of a tour we have already found, then we can discard the subproblem without any danger of missing a better tour.

Branch-and-bound is discussed further in Section 4.1. General treatments of branch-and-bound search can be found in books by Brusco and Stahl [97] and Chen and Bushnell [115].

HEURISTIC SEARCH

The TSP has played a role in the development of many of the most widely used paradigms for *heuristic-search algorithms*. Such algorithms are designed to run quickly and to return a hopefully good solution to a given problem. This theme does not directly match the TSP, which asks for the best possible tour, but the problem has nonetheless served as a basic model for developing and testing ideas in this important area. In this context, researchers look for methods that can detect high-quality tours while using only a modest amount of computational resources. Several of the techniques that have arisen from this work are the following.

- The general class of *local-search algorithms* has its roots in TSP papers from the 1950s by Morton and Land [403], Flood [182], Bock [69], and Croes [145]. These heuristics take as input an approximate solution to a problem and attempt to iteratively improve it by moving to new solutions that are in some sense close to the original. The basic ingredient is the notion of a "neighborhood" that implicitly defines the list of candidates that can be considered in a given iteration. A nice discussion of these algorithms can be found in the book by Aarts and Lenstra [1].

- The paper of Kirkpatrick, Gelatt, and Vecchi [316] that introduced the *simulated annealing* paradigm worked with the TSP, reporting on heuristic tours for a 400-city problem. A typical local-search algorithm follows the hill-climbing strategy of only moving to a neighboring solution if it is better than the one we have currently. In simulated annealing heuristics, this is relaxed to allow the algorithm to accept with a certain probability a neighbor that is worse than the current solution. At the start of the algorithm the probability of acceptance is high, but it is gradually decreased as the run progresses. The idea is to allow the algorithm to jump over to a better hill before switching to a steady climb. Kirkpatrick et al. write that the motivation comes from a connection with statistical mechanics, where annealing is the process of heating a material and then allowing it to slowly cool to obtain a ground state of minimal energy. A detailed discussion of simulated annealing can be found in the book of van Laarhoven and Aarts [330].

- *Neural network* algorithms follow a computational paradigm that is based on models of how the brain processes information. The basic components of a neural network are large numbers of simple computing units that communicate via a network of connections, simulating the parallel operation of neurons in the brain. The TSP was used as a primary example in the classic work of Hopfield and Tank [271] that introduced optimizational aspects of this model.

- *Genetic algorithms* work with a population of solutions that are combined and modified to produce new solutions, the best of which are selected for the

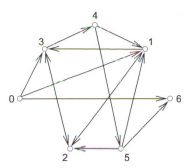

Figure 1.42 A directed Hamiltonian path problem.

next generation of the population. The general scheme is modeled after evolutionary processes in nature. The TSP has been used in numerous studies of this paradigm, including the early papers of Brady [91], Suh and Van Gucht [513], and Mühlenbein, Gorges-Schleuter, and Krämer [405]. A survey of this work can be found in Johnson and McGeoch [291].

In Chapter 15 we discuss in detail how some aspects of these general-purpose heuristic ideas have been combined to produce extremely good algorithms for finding approximate solutions to classes of TSP instances.

DNA COMPUTING

Over the past decade an intriguing field of research has sprung up, with the goal of carrying out large-scale computations with devices that work at the molecular level. The immense amount of information that can be stored in tiny amounts of DNA makes researchers speculate that one day we may be able to utilize molecular devices to attack classes of problems that are out of reach for standard electronic computers. The TSP has played a role in this development, providing the example that was attacked by Leonard Adleman [3] in the groundbreaking experiment that launched this research area in 1994. His solution of a seven-city version of the TSP using a DNA computation was described in popular articles in the *New York Times* [322], *Discover Magazine* [220], and elsewhere.

The variant of the TSP studied by Adleman consists of a set of points with a limited collection of links joining pairs of them. The goal is to find a path among the links to travel from a specified starting point to a specified finishing point, visiting every other point along the way. Such a route is often called a *Hamiltonian path*. In the seven-city problem considered by Adleman, no travel costs were involved; the challenge was simply to find a Hamiltonian path, which is still a very difficult problem in general.

The example used in the DNA experiment is drawn in Figure 1.42; the starting point is city 0 and the ending point is city 6. The Hamiltonian path through the points must obey the indicated directions; travel in either direction is permitted on the links between points 1 and 2 and between points 2 and 3, while travel is permitted in only one direction on the remaining links.

Adleman created a molecular encoding of the problem by assigning each of the seven points to a random string of 20 DNA letters. For example, point 2 was assigned to

$$TATCGGATCG|GTATATCCGA$$

and point 3 was assigned to

$$GCTATTCGAG|CTTAAAGCTA.$$

By considering the 20-letter tags for the points as the combination of two 10-letter tags, a link directed from point a to point b is represented by the string consisting of the second half of the tag for point a and the first half of the tag for point b. For example, a link from point 2 to point 3 would be

$$GTATATCCGA|GCTATTCGAG.$$

The only exceptions to this rule are links involving the starting and ending cities, 0 and 6, where the entire 20-letter tag is used for these points rather than just the first or second half. To capture the links where travel along either direction is possible, a pair of links is created, one in each direction. In the experiment, many copies of the DNA for the points and links were produced in the laboratory.

The next step in Adleman's approach is to provide a mechanism for joining links together into a path. To this end, for each point other than 0 and 6, the complementary DNA sequence for the tag is created. Such sequences act as "splints" to pair up links in their proper orientation. For example, the complementary sequence for point 3 could serve to join the links $(2, 3)$ and $(3, 4)$, since the first half of the sequence would pair up with the second half of the encoding of $(2, 3)$ and the second half of the sequence would pair up with the first half of the encoding of $(3, 4)$. Adding many copies of these splints, the mixture is able to generate double-stranded DNA that corresponds to paths in the problem. After careful work in the laboratory over a seven-day period, a double-strand yielding a Hamiltonian path was identified.

This method clearly is not suitable for large examples of the TSP, since the amount of required DNA components grows very rapidly with the number of cities (all paths in the problem are created in the experiment, not just a Hamiltonian path). But this fascinating solution technique is being explored in many directions to develop methods suitable for attacking other classes of difficult of problems. As in other cases, the simple nature of the TSP allowed it to play a leading role in the creation of a new computational paradigm. Surveys of work on DNA-based computing can be found in Lipton [359], Paun et al. [457], and Amos et al. [11].

1.6 IS THE TSP HARD?

When a description of the TSP appears in a popular setting, it is usually accompanied by a short statement of the notorious difficulty of the problem. Is it the case, however, that the TSP is actually hard to solve? The truth is that we do not really know.

In Menger's presentation of the *Botenproblem* he made the following comments on the complexity of possible solution methods [389], translated from the German original.

> This problem can naturally be solved using a finite number of trials. Rules which reduce the number of trials below the number of permutations of the given point set are not known. The rule that one should go from the starting point to the next nearest point, then to the next nearest point, and so on, does not always produce the shortest path.

So Menger observed that it is possible to solve the TSP by simply checking each tour, one after another, and choosing the cheapest. He immediately calls for better solution methods, however, not being satisfied with a technique that is finite but clearly impractical.

The notion of a "better than finite" solution method is a subtle concept and is not considered in the usual settings of classical mathematics. Hints of the subject appear in other TSP works, including the following statement in Flood's 1956 paper [182].

> It seems very likely that quite a different approach from any yet used may be required for successful treatment of the problem. In fact, there may well be no general method for treating the problem and impossibility results would also be valuable.

In 1964 Gilmore and Gomory wrote a paper [206] with the title "A solvable case of the traveling salesman problem," clearly implying that the simple algorithm of enumerating all tours is not a feasible solution method.

To put the issue of efficiency in mathematical terms, let us make a rough estimate of the running time for the "finite number of trials" algorithm mentioned by Menger. In computing the estimate we follow standard practice and count the number of elementary steps that are required when the algorithm is presented with an n-city TSP as input. This notion can be formalized using *Turing machines* or other models of computation. Excellent treatments of this subject can be found in books by Aho, Hopcroft, and Ullman [5] and Garey and Johnson [200].

There are many ways to implement a process to enumerate and check all tours; an easy-to-read discussion can be found in Jon Bentley's article [57] in *Unix Review*. Good implementations of the process all run in time that is proportional to the total number of tours. To count this number for an n-city TSP is an easy matter. Given a tour, we can choose any point as the starting city. Now from the start we have $n - 1$ choices for the second city, $n - 2$ choices for the third city, and so on. Multiplying these together we have that the total number of tours is equal to

$$(n - 1)! = (n - 1) \cdot (n - 2) \cdot (n - 3) \cdots 3 \cdot 2 \cdot 1.$$

So the direct method suggested by Menger would take time proportional to $(n-1)!$ to solve an n-city TSP.

The simple finite trials method is easy to compare with other possible algorithms for the TSP, since its running-time estimate is the same no matter what n-city instance of the TSP is solved. More complicated algorithms, including some of Bentley's [57] variants of the finite trials scheme, do not share this property. In many cases the running time of an algorithm can vary wildly with the structure of the travel costs between the cities.

How then should we compare two TSP solution methods? A simple judgment is to say that method A is superior to method B if A requires fewer elementary steps to solve every instance of the problem. This is a clean rule, but it makes direct rankings of methods next to impossible, since only closely related methods would yield such a simple comparison. It seems necessary to considerably relax this criterion. To this end, a judge with a more open mind might be willing to ignore results on very small instances, since these can be solved by all good solution techniques. Taking this further, for a given number of cities n, the judge might want to concentrate on those n-city instances that cause the most difficulty for a proposed method that he or she must evaluate. Adopting this approach, we would rank method A ahead of method B if for every large value of n the worst n-city example for A takes less time to solve than does the worst n-city example for B.

To make this comparison idea work in practice, we can analyze a given solution method to obtain a guarantee that it takes at most some amount of time $f(n)$ for any n-city TSP, where $f(n)$ is shorthand for some formula that depends only on n. Now to compare two solution methods we compare the best guarantees that we have found for them. This may of course produce misleading results, since a really good method might just be tough to analyze and therefore appear to be poor when compared to a method that leads to a good analysis. On many computational problems, however, the study of algorithms+guarantees has led to some beautiful mathematical results as well as important improvements in practical problems; this area is a primary subject of study in the field of computer science.

So what can we say about solution methods for the TSP? A running time proportional to $(n-1)!$ is clearly impractical, putting the direct solution of instances with 50 or so cities well out of reach of the combined computing power of all of the world's machinery. This, however, is incorrectly cited in popular articles as the reason the TSP is difficult to solve. The argument only shows that checking all tours is out of the question, it does not at all exclude the possibility that a "quite different approach," quoting Flood, could give a practical solution method.

This point can be made clear by considering a network design problem that is closely related to the TSP but is apparently much easier to solve. The problem is, given a set of cities with specified costs to build roads between each pair of them, find a minimum-cost plan to construct a road system so that travel is possible between all cities. Rather than roads, one could describe the problem as building a minimum-cost communications network between a given set of customers. In counting the number of possible solutions to this problem we should consider only those solutions that are minimal in the sense that if we remove some road from our solution, then the cities are no longer connected. Even with this restriction, however, the number of network design solutions is far greater than the number of tours. An easy way to see this is to note that each tour through n cities gives n solutions to the design problem, by removing one at a time each road that is used in the tour. And many more solutions that do not resemble tours are possible, like the minimum-cost design shown in Figure 1.43. To be more specific, the total number of design solutions for 10 cities is 100,000,000, while the total number of

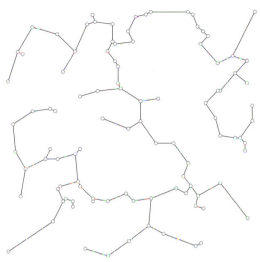

Figure 1.43 Optimal solution for a network design problem.

tours is a more modest 362,880. Moreover, the difference between the two counts grows rapidly as n increases, since the number of design solutions is n^{n-2}, which greatly exceeds $(n-1)!$ for large values of n. A nice proof of this classic result on counting network design solutions, or "spanning trees," can be found in Aigner and Ziegler [8].

If we are to accept the argument that the number of solutions indicates the difficulty of a combinatorial problem, then we can only conclude that the network design problem must be more difficult than the TSP. This is far from the case, however. For the network design problem, an optimal solution can be found by building the network step by step, each time selecting the cheapest road joining two cities that are not yet connected in the network that we have constructed thus far. Repeating this $n-1$ times, we obtain a minimum-cost network. We thus have a simple and efficient algorithm for the problem.

A second argument against the idea that the large number of tours ensures that the TSP is difficult to solve is the fact that there exists a method that is provably better than the direct solution algorithm! In 1962, Michael Held and Richard Karp [252] discovered an algorithm and a guarantee that is proportional to $n^2 2^n$. For any large value of n, the Held-Karp guarantee is much less than the $(n-1)!$ bound for the direct algorithm. We will discuss this result in Chapter 4.

Despite its improvement over the direct algorithm, the Held-Karp method can still not be classified as efficient in any common sense of the word. Mustering the world's computer resources may be able to solve 50-city instances with Held-Karp, but 100 cities is still far beyond its capabilities. This leads to the question of what does constitute an efficient algorithm or guarantee? The great mathematician Jack Edmonds championed the development of a theory to address this issue, making formal the notion that some solution methods are efficient while others are not. With such a theory we can classify problems as easy or hard, depending on whether or not an efficient method exists. The beautiful discussions in Edmonds [164] and elsewhere spurred the growth of the field of computational complexity that attempts to understand this classification.

The role played by the TSP in the development of complexity theory is emphasized by Johnson and Papadimitriou [294], who write the following about the classification of easy and hard problems.

> The TSP is probably the most important among the latter. It has served as a testbed for almost every new algorithmic idea, and was one of the first optimization problems conjectured to be "hard" in a specific technical sense.

The paper of Johnson and Papadimitriou [294] gives a detailed treatment of the computational complexity issues surrounding the TSP; we will limit our discussion to a short introduction.

POLYNOMIAL-TIME ALGORITHMS

Edmonds [164] proposed a precise criterion for a method to be called a *good algorithm*. To be in this category the algorithm should have a guarantee $f(n)$ that is a polynomial function of the problem size n, that is, for large values of n, the algorithm should run in time at most Kn^c for some constant numbers K and c. Note that K and c must be the same for all large n. The constant of proportionality K is often dropped when describing a guarantee, using the standard notation $O(n^c)$ to indicate the running-time bound.

The Held-Karp method is an $O(n^2 2^n)$ algorithm for the TSP and thus is not a good algorithm in the technical sense. The algorithm for solving the network design problem does, however, meet Edmonds' criterion, as we now observe. A simple implementation is to first sort the n^2 road costs, then go through the list in order to see if the ends of the roads are already connected. This checking condition can be carried out in time proportional to n, so we get a guarantee that is $O(n^3)$. Sophisticated implementations by Chazelle [113] and others get much better running-time guarantees, but this rough estimate is enough to establish that it is a good algorithm. Since good has many meanings, these days most researchers use the term *polynomial-time algorithm* to mean good in the sense of Edmonds.

Examining the values in Table 1.4, it is easy to see why polynomial-time bounds are desirable; they exhibit modest growth with increasing problem size. Of course,

Table 1.4 Values for n^3 and $n^2 2^n$.

n	5	10	20	40
n^3	125	1,000	8,000	64,000
$n^2 2^n$	800	102,400	419,430,400	1,759,218,604,441,600

there are some immediate objections to this general classification. First, one could hardly argue that an $O(n^{1,000})$ algorithm is good in a practical sense. Although this is clearly true, the search for polynomial-time algorithms has led to numerous important developments that have had a great impact on practical computations. Indeed, although it has happened that the first discovered polynomial-time algorithms for some problems have had high complexity, such as $O(n^{20})$ or more, usually over time these results have been improved to obtain running-time bounds that are also good in the practical sense. As a research guide, Edmonds' classification has been a major success in computer science. Second, it would be poor judgment to dismiss an algorithm that has performed well in practice just because it is not good in the technical sense. Concerning this issue, Edmonds [164] wrote the following.

> It would be unfortunate for any rigid criterion to inhibit the practical development of algorithms which are either not known or known not to conform nicely to the criterion. Many of the best algorithmic ideas known today would suffer by such theoretical pedantry.

He goes on to cite Dantzig's simplex algorithm as a prime example of an algorithm that is not known to be good, but yet is the basis for a large body of computational work, including our TSP study.

NP-HARD PROBLEMS

An important theoretical question raised by Edmonds in the 1960s is whether or not there is a good algorithm for the TSP. To date, unfortunately, this question has not been settled. Indeed, the discovery of a good algorithm for the TSP or a proof that no such algorithm exists would fetch a $1,000,000 prize from the Clay Mathematics Institute [127]. This is what we referred to at the start of the section when we wrote that we do not really know if the TSP is a hard problem.

Although the status of the TSP is unknown, over the past 40 years it has been placed in a general context within the subject of complexity theory. In this theory problems are set as decision questions, such as asking if there is a tour of cost less than K, rather than asking for a minimum-cost tour. The problems for which there exist good algorithms are known as the class $\mathcal{P}$, for *polynomial time*. A possibly more general class is known as $\mathcal{NP}$, for *nondeterministic polynomial time*. A problem is in $\mathcal{NP}$ if whenever the answer to the decision question is yes, then there exists a means to certify this yes answer in such a way that the certificate can be checked in polynomial time. For example, if the answer is yes to the TSP question, then this can be certified by exhibiting a tour that does indeed have cost less than K. So the TSP question is in $\mathcal{NP}$.

The cornerstone of this branch of complexity theory is the result of Stephen Cook [134] from 1971 showing that there exists a problem such that if it has a polynomial-time algorithm then every problem in $\mathcal{NP}$ has a polynomial-time algorithm. A problem with this property is called $\mathcal{NP}$-*hard*, or $\mathcal{NP}$-*complete* when it also belongs to $\mathcal{NP}$. Cook's initial $\mathcal{NP}$-hard problem served as an anchor for Richard Karp [308] to show that a number of prominent combinatorial problems are also $\mathcal{NP}$-hard, including the TSP. The technique employed by Karp is to transform an instance of Cook's hard problem to an instance of the TSP, while only increasing the size of the instance by a polynomial factor. This transformation method has subsequently been adopted by numerous researchers, leading to a collection of hundreds of problems that are known to be $\mathcal{NP}$-hard.

In the complexity theory setting, the main question is whether or not there exists a polynomial-time algorithm for an $\mathcal{NP}$-hard problem. If so, then every problem in $\mathcal{NP}$ can be solved in polynomial time, and thus the classes $\mathcal{P}$ and $\mathcal{NP}$ would be equal. This is one of the outstanding open problems in mathematics, and it is in this framework that Clay's $1,000,000 has been offered.

Each year there are new announcements of proofs that $\mathcal{P} = \mathcal{NP}$, usually by researchers providing a good algorithm for the TSP. This is natural, since the TSP is probably the most studied of all $\mathcal{NP}$-hard problems. So far none of these claims have held up under close examination, and many researchers feel that it is more likely the case that $\mathcal{P} \neq \mathcal{NP}$. Still, the question is relatively young as far as mathematics problems go and the resolution of $\mathcal{P}$ versus $\mathcal{NP}$ could turn out one way or the other. The attack on this problem has certainly made complexity theory a lively field, and we refer the reader to the book of Garey and Johnson [200] for a thorough introduction.

1.7 MILESTONES IN TSP COMPUTATION

The TSP algorithm developed by Held and Karp in 1962 still carries the best-known guarantee on the running time of a general solution method for the problem. Their $O(n^2 2^n)$ bound has survived over 40 years of challenges. Applying this method to Dantzig, Fulkerson, and Johnson's 49-city TSP would be a daunting challenge, however, where the guarantee of having the number of elementary steps only a multiple of 1,351,642,838,164,570,112 is of little comfort.

This lack of progress on algorithms with guarantees is disappointing, but Edmonds advised researchers not to let "any rigid criterion" inhibit the pursuit of practical algorithms. The TSP community has certainly not sat still in the face of the seemingly weak $O(n^2 2^n)$ bound. Following a tradition begun with Dantzig et al., computational researchers have focused on test suites of specific instances of the TSP. An easily recognized sign of progress in these efforts is the increasing size of the largest instances that have been solved over the years. We list these important milestones in TSP computation in Table 1.5, together with the researchers who led the computations. Many of the test problems studied over the past 50 years are now included in Reinelt's TSPLIB [472]. The short names provided in the last column of Table 1.5 are references to this library; the exceptions are the 57-city problem described in Karg and Thompson [304] and the three random geometric problems.

It is interesting to note that Held and Karp themselves played a large role in the chase for the solution of ever larger TSP instances, finally improving on Dantzig, Fulkerson, and Johnson's result after a span of 17 years. This computational work of Held and Karp does not make use of their $O(n^2 2^n)$ algorithm, but rather a method that comes without a competitive worst-case guarantee. Being the first paper that improves on the initial 49-city TSP record, they set an important prece-

Table 1.5 Milestones in the solution of TSP instances.

1954	G. Dantzig, R. Fulkerson, S. Johnson	49 cities	dantzig42
1971	M. Held and R. M. Karp	57 cities	[304]
1971	M. Held and R. M. Karp	64 cities	random points
1975	P. M. Camerini, L. Fratta, F. Maffioli	67 cities	random points
1975	P. Miliotis	80 cities	random points
1977	M. Grötschel	120 cities	gr120
1980	H. Crowder and M. W. Padberg	318 cities	lin318
1987	M. Padberg and G. Rinaldi	532 cities	att532
1987	M. Grötschel and O. Holland	666 cities	gr666
1987	M. Padberg and G. Rinaldi	1,002 cities	pr1002
1987	M. Padberg and G. Rinaldi	2,392 cities	pr2392

dent by not only solving two larger instances but also going back and resolving the Dantzig-Fulkerson-Johnson example. Taking care in this way to continue to hold on to past gains, the computational research community as a whole can make general progress in solution methods for the TSP, growing the collection of solvable test instances in both size and variety.

In Figure 1.44 we plot the TSP data from Table 1.5, giving the number of cities versus the year the instances were solved. Note that the size of the TSP is plotted on a log scale, underemphasizing the large improvements that were made in the 1980s. The big push during these years was headed by the research efforts of Martin Grötschel and Manfred Padberg, both jointly and with colleagues. Their work is responsible for the rapid switch in the slope of the curve that occurs around 1975, taking the TSP race to new heights.

The time period covered in Figure 1.44 saw rapid increases in the availability and power of computing machinery. While this hardware played a crucial role in shaping the research, it must be emphasized that advances in computing hardware have only a secondary role in the improved capability to solve problems such as the TSP. Indeed, problems that appear to require running time that grows exponentially with the data size are not suitable for solution using raw computing power. Consider, for example, the Held-Karp guarantee of $n^2 2^n$. If we double the amount of computing power at our disposal, we add at most a single city to the largest instance we can solve in the guaranteed number of steps. The wide availability of computing machinery has, however, allowed research groups to readily experiment with new solution approaches and thus channel their efforts in the most promising directions. This is without a doubt a factor in the rapid expansion of solution techniques for the TSP and in general computational mathematics. To see continued growth in these areas it is important for computational researchers to have easy access to advanced hardware platforms.

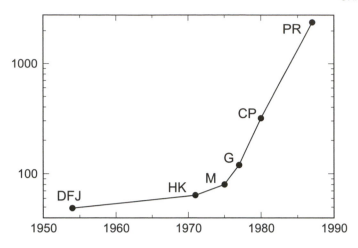

Figure 1.44 Progress in the TSP.

CONCORDE

When Reinelt first put together the TSPLIB in 1990, it included 24 instances having 1,000 cities or more. This number increased to 34 instances in 1995 as new challenge problems were added to the library. Two of these problems, pr1002 and pr2392, were solved previously in the breakthrough paper of Padberg and Rinaldi [447]. With the introduction of Concorde we have solved all 32 of the remaining instances.

Concorde is written in the C programming language, described in Kernighan and Ritchie [314], and consists of roughly 130,000 lines. The crucial modules build on earlier TSP research, following the basic line of attack outlined in Dantzig, Fulkerson, and Johnson's fundamental paper. In Chapters 3 and 4 we describe the work of researchers in the 1950s through the 1980s that led to the start of our project.

The highlights of the computational studies with Concorde are presented in Table 1.6, where we list the year of solution for new largest-sized instances solved with the code. Each of the problems cited in Table 1.6 is from the TSPLIB, with the exception of sw24978, which consists of all city locations in Sweden. A plot of the full set of record solutions for the TSP is given in Figure 1.45, again using a log scale for the number of cities. The progress made with the Concorde code is of course shown more vividly when plotted with a linear scale, as in Figure 1.46.

To give the reader a quick impression of the variety of problems solved, we display in Figure 1.47 the optimal tours for four of the largest instances. The tours are drawn as solid curves.

A close-up view of the 24,978-city tour of Sweden is given in Figure 1.48. In solving the Sweden TSP the Concorde code was heavily aided by the work of Danish computer scientist Keld Helsgaun, who applied his TSP heuristic algorithm LKH [260] to obtain what proved to be the optimal tour for the problem.

The largest two solved instances, pla33810 and pla85900, arose in a VLSI ap-

Table 1.6 Solution of TSP instances with Concorde.

1992	Concorde	3,038 cities	pcb3038
1993	Concorde	4,461 cities	fnl4461
1994	Concorde	7,397 cities	pla7397
1998	Concorde	13,509 cities	usa13509
2001	Concorde	15,112 cities	d15112
2004	Concorde	24,978 cities	sw24978
2004	Concorde with Domino-Parity	33,810 cities	pla33810
2006	Concorde with Domino-Parity	85,900 cities	pla85900

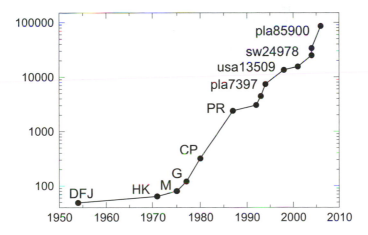

Figure 1.45 Further progress in the TSP, log scale.

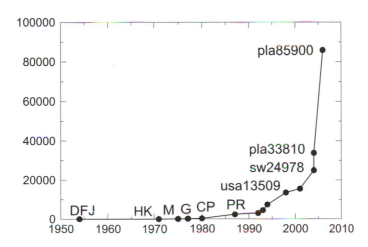

Figure 1.46 Further progress in the TSP, linear scale.

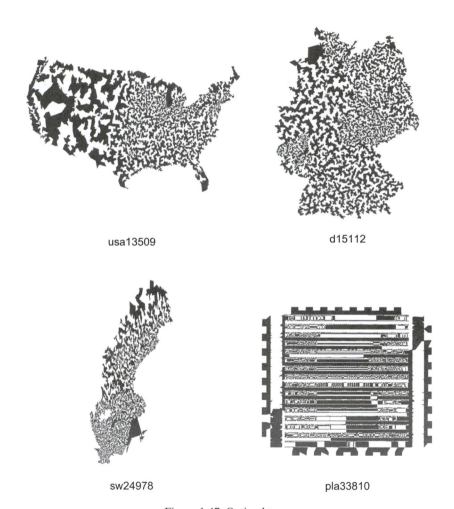

usa13509   d15112

sw24978   pla33810

Figure 1.47 Optimal tours.

Figure 1.48 Optimal tour of 24,978 cities in Sweden.

plication that is described in the next chapter. The solution to these problems was
accomplished in part by extending the basic Concorde code with new techniques
by Cook, Espinoza, and Goycoolea [136], described in Chapter 9. These efforts
build on algorithmic work of Letchford [347], and we hope it is a sign of future
improvements by other research teams, as Concorde is used as a bridge to bring
TSP theory to practical computation.

WORLD TSP

With the solution of the 85,900-city instance, the complete collection of TSPLIB
instances has now been solved. This should not signal the end of the line for TSP
computation, however. Much more can be learned by pursuing larger and even
more difficult instances. To take stock of what may be needed to face the challenge
of much larger problems, in 2001 we put together a 1,904,711-city instance of locations throughout the world. The solution of such a TSP is far too difficult for
Concorde, but many researchers have taken up the challenge of trying to produce
close approximations to an optimal tour. At the time of the writing of this book, the
best-known solution has been obtained by Keld Helsgaun with variants of his LKH
heuristic code. Remarkably, with lower bounds provided by Concorde, we know
that Helsgaun's tour is no more than 0.058% longer than an optimal world tour. A
picture of the tour is given in Figure 1.49. We discuss in Chapter 16 the current
attacks on this instance and how Concorde's solution modules behave when scaled
up to problems of this size.

1.8 OUTLINE OF THE BOOK

We mentioned at the outset that the primary goal of this book is to describe the
theory and algorithms adopted in the Concorde TSP code. Continuing some of the
discussions begun in this chapter, we also cover the history of TSP computation,
along with applications of TSP models.

In Chapter 2 we begin with a survey of applications, and follow this in Chapter
3 with a detailed discussion of Dantzig, Fulkerson, and Johnson's [151] seminal
work. The cutting-plane method they created for the solution of the 49-city TSP
set the stage for the leaps in performance that were made over the next 50 years.
Much of the further development and extension of the Dantzig et al. ideas was led
by Martin Grötschel and Manfred Padberg, and we survey this work in Chapter 4.

At the heart of the Concorde TSP code are the implementations of algorithms
and techniques for finding cutting planes. The methods that are utilized in these
implementations are presented in Chapters 5 through 11.

In Chapter 12 we lay out the steps that were necessary to manage the linear
programming problems that need to be solved when applying the cutting-plane
method to large-scale TSP instances. We follow this in Chapter 13 with a detailed
description of the workings of the ILOG-CPLEX linear programming solver that is
the engine for our cutting-plane implementation.

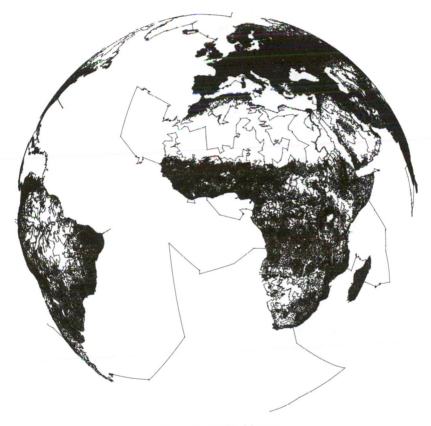

Figure 1.49 World tour.

The Concorde code uses an enumeration scheme to continue the TSP solution process when cutting planes themselves are no longer making significant progress. The design of this portion of the code is described in Chapter 14. Heuristic algorithms for finding tours (a component of the solution process) are presented in Chapter 15.

Results of computational tests with Concorde are given in Chapter 16. We present results on the TSPLIB and World TSP instances, as well as on random Euclidean problems, leading to a study of the Beardwood, Halton, and Hammersley [45] TSP constant β.

Finally, in Chapter 17 we discuss some of the research directions various groups have adopted in attempts to push beyond the limits of the TSP approach used in Concorde.

Chapter Two

Applications

Much of the earliest work on the TSP was motivated by direct applications. For example, Flood [182] considers the planning of school bus routes, Mahalanobis [368] and Jessen [286] work with crop surveys, and Morton and Land [403] mention the routing of a laundry van.

Although the best argument for continued interest in the TSP is its wild success as a general engine of discovery, a steady stream of new applications breathes life into the research area and keeps the mathematical community focused on the computational aspects of the problem. In this chapter we cover some of the general applications of the TSP, ranging from mapping genes to scheduling space-based telescopes.

2.1 LOGISTICS

A common application of the TSP is of course the movement of people, equipment, and vehicles around tours of duty, such as those treated in the early studies mentioned above. Here we give a sampling of the many applications of this type.

SALESMEN AND TOURISTS

Although people planning trips for business or pleasure usually do not check textbooks for TSP solution methods, they do often make use of software that does the work on their behalf. Personal mapping packages include a TSP solver for small instances having a dozen cities or so, and this is normally adequate for the tours traveled by most salesmen and tourists.

Occasionally larger problems involving a personal trip turn up, usually just for fun. For example, a fan of the sport of baseball used Concorde's solver to compute the quickest way to visit all 30 of the Major League stadiums. On a grander scale, Megain Dosher [158] reported in the *Wall Street Journal* that a member of the Extra Miler Club [171] proposed to eat a Big Mac in each of the over 13,000 McDonald's in North America. This would be a nice application of the TSP, but the club's motto, "The shortest distance between two points is no fun," suggests the members are not likely users of optimal tours.

SCHOOL BUS ROUTING

Scheduling a single run of a school bus to pick up waiting children is naturally a TSP. The fact that Merill Flood was drawn to the problem through this application

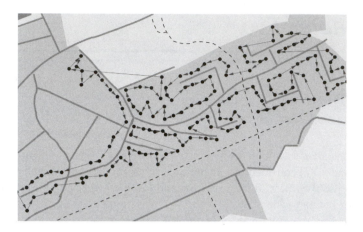

Figure 2.1 TSP tour for deliveries by Forbruger-Kontakt. Courtesy of Thomas Isrealsen.

gives it a special historical role, but the application is still around today. Indeed, a number of companies specialize in school bus routing software, offering school districts a quick way to optimize their pickup schedules via the solution of the TSP.

POSTAL DELIVERIES

Postal routing was mentioned by Menger [389] as a potential application of the TSP. In our description of his paper in Chapter 1 we noted that postman problems are often modeled these days as that of traversing a given set of streets in a city, rather than visiting a set of specified locations. Moreover, Edmonds [165] showed how to solve this version of the problem efficiently in the case where the collection of streets is connected. The TSP nevertheless plays a role in general postal problems, where the houses or streets are far apart or where only a subset of the houses needs to be visited, such as in the delivery of parcels. Indeed, a number of developers have recently adopted the TSP in software for use in these postal applications.

The firm Rapidis employs a heuristic module from Concorde to plot routes for their customer Forbruger-Kontakt, a distributor of advertising material and samples, operating in Denmark and several other countries. The image in Figure 2.1 is a drawing made from a screen dump of the routing software created by Rapidis. Note that the route in the drawing obeys one-way streets and other travel restrictions, making the cost to travel between two points depend on the direction that is chosen.

Another TSP-based software package was created by the firm Eurobios and adopted by Post Danemark, the Danish postal service. In this case the tours are constructed by a local-search heuristic, as reported in an article by Graham-Rowe [226] describing the application. A screen dump from the Eurobios software displayed in Figure 2.2 shows a set of locations that must be visited in the city of Roskilde. Vince Darley of Eurobios UK explained that each of the locations has a probability between 50% and 90% of being visited on any given day, and that multiple routes

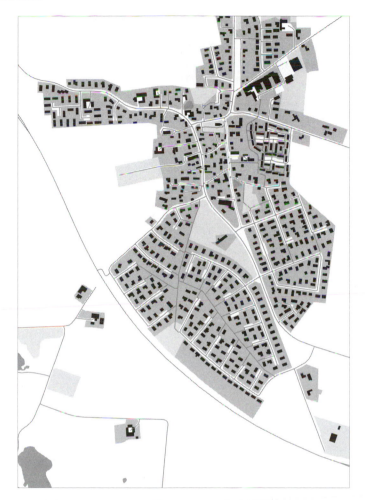

Figure 2.2 Set of locations in Denmark. Image courtesy of Vince Darley.

can be used to cover a given area. In this multi-salesman scenario, a component of the solution procedure is to dvide the customers among the individual tours.

MEALS ON WHEELS

To meet the needs of elderly and sickly persons, many urban areas have organizations that make regular visits to individual homes. In this context, Bartholdi et al. [42] describe a successful application of a fast TSP heuristic algorithm for constructing the routes of aid workers in a "Meals on Wheels" program in Atlanta, Georgia. Each driver working in the program delivers meals to 30 to 40 locations out of a total of 200 or so that are served daily. To construct the routes, Bartholdi et al. place all 200 locations in a tour and divide the tour into segments of the appropriate lengths. The overall tour is found with the aid of the spacefilling curve

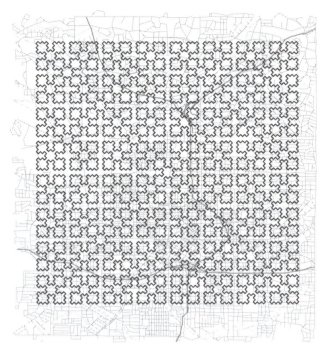

Figure 2.3 Spacefilling curve for Atlanta region. Image courtesy of John Bartholdi.

illustrated in Figure 2.3. Ever finer versions of the curve will eventually include any point in the city, and the heuristic tour through the 200 locations is obtained by taking the order in which the locations appear on the curve. This tour-finding method, proposed by Bartholdi and Platzman [41], worked well in the application. The simplicity of the method allowed the manager of the program to easily update the tour by hand as new clients joined the system and existing clients left the system.

INSPECTIONS

The crop survey studies by Mahalanobis [368] and Jessen [286] in the 1940s are early examples of the use of the TSP in the planning of inspections of remote sites. This type of logistical application occurs in many other contexts as well. For example, William Pulleyblank reports the use of TSP software to plan routes for an oil firm to visit a set of 47 platforms off the coast of Nigeria (see Cook et al. [138]). In this instance, the platforms are visited via helicopter flying from an onshore base. In another sea-based example, a group at the University of Maryland modeled the problem of scheduling boat crew visits to approximately 200 stations in the Chesapeake Bay. The purpose of the boat trips was to monitor the blue crab population in the bay; the researchers turned to the TSP after having difficulty completing trips quickly enough to permit frequent monitoring of all sites.

2.2 GENOME SEQUENCING

The mapping of the human genome is one of the great achievements in the history of science. The initial publication of the ordering of some 3 billion base pairs, by Lander et al. [336] and Venter et al. [534], has been followed by a wide range of efforts to improve and analyze the data, as well as to obtain information on the genetic material of other species. A focus of this work before finding the genome sequence is the accurate placement of *markers* that serve as landmarks for the genome maps. The TSP plays an important role here, providing a tool for building sequences from experimental data on the proximity of individual pairs of markers.

A genome map has for each chromosome or chromosome arm a sequence of markers with some estimate of the distance between adjacent markers. The markers in these maps are segments of DNA that appear exactly once in the genome under study and can be reliably detected in laboratory work. The ability to recognize these unique segments allows researchers to use them to verify, compare, and combine physical maps created across large numbers of different laboratories. In this context it is particularly useful to have accurate information on the order in which the markers appear on the genome, and this is where the TSP comes into play.

One of the primary techniques for obtaining laboratory data on the relative position of markers is known as *radiation hybrid* (RH) *mapping*, developed by Goss and Harris [223] and Cox et al. [142]. This process involves the exposure of a genome to high levels of X-rays to break it into fragments. The fragments are then combined with genetic material from rodents to form hybrid cell lines that can be analyzed for the presence of markers from the collection under study. A simple illustration of the two steps is given in Figure 2.4.

The central theme in RH mapping is that positional information can be gleaned from an analysis of which pairs of markers appear together in cell lines obtained in a series of independent experiments. If two markers A and B are close on the genome, then they are unlikely to be split apart in the radiation step. Thus, in this case, if A is present in a cell line, it is likely that B is present as well. On the other hand, if A and B are far apart on the genome, then we can expect to have cell lines that contain just A or just B, and only rarely a cell line containing both A and B. This positional reasoning can be crafted into a notion of the experimental distance between two markers, adopting one of two commonly used approaches we describe below.

Before we present the distance functions, note that, by considering markers as cities, a genome ordering can be viewed as a path traveling through each marker in the collection. As usual, such a Hamiltonian path is readily converted to a tour by adding an extra city to permit the ends of the path to be joined. Now, using the notion of distance between markers, we can model the problem of finding the genome order as a TSP.

The original paper of Cox et al. [142] outlines an implicit use of this TSP approach in the following passage.

To construct an RH map for this set of 14 markers, we first identified

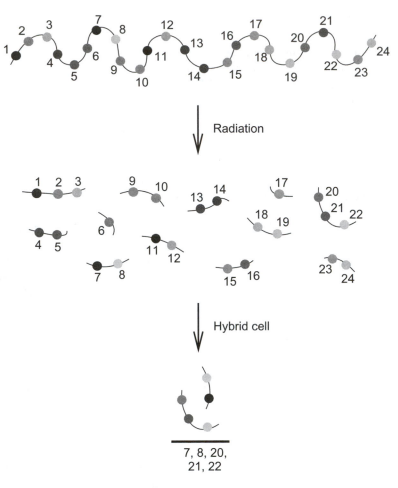

Figure 2.4 Radiation hybrid mapping.

those pairs of markers that are significantly linked. We then used only this set of linked marker pairs to determine the "best" map, defined as that which included the entire set of markers in an order such that the sum of distances between adjacent markers is minimized. This process of identifying the best map was carried out by trial and error, resulting in a map of 14 markers . . .

The distance functions used in this context are aimed at creating TSP instances that can translate high-quality experimental data into a close approximation to the actual ordering of the genome under study.

To set things up, we note that a cell line obtained in an RH experiment is typically recorded as an array of zeros and ones, indicating the set of markers that are present. The example in Figure 2.4 gives the array

$$[0, 0, 0, 0, 0, 0, 1, 1, 0, 0, 0, 0, 0, 0, 0, 0, 0, 0, 0, 1, 1, 1, 0, 0]$$

where the entries correspond to the markers, and the ones in positions

$$7, 8, 20, 21, 22$$

describe the subset in the cell line.

A collection of results from a series of experiments is called a *panel* and is recorded as a matrix of zeros and ones, with a row for each individual cell line. For instance, continuing the example in Figure 2.4 could lead to a panel

$$\begin{bmatrix}
0,0,0,0,0,0,1,1,0,0,0,0,0,0,0,0,0,0,0,1,1,1,0,0 \\
0,1,1,0,0,1,1,1,0,0,0,0,1,1,0,0,0,0,0,0,0,0,0,0 \\
1,0,0,1,1,0,0,0,0,0,0,0,0,1,1,0,1,1,0,0,0,0,0,1 \\
1,1,1,0,0,0,0,1,1,0,0,0,0,0,0,0,1,1,1,0,0,1,0,0 \\
0,1,0,0,0,1,1,1,0,0,0,0,0,0,0,0,0,0,1,0,1,1,0,0 \\
0,1,1,0,0,0,1,1,0,0,0,0,0,1,0,0,0,1,0,0,0,0,0,0 \\
0,0,0,1,1,0,0,1,0,0,0,0,0,0,0,0,0,0,0,0,1,1,0,0 \\
1,0,0,0,0,0,0,0,1,1,1,1,0,0,0,0,1,0,0,0,0,1,0,0 \\
1,0,0,0,0,0,0,0,1,1,1,1,1,1,1,1,0,1,0,0,0,1,1,1 \\
0,0,0,0,0,0,0,1,1,1,0,0,0,0,1,1,0,0,0,1,1,1,0,0
\end{bmatrix}$$

consisting of 10 experiments. From such a panel we can extract information, called an *RH vector*, for a single marker by taking the corresponding column of the matrix. In the example, the RH vectors for markers 1, 2, 3, and 4 are

$$v_1 = \begin{bmatrix} 0 \\ 0 \\ 1 \\ 1 \\ 0 \\ 0 \\ 0 \\ 1 \\ 1 \\ 0 \end{bmatrix}, \quad v_2 = \begin{bmatrix} 0 \\ 1 \\ 0 \\ 1 \\ 1 \\ 1 \\ 0 \\ 0 \\ 0 \\ 0 \end{bmatrix}, \quad v_3 = \begin{bmatrix} 0 \\ 1 \\ 0 \\ 1 \\ 0 \\ 1 \\ 0 \\ 0 \\ 0 \\ 0 \end{bmatrix}, \text{ and } v_4 = \begin{bmatrix} 0 \\ 0 \\ 1 \\ 0 \\ 0 \\ 0 \\ 1 \\ 0 \\ 0 \\ 0 \end{bmatrix}.$$

The positional arguments given above imply that markers that are near to one another on the genome should have similar RH vectors, and this suggests that distance measures for markers can be based on these vectors.

An immediate idea for such a measure is to define the distance between two markers as the number of positions where they differ in their corresponding RH vectors. This distance function captures the idea of minimizing the number of *obligate chromosome breaks* (OCB), that is, the least number of breaks in the genome that are necessary to produce the data observed in an experiment. For example, with markers in sequence from 1 up to 24, an experiment that produces the fragments $(7, 8)$ and $(20, 21, 22)$ in Figure 2.4 must at least have breaks between markers 6 and 7, between markers 8 and 9, between markers 19 and 20, and between markers 22 and 23. In this case the OCB number is 4, and this matches the length of the tour $(1, 2, \ldots, 24)$ under the proposed distance function, when we consider the panel consisting of the single experiment. For the full panel described above, the length of a tour is equal to the sum of the OCB numbers for each of the 10 experiments.

A second distance measure is derived from a probabilistic model of the RH process. Here a number of assumptions are made that permit a computation of the probability of observing the data in a panel given a specific ordering of the markers on the genome. The goal then is to find a *maximum-likelihood estimate* (MLE) for the genome ordering. Karp et al. [307] demonstrate how to convert this to a TSP, taking the logarithm of the estimator to replace a product of distances by a sum.

A detailed study of both the OCB and MLE versions of the genome TSP model was carried out by Amir Ben-Dor and Benny Chor [53], leading to a software package described in Ben-Dor et al. [54]. This work begins to address, within the TSP model, the important practical consideration of handling imperfect or missing data from the RH procedure.

In an RH panel each entry should be either 0 or 1, indicating whether or not a marker was detected in the given cell line. Unfortunately, repeated laboratory tests on a given line can give contradictory results, and these are recorded as a 2 in the panel. The distance functions can be modified to handle these incomplete panels, but a more difficult task is to deal with data that are recorded incorrectly or data that result from errors, a frequent occurrence in the operation of wet laboratories.

A group at the National Institutes of Health (NIH), led by Richa Agarwala and Alejandro Schäffer, has followed the Ben-Dor and Chor study with the development of methods and software for directly handling erroneous data. One of their tools is to apply variants of the OCB and MLE distance measures to obtain five different TSP instances for each genome problem, using differences in the resulting tours to spot potential problems with the input data. The initial work on their *rh_tsp_map* package is described in Agarwala et al. [4].

In the Ben-Dor and Chor study the TSP tours are found with a heuristic algorithm based on simulated annealing. The NIH package replaces this with Concorde to permit the software to find optimal tours for each of the selected distance functions. The algorithmic and software development by Agarwala and Schäffer has led to the adoption of the TSP approach with the Concorde solver in a number of important studies in genome sequencing, including work on the following species.

- **Human.** The paper of Agarwala et al. [4] contains a study of the human genome. In this work the software is used to place markers on a map that integrates data from studies by Gyapay et al. [245] and Stewart et al. [512]. In a separate study, Brüls et al. [96] develop a physical map of human chromosome 14.

- **Macaque.** Murphy et al. [411] create a map of the rhesus macaque genome, with a cross-reference to the human genome.

- **Horse.** Brinkmeyer-Langford et al. [93] build high-resolution maps of the two horse genome pieces corresponding to a partition of human chromosome 19. A further study is made by Wagner et al. [547].

- **Dog.** Guyon et al. [244] compute a map of the dog genome. Further maps are developed by Breen et al. [92] and by Comstock et al. [133].

- **Cat.** Menotti-Raymond et al. [391] and Menotti-Raymond et al. [390] create genetic maps of the domestic cat. An additional study is made by Murphy et al. [412].

- **Mouse.** Avner et al. [27] adopt the TSP approach to build a first-generation map of the mouse genome. Their paper contains a nice discussion of the practical issues involved in handling data errors using the strategy outlined in Agarwala et al. [4].

- **Rat.** Krzywinski et al. [327] create a map of the rat genome.

- **Cow.** Everts-van der Wind et al. [555] create a map of the cattle genome and make cattle-human comparisons.

The TSP approach competes with several other methods for building RH maps, but it has advantages in both the quality of the solutions that are constructed and in the speed of the software. Comparisons between the rh_tsp_map package with Concorde and other mapping techniques can be found in Agarwala et al. [4] and in Hitte et al. [265]. Further information on RH mapping and the TSP approach can be found in Ivansson [278], Ivansson and Lagergren [279], and Schäffer [491].

2.3 SCAN CHAINS

Post-manufacturing testing is a critical step in the production process of computer chips, that is, integrated circuits. The complexity of the chips and the circuit boards that are built with them make it extremely difficult to reliably produce working parts, often causing the loss of over half of the production.

Some years ago circuit boards were usually tested with a *bed-of-nails* system using a large set of probes to contact specific locations on a board. The bed-of-nails method allows data to be loaded and read from various components, permitting testing of the operation of the board. As computer chips became more complex and the mounting technologies permitted ever finer traces on the boards, bed-of-nails systems became impractical to construct and operate. To address this issue, *scan chains* were introduced in the 1980s to link components, or *scan points*, of a computer chip in a path having input and output connections on the boundary of the chip, as illustrated in Figure 2.5. A scan chain permits test data to be loaded into the scan points through the input end, and after the chip performs a series of test operations the data can be read and evaluated from the output end. The boundary connections allow circuit board designers to connect many chips into a single chain for testing the entire board. A nice overview of the use of scan chains can be found in Vermeulen and Goel [538], and descriptions of the chains in versions of the Pentium and UltraSparc processors can be found in Carbine and Feltham [105] and Levitt [354], respectively.

There are numerous technical issues involved in designing scan chains that we cannot hope to cover here, and we refer the interested reader to books by Bleeker et al. [67] and Parker [455] and to research papers by Gupta et al. [241], Nicolici

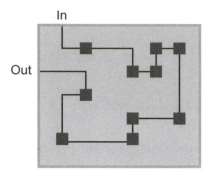

In

Out

Figure 2.5 Scan chain.

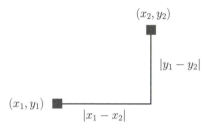

(x_2, y_2)

$|y_1 - y_2|$

(x_1, y_1)

$|x_1 - x_2|$

Figure 2.6 L_1-norm

and Al-Hashimi [434], and others. The aspect that concerns the TSP is, however, straightforward: we want to determine the ordering of scan points to make the chain as short as possible. Minimizing the chain length helps to meet a number of goals, including saving valuable wiring space on the chip and saving time in the testing phase by allowing the signals to be sent more quickly. This TSP approach has been proposed and studied by many groups, including Barbagallo [40], Gupta et al. [242], Hirech et al. [264], and Makar [370].

In most cases the distance between two points in a scan-chain TSP is measured using the L_1-norm, that is, if (x_1, y_1) and (x_2, y_2) describe the locations of two points on a chip, then the distance between them is $|x_1 - x_2| + |y_1 - y_2|$. This distance measure is also known as the Manhattan norm, since it corresponds to traveling using only horizontal and vertical lines, as we illustrate in Figure 2.6. The L_1 norm is used in work with computer chips owing to VLSI design and manufacturing technology that allows only horizontal and vertical connections.

A drawing of an optimal path for a 764-city scan-chain problem is given in Figure 2.7. This example was provided by Michael Jacobs and Andre Rohe from Sun Microsystems and was solved using the Concorde code (after converting the path problem to a TSP by adding an additional city). When we received the example in 2003 this was classified as a medium-sized scan-chain problem, with larger ones ranging in size up to 5,000 cities.

To reduce the time required for testing, a modern computer chip will typically have multiple scan chains (the 764-city example was just one of 25 or so chains on

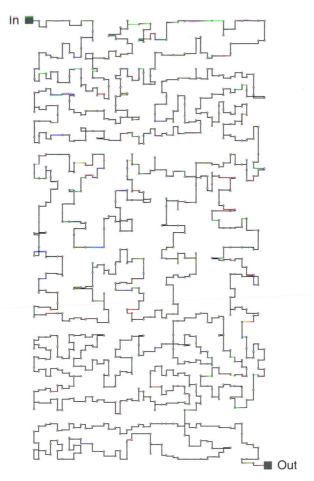

Figure 2.7 Scan-chain TSP with 764 cities.

the given chip). A standard approach is to first partition the scan points into sets, and then solve separately a TSP for each set to obtain the corresponding scan chain. The process of partitioning the scan points is discussed in Narayanan et al. [426] and Rahimi and Soma [468]. Even with multiple chains, it is expected that the size of the resulting TSP instances will continue to grow as the computer chips include an ever larger number of components.

2.4 DRILLING PROBLEMS

Printed circuit boards can be found in many common electronic devices. They are typically used to mount the integrated circuits discussed in the previous section,

Figure 2.8 Printed circuit board with 442 holes. Image courtesy of Martin Grötschel.

allowing chips to be combined with other hardware to obtain some specific functionality. While scan chains arise in chip design, a classic application of the TSP arises in the production of the basic boards.

A photograph by Martin Grötschel showing an example of a printed circuit board is displayed in Figure 2.8. As can be seen, the board has a relatively large number of holes used for mounting chips or for connecting among its layers. Such holes are typically produced by automated drilling machines that move between specified locations to drill one hole after another. The TSP in this application is to minimize the total travel time of the drill, where the cities correspond to the hole locations and the travel costs are estimates of the time to move between the holes. The particular set of locations for the board in the photograph is known in the research community as problem pcb442 in the TSPLIB, contributed by Grötschel, Michael Jünger, and Gerhard Reinelt. A drawing of the locations for this problem is given in Figure 2.9. Note that an additional city is added in the lower left-hand corner of pcb442: this is the place where the drill rests while the boards are loaded and unloaded from the machine.

The pcb442 instance is an often cited example in TSP studies, but a smaller problem on 318 cities is better known as the first instance of this type to appear in the literature. The problem, called lin318 in the TSPLIB, is described in the 1973 paper by Lin and Kernighan [357] that introduced their well-known heuristic algorithm. Concerning this drilling example, Lin and Kernighan write the following.

> It has often been observed that the problem of routing a numerically controlled drilling machine efficiently through a set of hole positions is a travelling-salesman problem, but if drilling time outweighs travel time, there is no particular advantage to any optimization. However, as we were experimenting we were given a problem in which drilling

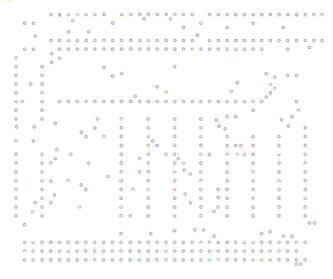

Figure 2.9 Points for pcb442.

is not done mechanically, but by a pulsed laser, and hence almost the entire cost *is* travel time.

They credit R. Habermann for providing the problem data and for sending a sample tour through the points. For a span of seven years in the 1980s the 318-city instance was the largest TSP instance that had been solved. A drawing of the hole locations for the problem is given in Figure 2.10.

The two drilling examples display several features common to many circuit board problems. The most notable of these are the large number of colinear points and the existence of repeated patterns. Despite this special structure, researchers have achieved significant improvements in industrial settings by using general-purpose TSP heuristics.

- Magirou [367] implemented a convex-hull method for the TSP that was adopted by Metalco, a circuit board manufacturer in Athens, Greece.

- Grötschel, Jünger, and Reinelt [235] describe a pilot project with the computer firm Siemens, leading to an industrial partnership to include TSP methods in everyday production of their circuit boards.

- Bernhard Korte led a study at the University of Bonn on circuit board production at IBM Germany. A range of methods were delivered to IBM's manufacturing devision.

Although positioning time is just one component of the drilling process, the use of TSP algorithms in these studies led to improvements of approximately 10% in the overall throughput of the production lines.

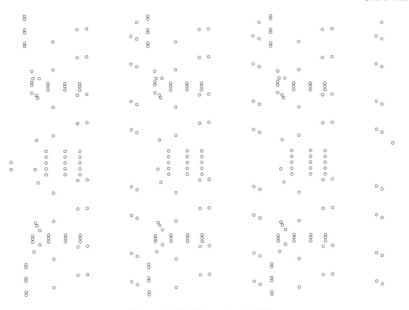

Figure 2.10 Points for lin318.

Typical problems in this class range in size from several hundred cities up to
several thousand cities. In Table 2.1 we display data provided by Ulf Radicke for
a board in the Bonn study. As in most applications, this example requires holes of
varying size, each drilled with a different spindle. The problem naturally decom-

Table 2.1 Circuit board drilling problem.

Drill Size (mm)	No. of Holes	Positioning Time(s)	Drilling Time(s)
18	5,132	1,107.4	1,385.6
43	2,722	581.5	680.5
44	35	29.4	13.7
50	130	45.3	50.7
67	10	20.0	5.0
81	2	13.0	1.0
96	32	26.8	18.2

poses, however, into separate TSP instances for each size, since the time needed to
change a spindle is relatively long compared with the time for drill movement.

In Table 2.1 we also report the total travel time used to position the drill head
in each of the locations, together with the time used for the actual drilling of the
holes. For most of the subproblems, the time is split evenly between movement
and drilling. This is not as attractive as the example cited by Lin and Kernighan,

but it still leaves ample opportunity for overall improvement via short tours, as demonstrated in the case studies mentioned above.

The travel times reported in Table 2.1 are the sum of the actual times used by the drilling machine. Industrial drills such as the one in the Bonn study often have separate motors for moving the drill head in the horizontal and vertical directions, allowing movement in both directions simultaneously. The travel times are therefore often modeled by the L_∞-norm, that is, if (x_1, y_1) and (x_2, y_2) describe the locations of two points on the board, then the travel cost is the larger of $|x_1 - x_2|$ and $|y_1 - y_2|$. This distance function is also known as the maximum norm. To handle the case where one of the two motors operates more quickly than the other, either the x or y coordinates can be scaled appropriately.

In actual operation, the movement of a drill head between two points involves acceleration at the start before it attains its peak speed and deceleration at the finish as it comes to a stop. In the Bonn study this aspect was modeled as a step-function, as shown in Table 2.2 for movement in the horizontal direction x. Similar values

Table 2.2 Drill movement time in horizontal direction.

Range of Movement (inches)	Time to Move t Inches (ms)
$0.00 \le t < 0.01$	$t * 15500.0$
$0.01 \le t < 0.05$	$155.0 + (t - 0.01) * 1062.5$
$0.05 \le t < 0.10$	$197.5 + (t - 0.05) * 300.0$
$0.10 \le t < 0.15$	$212.5 + (t - 0.1) * 300.0$
$0.15 \le t < 0.20$	$227.5 + (t - 0.15) * 250.0$
$0.20 \le t < 0.25$	$240.5 + (t - 0.2) * 300.0$
$0.25 \le t < 0.30$	$255.0 + (t - 0.25) * 1000.0$
≥ 0.30	$305.0 + (t - 0.3) * 154.6$

were used for the vertical direction y. The travel costs in the TSP were set to the maximum of the estimates for x and y times.

CUSTOMIZED INTEGRATED CIRCUITS

This same class of application, on a much smaller physical scale, arose in work at Bell Laboratories in the mid-1980s. Researchers at the Labs developed a technique for the quick production of customized integrated circuits. A news article by Joel Rosenkrantz [487] contains the following quote by Marty Gasper, one of the developers.

> We could go directly from circuit concept, to the layout and conversion to laser control coordinates, then to the laser machine and out with a working prototype in a matter of hours. The actual laser operation takes only 30 minutes for 1,000 logic gates.

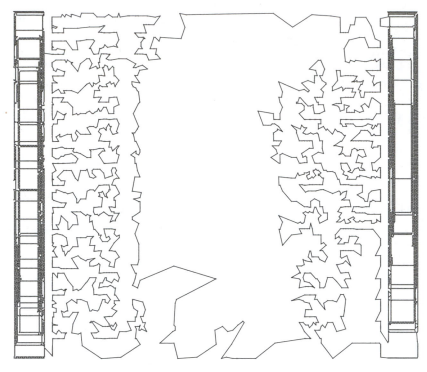

Figure 2.11 Optimal tour for pla7397.

The process starts with a basic chip having a network of simple gates. A set of the connections is then cut with a laser to build the logic for the customized chip. In this case, the TSP is to guide the laser through the locations that need to be cut.

David S. Johnson of AT&T Labs, and formally of Bell Labs, gathered three examples from this application and contributed them to the research community as part of the TSPLIB. The three instances are called pla7397, pla33810, and pla85900; the last two are the largest instances in the TSPLIB test suite. Johnson's own work on fast heuristic algorithms was able to provide very good solutions to the problems, and as part of the TSPLIB they became important challenge instances for both exact and heuristic solution methods. An optimal tour for the 7,397-city instance is displayed in Figure 2.11.

GUIDING LASERS FOR CRYSTAL ART

Pulsed lasers are used in many other manufacturing settings as well. A nice example is in the production of models and artwork burned into clear solid crystals, such as the "pla85900" object produced by Mark Dickens of Precision Laser Art displayed in Figure 2.12. The focal point of the laser beam is used to create fractures at specified three-dimensional points in the crystal, creating tiny points that are visible in the clear material; descriptions of the process can be found in Gross-

Figure 2.12 Crystal image drawn by laser.

man [227] and Troitski [525]. The TSP is again to route the laser through the points to minimize the production time.

Dickens has adopted heuristic methods from the Concorde TSP code to handle the massive data sets needed to obtain high-quality productions of elaborate images. A challenge here is to obtain appropriate estimates of the travel costs, modeling the combination of movement of servo motors and galvo mirrors that focus the laser.

This application holds a place of honor as having generated the largest industrial instances of the TSP that we have encountered to date, with some examples exceeding one million cities.

2.5 AIMING TELESCOPES AND X-RAYS

Although we normally associate the TSP with applications that require a physical visit to remote locations, the problem also arises when the sites can be observed from afar, without any actual traveling. A natural example is when the sites are planets, stars, and galaxies and the observations are to be made with some form of telescope. Here the process of rotating equipment into position is called *slewing*. For large-scale telescopes, slewing is a complicated and time-consuming procedure, handled by computer-driven motors. In this setting, a TSP tour that minimizes the total slewing time for a set of observations can easily be implemented as part of an overall scheduling process. The cities in the TSP are the objects to be imaged and the travel costs are estimates of the slewing times to move from one object to the next.

In a *Scientific American* article, Shawn Carlson [107] describes how a TSP heuristic came to his aid in scheduling a fragile, older telescope to image approximately 200 galaxies per night. Concerning the need for good TSP tours, Carlson writes the following.

> Because large excursions from horizon to horizon sent the telescope's 40-year-old drive system into shock, it was vital that the feeble old veteran be moved as little as possible.

Modern telescope installations are certainly not feeble, but good solutions to the TSP are vital for the efficient use of the very costly equipment. For example, the user information for the Australia Telescope Compact Array [26] contains the following item.

> For observations of many sources, use the ATMOS program in
> MIRIAD to solve the travelling salesman problem and optimise the
> order in which your sources are observed.

The ATMOS program is a Fortran computer code implementing a simulated annealing heuristic for the TSP. This tool is included in the MIRIAD package described in Sault et al. [489] and adopted at a number of telescope installations throughout the world.

Another example of the TSP in telescope scheduling is presented in Kaufer et al. [312] and Shortridge et al. [500]. This work is part of the Very Large Telescope project run by the European Southern Observatory. The computer codes in this case implement a collection of heuristics that are selected according to the number of observations that need to be made.

SPACE-BASED TELESCOPES

An interesting variation on this theme was considered by Bailey et al. [29] in planning work for a space-based telescope mission by NASA. In this study, the telescope is actually a pair of spacecraft operating in tandem as an interferometer, that is, the images are created by the interference patterns in the signals collected by the separated spacecraft. The details of the study focus on a mission called *StarLight*. The total number of stars to be imaged during the StarLight mission was only estimated to be 20, but an optimal solution to the corresponding TSP was important because of the limited fuel available for use in maneuvering the pair of spacecraft during slewing operations.

The StarLight mission was canceled in 2002, but NASA currently has a 2010 scheduled launch of a similar project called SIM PlanetQuest, consisting of a single spacecraft with collectors separated by 10 meters. An artist's image of the SIM spacecraft is shown in Figure 2.13. Documentation on SIM from the Jet Propulsion Laboratory [287] describes the targeting procedure for the interferometer telescope as follows.

> Pointing of the spacecraft will be performed using reaction wheels,
> with small thrusters used to desaturate the reaction wheels.

The JBL white paper on the project, edited by Edberg et al. [161], estimates that 12% of the flight time of the mission will be devoted to spacecraft slewing. Minimizing the time and/or energy needed for these operations is an obvious target for the TSP, following the work of Bailey et al.

X-RAY CRYSTALLOGRAPHY

The telescope TSP is quite similar to a study by Robert Bland and David Shallcross [66] in a different domain. Working with a team at the Cornell High Energy

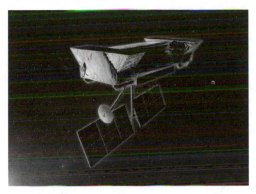

Figure 2.13 SIM PlanetQuest spacecraft. Image courtesy of NASA/JPL-Caltech.

Synchrotron Source in the mid-1980s, Bland and Shallcross adopt the TSP to aim a diffractometer in X-ray crystallography. The travel costs in this case are estimates of the time for the computer-driven motors to reposition the sample crystal and the X-ray equipment in between two observations.

One notable difference when compared with the telescope application is that in X-ray crystallography the observations to be taken number in the thousands. Indeed, Bland and Shallcross tested their heuristic methods on instances having between 2,762 cities and 14,464 cities. On these examples, obtained from five different crystals, an implementation of the Lin-Kernighan algorithm consistently produced tours of excellent quality. In all cases the length of the best tour was within 1.7% of a lower bound computed by a technique due to Held and Karp [254].

Bland and Shallcross contributed a set of test instances to the TSPLIB, described by a small amount of data defining the crystals, together with a computer code for producing the list of observations that are needed. These examples have not been widely studied, however, owing in part to differences in how to compute the travel costs between the cities.

2.6 DATA CLUSTERING

Organizing information into groups of elements with similar properties is a basic problem in data analysis and machine learning. Jain et al. [282] give a nice survey of this *data clustering* task, including a wide range of applications in object recognition, information retrieval, and data mining. The TSP has been adopted in a number of studies in this area, when there is a good measure $s(a, b)$ of the similarity between pairs (a, b) of data points. The idea is that using the $s(a, b)$ values as travel costs, a Hamiltonian path of maximum cost will place similar points near one another, and thus intervals in the path can be used as candidates for clusters.

This TSP approach was first adopted in a paper by McCormick et al. [386], where the data points are described by rows in a matrix. In their work, $s(a, b)$ is defined as the sum of the pairwise products of the entries in the rows corresponding to a and b.

The method is presented as one for reordering the matrix, but the connection to the TSP is made explicit in Lenstra [345]. To obtain clusters, the tours are examined by hand to select natural breakpoints in the ordering. This part can also be handled automatically in a variety of ways. For example, Alpert and Kahng [10] use a dynamic programming algorithm to split a tour into k intervals that serve as good clusters.

As an elegant alternative to a two-stage approach of creating a tour and then splitting it into intervals, Sharlee Climer and Weixiong Zhang [128] proposed to add $k + 1$ dummy cities in creating the TSP, rather than just the single city to convert from a path to a tour. Each of the dummy cities is assigned a travel cost of zero to all other cities. The additional cities serve as breakpoints to identify k clusters, since a good tour will use the zero-cost connections to the dummy cities to replace the large travel costs between clusters of points.

Climer and Zhang use their TSP+k method as a tool for clustering gene expression data, adopting Concorde to compute optimal tours and varying k to study the impact of different cluster counts. The image in Figure 2.14 was produced with their software. The data set displayed in the figure consists of 499 genes from the plant *Arabidopis* under five different environmental conditions; the shades of gray represent the gene expression values and the clusters are indicated by the solid white borders.

2.7 VARIOUS APPLICATIONS

We have covered some of the prominent applications of the TSP, but the problem also turns up in many other scenarios. We mention a few of these that have appeared in the literature.

- Gilmore and Gomory [207] show that a machine-scheduling problem can be modeled as a TSP. An interesting feature is that the application gives a case of the TSP that can be solved efficiently.

- Garfinkel [202] considers the TSP to model a problem in minimizing wallpaper waste.

- Ratliff and Rosenthal [470] solve a problem of picking items in a rectangular warehouse as a TSP.

- Madsen [366] uses the TSP to plan production in a pattern-cutting application in the glass industry.

- Grötschel, Jünger, and Reinelt [235] use the TSP to control a photo plotter in the drawing of masks for printed circuit boards. In an industrial application at Siemens, Grötschel et al. report savings of up to 17% in the total drawing time.

- Jacobson et al. [281] construct shortest possible universal nucleotide linkers using an early version of the Concorde TSP code. Here a universal linker

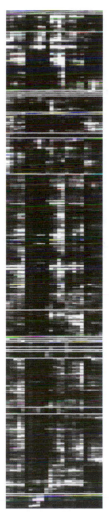

Figure 2.14 Gene expression data. Image courtesy of Sharlee Climer and Weixiong Zhang.

of order n is a sequence of DNA containing all possible co-palindromes of length $2n$ as subsequences, where a co-palindrome is a string of nucleotides that is identical to the string obtained by reading backward and complementing the base pairs.

- Cardiff et al. [106] find optimal routes through a theme park using a TSP model.

- Bock [68] uses the TSP to estimate the trenching costs for connecting the antennas of a large ground-based telescope array.

- Moret et al. [402] study problems in evolutionary change using TSP algorithms.

- Csürös et al. [148] use the TSP to assemble a genome map from a library of known subsequences, using a method called "clone-array pooled shotgun mapping." The Concorde code is used in this study to compute optimal orders for the models.

- Gutin et al. [243] use a TSP model in a problem of gathering geophysical seismic data. In this setting the items to be surveyed are line segments and the TSP is used to minimize the travel time between the traversals of the segments.

- Johnson et al. [290] test the TSP as a means for compressing large data sets of zero-one-valued arrays. The cost function in this case is the Hamming distance between two arrays, that is, the number of places where the entries in the arrays differ.

- Massberg and Vygen [382] use the TSP in the construction of an algorithm for a location problem in VLSI design.

- Pohle et al. [462] develop a TSP-based method for organizing the musical playlists on portable devices such as the iPod. Here the distance function measures the similarity of the timbre of the musical pieces.

These examples, and ones cited earlier, give a good picture of the range of disciplines where the TSP has appeared in one form or another. The breadth of the applications is a healthy sign for a problem with applied roots mainly in the area of logistics.

Other surveys of models and applications using the TSP can be found in papers by Garfinkel [201], Lenstra and Rinnooy Kan [346] and Punnen [466], and in the book by Reinelt [475].

Chapter Three

Dantzig, Fulkerson, and Johnson

The origins of the study of the TSP as a mathematical problem are somewhat obscure. Karl Menger's [389] *Botenproblem* was presented in Vienna in 1930, and later in the decade the TSP was widely discussed at Princeton University. There is little mention of the problem, however, in the mathematical literature from this period.

In one of the earliest papers on the TSP, Merrill Flood [182] made the following statement.

> This problem was posed, in 1934, by Hassler Whitney in a seminar talk at Princeton University.

This points to Whitney as the founding father of the TSP at Princeton. Hoffman and Wolfe [267] suggest that Whitney may have served "possibly as a messenger from Menger" in regards to the TSP. In a fascinating historical treatment, Schrijver [493] provides evidence of this, documenting that Menger and Whitney met at Harvard in 1931 and exchanged ideas related to the problem.

From Whitney, enthusiasm for the TSP passed on to Flood, about whom Hoffman and Wolfe [267] write the following.

> We do not know who brought the name TSP into mathematical circles, but there is no question that Merrill Flood is most responsible for publicizing it within that community and the operations research fraternity as well.

In the late 1940s, Flood moved from Princeton to the RAND Corporation, a hotbed of early work in mathematical optimization, bringing along his interest in the TSP.

It was at RAND that a breakthrough occurred in 1954, when George Dantzig, Ray Fulkerson, and Selmer Johnson [151] published a description of a method for the TSP and illustrated its power by solving an instance with 49 cities, an overwhelming size at that time. Riding the wave of excitement over the numerous applications of the simplex algorithm (designed by Dantzig in 1947), the three researchers attacked the salesman with linear programming, as we describe below.

3.1 THE 49-CITY PROBLEM

The TSP work at the RAND Corporation appears to have been motivated, in part, by the challenge of solving a particular example of the TSP through the United States. Julia Robinson wrote the following in a 1949 report [483].

One formulation is to find the shortest route for a salesman starting from Washington, visiting all the state capitals and then returning to Washington.

Also, Merrill Flood [183] referred to the "'48 states problem' of Hassler Whitney" when recalling early TSP work in an interview in 1984.

Dantzig, Fulkerson, and Johnson took up the challenge, creating their test instance by picking one city from each of the 48 states in the USA (Alaska and Hawaii became states only in 1959) and adding Washington, D.C.; the costs of travel between these cities were defined by road distances. Rather than solving this problem, they solved the 42-city problem obtained by removing Baltimore, Wilmington, Philadelphia, Newark, New York, Hartford, and Providence. As it turned out, an optimal tour through the 42 cities used the link joining Washington, D.C., to Boston; since the shortest route between these two cities passes through the seven removed cities, this solution of the 42-city problem yields a solution of the 49-city problem. The remaining 42 cities are listed in Table 3.1; the locations of the cities are indicated in Figure 3.1.

Table 3.1 The 42 cities in the reduced problem.

1. Manchester, NH	15. Portland, OR	29. Dallas, TX
2. Montpelier, VT	16. Boise, ID	30. Little Rock, AR
3. Detroit, MI	17. Salt Lake City, UT	31. Memphis, TN
4. Cleveland, OH	18. Carson City, NV	32. Jackson, MS
5. Charleston, WV	19. Los Angeles, CA	33. New Orleans, LA
6. Louisville, KY	20. Phoenix, AZ	34. Birmingham, AL
7. Indianapolis, IN	21. Santa Fe, NM	35. Atlanta, GA
8. Chicago, IL	22. Denver, CO	36. Jacksonville, FL
9. Milwaukee, WI	23. Cheyenne, WY	37. Columbia, SC
10. Minneapolis, MN	24. Omaha, NE	38. Raleigh, NC
11. Pierre, SD	25. Des Moines, IA	39. Richmond, VA
12. Bismarck, ND	26. Kansas City, MO	40. Washington, DC
13. Helena, MT	27. Topeka, KS	41. Boston, MA
14. Seattle, WA	28. Oklahoma City, OK	42. Portland, ME

The data for an n-city TSP instance consist of $n(n-1)/2$ travel costs between the pairs of cities, giving 861 items for the reduced 42-city problem. These costs were recorded by Dantzig, Fulkerson, and Johnson in a distance matrix like those found in atlases, but for our purposes it is useful to consider the data assembled as a vector c indexed by the pairs of cities. Each component of c corresponds to an *edge* e indicating a potential tour leg between two cities; we refer to the two cities as the *ends* of e. (We introduce standard graph theory terminology in Chapter 5 for dealing with this representation of the TSP.)

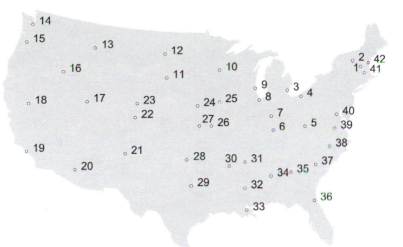

Figure 3.1 Locations of the 42 cities.

A tour through n cities can be represented as its incidence vector of length $n(n-1)/2$, with each component of the vector set at 1 if the corresponding edge is a part of the tour, and set at 0 otherwise. If x is such an incidence vector for the 42-city instance, then $c^T x$ gives the cost of the tour (since the product of the two vectors adds the distances used by the legs of the tour). So, letting $\mathcal{S}$ denote the set of the incidence vectors of all the tours, the problem is to

$$\text{minimize } c^T x \text{ subject to } x \in \mathcal{S}. \tag{3.1}$$

Like the man searching for his lost wallet not in the dark alley where he actually dropped it but under a street lamp where he can see, Dantzig, Fulkerson and Johnson begin not with the problem they *want to* solve but with a related problem they *can* solve, as we now describe.

For each $x \in \mathcal{S}$ and each edge e, let x_e denote the component of x corresponding to e. Trivially, each x in $\mathcal{S}$ satisfies

$$0 \le x_e \le 1 \text{ for all edges } e, \tag{3.2}$$

since each component of an incidence vector is either 0 or 1. Furthermore,

$$\sum (x_e : v \text{ is an end of } e) = 2 \text{ for all cities } v, \tag{3.3}$$

since in any tour $x \in \mathcal{S}$ and for any city v, we must have $x_e = 1$ for exactly two edges having v as an end (the tour enters v and then leaves v). Therefore, the problem

$$\text{minimize } c^T x \text{ subject to } (3.2), (3.3) \tag{3.4}$$

is a *relaxation* of the TSP in the sense that every solution to the TSP is also a feasible solution to (3.4). Although it is true that an optimal solution x^* to (3.4) need not be a tour, from the above discussion it follows that no tour can have cost less than $c^T x^*$. Thus, solving (3.4) provides us with a *lower* bound for the TSP; the optimal value $c^T x^*$ can be used to measure the quality of any proposed tour.

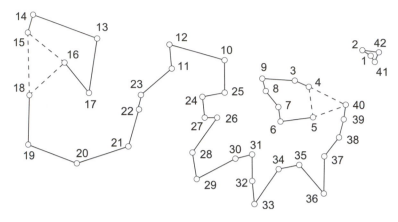

Figure 3.2 Solution of the initial LP relaxation.

At the time of Dantzig, Fulkerson, and Johnson's study, problems such as (3.4) were the center of attention: it is an example of a linear programming problem, or LP problem, for short. In general, an LP problem is to optimize a linear function subject to linear equality and inequality constraints. The excitement in the late 1940s and early 1950s followed Dantzig's [149] development of the simplex algorithm for solving LP problems, and the realization that many planning models could be cast in this form. (Today we have a variety of efficient methods and computer codes for solving LP problems, as we describe in Chapter 13.)

The idea of using an LP relaxation to study the TSP was proposed earlier in a paper by Robinson [483], and Dantzig, Fulkerson, and Johnson [151] wrote the following in a historical note to their seminal paper.

> The relations between the traveling-salesman problem and the transportation problem of linear programming appear to have been first explored by M. Flood, J. Robinson, T. C. Koopmans, M. Beckmann, and later by I. Heller and H. Kuhn.

The groundbreaking idea of Dantzig, Fulkerson, and Johnson was that solving the LP relaxation (3.4) can help with solving (3.1) in a far more substantial way than just by providing a lower bound: having satisfied oneself that the wallet is not under the street lamp, one can pick the street lamp up and bring it a little closer to the place where the wallet was lost. We will introduce their general idea by giving the details of its application to the 42-city TSP.

IMPROVING THE LP RELAXATION

The optimal solution to problem (3.4) is shown in Figure 3.2. In this drawing, the solid edges e carry $x_e^* = 1$, the dashed edges e carry $x_e^* = 1/2$, and the edges e that are not drawn carry $x_e^* = 0$. The value of the solution is 641, a respectable lower bound, given that the optimal tour turns out to have cost 699.

Notice, however, that the solution x^* presented in Figure 3.2 is disconnected; one of its two components has cities $1, 2, 41, 42$ and the other component has cities

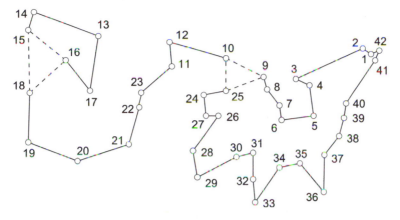

Figure 3.3 LP solution after three subtour constraints.

$3, 4, \ldots, 40$. This structural fault suggests that we can improve the LP relaxation by adding a suitably chosen linear inequality to the list of constraints. Indeed, since every tour must cross every demarcation line separating the set of cities into two nonempty parts at least twice, every x in $\mathcal{S}$ satisfies

$$\sum (x_e : e \text{ has one end in } S \text{ and one end not in } S) \geq 2 \qquad (3.5)$$

for all nonempty proper subsets S of cities; however, x^* in place of x violates (3.5) with $S = \{1, 2, 41, 42\}$. The addition of this inequality raises the optimal value of the LP relaxation to 676, getting us more than halfway toward the 699 target in a single stroke.

Constraints (3.5), called "loop conditions" by Dantzig, Fulkerson, and Johnson, are nowadays called *subtour elimination constraints* or simply *subtour constraints*: from the set of integer solutions of (3.2), (3.3), they eliminate those that are disconnected (consisting of two or more disjoint "subtours") and leave only the incidence vectors of tours. There are far too many of these inequalities to simply add them all to the LP relaxation. Instead, Dantzig, Fulkerson, and Johnson propose to utilize subtour constraints in an iterative fashion, directly attacking any additional faults that arise in the solution vectors that are obtained.

Following this procedure, the next two iterations are similar to the first: the LP solutions x^* are disconnected and we add two more subtour constraints, one with $S = \{3, 4, \ldots, 9\}$, the other with $S = \{24, 25, 26, 27\}$. Together, these constraints raise the LP value to 682.5.

A drawing of the new LP solution is given in Figure 3.3. At this point the solution x^* becomes connected but not 2-connected: the removal of city 18 splits it into two connected components, one with cities $13, 14, \ldots, 17$ and the other with cities $19, 20, \ldots, 42, 1, 2, \ldots, 12$. Again, this structural fault points out a violated subtour constraint: more generally, if the removal of a single city splits the rest of the solution into connected components with city sets $S_1, S_2, \ldots, S_k$ ($k \geq 2$), then

$$\sum (x_e^* : e \text{ has one end in } S_i \text{ and one end not in } S_i) \leq 1$$

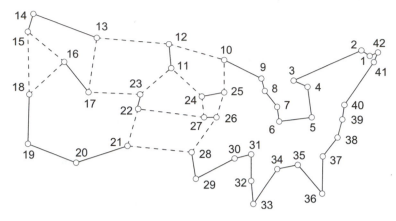

Figure 3.4 LP solution satisfying all subtour constraints.

for at least one set S_i, which violates equation (3.5) for this set. We add the subtour constraint with $S = \{13, 14, \ldots, 17\}$ and continue through three similar iterations (adding subtour constraints with $S = \{10, 11, 12\}$, $S = \{11, 12, \ldots, 23\}$, and $S = \{13, 14, \ldots, 23\}$) until we obtain a relaxation of value 697, whose optimal solution x^* is shown in Figure 3.4. This solution is 2-connected and no violated subtour constraints are apparent; in fact (using techniques we shall describe in Chapter 6), it can be checked that there are none.

BEYOND SUBTOUR CONSTRAINTS

Dantzig, Fulkerson, and Johnson's approach does not end here, however. As long as the optimal solution x^* found by the simplex algorithm is not a tour, by geometric arguments (discussed in the next section) we know that there must be a linear inequality that is satisfied by all x in $\mathcal{S}$ and is violated by x^*. Any such inequality is a candidate for inclusion in our LP relaxation; there is no need to restrict ourselves to subtour constraints.

Continuing the solution process, Dantzig, Fulkerson, and Johnson focus their attention on the triangle $\{15, 16, 18\}$ and the adjacent edges $\{14, 15\}$, $\{16, 17\}$, $\{18, 19\}$. The values of x_e^* for these edges sum to 4.5, while any tour can use at most 4 of them. This is a structural fault that can again be attacked with a linear inequality; we will present this inequality in a general form that will also be exploited in the next iteration.

Suppose we have four proper subsets of cities S_0, S_1, S_2, S_3 whose Venn diagram can be drawn as in Figure 3.5, that is, S_1, S_2, S_3 are pairwise disjoint and for each $i = 1, 2, 3$, S_i has at least one city in S_0 and at least one city not in S_0. Such a configure is called a *comb* (we will study these in a more general form in Chapter 5). We give below a short argument showing that every $x \in \mathcal{S}$ satisfies

$$\sum_{i=0}^{i=3} \left(\sum (x_e : e \text{ has one end in } S_i \text{ and one end not in } S_i) \right) \geq 10 \qquad (3.6)$$

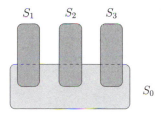

Figure 3.5 Venn diagram for a comb.

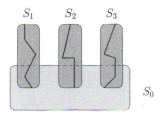

Figure 3.6 Three paths entering S_0.

for any comb. Now, if we take the comb consisting of the four sets

$$S_0 = \{15, 16, 18\}, S_1 = \{14, 15\}, S_2 = \{16, 17\}, S_3 = \{18, 19\}$$

then the left-hand side of (3.6) has value only 9, rather than the required 10. This is the inequality that is added to the LP relaxation, raising the lower bound to 698. (Dantzig, Fulkerson, and Johnson use a variation of this inequality, involving the additional edge $\{19, 20\}$.)

Let us show that (3.6) holds for any tour $x \in \mathcal{S}$. Given a nonempty proper subset of cities S, notice that there must be an even number of tour edges crossing S's border, that is, there is an even number of edges e such that $x_e = 1$ and e has one end in S and one end not in S. Furthermore, if exactly two tour edges cross S's border, then the tour x restricted to the cities S consists of a single path through S. Now, if x restricted to each of S_1, S_2, and S_3 consists of such a single path, then at least three tour edges cross S_0's border (since the three paths must enter S_0; see Figure 3.6). It follows that the left-hand side of (3.6) must be at least 9 for the tour x, and therefore at least 10 (since it must be an even number).

The updated LP solution x^*, after adding the specified inequality (3.6), is shown in Figure 3.7. Although it is not easy to spot, this vector again has a fault that can be attacked with a comb inequality. In this case, we take

$$S_0 = \{11, 12, \dots, 23\},$$
$$S_1 = \{1, 2, \dots, 12, 24, 25, \dots, 42\},$$
$$S_2 = \{20, 21\},$$
$$S_3 = \{22, 23\},$$

as indicated in Figure 3.8. Again, the sum of the left-hand side of (3.6) has value only 9, rather than the required 10. Adding this inequality to the LP relaxation produces a faultless solution: the optimal x^* is a tour.

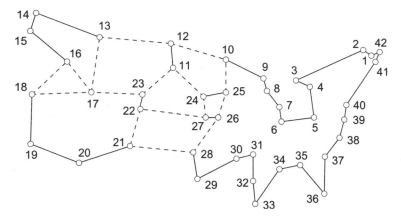

Figure 3.7 What is wrong with this vector?

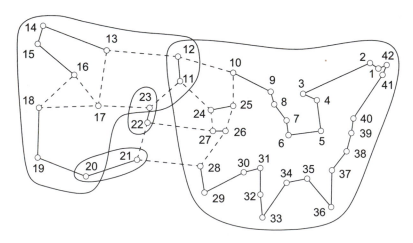

Figure 3.8 A violated comb.

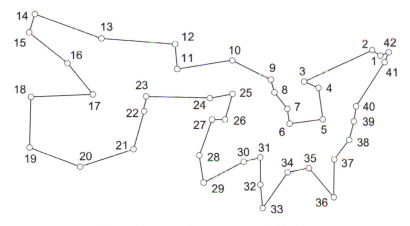

Figure 3.9 An optimal tour through 42 cities.

In place of this second comb, Dantzig, Fulkerson, and Johnson use an ingenious argument to show that an alternative inequality holds for all tours. In a footnote they write, "We are indebted to I. Glicksberg of Rand for pointing out relations of this kind to us." Like the second comb, this extra inequality also finishes the job. By adding this final constraint, Dantzig, Fulkerson and Johnson put the street lamp next to the wallet: they obtained a relaxation, whose optimal solution is the incidence vector of a tour (passing through the 42 cities in the order of their labels; a drawing is given in Figure 3.9). And that was the end of the 49-city problem.

3.2 THE CUTTING-PLANE METHOD

The linear programming approach adopted by Dantzig, Fulkerson, and Johnson grew out of earlier work by Heller [256] and Kuhn [328] on linear inequalities for the TSP. The main conclusion of the Heller-Kuhn studies is that although it is possible to give beforehand a finite list of all of the inequalities that might possibly be needed in the TSP model (and so a complete description of the convex hull of the vectors S), the length of such a list would necessarily be far too great for any linear programming solver to handle directly. The beauty of the Dantzig-Fulkerson-Johnson method is that it can proceed by generating these inequalities only as they are needed. This has great practical consequences, as Dantzig, Fulkerson, and Johnson [151] suggest in their modest conclusion:

> It is clear that we have left unanswered practically any question one might pose of a theoretical nature concerning the traveling-salesman problem; however, we hope that the feasibility of attacking problems involving a moderate number of points has been successfully demonstrated, and that perhaps some of the ideas can be used in problems of similar nature.

Indeed, the influence of Dantzig, Fulkerson, and Johnson's work has reached far beyond the narrow confines of the TSP.

On the one hand, the method can be used to attack problems of the form

$$\text{minimize } c^T x \text{ subject to } x \in \mathcal{S} \tag{3.7}$$

for any finite set $\mathcal{S}$ (or certain infinite sets as well) of m-dimensional vectors, not just when $\mathcal{S}$ is a set of tour vectors. In this general case, we begin with an LP problem

$$\text{minimize } c^T x \text{ subject to } Ax \leq b \tag{3.8}$$

with some suitably chosen system $Ax \leq b$ of linear inequalities satisfied by all x in $\mathcal{S}$; here A is a matrix with m columns and b is a vector with a component for each row of A. Since (3.8) is a relaxation of (3.7) in the sense that every feasible solution of (3.7) is a feasible solution of (3.8), the optimal value of (3.8) provides a lower bound on the optimal value of (3.7). Moreover, it is a characteristic feature of the simplex method that the optimal solution x^* it finds is an extreme point of the polyhedron defined by $Ax \leq b$; in particular, if x^* is not one of the points in $\mathcal{S}$, then it lies outside the convex hull of $\mathcal{S}$. In that case, x^* can be separated from $\mathcal{S}$ by a hyperplane: some linear inequality is satisfied by all the points in $\mathcal{S}$ and violated by x^*. Such an inequality is called a *cutting plane* or simply a *cut*. Having found a cut, one can add it to the system $Ax \leq b$, solve the resulting tighter relaxation by the simplex method, and iterate this process until a relaxation (3.8) with an optimal solution in $\mathcal{S}$ is found. This procedure is called the *cutting-plane method*.

Many problems in combinatorial optimization can be described as (3.7) for a particular choice of $\mathcal{S}$; these are all targets for the cutting-plane method. For example, in the maximum clique problem, $\mathcal{S}$ consists of the incidence vectors of all cliques in the input graph G (the components of these vectors are indexed by the vertices of G); in the maximum cut problem, $\mathcal{S}$ consists of the incidence vectors of all edge cuts in the input graph G (the components of these vectors are indexed by the edges of G); and so on. In each example, the challenge is to come up with linear inequalities satisfied by all the points in $\mathcal{S}$ and violated by the optimal solution x^* of the current LP relaxation. To meet this challenge, one often begins with common sense to establish some combinatorial property of $\mathcal{S}$ and only then (as we have seen in the 49-city example) one expresses this property in terms of linear inequalities. This line of attack has led to the development of the flourishing field of *polyhedral combinatorics*.

Second, the approach used in the solution of the 49-city TSP can be adopted to attack any mixed-integer linear programming problem, that is, an LP problem where some of the variables are required to take on integer values. To cast the TSP in this form, note that its $\mathcal{S}$ consists of all integer solutions of (3.2), (3.3), and (3.5), taken for all proper nonempty subsets of cities S. Again, the challenge here is to come up with linear inequalities satisfied by all the mixed-integer feasible solutions of the current relaxation (3.8) and violated by its optimal solution x^*. In the late 1950s and early 1960s, Ralph Gomory [214] answered this challenge with breathtaking elegance, designing automatic procedures to generate cutting planes in this context.

In this way, the work of Dantzig, Fulkerson, and Johnson became the prototype of two different methodologies: polyhedral combinatorics in combinatorial optimization and cutting-plane algorithms in mixed-integer linear programming.

3.3 PRIMAL APPROACH

Our description of the cutting-plane method and its application to the 49-city TSP differs from those originally proposed by Dantzig, Fulkerson, and Johnson. Given the initial LP relaxation, we write casually that one goes ahead and finds the optimal solution x^*. This is a simple matter today, given the wide availability of LP software, but Dantzig, Fulkerson, and Johnson carried out their computations by hand, as Fulkerson noted in a letter to Isidor Heller, dated 11 March 1954.[1]

> Recently G. Dantzig, S. Johnson, and I have been working on computational aspects of the problem via linear programming techniques even though we don't know, of course, all the faces of the convex C_n of tours for general n. The methods we have been using seem hopeful, however; in particular, an optimal tour has been found by hand computation for a large scale problem using 48 cities, one in each state, and also several smaller problems have yielded rather quickly.

This size of LP instances in question makes this a daunting task. As a first step, Dantzig, Fulkerson, and Johnson developed methods for implicitly handling the great number of variables in the LP relaxation; these ideas are still used today, as we describe in Chapter 12. Moreover, they proceed *without* actually solving the LP problems, working instead by taking only a single step of the simplex algorithm after adding a cutting plane.

The approach adopted by Dantzig, Fulkerson, and Johnson is summarized in a preliminary version [150] of their paper, in a section titled "The method."

> The title of this section is perhaps pretentious, as we don't have a method in any precise sense. What we do is this: Pick a tour x which looks good, and consider it as an extreme point of C_1; use the simplex algorithm to move to an adjacent extreme point e in C_1 which gives a smaller value of the functional; either e is a tour, in which case start again with this new tour, or there exists a hyperplane separating e from the convex of tours; in the latter case cut down C_1 by one such hyperplane that passes through x, obtaining a new convex C_2 with x as an extreme point. Starting with x again, repeat the process until a tour $\hat{x}$ and a convex $C_m \supset T_n$ are obtained over which $\hat{x}$ gives a minimum of $\sum d_{ij}x_{ij}$.

In this description, C_1 is the solution set of the LP relaxation and d_{ij} is the cost of the edge having ends i and j.

---

[1] We thank Robert Bland and David Shmoys for giving us access to the Fulkerson archives at Cornell University.

For the reduced 42-city problem, Dantzig, Fulkerson, and Johnson start their procedure with what turned out to be the optimal tour of length 699. (Hoffman and Wolfe [267] write that this tour was "worked out with strings on a model.") The incidence vector, $\bar{x}$, of the tour is a basic feasible solution (see Chapter 13) of the initial LP relaxation and it can be used as a starting point for the simplex algorithm. If $\bar{x}$ is not optimal, then the simplex algorithm selects a direction to move to a new basic solution x' via a pivot operation. If x' is not a tour, then a cutting plane is added to the LP. In choosing such a cut, Dantzig, Fulkerson, and Johnson restrict their attention to those that are both violated by x' and hold as an equation for $\bar{x}$; this extra condition allows them to cut off the entire line segment from $\bar{x}$ to x' and thus force the simplex algorithm to move in a different direction in the next step.

This process is known as a *primal cutting-plane method* since it attempts to move from tour to tour, rather than moving from lower bound to lower bound. The primal method has not been explored by the research community to nearly the same degree as the (dual) cutting-plane method we described in the previous section. In this book we will not treat further the primal method, except for some short remarks on the work of Padberg and Hong [451], who created a computer implementation for the TSP in the late 1970s.

A general discussion of the primal cutting-plane method can be found in Letchford and Lodi [349], where some trade-offs between the primal and dual methods are considered. The choice of one method over the other is not discussed in the paper of Dantzig, Fulkerson, and Johnson, but the ability to proceed with relatively few simplex pivots may have been decisive, given their reliance on by-hand computations.

Chapter Four

History of TSP Computation

The triumph of Dantzig, Fulkerson, and Johnson in 1954 set off a flurry of activity in mathematics and operations research circles, building toward Gomory's fundamental work in mixed-integer linear programming at the end of the decade. Interestingly, however, the TSP literature from the 1950s does not include any attempts to extend or automate the Dantzig et al. solution procedure; from the narrow perspective of the TSP, this appears to be the dark ages of the cutting-plane method. This inactivity slowed the overall progress on the problem, and it would be another 16 years before a larger TSP instance was solved, by Michael Held and Richard Karp.

We can only speculate as to why early TSP researchers did not pursue the use of cutting planes in their computational research. One factor may have been the frightening task of trying to apply by hand the simplex algorithm for LP problems of the size studied by Dantzig, Fulkerson, and Johnson; neither computing equipment nor LP software was widely available during this period. We discuss this point in detail in Chapter 13, where we present a history of LP computation, covering the influence of the TSP on the development of LP software.

A second factor may have been simply that LP modeling was a new subject, and aspects of the cutting-plane work were not fully digested by other TSP researchers. This point led Dantzig, Fulkerson, and Johnson [153] to write a second paper, in 1959, illustrating the cutting-plane method step by step on a 10-city example from Barachet [36]. They write, "...judging from the number of queries we have received from readers, this method was not elaborated sufficiently to make the proposal clear."

A third factor may have been an overemphasis on solution methods suitable only for small TSP instances (say, up to 20 cities or so). In this domain, the cutting-plane method is not the solution method of choice, and a small study by J. T. Robacker [481] at RAND in 1955 may have inadvertently created optimism for the power of alternative techniques. Robacker carried out computational tests of the cutting-plane method by hand on 10 instances, each having 9 cities. His report that "the average time to work one of the problems was about 3 hours" was cited as a benchmark for the next few years of computational work on the TSP.

Whether due to a combination of these factors or to others of which we are not aware, besides the study of Robacker and the explanatory paper of Dantzig et al., no mention of TSP computation with the cutting-plane method is given in the years immediately following the paper of 1954. Commenting on this in 1966, Ralph Gomory [217] made the following statement in a survey paper presented at a workshop at IBM.

I do not see why this particular approach stopped where it did. It should be possible to use the same approach today, but in an algorithmic manner.

Shortly after this, the theme was finally picked up again in the Ph.D. work of Saman Hong, studying with Mandell Bellmore at Johns Hopkins University. Hong automated the search for subtour inequalities and introduced a combination of cutting planes and branch-and-bound methods (see Section 4.1). Hong's work was followed by the breakthrough results of Martin Grötschel and Manfred Padberg, who championed the use of cutting planes in the 1970s and 1980s. With Grötschel and Padberg in the lead, the cutting-plane method went on to become the dominant solution technique for the TSP, and for other combinatorial problems as well.

In the sections below, we present a short history of TSP computation. More detailed coverage of early work on particular topics can be found in individual chapters later in the book.

4.1 BRANCH-AND-BOUND METHOD

Although the years following the 1954 paper were quiet times for cutting planes, during this period an entirely different method for solving the TSP emerged and eventually gained wide popularity. The algorithms designed by Bock [69], Croes [145], Eastman [159], and Rossman and Twery [488] seem to be its earliest instances; Land and Doig [334] pointed out that its use extends beyond the TSP; Little, Murty, Sweeney, and Karel [360] dubbed it the *branch-and-bound method*.

Hints of branch-and-bound techniques are actually contained in the papers of Dantzig, Fulkerson, and Johnson [151], where they are described as an extension of the cutting-plane method. We introduce the topic in this way (although the general process we describe was not laid out until after the Dantzig et al. work).

To begin, progress of the cutting-plane method toward solving a particular instance of the problem

$$\text{minimize } c^T x \text{ subject to } x \in \mathcal{S} \tag{4.1}$$

for some set $\mathcal{S}$ is often estimated by the increase in the optimal value of its LP relaxation; as more and more cuts are added, these increases tend to get smaller and smaller. When they become unbearably small, the sensible thing to do may be to *branch*: having chosen a vector α and numbers β', β'' with $\beta' < \beta''$ such that each $x \in \mathcal{S}$ satisfies either $\alpha^T x \leq \beta'$ or $\alpha^T x \geq \beta''$, we solve the two *subproblems*,

$$\text{minimize } c^T x \text{ subject to } x \in \mathcal{S} \text{ and } \alpha^T x \leq \beta'$$

and

$$\text{minimize } c^T x \text{ subject to } x \in \mathcal{S} \text{ and } \alpha^T x \geq \beta'',$$

separately. At some later time, one or both of these two subproblems may be split into sub-subproblems, and so on. The subproblems are in a natural one-to-one correspondence with the nodes of a *branch-and-bound tree*: the original problem (4.1) is the root of the tree, each subproblem that leads to a branching step is the

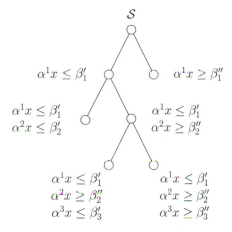

Figure 4.1 Branch-and-bound tree.

parent of two sub-subproblems, and all the remaining subproblems are the leaves of the tree; see Figure 4.1. Each node in the tree has the form

$$\text{minimize } c^T x \text{ subject to } x \in \mathcal{S} \text{ and } Cx \leq d \qquad (4.2)$$

for some system $Cx \leq d$ of linear inequalities, where C is a matrix and d is a vector. Each leaf will have been either solved without recourse to branching or else found irrelevant (*pruned*), since the optimal value of its LP relaxation turned out to be at least as large as $c^T x$ for some previously known element x of $\mathcal{S}$.

In the special case where all points in $\mathcal{S}$ are integer and x^* is not, branching can be implemented by choosing a subscript e such that x_e^* is not an integer and using the constraints

$$x_e \leq \lfloor x_e^* \rfloor \text{ and } x_e \geq \lceil x_e^* \rceil,$$

where, as usual, $\lfloor t \rfloor$ is t rounded down to the nearest integer and $\lceil t \rceil$ is t rounded up to the nearest integer. More generally, if all points in $\mathcal{S}$ are integer, branching can be implemented by choosing an integer vector α such that $\alpha^T x^*$ is not integer and letting

$$\beta' = \lfloor \alpha^T x^* \rfloor \text{ and } \beta'' = \lceil \alpha^T x^* \rceil.$$

This scheme is one of the many variants of the branch-and-bound method. The term "branch-and-bound," coined by Little, Murty, Sweeney, and Karel, refers to a general class of algorithms, where relaxations of (4.1) may come from a universe far wider than that of LP relaxations, and each subproblem may be split into more than two sub-subproblems.

The early algorithms of Bock [69], Croes [69], and Rossman and Twery [488] are based on enumeration techniques that do not easily fit into the branch-and-bound scheme, but use the same idea of creating subproblems and pruning them when possible. Bock's algorithm was programmed for an IBM computer and was used to solve two 10-city instances, including the example of Barachet [36] mentioned earlier. Croes' algorithm was not implemented for a computer, but he reports solving the reduced 42-city problem of Dantzig, Fulkerson, and Johnson with 70 hours of by-hand computation. Regarding a computer implementation, Croes writes the following.

> Programming of the rest will present some difficulties. Though these are certainly not insurmountable, it remains dubious whether it is efficient to use an electronic computer for these calculations, as they involve mostly inspectional work.

The Rossman and Twery algorithm also appears not to have been implemented (little information is given in the conference abstract [488], other than reporting the solution of a 13-city instance by hand). These three papers from 1958, as well as the comments in Dantzig, Fulkerson, and Johnson [151], should be viewed as precursors to the branch-and-bound method. Later papers along similar lines include Müller-Merbach [406] and Thüring [522].

At about the same time as the Bock, Croes, and Rossman-Twery work, Eastman [159] wrote a doctoral thesis containing a TSP algorithm that may be the earliest description of a complete branch-and-bound method. Eastman worked with a variation of the LP problem

$$\text{minimize } c^T x \tag{4.3}$$

$$\text{subject to}$$

$$\sum(x_e : v \text{ is an end of } e) = 2 \quad \text{for all cities } v$$

$$0 \le x_e \le 1 \quad \text{for all edges } e$$

that starts the Dantzig, Fulkerson, and Johnson scheme. (He adopts the asymmetric version, with each variable x_e replaced by a pair of variables x_{uv} and x_{vu}, where u and v are the ends of the edge e.) He avoids using a general LP method, instead solving the relaxations using network-flow algorithms developed by Ford and Fulkerson [185] several years earlier. In Eastman's formulation, the LP solutions are integer vectors, but they may contain subtours. In the branching step, a subtour having k edges is chosen, and k subproblems are created by setting, in turn, $x_e = 0$ for each of the subtour's edges. Thus, in typical fashion for branch-and-bound algorithms, the relaxations provide lower bounds, as well as providing a guide to carrying out the branching step.

Eastman [159] gives a step-by-step description of the solution of a 10-city TSP, but he does not report on a computer implementation of his algorithm. Refined versions of Eastman's algorithm were implemented in 1966 by Shapiro [498] and in 1971 by Bellmore and Malone [49], reporting computational tests on instances with up to 20 cities.

In between the studies of Eastman and Shapiro, the research team of Little, Murty, Sweeney, and Karel [360] developed a different approach to working with the LP relaxation. In their branch-and-bound algorithm, they do not actually solve the LP problem at each step. Instead, Little et al. compute an approximate dual solution that serves as a lower bound on the LP problem (via weak LP duality) and thus as a lower bound for the TSP. The branching step creates two subproblems by selecting a variable x_e and adding the constraints $x_e = 0$ in one case and $x_e = 1$ in the other case; the edge e is selected so as to increase as much as possible the lower bound on the $x_e = 0$ subproblem. This work was carried out in 1963, and their paper is the first to report extensive computational tests of a proposed TSP solution algorithm. Among other examples, Little et al. include the solution of a 25-city instance from Held and Karp [252].

A further TSP study was made by Fleischmann [179] in 1967, using a branch-and-bound algorithm of Balas [30], designed to handle integer LP problems with zero-one variables. Fleischmann reports on the solution of two small TSP instances, one having six cities and the other having seven cities.

HELD-KARP BOUND

The general weakness of the above branch-and-bound algorithms, in terms of the size of problem instances that can be solved, is largely due to the poor quality of the lower bound provided by the LP relaxation. Recall from the previous chapter that the LP relaxation (4.3) for the 42-city problem gives a bound of 641, versus the 699 cost of the optimal tour. This 9% gap between lower bound and tour is difficult to close by branching techniques alone, preventing researchers from pushing beyond the size of problems studied years earlier by Dantzig, Fulkerson, and Johnson.

A great improvement came with the 1970 work of Held and Karp [253], who directly attacked the lower bound issue with a new iterative scheme for approximating the LP relaxation

$$\text{minimize } c^T x \tag{4.4}$$

$$\text{subject to}$$

$$\sum(x_e : v \text{ is an end of } e) = 2 \quad \text{for all cities } v$$

$$\sum(x_e : e \text{ has exactly one end in } S) \geq 2 \quad \text{for all proper subsets } S \neq \emptyset$$

$$0 \leq x_e \leq 1 \quad \text{for all edges } e$$

obtained by adding subtour inequalities for all nonempty proper subsets S of cities. Solving (4.4) can provide a much stronger lower bound than the one given by the solution of (4.3). For example, it gives a bound of 697 for the 42-city TSP, closing 96% of the gap between the value of (4.3) and the cost of the optimal tour. Even today, however, the only efficient way to solve (4.4) is via the cutting-plane method, and this was not a process Held and Karp wanted to adopt, given the state of LP software at the time of their study.

The bound proposed by Held and Karp is based on computing a minimum-cost spanning tree for the set of cities of a TSP instance. A *spanning tree* is a minimal set of edges that joins all of the cities into a single connected component; by "minimal" we mean that if any edge is removed, then the set of cities is no longer connected. A TSP tour is not a spanning tree (since deleting any edge keeps the cities connected), but if we remove a single city v and the two tour edges meeting v, then the resulting path is a spanning tree for the remaining nodes. In this way, a lower bound for the TSP can be computed by adding the costs of the cheapest two edges meeting city v together with the minimum cost of a spanning tree connecting the remaining cities; Held and Karp call this collection of edges a *1-tree*, labeling v as city 1.

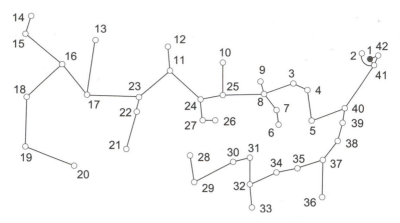

Figure 4.2 Optimal 1-tree for the Dantzig-Fulkerson-Johnson 42-city TSP.

What makes this useful is the fact that minimum-cost spanning trees (and hence 1-trees) can be computed very efficiently with a number of different methods (studies on the history of spanning-tree algorithms can be found in Graham and Hell [224], Nešetřil [429], and Schrijver [495]). A drawing of the optimal 1-tree for the reduced 42-city TSP of Dantzig, Fulkerson, and Johnson is given in Figure 4.2.

The connection between spanning trees and the TSP was well known in the research community. In fact, a 1956 paper by Kruskal [326], giving one of the first efficient algorithms for computing minimum-cost spanning trees, is titled "On the shortest spanning subtree of a graph and the traveling salesman problem." Furthermore, in a 1965 survey paper, Lawler and Wood [339] suggest that the spanning-tree bound could be used in a branch-and-bound algorithm for the TSP.

This idea of Lawler and Wood was picked up by Held and Karp, but with a crucial twist that allowed for a great improvement in the 1-tree lower bound. The difficulty with 1-trees is that their structure can be far removed from that of a tour, as evidenced by the many cities meeting either one or three edges in the optimal 1-tree in Figure 4.2. This extra freedom often means that the optimal 1-tree has much lower cost than does the optimal tour: the cost of the 1-tree for the 42-city TSP is only 600, while the optimal tour has cost 699. In Held and Karp's scheme, however, a 1-tree that does not look like a tour presents an opportunity to possibly improve the lower bound via a transformation of the edge costs, as we describe below.

The Held-Karp idea starts with Flood's [182] observation that if we choose a city v and subtract a number y_v from the cost of each edge meeting v, then the TSP remains the same, in the sense that one tour is cheaper than another after the transformation if and only if it was cheaper before the transformation: all we have done is subtract $2y_v$ from the cost of every tour. Expanding this to all of the cities, we can assign a number y_v to each city v and replace each edge cost c_e by $\bar{c}_e = c_e - y_u - y_v$, where u and v are the ends of e. Again, the TSP has not been changed, since the cost of each tour is just its old cost minus $\sum(2y_v : v \in V)$, where V denotes the set of all cities. It follows that if we have a lower bound B on

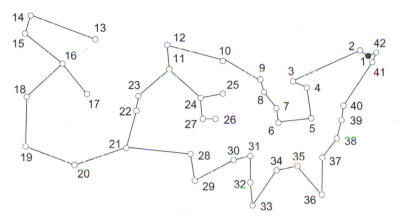

Figure 4.3 Held-Karp 1-tree for the Dantzig-Fulkerson-Johnson 42-city TSP.

the cost of a tour with the transformed edge costs, then $B + \sum(2y_v : v \in V)$ is a lower bound for the original TSP.

The crucial point is that although the above transformation does not change the TSP, it *can* change the 1-tree problem in a fundamental way. If the city weights $(y_v : v \in V)$ are chosen carefully, then the value of the new optimal 1-tree can provide a much stronger lower bound than that given by the original optimal 1-tree. Held and Karp proposed an iterative scheme for making this careful choice. First, note that to obtain an improvement in the bound, we want the value of the 1-tree to go down by less than the "correction factor" $\sum(2y_v : v \in V)$. So if a city v meets only one edge in the original optimal 1-tree, then it makes sense to assign y_v a positive value to make v's edges cheaper. On the other hand, if a city v meets more than two edges in the original optimal 1-tree, then the sensible thing to do is to assign y_v a negative value. Held and Karp do just this, moving from 1-tree to 1-tree, repeatedly adjusting the city weights according to these two rules.

Held and Karp [254] and Held, Wolfe, and Crowder [255] present methods for choosing the amounts by which to increase and decrease the city weights $(y_v : v \in V)$ in such a way that the best value in the sequence of lower bounds will converge to the *best possible bound* that can be obtained as a 1-tree with a correction factor. In a beautiful connection with the Dantzig, Fulkerson, and Johnson study, Held and Karp [253] show that this best possible bound is precisely equal to the optimal value of the LP relaxation (4.4) obtained by adding all subtour constraints.

In practice, it is not feasible to allow the Held-Karp process to run long enough to converge to the best possible bound. Quite good bounds, however, can often be obtained with only modest runs of their procedure. For example, a short run of Held-Karp produced the 1-tree displayed in Figure 4.3 for the 42-city TSP. Together with the corresponding correction factor, this 1-tree gives a bound of 696.875 (quite close to the 697 optimal value for (4.4)).

Held and Karp [254] combine this lower bound with a branching rule designed to increase the bounds of the subproblems by forcing certain edges to be included or excluded. The strength of their bound, coupled with the speed with which they could execute the iterative process, put them in a good position to attack large instances of the TSP.

In his 1985 Turing Award Lecture, Karp [310] made it clear that the intention of the Held-Karp study was indeed to surpass the computational results of Dantzig, Fulkerson, and Johnson.

> A few years earlier, George Dantzig, Raymond Fulkerson, and Selmer Johnson at the RAND Corporation, using a mixture of manual and automatic computation, had succeeded in solving a 49-city problem, and we hoped to break their record.

And this they did. Their successes included the solution of the following examples:

- The 42-city reduced TSP from Dantzig, Fulkerson, and Johnson [151].

- A 48-city instance from an earlier paper by Held and Karp [252].

- A 57-city instance from Karg and Thompson [304].

- A 64-city random Euclidean instance, obtained by randomly placing cities in a square (of side 1,500) according to a uniform distribution, and using the Euclidean distance between pairs of cities as the travel costs.

- A second 64-city instance obtained by joining the 42-city instance with a random Euclidean 22-city instance, using a construction of Lin [356].

Seventeen years after Dantzig, Fulkerson, and Johnson, there was now a new standard in TSP computation. Each of the problems was solved in under 15 minutes of computing time on an unspecified computer; Held and Karp thank Linda Ibrahim "who programmed the algorithm reported in this paper." A drawing of the 57-city tour is given in Figure 4.4.

The great success of the Held-Karp solution process encouraged a number of researchers to pursue their branch-and-bound technique; refinements were reported by Helbig Hansen and Krarup [251], Camerini, Fratta, and Maffioli [102], Bazaraa and Goode [44], Smith and Thompson [505], Volgenant and Jonker [544], and Smith, Meyer, and Thompson [504]. The ideas presented in these papers include the use of heuristic algorithms to search for tours while processing subproblems, new branching strategies, and speedups for the iterative bounding procedure. Computational tests were carried out on sets of problems similar to those used by Held and Karp, with the aim of reducing the computing time required to solve the instances.

Notable among these post-Held-Karp studies are the experiments of Camerini, Fratta, and Maffioli [102]. In 1975, this team from Milano, Italy, solved a 67-city random Euclidean instance, as well as a 100-city instance that was composed from a 48-city problem of Held and Karp [252] and several smaller problems. The 100-city instance is unusual in that it glues together four small geometric problems in a nongeometric way (using the construction of Lin [356]), and it is not clear how we should compare this with other TSP computations. There is no such debate with the 67-city instance, however, and this again pushed the TSP to a new record.

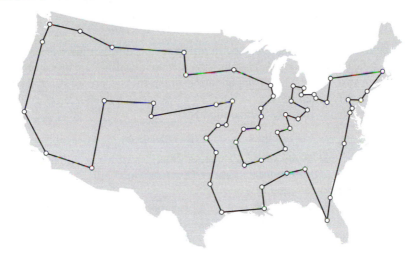

Figure 4.4 Optimal tour for the Karg-Thompson 57-city TSP.

4.2 DYNAMIC PROGRAMMING

A third class of TSP algorithms arose in the early 1960s, utilizing Richard Bell-man's [46] theory of *dynamic programming*. The idea here is that in an optimal tour, after we traverse a number of cities, the path through the remaining subset must itself be optimal. This allows us to build the tour step by step: using a list of all minimum-cost paths through subsets of k cities (with specified starting and end-ing points), we can create a list of all such paths through sets of $k + 1$ cities. This strategy was taken up by Bellman [47], Gonzalez [221], and Held and Karp [252] (in work that preceded their record-setting computations at the end of the decade).

The Held-Karp paper reports on a computer implementation that solved instances with up to 13 cities. This modest problem size is due to the rapid increase in the amount of data that needs to be processed and recorded in order to keep the list of all k-city paths. On the positive side, an analysis by Held and Karp shows that the dynamic programming algorithm can solve any n-city TSP instance in time that is at most proportional to $n^2 2^n$. Although at first glance this may appear to be a weak time bound, it is significantly less than the n factorial time it would take to enumerate all tours. In fact, this analysis of Held and Karp holds a place of honor in the TSP literature: it has the best time complexity for any known algorithm capable of solving all instances of the TSP. A discussion of this point can be found in Woeginger [556].

We will not treat further this dynamic programming approach to the TSP. De-spite the nice worst-case time bound obtained by Held and Karp, the inherent growth in the practical running time of the method restricts its use, even today, to tiny instances of the problem.

4.3 GOMORY CUTS

Returning to the cutting-plane method, a milestone occurred in the work of Ralph Gomory at the end of the 1950s, applying the idea to mixed-integer linear programming. In our discussion, we restrict attention to the case where all variables in an LP problem are required to take on integer values; these models are referred to as *integer LP problems*. Here the formulation is

$$\text{minimize } c^T x \text{ subject to } x \in \mathcal{S} \qquad (4.5)$$

where $\mathcal{S}$ is specified as the set of all integer solutions of a prescribed system of linear inequalities. For this class, Gomory [214] designed efficient procedures to automatically generate cutting planes.

Let us describe Gomory's idea under the assumptions that the LP relaxation is defined by a system $Ax \leq b$ of linear inequalities such that all the entries of matrix A and vector b are integers and A has full column rank. We will also assume that the LP relaxation is solved by the simplex method.

Along with an optimal solution x^* of the current LP relaxation, the simplex method finds a subsystem $A'x \leq b'$ of $Ax \leq b$ such that A' is a square nonsingular matrix and $A'x^* = b'$. Since x^* is not an integer vector, there is an integer vector w such that $w^T x^*$ is not an integer. Every such w yields the cut

$$(w^T - \lfloor v \rfloor^T A')x \leq \lfloor w^T x^* \rfloor - \lfloor v \rfloor^T b' \qquad (4.6)$$

where v is obtained from w by solving the system $v^T A' = w^T$ and $\lfloor v \rfloor$ is obtained from v by rounding each component down to the nearest integer.

To see that (4.6) is satisfied by all points of $\mathcal{S}$, note first that every vector x satisfies

$$w^T x + v^T (b' - A'x) = w^T x^*,$$

then that every solution x of $A'x \leq b'$ satisfies

$$w^T x + \lfloor v \rfloor^T (b' - A'x) \leq w^T x^*,$$

and finally that every integer solution x of $A'x \leq b'$ satisfies

$$w^T x + \lfloor v \rfloor^T (b' - A'x) \leq \lfloor w^T x^* \rfloor.$$

Gomory showed how to generate cuts of this type in such a way that guarantees the cutting-plane method's termination. His methods were quickly adopted by a number of TSP researchers, who formulated the TSP directly as an integer LP problem and applied Gomory's algorithm:

- Lambert [332] solves a 5-city instance of the TSP using Gomory cuts.

- Miller, Tucker, and Zemlin [395] present an integer LP formulation of the TSP and report their computational experience with Gomory cuts. The study includes the solution to a 4-city example, and an unsuccessful attempt to solve the 10-city example of Barachet [36].

- Mudrov [404] presents an integer LP formulation of the TSP and proposes to use Gomory cuts to solve TSP instances.

Notice that this TSP work was restricted to tiny instances, owing to the size and complexity of the formulations.

The experience of the above researchers, however, is not the end of the story for Gomory cuts in the world of the TSP. An alternative to using direct integer LP formulations was proposed by Glenn Martin [378] in an unpublished manuscript from 1966. Martin's algorithm works roughly as follows. Start with the degree constraints

$$\sum(x_e : v \text{ is an end of } e) = 2 \quad \text{for all cities } v \tag{4.7}$$

and one subtour constraint for the ends of the cheapest edge incident to each city (these 2-city subtour constraints are equivalent to the upper bounds $x_e \leq 1$ on the selected edges e). Apply Gomory's algorithm to this problem. If the solution is a tour, it must be optimal; otherwise, find violated subtour constraints (by hand in this case), add them, call Gomory's algorithm again, and so on. In this way, Martin was able to solve the Dantzig, Fulkerson, and Johnson 42-city instance, calling Gomory's algorithm a total of three times.

In a 1978 paper, Panayiotis Miliotis [394] fully automated Martin's approach and solved a collection of TSP instances that had appeared earlier in the literature, with sizes ranging from 42 to 64 cities, and an additional set of random Euclidean instances having up to 80 cities; the largest number of Gomory cuts that he was forced to generate in solving these 14 instances was 27. The solution of the 80-city problem surpasses the earlier 67-city Euclidean problem of Camerini, Fratta, and Maffioli [102], but as we shall see, by 1978 a new record of 120 cities had been established by Grötschel [228] using a different approach. Nevertheless, given the delay in the publication of Miliotis' paper (it was submitted in August 1975), it seems likely that for a short time Miliotis held the record for the largest solved Euclidean TSP instance.

The successful studies of Martin and Miliotis were continued in the work of Land [333] and Fleischmann [180], adding versions of comb inequalities to the mix of cutting planes, as we report in Section 4.5.

4.4 THE LIN-KERNIGHAN HEURISTIC

The excitement in TSP research through the 1950s and 1960s was centered on the mathematical challenge of finding provably optimal solutions to the problem. A number of researchers, however, began to explore the relaxed challenge of finding tours that, while perhaps not optimal, are hopefully at least of low cost. This eventually became a very popular topic of study, including several hundred research papers in later decades; detailed surveys can be found in Johnson and McGeoch [291] and Reinelt [475]. In Chapter 15, we cover aspects of this tour-finding work that have directly impacted exact solution techniques, providing starting tours for branch-and-bound and cutting-plane algorithms. Chief among these heuristic methods is an algorithm proposed by Shen Lin and Brian Kernighan [357] and adopted in most subsequent studies of exact methods for the TSP.

Lin and Kernighan's paper appeared in 1973, making a great improvement in the

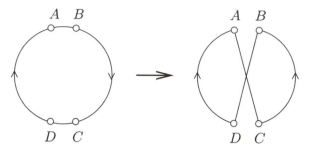

Figure 4.5 A 2-opt move.

quality of tours provided by heuristic methods. Even today, their algorithm remains
the key ingredient in the most successful approaches for finding high-quality tours.
This is remarkable, given the amount of attention tour finding has received and
the fact that, due to limited computational resources, Lin and Kernighan's study
included only instances with at most 110 cities.

The Lin-Kernighan algorithm is a *tour improvement method* in that it takes a
given tour and attempts to modify it in order to obtain an alternative tour of lower
cost. The basic step goes back to Flood [182], who observed that often a good tour
can be obtained by repeatedly replacing pairs of tour edges by cheaper alternative
pairs where possible. To describe the idea, suppose we have a tour such that for a
specified orientation (that is, a direction of travel around the tour), we visit city A
immediately before city B and we visit city C immediately before city D. Now if
the sum of the costs of the tour edges $\{A, B\}$ and $\{C, D\}$ is greater than the cost of
$\{A, C\}$ and $\{B, D\}$, then we can improve the tour by deleting $\{A, B\}$ and $\{C, D\}$
and adding $\{A, C\}$ and $\{B, D\}$, as in Figure 4.5. This exchange of Flood is known
as a *2-opt move* and was studied by other early TSP researchers, including Morton
and Land [403], Bock [69], Croes [145], and Lin [356].

To see how 2-opt moves work on a real example, consider the tour indicated in
Figure 4.6. This tour through Dantzig, Fulkerson, and Johnson's set of 42 cities has
length 1,013. It was found with the *nearest-neighbor algorithm* described by Flood
[182]: we choose a starting city and then repeat the process of visiting the closest
city not yet visited, returning to our starting point after we have reached all of the
cities.

Although the nearest-neighbor approach sounds appealing, the algorithm can
sometimes cause you to "paint yourself into a corner," requiring long edges to
get back to the unvisited cities. This can be seen in Figure 4.6, where tour edges
$\{15, 36\}$ and $\{2, 20\}$ cut across the entire length of the problem. Of course, this
pair of edges is an obvious candidate for a 2-opt move. If we delete $\{15, 36\}$ and
$\{2, 20\}$, we need to add $\{15, 2\}$ and $\{36, 20\}$ to get back to a tour, as indicated in
Figure 4.7. This exchange saves 31 units, bringing the total cost of the tour down
to 982. And many more such exchanges are available.

By repeatedly making 2-opt moves (27 of them altogether), we arrive at the tour
of cost 758 given in Figure 4.8. In this case, there exist no further 2-opt moves that
improve the tour. Recall that the optimal value for this problem is 699, so 2-opt

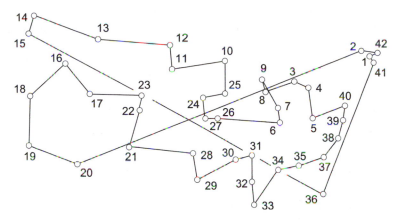

Figure 4.6 Nearest-neighbor tour for the Dantzig-Fulkerson-Johnson 42-city TSP.

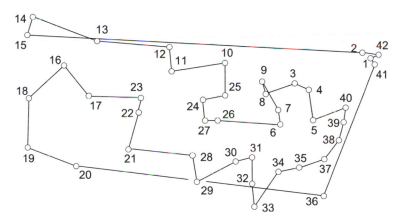

Figure 4.7 Tour after 2-opt move.

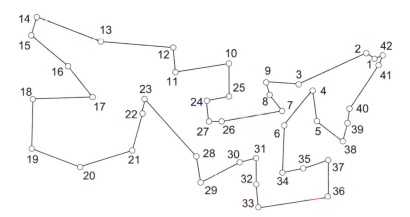

Figure 4.8 Tour with no further improving 2-opt moves.

alone has produced a result that is 8% more costly than the best possible tour. Lin and Kernighan, however, can get all the way down to 699, as we describe below.

A natural idea to improve upon 2-opt tours is to consider k-opt moves for larger values of k, that is, we remove k edges and reconnect the resulting tour fragments with an alternative set of k edges. Some success with 3-opt was reported by Lin [356] and others, but it is clear that the computational burden of directly searching for improving k-opt moves makes the k-opt algorithm impractical for k much larger than 2 or 3. Lin and Kernighan accomplish this, however, in a beautifully engineered algorithm that does not set k to a predetermined value but can, in fact, discover moves with large values of k. The key to the Lin-Kernighan algorithm is to restrict the set of k-opt moves to allow for a strong heuristic search process for detecting an improving move. The restriction is to consider only k-opt moves that can be built from a sequence of specially constructed 2-opt moves. Thus, rather than searching for a single improving move, the Lin-Kernighan algorithm attempts to build a sequence of 2-opt moves that, taken together, one after another, end up at an improved tour. The point is that by allowing some of the intermediate tours to be more costly than the initial tour, the Lin-Kernighan algorithm can go well beyond the point where 2-opt would terminate.

In Figures 4.9 and 4.10 we illustrate the algorithm, starting with the 2-opt tour from Figure 4.8. In this example, the algorithm finds five improving moves that together reach the optimal 699 tour; the drawings in the figures represent the five moves, in order, and the final tour. The highlighted edges in the drawings indicate those that are removed from the tour by each of the moves; the moves improve the tour by 4, 39, 8, 3, and 5 units, respectively.

The success of Lin-Kernighan on the Dantzig, Fulkerson, and Johnson instance is not a surprise. Indeed, in their original paper, Lin and Kernighan write:

> . . . the probability of obtaining optimal solutions in a single trial is close to 1 for small-to-medium problems, say up to the 42-city problem.

They also report finding the optimal tour for the 57-city problem of Karg and Thompson [304], and new best-known tours for three of five 100-city random Euclidean instances described in Krolak, Felts, and Marble [325].

Over the past 20 years, a number of innovations, most notably by Martin, Otto, and Felten [379] and Helsgaun [260], have pushed the success of Lin-Kernighan on to much larger problems. We describe in detail the basic algorithm and its modern variants in Chapter 15.

4.5 TSP CUTS

Gomory's algorithm for integer LP problems fits the general cutting-plane scheme proposed by Dantzig, Fulkerson, and Johnson, but its application to the TSP involves an entangled formulation of an LP relaxation and an integer LP relaxation. By the mid-1960s, the research community still had not returned to the direct

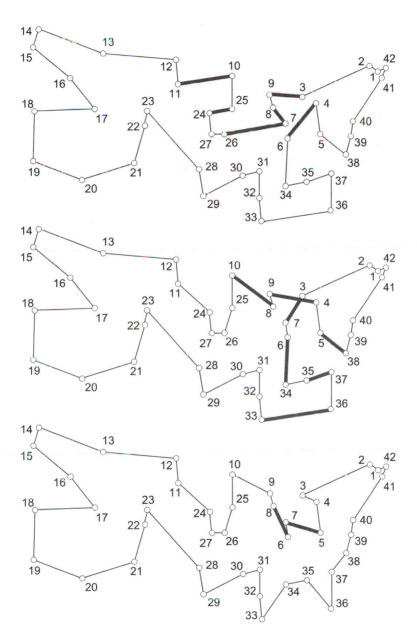

Figure 4.9 First three Lin-Kernighan moves.

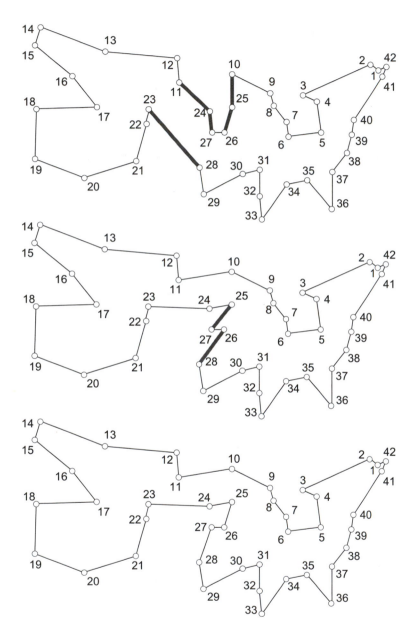

Figure 4.10 Final two Lin-Kernighan moves.

Dantzig-Fulkerson-Johnson approach of using cutting planes to solve the TSP. Again, quoting Gomory [217] from his 1966 survey paper:

> So it should be possible to do the whole thing now systematically. This is an approach one might not expect to work, but we already know that it does.

Indeed, the stage was set for dramatic improvements in the computation of optimal solutions to the TSP.

SAMAN HONG'S DOCTORAL THESIS

The first person to turn back to the TSP cutting-plane method was Saman Hong [270] in his doctoral thesis from 1972. Hong's work is notable in many respects. First, his computer code was the first to fully automate the Dantzig-Fulkerson-Johnson approach, including the use of network-flow methods to spot violated subtour constraints in the case when x^* is not necessarily an integer vector. He was also the first to include a general class of TSP inequalities beyond subtour constraints, using the *2-matching inequalities* that were developed by Jack Edmonds [163] (these are comb inequalities with a general odd number of teeth, but with each tooth containing only two cities). Furthermore, his is also apparently the first code to integrate cutting planes within a branch-and-bound framework, an approach later dubbed *branch-and-cut* by Padberg and Rinaldi [447], as we describe in the next section.

Hong's hybrid of branch-and-bound techniques and cutting planes makes use of a truncated version of the cutting-plane method. There, rather than insisting on finding cuts whenever $x^* \notin S$, we settle for some algorithm FINDCUTS (S, x^*), which sometimes finds cuts and sometimes does not. When it does not, we simply give up and return the current LP relaxation and x^* as a consolation prize for not having solved the problem: the value $c^T x^*$ can at least provide a hopefully good lower bound on the TSP. This is the lower bound Hong employs for subproblems in his branch-and-bound tree.

Despite being the first code to contain many of the components of modern TSP solvers, Hong's computational study was not notable in its success: the bottleneck in Hong's computer code was his LP subroutine, which allowed him to solve only instances with up to 20 cities.

COMB INEQUALITIES AND GRÖTSCHEL'S 120-CITY PROBLEM

As we saw in Chapter 3, comb inequalities (with three teeth) can be combined with subtour constraints to give a quick solution to the 42-city instance of Dantzig, Fulkerson, and Johnson. The solution used two combs. The first of these was a 2-matching constraint, equivalent to the cutting plane used by Dantzig, Fulkerson, and Johnson, while the second was a substitute for I. Glicksberg's ad hoc constraint. General combs (with any odd number k of teeth, where k is at least three) were introduced in the mid-1970s by Chvátal [125] and Grötschel and Padberg [236], and the corresponding inequalities were quickly shown to be a powerful source of cutting planes for the TSP.

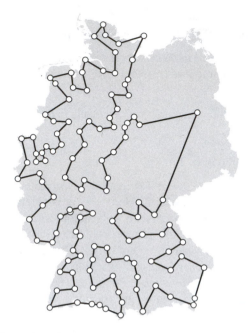

Figure 4.11 Optimal tour through 120 cities in Germany.

The first demonstration of the strength of general combs was made in a computation by Grötschel, finding the optimal tour through 120 German cities (with distances taken from a road atlas), a new record in the solution of the TSP. A picture of his optimal tour is given in Figure 4.11. Grötschel adopts a pure cutting-plane approach, using computer software to solve the LP relaxations and by-hand inspections to determine cutting planes. His work was reported in his 1977 doctoral thesis [228] from the University of Bonn, as well as in a 1980 research paper [230]. In this record-setting computation, 13 LP relaxations were solved, using a total of 36 subtour constraints, 25 2-matching inequalities, and 35 general comb inequalities (a full list is given in [230]). Each of the 13 rounds took between 30 minutes and 3 hours of by-hand computations, and between 30 seconds and 2 minutes of computer time.

A by-hand drawing made by Grötschel in June 1976 giving the LP solution and cutting planes he identified in the second round is displayed in Figure 4.12. The cuts are indicated by circular regions; there are a number of large regions corresponding to violated subtour constraints as well as several combs having three teeth. We thank Martin Grötschel for kindly providing us with a copy of this drawing from his archives.

PADBERG AND HONG STUDY

An automated use of general comb inequalities was reported by Manfred Padberg and Saman Hong [451]. Their paper appeared in 1980 but was based on earlier

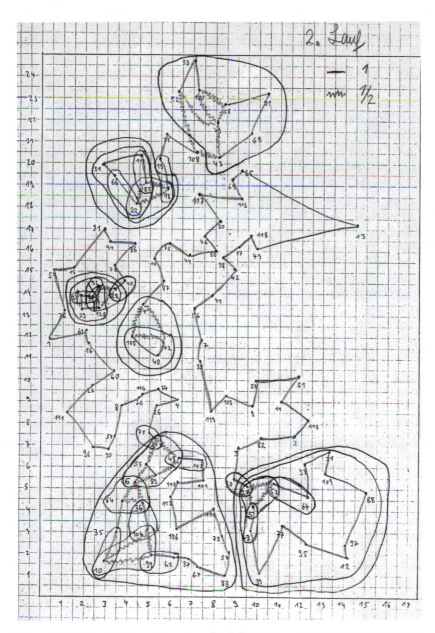

Figure 4.12 Grötschel's hand drawing of an LP solution and identified cutting planes.

work (the manuscript was originally submitted in 1977). Interestingly, they do not follow the approached adopted by Hong [270], choosing instead to use the primal cutting-plane method employed by Dantzig, Fulkerson and Johnson [151]. Padberg and Hong implement a truncated version of this method with FINDCUTS consisting of a routine for finding subtour inequalities, together with heuristic routines for finding comb inequalities and certain "chain constraints" (the heuristic routines may fail to spot a violated comb inequality or chain constraint, even if one exists; this point is discussed in Chapter 5).

Using the primal cutting-plane method, Padberg and Hong restrict their FIND-CUTS to return inequalities that are satisfied as an equation by the incidence vector $\bar{x}$ of their active tour (see the discussion in Section 3.3). They take advantage of this in their routine for subtour inequalities, by simply searching through the sets of cities corresponding to subpaths of $\bar{x}$, since these are the only sets whose corresponding subtour inequality holds as an equation for $\bar{x}$. Any given city is the starting point of only n subpaths, where n is the total number of cities, so this direct search can easily be carried out in time proportional to n^2. This is a more efficient approach than the use of network-flow methods proposed by Hong [270]; we discuss this in a more general context in Section 6.3.

With a truncated version of the primal cutting-plane method, the Padberg-Hong procedure can terminate without solving a TSP instance. In this case, a lower bound is obtained by solving the final LP relaxation (recall that in the primal method only single simplex pivots are made after the addition of a cutting plane, and so we do not typically have an optimal LP solution). The costs of the tours produced in their computational study, either by the heuristic of Lin and Kernighan [357] or by the simplex method in the process of working with the sequence of LP relaxations, were consistently within 0.5% of the optimum value of the final LP relaxation. In fact, for several of the smaller instances they established LP bounds equal to the cost of the optimal tour, including the 42-city problem of Dantzig, Fulkerson, and Johnson [151], two 48-city problems from Held and Karp [252] and Grötschel [228], and three 75-city random Euclidean problems.

The largest instance considered by Padberg and Hong is a 318-city TSP described by Lin and Kernighan [357]. This instance arose in a drilling application for a printed circuit board, described in Section 2.4. In the application, the starting and ending locations are specified, so the problem is to find a path through the 318 points; this can easily be converted to a standard TSP, for example, by assigning a large negative travel cost between the start and end points. For this problem, Padberg and Hong obtained a solution of cost 41,349 and a lower bound of value 41,236.75; furthermore, they noted that most of the initial 50,403 variables could be eliminated through value fixing (a trick devised in the paper of Dantzig, Fulkerson, and Johnson [151]), leaving a problem in 1,532 zero-one variables, 318 equations, and 171 inequalities. Their conclusion

> If the economics of this particular application demanded a true optimum solution, one would have—in view of the small remaining gap of 112—a better than even chance to solve this comparatively small problem exactly by any good branch-and-bound code.

seems to anticipate the subsequent success of Crowder and Padberg [147] described below.

CROWDER-PADBERG CODE

The excellent LP lower bounds found by Padberg and Hong [451] called out for further work to continue the solution process. The call was answered by Harlan Crowder and Manfred Padberg [147], who, in 1980, solved a set of 10 previously published TSP instances; most notably, this set included the 318-city drilling problem of Lin and Kernighan [357]. The Crowder-Padberg work was easily the greatest computational achievement of this period; the 318-city solution nearly tripled the size of the largest instance solved up to that point.

Crowder and Padberg employed the truncated primal cutting-plane method as implemented by Padberg and Hong [451]. Rather than stop with the LP bound, however, they go on to solve the integer LP problems

$$\text{minimize } c^T x \text{ subject to } Cx \le d, x \text{ integer,} \tag{4.8}$$

using IBM's package MPSX-MIP/370, which they treat as a black box. If the black box returns a tour vector, then the TSP is solved. If the returned integer solution is not a tour, then violated subtour constraints are added to the LP and the black box is called again. This hybrid approach is similar to Martin's [378] combined use of subtour constraints and Gomory's algorithm to solve the TSP. In the Crowder-Padberg study, the integer LP black box was used no more than three times in the solution of any of their test instances.

In between the cutting-plane phase and the integer LP phase of their method, Crowder and Padberg interpose several steps that have great importance in practice. First, when the primal method terminates, there may be simplex pivots that lead to integer solutions that are not tours; Crowder and Padberg add the corresponding subtour constraints to the LP, even if they are not satisfied as an equation by the active tour $\bar{x}$. The purpose of this action is to improve the chances that the black box returns a tour, by removing from the LP solution set subtour solutions having better objective value than the tour $\bar{x}$. Taking this another step, at this point Crowder and Padberg switch to the dual cutting-plane method, employing a limited search for subtour constraints using network-flow techniques; the added constraints further restrict the set of integer LP solutions as well as possibly increasing the value of the LP bound.

As a final step before turning to the black box, Crowder and Padberg fix the values of many variables using the ideas from Dantzig, Fulkerson, and Johnson [151] and Padberg and Hong [451]. In the Crowder-Padberg computations, variable fixing reduces their total number to less than $4.5n$, where n is the number of cities. We discuss this in detail in Chapter 12.

As we have mentioned, the computational results of Crowder and Padberg are impressive. Their set of 10 solved TSP instances ranged in size from 48 cities to 318 cities, including the 48-city and 120-city problems of Grötschel [228] and the 57-city problem of Karg and Thompson [304]. A picture of the optimal 318-city tour is given in Figure 4.13. Crowder and Padberg's paper includes an interesting detailed discussion of the solution computations for this large TSP.

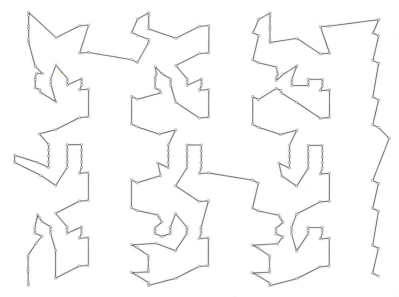

Figure 4.13 Optimal path for the 318-city Lin-Kernighan problem.

Crowder and Padberg conclude their paper by stating their confidence that further progress in the TSP could be expected with the proposed solution method.

> Specifically, as far as the symmetric travelling salesman problem is concerned, we are confident that problems involving 1,000 cities and more are amenable to exact solution by today's technology—the hardware exi[s]ts and the software can be developed to handle such problems.

As we will see, further progress was indeed made later in the 1980s, although several new algorithmic ideas were required to move computer codes for the TSP beyond the 318-city record.

CODES OF LAND AND FLEISCHMANN

At about the time of the Crowder-Padberg study, an important unpublished paper by Ailsa Land [333] appeared. The manuscript is dated 1979 from the London School of Economics. The stated purpose of the paper was to report on work aimed at removing the limitations that held Miliotis [394] to examples with at most 80 cities. Land continues the approach of Miliotis, adopting Gomory cuts as a means to find integer solutions to the LP relaxations. Like Miliotis, Land begins with a heuristic for subtour constraints, in her case using a method based on shrinking sets of cities joined by edges assigned the value 1.0 in the LP relaxation (this idea is also used by Crowder and Padberg [147], before applying network-flow techniques). This subtour heuristic improves upon Miliotis' code for spotting subtour constraints, which only handles the case when the LP solution is not connected. Furthermore, Land adds to the mix a heuristic algorithm for 2-matching inequalities, before turning to Gomory's procedure.

The richer set of cutting planes used by Land is a considerable improvement on the algorithms of Martin [378] and Miliotis [394], although not as powerful as the comb-finding methods used by Padberg and Hong [451]. A major contribution of the paper, however, is a description of a mechanism for automatically dealing with the large number of variables that can appear in the LP relaxations. Dantzig, Fulkerson, and Johnson [151] had to address this issue in their by-hand computations, and Padberg and Hong [451] used an ad hoc method for dealing with the 50,403 edges in their work on the 318-city TSP. Land took the ideas used by these researchers and molded them into an algorithm that maintains a small core set of LP variables and periodically updates the core by scanning the remaining set of edges. Variants of Land's technique are adopted in all recent studies of the cutting-plane method for large-scale problems. We discuss this point in detail in Chapter 12.

Land's test set consisted of 10 random Euclidean problems, each having 100 cites. Her TSP code was able to solve 9 of these instances, and it established a strong lower bound on the 10th problem before exceeding a computation time limit. Concerning the unsolved instance, Land [333] writes the following.

> Problem 10 is the only one which failed to reach an unambiguous answer. The 3-opt tour costs 769. After 1096 iterations (and 398.2 seconds) it was established that the optimal function was definitely greater than 767. After 625.5 seconds the optimal LP was 768 exactly, and remained at that value (with solution values of halves, quarters, occasionally twelfths) as constraints were added and purged, but never either found an integer solution at 768 nor proved the cost to be greater than 768 in the remaining time before the computation was abandoned. Probably a modest amount of "branching and bounding" would quickly establish whether there is or is not a solution at 768 for this problem.

This gives a good indication of the difficulty that can arise when relying on Gomory cuts alone to solve the integer IP relaxations for the TSP.

After a break of several years, this Gomory-based approach was again picked up by Bernhard Fleischmann [180] in 1985. Fleischmann was able to improve on the work of Land by pressing ahead and adding even more TSP-specific cutting planes before moving to Gomory cuts. In his work, Fleischmann incorporates a heuristic for combs with three teeth. In fact, Fleischmann describes a generalization of this class of inequalities that are valid for the *graphical traveling salesman problem*, where tours are permitted to revisit cities that have already been visited. This model is relevant in applications where direct travel is only permitted between certain pairs of cities (this can be converted to a standard TSP by computing the minimum-cost travel distance between each excluded pair, but it can be more efficient to directly tackle the original model).

In his study, Fleischmann solved 17 instances with sizes ranging from 42 to 292 cities; the largest number of Gomory cuts that he was forced to generate in solving these any of these instances was 42. The three largest instances, having 169, 172, and 292 cities, are graphical TSPs derived from road maps. In solving his 292-city example, Fleischmann used an unspecified number of subtour inequalities, 73 comb inequalities, and 2 Gomory cuts; in solving Grötschel's 120-city instance, he used an unspecified number of subtour inequalities, 128 comb inequalities, and no Gomory cuts at all.

GRÖTSCHEL AND HOLLAND'S 666-CITY TSP

In 1987, major studies by Martin Grötschel and Olaf Holland [232] and Manfred Padberg and Giovanni Rinaldi [447] again pushed TSP computation to a new level. We begin our discussion of the 1987 breakthroughs with a short description of Grötschel and Holland's work, and continue with the Padberg and Rinaldi study in the next section.[1]

Following the successful approach of Crowder and Padberg [147], Grötschel and Holland [232] again adopt the hybrid scheme with MPSX-MIP/370 used as a black box for solving the integer LP problem (4.8). Grötschel and Holland take a different tack, however, choosing the (dual) cutting-plane method rather than continuing with the primal method used by Dantzig, Fulkerson, and Johnson [151] and Crowder and Padberg [147]. This choice allowed Grötschel and Holland to use a commercial code to solve the LP relaxations, since no special simplex pivoting rules are required. This is an advantage, given the increasing size and difficulty of the LP relaxations associated with larger TSP instances, but Grötschel and Holland [232] point out that there are drawbacks as well.

> A basic design error we made was to use a black box LP-solver. Commercial LP-solvers like MPSX are certainly much faster than any such code we can come up with. This is why we used it. But from a certain size on, in using MPSX most time is wasted by communicating between the various parts of the code, setting up and revising data structures. LP-solving, cutting-plane addition, and branch & bound have to be married to be really successful.

We return to this point in Chapter 13, when we discuss the design of the CPLEX LP solver.

The main innovations in Grötschel and Holland's code are, quite naturally, in their version of the FINDCUTS routine. First, they adopt a network-flow-based algorithm (as proposed by Hong [270]) that is guaranteed to find a violated subtour constraint if one exists, rather than the limited system used in the final stage of the Crowder-Padberg cutting-plane procedure. Second, they implement an algorithm of Padberg and Rao [453] that guarantees that they will also find a violated 2-matching constraint if one exists. This algorithm of Padberg and Rao is a much more powerful weapon than the simple heuristics for 2-matchings used by Hong [270], Padberg and Hong [270], and Land [333], and Grötschel and Holland utilize it also as a means to search for more general comb constraints. The idea of the comb

---

[1] Although the two main papers appeared in 1991, the Grötschel and Holland work is also described in the 1987 doctoral thesis of Holland [268], and a brief description of the Padberg and Rinaldi work appeared in the 1987 paper [444].

heuristics is to selectively shrink sets of cities down to single supercities and then apply the Padberg-Rao algorithm in the shrunk graph. If a 2-matching inequality is detected in the shrunk graph, then expanding the supercities back to the original sets can produce a general comb (if one of the supercities is contained in a tooth of the 2-matching inequality). We describe this method in detail in Chapter 7.

The Grötschel-Holland computer code solved a large collection of TSP instances, including 10 new geographical problems, a new 442-city drilling problem, and 12 previously published instances. The geographical problems range in size from 17 to 666 cities, with the largest problem asking for a tour around the world. Drawings of optimal tours for the 442-city and 666-city instances are given in Figure 4.14 and Figure 4.15, respectively. For the 666-city instance, three runs of the Grötschel-Holland code (with different parameter settings) used totals of 16, 19, and 23 calls to the integer LP black box.

The solution of the 666-city world TSP was one of the new TSP records set in 1987. Grötschel and Holland's important work is somewhat overshadowed by the Padberg-Rinaldi study described below, but it was a major accomplishment, and many of their techniques are used in work on the TSP described later in the book. Grötschel and Holland [231] conclude their paper with the simple statement, "There is still much to do."

4.6 BRANCH-AND-CUT METHOD

The computer codes of Crowder and Padberg [147] and Grötschel and Holland [232] are hybrids of the cutting-plane and branch-and-bound schemes, in the sense that the MPSX-MIP/370 package is itself an implementation of the branch-and-bound method. It seems clear, however, that the 1972 algorithm of Hong [270] is a better way of making the two components cooperate, as we describe below.

In the course of the branch-and-bound method, we have an upper bound u set to either ∞ or to the best value of a solution $\bar{x} \in S$ we have found thus far in the search. Now, branching is a last resort in dealing with a subproblem (4.2): it is preferable either to establish that $\min\{c^T x : x \subset S \text{ and } Cx \leq d\} \geq u$, in which case the subproblem can be dismissed, or to find an optimal solution x^* of (4.2) with $c^T x^* < c^T \bar{x}$, in which case u and $\bar{x}$ are updated. Solving the LP relaxation of (4.2) is an attempt to force one of these two desirable outcomes; a finer instrument serving the same purpose is the truncated cutting-plane method applied to (4.2). This instrument becomes even more attractive if FINDCUTS (S, x^*), rather than FINDCUTS $(\{x \in S \text{ and } Cx \leq d\}, x^*)$, is used in tackling the subproblem: all of the cuts returned by FINDCUTS (S, x^*) may be used as potential cutting planes to tighten up the LP relaxations of other subproblems as the branch-and-bound process chugs along.

The name *branch-and-cut* for this scheme was coined by Padberg and Rinaldi [444]; the earliest branch-and-cut computer code that we know of is the one written by Hong for the TSP. Hong's FINDCUTS varies with the position of the subproblem in the branch-and-cut tree: he uses both an algorithm for subtour inequalities and a heuristic for 2-matching inequalities at the root of the tree, and only the former al-

Figure 4.14 Optimal tour of a 442-hole printed circuit board.

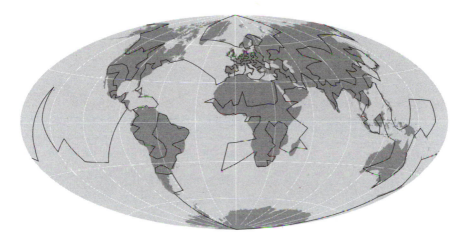

Figure 4.15 Solution to 666-city tour of the world.

gorithm at all other nodes. As noted in the previous section, Hong's weak LP solver
prevented him from solving instances with more than 20 cities. Consequently, the
full power of his branch-and-cut algorithm could hardly manifest itself: of the 60
instances he solved, 59 required no branching at all, and the remaining instance
required just one execution of the branching instruction.

Another branch-and-cut computer code for use in the TSP was written by Miliotis [393] in 1976. Ironically, he concludes his paper by saying

> The modest solution times achieved for problems which are recog-
> nized as testing ones give us encouragement to believe that such an LP
> approach to combinatorial problems deserves further consideration ...

and yet he abandons the branch-and-cut approach two years later in favor of the
cutting-plane method with subtour cuts and Gomory cuts, as we described in Section 4.3. This change of direction may be related to the fact that Miliotis starts out
with the TSP forced into integer LP format as problem (4.3) with each x_e restricted
to integer values; in this formulation, subtour inequalities acquire an aura of special
importance. Miliotis' FINDCUTS returns only subtour cuts and it returns them if
and only if x^* is integer; this restriction makes it weak; on a set of test instances, the
resulting branch-and-cut code was consistently slower than the implementations of
the cutting-plane method reported in Miliotis [394].

An application of the branch-and-cut method to a problem outside the domain
of the TSP was described in 1984 by Grötschel, Jünger, and Reinelt [234]. This
research team designed and implemented an algorithm for *linear ordering problems*
(another model in combinatorial optimization). The strength of their initial cutting-
plane phase, however, allowed them to solve all 44 of their real-world test instances
without recourse to branching. Their code, however, can rightly be viewed as the
first model of a modern implementation of the branch-and-cut method.

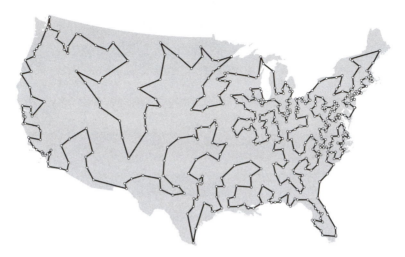

Figure 4.16 Optimal tour through 532 USA cities.

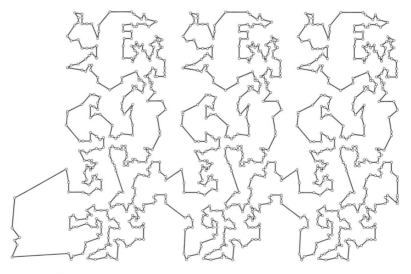

Figure 4.17 Optimal tour of a 1,002-hole printed circuit board.

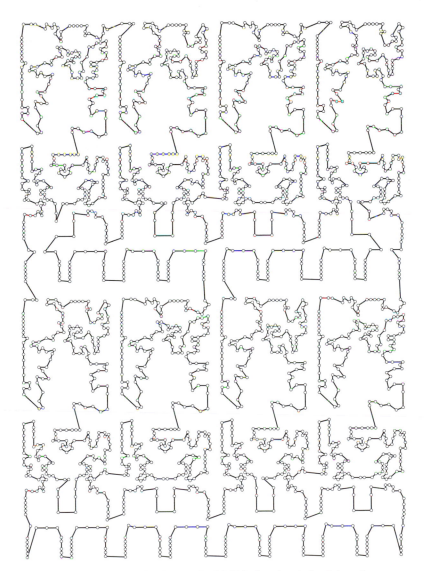

Figure 4.18 Optimal tour of a 2,392-hole printed circuit board.

THE PADBERG-RINALDI CODE

The first computational success of the branch-and-cut method in the context of the TSP was by Padberg and Rinaldi [447], in the second of the two major TSP studies in 1987. The Padberg-Rinaldi computer code solved a set of 42 instances with sizes ranging from 48 to 2,392 cities; this set included a tour of 532 cities in the United States (Figure 4.16), the Grötschel-Holland 666-city problem, a 1,002-city drilling problem (Figure 4.17), and a 2,392-city drilling problem (Figure 4.18). The solution of the 2,392-city TSP was a stunning accomplishment; the computer code of Padberg and Rinaldi was easily the most complex attack on a combinatorial optimization problem up to that time.

Padberg and Rinaldi's version of FINDCUTS introduced many new ideas, including a shrinking routine to greatly speed up the search for subtour inequalities and new heuristics for finding comb inequalities and clique-tree inequalities (a generalization of combs); these algorithms are described in Padberg and Rinaldi [445]. Their innovations went beyond finding cuts, however, impacting nearly all aspects of the implementation of a branch-and-cut code.

As in other studies of the TSP, the Lin-Kernighan heuristic is used to find an initial tour for the branch-and-cut search, but here also a new idea is introduced by Padberg and Rinaldi. Concerning the implementation of the heuristic, Padberg and Rinaldi [447] write:

> It is a substantial revision of the program used by Padberg and Hong (1980), the original (1975) version of which was a FORTRAN IV program due to Shen Lin.

tying this aspect of their work all the way back to Lin himself. The new idea is to take advantage of the common strategy of making repeated runs of the heuristic, collecting the union of the edge sets of the generated tours to use as the initial core set of edges for the LP relaxation. This tour union gives a much better representation of the full edge set than does the initial core edges considered in Land [333] (consisting of the edge set of a single tour together with the four cheapest edges meeting each city); this proposal of Padberg and Rinaldi is still used in current TSP codes, as we describe in Section 12.1.

In the storage and handling of cutting planes, Padberg and Rinaldi pioneered techniques for dealing with large-scale applications. A technical difficulty common to both the cutting-plane method and the branch-and-cut method arises from the fact that the system of linear inequalities defining the LP relaxation may eventually grow to contain too many inequalities, which puts a burden on both the computer memory and the LP solver. The obvious and standard remedy in the context of the cutting-plane method is to purge this system from time to time by deleting inequalities that no longer constrain the minimum of the LP relaxation (see, for instance, Miliotis [394]). Padberg and Rinaldi [444] transfer this remedy to the context of the branch-and-cut method with the additional twist we hinted at earlier: all of the purged inequalities are removed from the LP relaxation, but some of them may be stored away in a memory bank called the *cut pool* and subsequently inspected before the time-consuming, and possibly fruitless, calls of FINDCUTS while processing subproblems in the branch-and-cut tree.

The cut pool is an integral part of the Padberg and Rinaldi strategy for storing subproblems as they move from working on one part of the tree to another. In their scheme, the set of branching decisions $Cx \leq d$ is stored for a subproblem, but not the LP relaxation that was found by the cutting-plane method. Using the pool, an LP relaxation can quickly be reconstructed with the pool as the source of cuts (together with their algorithm for subtour constraints, since these inequalities are not included in the pool).

The use of a cut pool is just one of the many components described in detail in [447]; these components are fully exercised in the reported computational tests. Recall that in the earlier studies of Hong [270] and Grötschel, Jünger, and Reinelt [234], the branching phase of the branch-and-cut scheme was rarely used. This is not the case in the Padberg and Rinaldi study, where 28 of their 42 test instances branched into subproblems; the number of subproblems used by Padberg and Rinaldi in solving each of the four large problems mentioned earlier is reported in Table 4.1. Of the entire set of 42 instances, the largest number of subproblems used

Table 4.1 Number of subproblems in Padberg-Rinaldi study.

TSP Problem Name	Subproblems
532-city USA tour	107
666-city world tour	21
1,002-city drilling problem	13
2,392-city drilling problem	3

was 121; this largest search tree occurred in the solution of the 439-city drilling problem shown in Figure 4.19.

The study of Padberg and Rinaldi provided a quantum leap in the computation of solutions for the TSP and it remains a high point of computation for combinatorial optimization in general. Similar in spirit to the concluding remark of Gröstchel and Holland [232], Padberg and Rinaldi [447] end their paper with the statement "... the problem with 2392 cities should not be the end of the ongoing saga of the symmetric traveling salesman."

FOLLOWING PADBERG AND RINALDI

Much of the remainder of the book is devoted to a description of our work toward continuing the line of research that led from Dantzig, Fulkerson, and Johnson up to Padberg and Rinaldi. We of course are not alone in the continued pursuit of the traveling salesman problem; important recent computational TSP studies include the following:

- In a 1993 paper, Clochard and Naddef [130] demonstrate the effectiveness of path inequalities for the TSP (another generalization of combs). They also

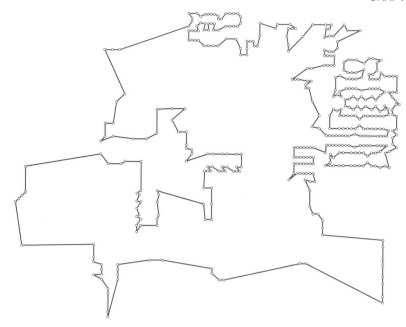

Figure 4.19 Optimal tour of a 439-hole printed circuit board.

introduce a branching rule that is based on the fact that each tour must cross the boundary of a subset of cities an even number of times.

- In a 1994 paper, Jünger, Reinelt, and Thienel [300] describe a new implementation of the branch-and-cut method for the TSP. They focus attention on some of the important design questions that arise when implementing the method, and introduce the idea of utilizing the LP solution vector x^* in heuristic search algorithms for high-quality tours.

- In a 1995 survey paper, Jünger, Reinelt, and Rinaldi [296] carry out further tests of the branch-and-cut implementations of Padberg and Rinaldi [447] and Jünger, Reinelt, and Thienel [300]. They introduce a new strategy for obtaining good bounds for large geometric TSP instances, preselecting a collection of subtour inequalities to add to the initial LP relaxation (4.3). The preselection is based on a clustering algorithm and the resulting large LP relaxation is solved by a non-simplex solver.

- In a series of papers in 2002, Naddef and Thienel [423, 424] introduce a new class of heuristic methods for spotting violated TSP inequalities, growing components of the underlying structures (for example, the handle and teeth of a comb) step by step.

We cover details of these contributions and others as we describe our computational approach to the TSP in later chapters.

4.7 NOTES

A great source for further reading is Alan Hoffman and Philip Wolfe's [267] chapter in the book *The Traveling Salesman Problem: A Guided Tour of Combinatorial Optimization*, edited by Lawler, Lenstra, Rinnooy Kan, and Shmoys [338]. Hoffman and Wolfe were on the scene of the work by Dantzig, Fulkerson, and Johnson and provide a very interesting commentary.

A detailed treatment of the origins of the TSP in mathematics can be found in the TSP chapter of Schrijver [495]. Of particular interest is Schrijver's research on the possible passing of the TSP from Menger to Whitney to Flood in the 1930s. Other studies of early work on the TSP can found in Flood [182], Arnoff and Sengupta [23], Gomory [217], Bellmore and Nemhauser [50], Isaac and Turban [277], Burkard [99], Parker and Rardin [456], and Laporte [337].

Computational TSP work that we did not cover in the chapter includes studies by Morton and Land [403], Riley [479], Bock [70], Suurballe [517], Pfluger [461], Steckhan and Thome [508], Houck, Picard, Queyranne, and Vemuganti [273], and Carpaneto, Fischetti, and Toth [108].

Random-Cost TSPs

The test instances we discussed in the chapter were typically geometric problems, where the travel cost between two cities is based on the distance between the associated points. For the most part, we did not mention problems that were obtained by assigning random travel costs to pairs of cities. These random-cost instances do not provide good tests for TSP codes, since it is often easy to obtain very strong lower bounds.

One random-cost TSP test we did discuss is the set of 9-city instances studied by Robacker [481], where the travel costs were "chosen from a table of two-digit random numbers." These instances were solved in a straightforward manner in by-hand computations.

Land [333] reported the solution of three 100-city instances with edge costs generated uniformly between 0.0 and 100.0, together with her Euclidean examples. Concerning the ease with which the random-cost instances were solved, Land wrote, "A secondary conclusion must be a warning against using random cost matrices when assessing the value of a TSP algorithm." Grötschel and Holland [232] solved a 1,000-city instance of this type, having integer edge costs generated uniformly between 1 and 5,000. Despite its larger size, this example did not present as great a challenge as the solution to the 666-city TSP that we described earlier. Regarding the time in the cutting-plane phase of their algorithm, Grötschel and Holland report that one run of the 666-city TSP took over 3 hours, while all runs on the 1,000-city TSP were completed in under 10 minutes.

The largest solved random-cost instances reported in the literature is a collection of 10,000-city examples by Jünger, Reinelt, and Rinaldi [296]. Their test set consisted of 10 problems with integer edges costs between 0 and 50,000. Of these, 7 problems were solved within a 5-hour limit on the computation time. The con-

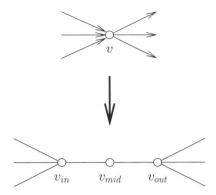

Figure 4.20 Reducing an asymmetric TSP to a symmetric TSP.

clusion of this study was again a warning against using random-cost instances in computational tests.

ASYMMETRIC TSP

In our historical treatment we have not covered the development of methods for the asymmetric TSP, since in this book we treat only the case where the cost of travel between two cities does not depend on whether we move from x to y or from y to x. Surveys of work on the asymmetric TSP can be found in Balas and Toth [34], Balas and Fischetti [32], and Fischetti, Lodi, and Toth [175].

Many of the computational tools for the TSP carry over to the asymmetric case by a standard construction for converting any asymmetric TSP to one with symmetric travel costs. The idea is to split each city v into a pair cities v_{in} and v_{out}, with v_{in} representing the end of each edge heading into v, and v_{out} representing the end of each each heading out of v; the cities v_{in} and v_{out} are then joined by an edge of zero cost that is required to be in the tour, for example, by inserting an additional city v_{mid} into the middle of the edge. This construction is illustrated in Figure 4.20.

The early branch-and-bound algorithms of Eastman [159], Shapiro [498], and others were designed to handle both symmetric and asymmetric TSP instances. Unfortunately, the asymmetric test instances were typically constructed by assigning uniformly distributed, random costs to each ordered pair of cities. As in the (symmetric) TSP case, these random instances do not provide a good test of the proposed methods. Indeed, Shapiro [498] made the following statement regarding his asymmetric tests.

> Another interesting aspect is that the assignment and tour solutions agree globally for problems as large as (in this case) 40 cities and there is no reason to suggest that larger problems will not exhibit this property.

Here the "assignment" solution is the optimal value of the initial LP relaxation (the directed version of the degree constraints). Further empirical support for Shapiro's observation is provided by the large-scale study of Miller and Pekny [396], including instances with up to 500,000 cities. Analytical support was provided by Frieze, Karp, and Reed [197], who showed that, for these random-cost instances, the optimal values of the LP relaxation and the asymmetric TSP will agree with probability approaching 1 as the number of cities increases.

Chapter Five

LP Bounds and Cutting Planes

In the next six chapters we describe methods for finding cutting planes for the TSP. To set the stage for these discussions, we gather together in this chapter some common notation and themes concerning linear programming, separation routines, and the cutting-plane method.

5.1 GRAPHS AND VECTORS

A general instance of an n-city TSP is specified by the $n(n-1)/2$ travel costs between the pairs of cities. When dealing with these data it is often convenient to adopt the language of graph theory. In this section we give only basic definitions that are needed later in the book; for further information we refer the reader to the many excellent texts on the subject, for example, Berge [58], Bondy and Murty [78], Diestel [156], and West [553].

A *graph* G consists of a finite set V of *vertices*, a finite set E of *edges*, and a relationship associating with each edge e a pair of vertices, called the *ends* of e; an edge is said to *join* its two ends. We treat the edges of a graph as 2-point subsets of its vertex set: $v \in e$ means that vertex v is an end of edge e; $e \cap Q \neq \emptyset$ means that edge e has an end in set Q; $e - Q \neq \emptyset$ means that edge e has an endpoint outside set Q; and so on.

Graphs can be visualized by drawing vertices as points in the plane, with each edge represented as a line segment between the points corresponding to its ends. A typical drawing of a graph is given in Figure 5.1, with vertices $\{a, b, c, d\}$ and edges $\{a, c\}, \{a, d\}, \{b, c\},$ and $\{c, d\}$.

A graph $H = (W, F)$ is a *subgraph* of $G = (V, E)$ if $W \subseteq V$ and $F \subseteq E$. If $W \subseteq V$ and $F = \{e : e \in E, e \subseteq W\}$, then $H = (W, F)$ is called the subgraph *induced* by W.

A *walk* in G is a sequence $v_0, e_1, v_1, \ldots, e_k, v_k$, where each v_i is a vertex and

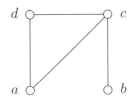

Figure 5.1 Drawing of a graph.

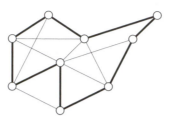

Figure 5.2 Hamiltonian circuit.

each e_i is an edge joining v_{i-1} and v_i, for $i = 1, \ldots, k$. A walk is called a *path* if each of the vertices $v_0, \ldots, v_k$ is distinct, and it is called a *circuit* if $v_0, \ldots, v_{k-1}$ are distinct and $v_0 = v_k$. We will usually identify a circuit or path with its edge set $\{e_1, \ldots, e_k\}$. A circuit or path containing all vertices in V is called *Hamiltonian*. An example of a graph with a Hamiltonian circuit is drawn in Figure 5.2; the circuit is indicated by the heavy lines in the drawing.

A graph $G = (V, E)$ is *connected* if for each pair of vertices $u, v \in V$ there exists a path starting at u and ending at v. A *tree* is a connected graph having no circuits.

Let $V_1, \ldots, V_k \subseteq V$ be the maximal sets that induce connected subgraphs of G; the subgraphs induced by $V_1, \ldots, V_k$ are called the *connected components* of G. A vertex v is called a *cut vertex* if the graph induced by $V - \{v\}$ has more connected components than does G. A graph is called *2-connected* if it contains no cut vertex; a *block* of a graph is a maximal subgraph that is 2-connected.

A *complete graph* is one with edges consisting of all 2-point subsets of V. A TSP can be described by an edge-weighted complete graph $G = (V, E)$, where the edge weights $(c_e : e \in E)$ give the travel costs. In this context, Hamiltonian circuits are called *tours* of V; the cost of a tour T is $\sum(c_e : e \in T)$.

For a proper subset S of vertices, $\delta(S)$ denotes the set of edges having one end in S and one end not in S, that is,

$$\delta(S) = \{e \in E : |e \cap S| = 1\}.$$

A set of edges of the form $\delta(S)$ called a *cut* of G; an example of a cut is indicated in Figure 5.3, where the shaded vertices are those in the set S and the heavy edges are those in the cut. For a single vertex v we will usually write $\delta(v)$ instead of $\delta(\{v\})$.

Given a vector $w = (w_e : e \in E)$ indexed by the edges of G and a set $F \subseteq E$, let

$$w(F) = \sum(w_e : e \in F).$$

For example, the cost of a circuit T can be written as $c(T)$.

Given disjoint subsets $S, T \subseteq V$, we write

$$w(S, T) = \sum(w_e : e \in E, e \cap S \neq \emptyset, e \cap T \neq \emptyset).$$

This notation is adopted from Ford and Fulkerson [189]; it is handy when describing linear inequalities for the TSP, where w is replaced by the vector of variables $x = (x_e : e \in E)$.

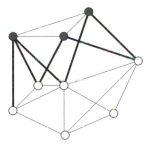

Figure 5.3 A cut in a graph.

5.2 LINEAR PROGRAMMING

We gave a brief introduction to linear programming in Section 3.1; here we give
a nearly so briefly, but more formal, summary of the theory. A detailed treatment
of the simplex method for solving LP problems can be found in Chapter 13. For
general coverage of LP theory and algorithms, we refer the reader to the texts by
Bertsimas and Tsitsiklis [60], Chvátal [126], Dantzig [152], Schrijver [494], and
Vanderbei [532].

An LP problem is to maximize or minimize a linear function subject to specified
linear equality and linear inequality constraints. A general form of an LP problem
is

$$\text{maximize } c_1 x_1 + c_2 x_2 + \cdots + c_n x_n$$
$$\text{subject to}$$
$$a_{11} x_1 + a_{12} x_2 + \cdots + a_{1n} x_n \le b_1$$
$$a_{21} x_1 + a_{22} x_2 + \cdots + a_{2n} x_n \le b_2$$
$$\vdots$$
$$a_{m1} x_1 + a_{m2} x_2 + \cdots + a_{mn} x_n \le b_m$$
$$l_1 \le x_1 \le u_1$$
$$l_2 \le x_2 \le u_2$$
$$\vdots$$
$$l_n \le x_n \le u_n$$

where $x_1, x_2, \ldots, x_n$ are variables and the remaining elements are input data. Each
of the input data can be any (rational) number; the $l_1, l_2, \ldots, l_n$ can also be assigned
the value of $-$infinity $(-\infty)$ and the $u_1, u_2, \ldots, u_n$ can also be assigned the value
of $+$infinity $(+\infty)$. The "maximize" can alternatively be "minimize," and each "$\le$"
relation can alternatively be an "$=$" or a "$\ge$" relation. It is straightforward to write
any LP problem in the above form.

It is often convenient to express an LP problem in matrix notation. To do this, we define a matrix

$$A = \begin{pmatrix} a_{11} & a_{12} & \cdots & a_{1n} \\ a_{21} & a_{22} & \cdots & a_{2n} \\ \vdots & & & \\ a_{m1} & a_{m2} & \cdots & a_{mn} \end{pmatrix}$$

and vectors

$$x = \begin{pmatrix} x_1 \\ x_2 \\ \vdots \\ x_n \end{pmatrix}, \quad c = \begin{pmatrix} c_1 \\ c_2 \\ \vdots \\ c_n \end{pmatrix}, \quad b = \begin{pmatrix} b_1 \\ b_2 \\ \vdots \\ b_m \end{pmatrix},$$

$$l = \begin{pmatrix} l_1 \\ l_2 \\ \vdots \\ l_n \end{pmatrix}, \quad u = \begin{pmatrix} u_1 \\ u_2 \\ \vdots \\ u_n \end{pmatrix}.$$

The general LP problem can then be written as

$$\text{maximize } c^T x$$
$$\text{subject to}$$
$$Ax \leq b$$
$$l \leq x \leq u$$

where c^T is the transpose of the c-vector. Matrix A is called the *constraint matrix*; vector b is the *right-hand-side vector*; vectors l and u are the *lower and upper bounds*; $c^T x$ is the *objective function*. An assignment of values to x that satisfies all of the constraints is called a *feasible solution* to the LP problem. An *optimal solution* x^* is a feasible solution such that $c^T x^* \geq c^T \bar{x}$ for all other feasible solutions $\bar{x}$.

In describing the basic theory of linear programming, let us take the simpler form

$$\text{maximize } c^T x \tag{5.1}$$
$$\text{subject to}$$
$$Ax \leq b$$

without the explicit lower and upper bounds on the variables (these can be added to the constraints $Ax \leq b$).

POLYHEDRA

A set of the form $\{x \in \mathbf{R}^n : Ax \leq b\}$ for some linear system $Ax \leq b$ is called a *polyhedron*; the set of feasible solutions to an LP problem is such a set. When studying linear programming, it is sometimes helpful to visualize the feasible solutions for an LP problem in two variables by drawing a polyhedron in the plane, as in Figure 5.4.

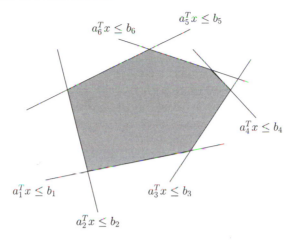

Figure 5.4 A polyhedron.

A linear inequality $\alpha^T x \leq \beta$ is *valid* for a polyhedron $P = \{x : Ax \leq b\}$ if $\alpha^T \bar{x} \leq \beta$ holds for all $\bar{x} \in P$. The set $\{x : \alpha^T x = \beta\}$ is a *supporting hyperplane* of P if $\alpha^T x \leq \beta$ is a valid inequality and at least one vector $\bar{x} \in P$ satisfies $\alpha^T \bar{x} = \beta$. A *face* of P is either P itself or a set of the form $F = P \cap \{x : \alpha^T x = \beta\}$ for some supporting hyperplane $\{x : \alpha^T x = \beta\}$; the inequality $\alpha^T x \leq \beta$ is said to *induce* the face F.

A subset S of some $\mathbf{R}^n$ is said to be *affinely independent* if there are no scalars λ_x $(x \in S)$ such that

$$\sum_{x \in S} \lambda_x x = 0, \quad \sum_{x \in S} \lambda_x = 0, \quad \text{and at least one } \lambda_x \text{ is nonzero;}$$

the *dimension* of S, denoted $\dim S$, is the largest number of affinely independent points in S minus one. The *affine hull* of S is the set of all vectors $y \in \mathbf{R}^n$ such that there exist scalars λ_x $(x \in S)$ satisfying

$$\sum_{x \in S} \lambda_x x = y \quad \text{and} \quad \sum_{x \in S} \lambda_x = 1.$$

A polyhedron P in $\mathbf{R}^n$ is of *full dimension* if $\dim P = n$; in this case there is no equation $\alpha^T x = \beta$ with α not equal to the zero vector and $P \subseteq \{x : \alpha^T x = \beta\}$. In general, if $P = \{x : Ax \leq b\}$ and if $A^= x \leq b^=$ are those inequalities in $Ax \leq b$ that hold as an equation for all vectors in P, then $\dim P = n - t$, where t is the rank of the matrix $A^=$; the set $\{x : A^= x = b^=\}$ is identical to the affine hull of P, and thus depends only on P, not on the choice of the system $Ax \leq b$.

A face F of a polyhedron P is a *facet* of P if $\dim F = \dim P - 1$. Within the affine hull of P, the facets correspond to the inequalities that must be included in any linear system that defines P. To make this statement precise, call a linear system $A'x = b'$, $A''x \leq b''$ a *minimal defining system* for the polyhedron $P = \{x : A'x = b', A''x \leq b''\}$ if no inequality can be made into an equation and no inequality or equation can be dropped, while still defining P. The following theorem is a classic result in polyhedral theory.

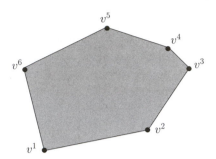

Figure 5.5 Extreme points.

THEOREM 5.1 *Let $P = \{x : A'x = b', A''x \le b''\}$ with $P \ne \emptyset$. The defining system is minimal if and only if the rows of matrix $A^=$ are linearly independent and each inequality in $A''x \le b''$ induces a distinct facet of P.*

An illustration of this result is given in Figure 5.4, where the facets of the indicated polyhedron are induced by the inequalities $a_i^T x \le b_i$, $i = 1, \ldots, 6$; these six inequalities form a minimal defining system for the polyhedron.

The facets of a polyhedron P are its maximal, with respect to inclusion, faces other than P itself. At the other end of the spectrum are the *extreme points* of a polyhedron, defined as faces consisting of single vectors. A vector $v \in P$ is an extreme point if and only if it cannot be written as $v = \lambda p + (1 - \lambda)q$ for distinct vectors $p, q \in P$ and a scalar λ satisfying $0 < \lambda < 1$. Such an expression of v is a *convex combination* of p and q. In general, a vector $y \in \mathbf{R}^n$ is a convex combination of $x^1, \ldots, x^k \in \mathbf{R}^n$ if there exist scalars $\lambda_1, \ldots, \lambda_k$ such that

$$\sum_i \lambda_i x^i = y, \quad \sum_i \lambda_i = 1, \quad \text{and } \lambda_i \ge 0 \text{ for all } i.$$

The *convex hull* of a set $S \subseteq \mathbf{R}^n$ is the set of all vectors that can be written as convex combinations of vectors in S.

Not all polyhedra have extreme points, for example $P = \{x \in \mathbf{R}^2 : x_1 + x_2 \le 0\}$ has as its only face, other than P itself, the hyperplane $\{x \in \mathbf{R}^2 : x_1 + x_2 = 0\}$. A sufficient condition for a polyhedron P to have at least one extreme point is that P be bounded, that is, P contains no infinite half-line. A bounded polyhedron is called a *polytope*.

The extreme points of the polytope drawn in Figure 5.5 are indicated by the dark points labeled $v^1, \ldots, v^6$. In this example, it is easy to see that the polytope is identical to the convex hull of its extreme points. A classic theorem of Minkowski [398] implies that this is always the case.

THEOREM 5.2 *A set $P \subseteq \mathbf{R}^n$ is a polytope if and only if P is the convex hull of a finite set $S \subseteq \mathbf{R}^n$.*

LP DUALITY

A crucial part of linear programming theory is the relationship between an LP problem (5.1) and its associated *dual LP problem*:

$$\text{minimize } y^T b \qquad\qquad (5.2)$$

$$\text{subject to}$$

$$y^T A = c^T$$

$$y \geq 0.$$

When discussing duality, the original LP problem is usually referred to as the *primal* problem. The dual problem has a constraint $y^T A^j = c_j$ for each primal variable x_j, where A^j denotes the jth column of A and c_j is the jth component of c; the quantity $c_j - y^T A^j$ is called the *reduced cost* of x_j.

By converting any LP problem to the form (5.1) we obtain a corresponding dual problem; it is straightforward to check that the dual of the dual problem is equivalent to the original primal LP problem.

A feasible solution to the dual LP problem provides an upper bound on the objective function value for the primal problem, and, conversely, a feasible solution to the primal problem provides a lower bound on the objective value for the dual problem.

THEOREM 5.3 *(Weak Duality Theorem) Let A be an m by n matrix, $b \in \mathbf{R}^m$, and $c \in \mathbf{R}^n$. If $\bar{x}$ is a feasible solution to the LP problem (5.1) and $\bar{y}$ is a feasible solution to the dual LP problem (5.2), then $c^T x \leq y^T b$.*

Proof. Note that $c^T \bar{x} = (\bar{y}^T A)\bar{x} = \bar{y}^T(A\bar{x}) \leq \bar{y}^T b$. □

We will turn to this basic result many times while describing our implementation of the cutting-plane method; it is the key to the validity of the lower bounds obtained by the LP relaxations.

A much stronger version of Theorem 5.3 also holds, as first shown by von Neumann in 1947.

THEOREM 5.4 *(Strong Duality Theorem) If both the primal problem (5.1) and the dual problem (5.2) have feasible solutions, then there exist optimal solutions x^* to (5.1) and y^* to (5.2) with $c^T x^* = y^{*T} b$.*

There are a number of different proofs of this theorem, including one obtained by analyzing the simplex algorithm for solving LP problems; we refer to the LP texts cited above. In the following we will not make explicit use of Theorem 5.4, but it is the driving force behind the simplex algorithm that is the basis of our LP solver, as we describe in Chapter 13.

LP RELAXATIONS FOR THE TSP

The notation presented in the previous section allows us to easily describe the LP relaxations for the TSP that were discussed in Chapters 3 and 4. Given a complete graph $G = (V, E)$ with edge costs $c = (c_e : e \in E)$, the relaxations have variables $x = (x_e : e \in E)$.

Recall that the starting point of the Dantzig-Fulkerson-Johnson model included an equation for each vertex v that requires the variables corresponding to edges having v as an end to sum to 2. These *degree equations* can be written as

$$x(\delta(v)) = 2 \quad \text{for all} \ v \in V, \tag{5.3}$$

since $\delta(v)$ is precisely the set of edges that meet vertex v, and $x(\delta(v))$ sums over all variables in this set. The LP problem

$$\text{minimize} \ c^T x \tag{5.4}$$

subject to

$$x(\delta(v)) = 2 \quad \text{for all vertices} \ v$$

$$0 \le x_e \le 1 \quad \text{for all edges} \ e$$

is called the *degree LP relaxation*, or sometimes the *assignment LP*.

Given a nonempty proper subset S of V, the *subtour inequality* for S requires that the variables corresponding to edges joining vertices in S to vertices in $V - S$ sum to at least 2. This inequality can be written as

$$x(\delta(S)) \ge 2 \tag{5.5}$$

or, alternatively, using the Ford-Fulkerson notation,

$$x(S, V - S) \ge 2. \tag{5.6}$$

Since variants of this inequality arise so often, we sometimes adopt the shorthand $\eta(S, x) \equiv x(\delta(S)) - 2$, and write the subtour inequality as

$$\eta(S, x) \ge 0. \tag{5.7}$$

The *subtour relaxation* of the TSP is

$$\text{minimize} \ c^T x \tag{5.8}$$

subject to

$$x(\delta(v)) = 2 \quad \text{for all vertices} \ v$$

$$x(\delta(S)) \ge 2 \quad \text{for all} \ S \subset V, S \ne V, |S| \ge 3$$

$$0 \le x_e \le 1 \quad \text{for all edges} \ e.$$

Note that this LP problem has an exponential number of constraints and cannot be solved with an explicit formulation, but as we have discussed, this is often solved as part of the cutting-plane method for the TSP.

In general, we study TSP relaxations of the form

$$\text{minimize} \ c^T x \tag{5.9}$$

subject to

$$x(\delta(v)) = 2 \quad \text{for all vertices} \ v$$

$$Cx \le d$$

$$0 \le x_e \le 1 \quad \text{for all edges} \ e,$$

where $Cx \le d$ is a system of m inequalities satisfied by all tours. It is a property of the simplex algorithm that it returns solutions x^* to (5.9) having at most $2n + m$ nonzero components. If m is small, this implies that the *support graph* G^* of x^*, having vertices V and edges $\{e \in E : x_e^* > 0\}$, will be sparse, that is, G^* will have few edges. An example of a support graph is displayed in Figure 5.6, where x^* is a solution to the subtour relaxation for a 442-city TSP.

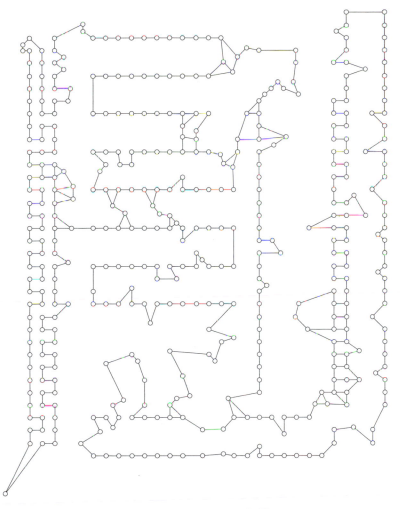

Figure 5.6 A support graph G^*.

5.3 OUTLINE OF THE CUTTING-PLANE METHOD

In the cutting-plane method of Dantzig, Fulkerson, and Johnson [151], a sequence of LP relaxations is created for the problem

$$\text{minimize } c^T x \text{ subject to } x \in \mathcal{S}$$

where $\mathcal{S}$ is a finite subset of some $\mathbf{R}^n$ and $c \in \mathbf{R}^n$ is a cost vector. We introduced this process by means of a TSP example in Chapter 3. The fundamental idea is that an LP relaxation can be improved during a solution procedure by adding as constraints selected linear inequalities $\alpha^T x \leq \beta$ that are satisfied by all points in $\mathcal{S}$. If $\alpha^T x^* > \beta$, where x^* is an optimal LP solution, then such an $\alpha^T x \leq \beta$ is called a *cutting plane* and its addition will eliminate x^* from the LP solution set, possibly leading to an improvement in the bound provided by the relaxation.

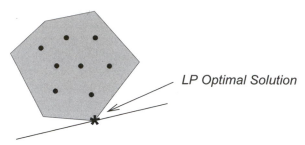

Figure 5.7 Optimal solution to LP relaxation.

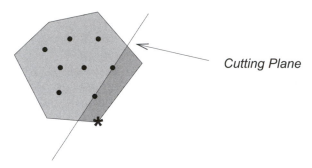

Figure 5.8 Cutting plane.

In Figures 5.7 and 5.8 we give standard textbook drawings of a cutting plane for a problem in $\mathbf{R}^2$. In the first figure, the polyhedron is an LP relaxation for the set S indicated by the dark points; the optimal solution to the LP relaxation, for the objective function corresponding to the drawn supporting hyperplane, is the extreme point indicated at the bottom of the polyhedron. The cutting plane in the second figure is the inequality defining the half-space indicated by the drawn line; the inequality cuts off a portion of the polyhedron that contains the optimal LP solution, but does not exclude any of the points in S.

The iterations of the cutting-plane method are summarized in Algorithm 5.1. Here, rather than adding a single cutting plane at each step, we follow the advice of the many practical implementations of the method and add a family of cuts, represented as a system of inequalities $A'x \leq b'$. Note that in the general form we have described, it may not be possible to carry out the method, since the optimal LP solution x^* could possibly be a convex combination of points in S, and so no cutting planes exist. In this case, the LP bound $c^T x^*$ is equal to the optimal value of $c^T x$ over all points in S, but we have not produced an optimal solution. This pitfall can be overcome by insisting that x^* be an extreme point of the polyhedron P (since S is finite, we can assume that P is a polytope and thus that P has extreme points; for such LP problems the simplex algorithm will return an extreme point of P).

In Chapter 4 we cite the frequent use of the *truncated cutting-plane method*, where it is accepted that a subroutine FINDCUTS (S, x^*), for finding cutting planes

Algorithm 5.1 Cutting-plane method.

choose an initial linear system $Ax \leq b$ satisfied by all $x \in S$
let $P = \{x : Ax \leq b\}$;
while $P \neq \emptyset$
do find a point x^* of P that minimizes $c^T x$;
    if $x^* \in S$
    then return x^*;
    else find a nonempty set $A'x \leq b'$ of inequalities such that
        $S \subset \{x : A'x \leq b'\}$ and $A'x^* > b'$;
        $P = P \cap \{x : A'x \leq b'\}$;
    end
end
return the message "$S = \emptyset$";

for S that are violated by the LP solution x^*, may not produce any cuts, even though they exist. This truncated method is summarized in Algorithm 5.2. In this case the procedure may not produce an optimal point in S, but will still return a lower bound and an LP relaxation that can be used as part of a full solution procedure, such as the branch-and-cut scheme.

Algorithm 5.2 Truncated cutting-plane method.

choose an initial linear system $Ax \leq b$ satisfied by all $x \in S$
let $P = \{x : Ax \leq b\}$;
while $P \neq \emptyset$
do find a point x^* of P that minimizes $c^T x^*$;
    if $x^* \notin S$
    then if FINDCUTS (S, x^*)
        returns a nonempty set $A'x \leq b'$ of cuts
        then $P = P \cap \{x : A'x \leq b'\}$;
        else return P and x^*;
        end
    else return P and x^*;
    end
end
return the message "$S = \emptyset$";

5.4 VALID LP BOUNDS

Suppose we have a TSP defined on the complete graph $G = (V, E)$. Using the truncated cutting-plane method with a FINDCUTS subroutine that returns subtour inequalities, we can obtain an LP relaxation

$$\text{minimize } c^T x \tag{5.10}$$

$$\text{subject to}$$

$$x(\delta(v)) = 2 \quad \text{for all } v \in V$$

$$x(\delta(S)) \geq 2 \quad \text{for all } S \in \mathcal{U}$$

$$0 \leq x_e \leq 1 \quad \text{for all } e \in E.$$

where $\mathcal{U}$ is a collection of sets $S \subseteq V$ with $S \neq V, |S| \geq 3$. The optimal solution to this relaxation is a lower bound on the cost of any tour T, that is, if $t = c^T x^*$ for an optimal LP solution x^*, then $c(T) \geq t$.

Although we have relied on the above fact in our discussions of the cutting-plane method, there is a great practical difficulty: for numerical reasons, modern LP computer codes do not always produce a correct optimal x^* when solving the relaxation. This is not a criticism of the current generation of LP solvers; numerical error is a fact of life when dealing with software built around floating-point libraries (see Knuth [321]). It is particularly relevant in solution methods for large-scale systems of equations, one of the core subproblems of LP solution techniques; for a discussion, see Murtagh [415]. To solve a TSP with an LP-based procedure, such as the cutting-plane method or the branch-and-cut scheme, we must confront the numerical issues involved in the solution of LP problems.

In the original study of Dantzig, Fulkerson, and Johnson [151], numerical errors were avoided by using exact rational arithmetic in the by-hand computations that were used to solve the LP relaxations. Early computer implementations of the cutting-plane method for the TSP also adopted this approach:

- Miliotis [394] writes, "The computation is performed throughout in integer arithmetic."

- Land [333] writes, "Here and elsewhere all LP calculations are carried out in exact arithmetic."

- Padberg and Hong [451] write, "The entire program uses fixed-point arithmetic so as to avoid any problems connected with round-off errors."

More recent studies of Padberg and Rinaldi [447] and others use a different tactic, setting tolerances to make their codes insensitive to small numerical errors and depending on the LP software to produce solutions within a specified range of the true optimal value. The particular methods used in the Padberg-Rinaldi study are not discussed in detail in [447], but the following passage, concerning the choice of a parameter for the minimum violation of a cutting plane, conveys the idea of this approach.

> We declare a constraint violated if its slack is negative and has an absolute value bigger than a parameter XTOLCT. Assume that the LP solver guarantees a *precision* of p decimal digits in the floating point representation of each component of $\bar{x}$. To be conservative the parameter XTOLCT is set to a value greater than $2q10^{-p}$, where q is the size of the current basis.

This is a practical strategy, allowing for the use of highly efficient, general-purpose LP software. When applied to the lower bound itself, however, it has the drawback of creating a dependence on the expected accuracy of the results provided by the solver.

To obtain a robust solution method that does not share this dependence, we turn to the Weak Duality Theorem and consider dual LP solutions. In the case of the TSP relaxation (5.10) we have dual variables y_v for each vertex v, corresponding to the degree equations; Y_S for each set $S \in \mathcal{U}$, corresponding to subtour constraints; and z_e for each edge e, corresponding to the upper bound constraints $-x_e \geq -1$. The dual LP problem for (5.10) is

$$\text{maximize } 2\sum(y_v : v \in V) + 2\sum(Y_S : S \in \mathcal{U}) - \sum(z_e : e \in E) \quad (5.11)$$

$$\text{subject to}$$

$$y_u + y_v + \sum(Y_S : S \in \mathcal{U}, e \in \delta(S)) - z_e \leq c_e \quad \text{for } e = \{u, v\} \in E$$

$$Y_S \geq 0 \quad \text{for all } S \in \mathcal{U}$$

$$z_e \geq 0 \quad \text{for all } e \in E.$$

Any feasible solution $(\bar{y}, \bar{Y}, \bar{z})$ to (5.11) provides the lower bound

$$\bar{t} = 2\sum(\bar{y}_v : v \in V) + 2\sum(\bar{Y}_S : S \in \mathcal{U}) - \sum(\bar{z}_e : e \in E).$$

From the Strong Duality Theorem, we know that there exists a dual solution such that $\bar{t} = t$, where t is the optimal value of (5.10). Now, to obtain a trustworthy bound $\bar{t}$ of value approximately t, we need only equip ourselves with a near-optimal dual solution provided by the LP solver. (The simplex algorithm produces dual solutions as a matter of course.)

This use of duality does not quite get us home, however. For general LP problems the feasibility of solutions is again subject to numerical errors; the dual values $(\bar{y}, \bar{Y}, \bar{z})$ returned by a solver may not satisfy all of the dual constraints and thus do not directly give a valid bound $\bar{t}$. To finish the bounding technique we take advantage of the form of the dual LP (5.11), where each constraint has a variable z_e that appears only in that constraint. Using these variables, any values $(\bar{y}, \bar{Y}, \bar{z})$ can be transformed into a feasible dual solution $(\bar{y}, \hat{Y}, \hat{z})$ as follows. First, we enforce the condition $Y_S \geq 0$ for all $S \in \mathcal{U}$ by setting

$$\hat{Y}_S = \begin{cases} \bar{Y}_S & \text{if } \bar{Y}_S \geq 0, \\ 0 & \text{otherwise.} \end{cases}$$

Next, writing

$$\lambda_e = c_e - \bar{y}_u - \bar{y}_v - \sum(Y_S : S \in \mathcal{U}, e \in \delta(S))$$

for each edge $e = \{u, v\} \in E$, we enforce the constraint

$$y_u + y_v + \sum(Y_S : S \in \mathcal{U}, e \in \delta(S)) - z_e \leq c_e$$

by letting

$$\hat{z}_e = \begin{cases} 0 & \text{if } \lambda_e \geq 0, \\ -\lambda_e & \text{otherwise.} \end{cases}$$

Since $\hat{z}_e \geq 0$ for all e, the values $(\bar{y}, \hat{Y}, \hat{z})$ satisfy all constraints in (5.11). The lower bound $\hat{t}$ provided by this solution may be less than $\bar{t}$, but only by at most the total amount by which $(\bar{y}, \bar{Y}, \bar{z})$ violates the dual constraints.

The computations to obtain $\hat{t}$ from $(\bar{y}, \bar{Y}, \bar{z})$ can be carried out in fixed-point arithmetic to ensure that no rounding errors are introduced. In the Concorde code we do this by representing each number by 32 bits (including the sign bit) before the decimal point and 32 bits after the decimal point. Addition and subtraction of any two such numbers as well as multiplication of any such number by an integer can be carried out without rounding errors; however, any of these operations may result in overflow. We start out with the best fixed-point approximations to the floating-point dual solution returned by the LP solver, then compute $\hat{t}$ while checking that no overflow occurs (in which case we report an error message).

With this approach we obtain valid lower bounds for the TSP, typically of quality very close to the original value reported by the LP solver. This is of critical importance in our study of large-scale instances, where we depend on the efficiency of floating-point computations to handle the large and difficult LP problems that we create.

This technique for obtaining valid bounds can be adopted while solving any LP problems having explicit upper bounds on the primal variables; a general discussion can be found in Neumaier and Shcherbina [431].

5.5 FACET-INDUCING INEQUALITIES

Some cuts are better than others. Ultimately, what makes a cut good is its contribution to reducing the total running time of the cutting-plane method; the two factors that make up the total are (i) the number of iterations of adding cuts and (ii) the average time required to execute each iteration. As for (ii), the simplex method slows down when the LP relaxation is dense in the sense that its constraint set $Ax \leq b$ has a relatively large number of nonzero entries; this fact makes sparse cuts preferable to dense ones. However, in this section we shall disregard criterion (ii) and concentrate on criterion (i).

To begin, consider an LP relaxation with feasible region P. If $\alpha^T x \leq \beta$ is a cut and if $Ax \leq b$ is a system of inequalities such that

$$\mathcal{S} \subset P \cap \{x : Ax \leq b\} \subseteq P \cap \{x : \alpha^T x \leq \beta\}, \qquad (5.12)$$

then the entire set of inequalities $Ax \leq b$ is preferable to the single cut $\alpha^T x \leq \beta$ in the sense that substituting $Ax \leq b$ for $\alpha^T x \leq \beta$ in all subsequent LP relaxations either reduces the number of iterations of cut addition or leaves this number unchanged.

Next, let us say that an inequality $\alpha^T x \leq \beta$ is a *nonnegative linear combination* of inequalities $Ax \leq b$ if there is a nonnegative vector y such that $y^T A = \alpha^T$ and $y^T b = \beta$. If $\alpha^T x \leq \beta$ is a nonnegative linear combination of inequalities $Ax \leq b$, then $\{x : Ax \leq b\} \subseteq \{x : \alpha^T x \leq \beta\}$ holds trivially; if, in addition, all inequalities in $Ax \leq b$ are satisfied by all x in $\mathcal{S}$, then (5.12) holds.

These observations motivate our interest in decomposing cuts into nonnegative linear combinations. We are about to quote a well-known theorem asserting that every cut is a nonnegative linear combination of special linear inequalities that are satisfied by all x in $\mathcal{S}$; this theorem is a consequence of Theorem 5.1 on minimal defining systems for polyhedra.

THEOREM 5.5 *Let $\mathcal{S}$ be a nonempty finite subset of $\mathbf{R}^n$ and let α be a nonzero vector in $\mathbf{R}^n$. Then there is a system $Ax \leq b$ of linear inequalities such that*

(i) *$\alpha^T x \leq \max\{\alpha^T x : x \in \mathcal{S}\}$ is*
   a nonnegative linear combination of $Ax \leq b$,

 (ii) *each inequality in $Ax \leq b$ is satisfied by all x in $\mathcal{S}$,*

 (iii) *each inequality, $g^T x \leq t$, in $Ax \leq b$ has the property*
$$\dim\{x \in \mathcal{S} : g^T x = t\} \geq \dim \mathcal{S} - 1.$$

The inequalities $g^T x \leq t$ that constitute the system $Ax \leq b$ featured in Theorem 5.5 come in two flavors: either
$$\dim\{x \in \mathcal{S} : g^T x = t\} = \dim \mathcal{S} - 1$$
or
$$\dim\{x \in \mathcal{S} : g^T x = t\} = \dim \mathcal{S}.$$
In the former case, $g^T x \leq t$ induces a facet of the convex hull of $\mathcal{S}$; for short, we will say that the inequality induces a facet of $\mathcal{S}$. In the latter case, we have
$$\mathcal{S} \subset \{x : g^T x = t\}. \tag{5.13}$$
Equations $g^T x = t$ satisfying (5.13) are easy to characterize: if $A'x = b'$ is a system of linear equations that defines the affine hull of $\mathcal{S}$, then g and t satisfy (5.13) if and only if $g^T = y^T A'$ and $t = y^T b'$ for some vector y. Theorem 5.5 has the following corollary.

THEOREM 5.6 *Let $\mathcal{S}$ be a nonempty finite subset of $\mathbf{R}^n$, let P be a subset of the affine hull of $\mathcal{S}$, and let $\alpha^T x \leq \beta$ be a linear inequality satisfied by all x in $\mathcal{S}$. Then there is a system $Ax \leq b$ of linear inequalities such that*
$$P \cap \{x : Ax \leq b\} \subseteq P \cap \{x : \alpha^T x \leq \beta\},$$
and such that each inequality in $Ax \leq b$ induces a facet of $\mathcal{S}$.

To summarize, if the initial P in the cutting-plane method is a subset of the affine hull of $\mathcal{S}$, then substituting for each cut a suitably chosen set of inequalities that induce facets of $\mathcal{S}$ either reduces the number of iterations of cut addition or leaves this number unchanged. In this sense, cuts that induce facets of $\mathcal{S}$ are preferable to cuts that do not.

Nevertheless, it is an entire collection $Ax \leq b$ of facet-inducing inequalities that we are pitting here against a single cut $\alpha^T x \leq \beta$. Furthermore, some of the inequalities in $Ax \leq b$ may be satisfied by x^*, and therefore not qualify as cuts; if these inequalities are removed from $Ax \leq b$, then the set inclusion
$$P \cap \{x : Ax \leq b\} \subseteq P \cap \{x : \alpha^T x \leq \beta\}$$

of Theorem 5.6 may no longer hold. In fact, restricting the choice of cuts in the cutting-plane method to inequalities that induce facets of S may bring about an increase in the number of iterations of cut addition. To take a miniature example, consider the integer linear programming problem

$$
\begin{aligned}
\text{minimize} \quad & 3x_1 + 5x_2 \\
\text{subject to} \quad & x_1 + x_2 \leq 1, \\
& x_1 - 2x_2 \leq 1, \\
& -2x_1 + x_2 \leq 1, \\
& -3x_1 - 4x_2 \leq 1, \\
& x_1, x_2 \text{ integer.}
\end{aligned}
$$

Choosing

$$
P = \{x : x_1 + x_2 \leq 1, \ x_1 - 2x_2 \leq 1, \ -2x_1 + x_2 \leq 1, \ -3x_1 - 4x_2 \leq 1\},
$$

we enter the first iteration of cut addition with $x^* = [1/5, -2/5]^T$. To find cuts, we may use the variation on Gomory's theme that was described in Section 4.3. Here,

$$
A' = \begin{bmatrix} 1 & -2 \\ -3 & -4 \end{bmatrix}, \quad b' = \begin{bmatrix} 1 \\ 1 \end{bmatrix},
$$

and every integer vector g such that $g^T x^*$ is not an integer yields the cut (4.6):

$$
\begin{aligned}
g^T = [0, 3] \text{ yields the cut} \quad -2x_1 - 3x_2 \leq 0, \\
g^T = [3, 0] \text{ yields the cut} \quad -x_1 - 2x_2 \leq 0,
\end{aligned}
$$

and so on. Having tightened the relaxation by cuts $-2x_1 - 3x_2 \leq 0$ and $-x_1 - 2x_2 \leq 0$, we arrive at $x^* = [0, 0]^T$ and are done. However, if our choice of cuts is restricted to inequalities that induce facets of S, then two iterations of cut addition are required to solve the problem. Inequalities

$$
-x_1 \leq 0, \quad -x_2 \leq 0, \quad x_1 + x_2 \leq 1
$$

are, up to scaling, the only three inequalities that induce facets of S; of these three, only $-x_2 \leq 0$ is violated by the initial x^*; having tightened the initial relaxation by cut $-x_2 \leq 0$, we arrive at $x^* = [-1/3, 0]^T$ and have to proceed to another iteration of cut addition.

Still, examples like this seem to illustrate exceptions rather than the rule. Let us return to the case where S consists of all tours through a set V: Grötschel and Padberg [236] proved that subtour inequalities induce facets of S, and empirical evidence gathered by Miliotis [394] suggests that subtour cuts are preferable to Gomory cuts. Miliotis designed two different implementations of the cutting-plane method for the TSP. In the "straight" algorithm, he tightened the current LP relaxation by subtour cuts whenever x^* was integer and by Gomory cuts whenever x^* was not integer. In the "reverse" algorithm, he used heuristic procedures for finding subtour cuts that are guaranteed to succeed for all integer x^* and often succeed even when x^* is not integer; he tightened the LP relaxation by Gomory cuts only when these heuristics failed to find subtour cuts. On the more difficult instances that Miliotis solved, the reverse algorithm was more efficient: it generated fewer cuts in order to solve the problem, it required fewer iterations in the cutting-plane method, and it ran faster.

In closing this section, let us take note of the fact that, when S consists of all tours through a set V, the polyhedron that Dantzig, Fulkerson, and Johnson choose to initialize P in the cutting-plane method is a subset of the affine hull of S.

THEOREM 5.7 (Maurras [385], Grötschel and Padberg [236].) *The affine hull of the set of all tours through V consists of all solutions x of the degree constraints.*

5.6 THE TEMPLATE PARADIGM FOR FINDING CUTS

If the cutting-plane method is specialized to inputs with a fixed S, then one may exploit peculiarities of this S in design of specialized algorithms for finding cuts. One idea is to identify once and for all a class $\mathcal{C}$ of linear inequalities satisfied by all points of S and then, in each iteration of cut addition, look for cuts that belong to $\mathcal{C}$. This idea is the starting point of the present section.

A *separation algorithm* for a class $\mathcal{C}$ of linear inequalities is an algorithm that, given any x^*, returns either an inequality in $\mathcal{C}$ that is violated by x^* or a failure message. Separation algorithms that return a failure message only if all inequalities in $\mathcal{C}$ are satisfied by x^* are called *exact*; separation algorithms that may return a failure message even when some inequality in $\mathcal{C}$ is violated by x^* are called *heuristic*.

Combinatorial optimization problems such as the TSP are characterized not by single sets S but by families of mutually related sets in Euclidean spaces of various dimensions. In particular, the TSP is characterized by a sequence of sets $\mathcal{S}_n$ such that $\mathcal{S}_n \subset \mathbf{R}^{n(n-1)/2}$ for all n. By a *template* of cuts, with respect to a family $\mathcal{F}$ of sets S, we mean any class $\bigcup(\mathcal{C}_S : S \in \mathcal{F})$ of linear inequalities such that, for every S in $\mathcal{F}$, all inequalities in $\mathcal{C}_S$ are satisfied by all points of S. For instance, the subtour inequalities constitute a template of cuts in the TSP.

The *template paradigm* for finding cuts goes as follows. Given a family $\mathcal{F}$ of finite subsets of Euclidean spaces of various dimensions, identify once and for all one or more templates of cuts with respect to $\mathcal{F}$ and design separation algorithms for each of these templates. Then, in solving any problem

$$\text{minimize } c^T x \text{ subject to } x \in S$$

with $S \in \mathcal{F}$ by the cutting-plane method, use your separation algorithms in an attempt to find cuts.

The earliest polynomial-time exact separation algorithm that we know of comes from the doctoral thesis of Hong [270], who observed that the problem of finding a subtour cut or establishing that there is none reduces to $|V| - 1$ max-flow min-cut computations on networks with $|V|$ vertices. The same thesis is also the earliest reference we know of where the template paradigm is employed in the TSP with a template other than subtour inequalities: Hong uses his exact separation algorithm for the class of subtour inequalities and a heuristic separation algorithm for another template of cuts.

The other template used by Hong is the class of *2-matching inequalities*, also known as *blossom inequalities*,

$$x(\delta(H)) + \sum_{j=1}^{s} x(\delta(T_j)) \geq 3s + 1,$$

with $H, T_1, T_2, \ldots, T_s$ a family of subsets of V such that
- each T_j consists of a point in H and a point in $V - H$,
- $T_1, T_2, \ldots, T_s$ are pairwise disjoint,
- s is odd and at least three.

We pointed out in the previous chapter that these inequalities originate in the work of Edmonds [163], who used them in a different context. It follows from this work, and is easy to see directly, that all blossom inequalities are satisfied by the incidence vectors of all tours through V.

A triumph of the template paradigm came with the doctoral thesis of Grötschel [213] and his solution of a 120-city tour of Germany, as we described in Section 4.5. Grötschel's two templates were subtour inequalities and comb inequalities, which generalize blossom inequalities. Rather than designing explicit separation algorithms, Grötschel resorted to a man-machine interaction, where the machine found optimal solutions x^* of LP relaxations and the man found cuts by inspecting x^*.

We have mentioned comb inequalities several times in earlier chapters. A *comb* is any family of subsets of V that consists of a single "handle" and a set of "teeth" with the following properties:
- the handle meets each tooth but does not contain any,
- the teeth are pairwise disjoint,
- the number of teeth is odd and at least three.

Every comb with handle H and teeth $T_1, T_2, \ldots, T_s$ gives rise to the *comb inequality*

$$x(\delta(H)) + \sum_{j=1}^{s} x(\delta(T_j)) \geq 3s + 1. \tag{5.14}$$

Using the $\eta()$ notation, this inequality is written as

$$\eta(H, x) + \sum_{j=1}^{s} \eta(T_j, x) \geq s - 1. \tag{5.15}$$

The name comes from Chvátal [125], who introduced a variant of (5.14) with the subsets $T_1, \ldots, T_s$ not required to be pairwise disjoint but for each $i = 1, \ldots, s$, subset T_i is required to intersect H in exactly one vertex. The present version is due to Grötschel and Padberg [236], who have shown that it properly subsumes the original theme.

THEOREM 5.8 *For any comb with handle $H \subseteq V$ and teeth $T_1, \ldots, T_s \subseteq V$, every tour through V satisfies the comb inequality (5.14).*

Proof. Let x be the incidence vector of a tour. For each $i = 1, \ldots s$, write

$$c_i = \begin{cases} 1 & \text{if } x \text{ includes an edge from } H \cap T_i \text{ to } T_i - H, \\ 0 & \text{otherwise.} \end{cases}$$

Since the teeth are pairwise disjoint, we have $x(\delta(H)) \geq \sum_i c_i$; by definition we have $\sum_i c_i \leq s$; since $x(\delta(H))$ is even and s is odd, we conclude that

$$x(\delta(H)) \geq 2\sum_i c_i - s + 1. \tag{5.16}$$

The restriction of x on tooth T_i consists of $x(\delta(T_i))/2$ segments; one of these segments passes through a point of T_i that does not belong to H; if $c_i = 0$, a second segment passes through a point in $T_i \cap H$; we conclude that

$$x(\delta(T_i)) \geq 4 - 2c_i. \tag{5.17}$$

Combining (5.16) with (5.17) for all $i = 1, \dots, s$, the comb inequality (5.14) follows.   □

Following Grötschel's study, combs became a central theme in implementations of the template paradigm for the TSP; their repeated success in computational studies, moreover, became a strong force in promoting the paradigm in general.

Grötschel and Padberg [237] showed that all comb inequalities induce facets for the convex hull of tours, and championed the template paradigm with the axiom that cuts inducing facets are preferable to other cuts. They argued for choosing templates $\bigcup C_S$ in such a way that, for each S in $\mathcal{F}$, all inequalities in C_S induce facets of S. Subsequent work following these principles includes

- separation algorithms for subtour inequalities, designed by Crowder and Padberg [147] and by Padberg and Rinaldi [445];

- a polynomial-time exact separation algorithm for blossom inequalities, designed by Padberg and Rao [453];

- heuristic separation algorithms for comb inequalities, designed by Padberg and Rinaldi [446], by Grötschel and Holland [232], and by Applegate et al. [15];

- heuristic separation algorithms for *clique-tree inequalities* (introduced and proved to be facet-inducing by Grötschel and Pulleyblank [239]), designed by Padberg and Rinaldi [446];

- a polynomial-time exact separation algorithm for clique-tree inequalities with a fixed number of handles and teeth, and for certain more general templates, designed by Carr [109];

- heuristic separation algorithms for *path inequalities* (introduced by Cornuéjols, Fonlupt, and Naddef [140]), designed by Clochard and Naddef [130];

- a heuristic separation algorithm for a template arising from a complete list of inequalities that induce facets of all tours through eight cities (compiled by Christof, Jünger, and Reinelt [118]), designed by Christof and Reinelt [119] and Wenger [551];

- a family of greedy-type separation heuristics for comb inequalities, path inequalities, and a variety of other templates, designed by Naddef and Thienel [423].

We describe these lines of research in the chapters that follow.

An obvious advantage of the template paradigm is that all the cuts it provides are of a quality approved in advance; an obvious disadvantage is that it may provide no cuts at all. When the cutting-plane method demands finding cuts but the template paradigm does not provide any, either (A) the paradigm has to be augmented, at least temporarily, or (B) the method has to be given up, at least temporarily, in order to break the stalemate.

Option (A) is chosen in the "reverse algorithm" of Miliotis [394]: when his heuristics fail to provide subtour cuts, he generates Gomory cuts as a last resort. Land [333] adopts the same strategy with a refined implementation of the template paradigm: she uses heuristic separation algorithms for both subtour inequalities and blossom inequalities. Fleischmann [180] follows this strategy once again, but with yet another implementation of the template paradigm: he uses an exact separation algorithm for subtour inequalities and heuristic separation algorithms for comb inequalities with three teeth.

Option (B) was strongly advocated by Padberg and Grötschel [450], and variations were adopted in the work of Grötschel [228], Crowder and Padberg [147], Grötschel and Holland [232], and Padberg and Rinaldi [447], leading to the branch-and-cut scheme that we describe in Section 5.7.

5.7 BRANCH-AND-CUT METHOD

The truncated cutting-plane method formalizes the idea that at some point in a computation we may no longer be satisfied with the quality of the cutting planes we are obtaining. This could be when the template paradigm does not provide any cuts, or, in general, if the amount of increase in the LP lower bound is insignificant when compared to the remaining gap to the value of the best-known solution to our problem. In our study we adopt the *branch-and-cut* scheme as a means for turning the truncated cutting-plane method into a full solution procedure for the TSP.

An introduction and history of the branch-and-cut scheme is given in Section 4.6. We will not repeat the earlier material, but to have a point of reference we give below a formal outline of the method via pseudocode.

The setting is again that we have a finite subset S of points in some $\mathbf{R}^n$, and our problem is

$$\text{minimize } c^T x \text{ subject to } x \in S$$

for some cost vector $c \in \mathbf{R}^n$. It is trivial to create an initial linear system $Ax \leq b$ that is satisfied by all points in S; in our presentation we assume that the polyhedron $P = \{x : Ax \leq b\}$ is bounded. Since S is finite it is a simple matter to choose a system that meets this requirement.

The LP-based *branch-and-bound* method we outlined in Section 4.1 is summarized in the pseudocode given in Algorithm 5.3. The bounding process here is to

Algorithm 5.3 LP-based branch-and-bound method.

choose an initial linear system $Ax \leq b$ satisfied by all $x \in S$
set $\mathcal{L} = \{(Ax \leq b)\}, u = +\infty$;
while $\mathcal{L} \neq \emptyset$
do   remove a system $(Cx \leq d)$ from $\mathcal{L}$;
     if $\{x : Cx \leq d\} = \emptyset$ **then** set $t = +\infty$
     else   find x^* that minimizes $c^T x$ subject to $Cx \leq d$
           and set $t = c^T x^*$;
     end
     if $t < u$
     then if $x^* \in S$
          then set $u = t, \bar{x} = x^*$;
          else choose a vector α and numbers $\beta' < \beta''$ so that
               each $x \in S$ satisfies either $\alpha^T x \leq \beta'$ or $\alpha^T x \geq \beta''$
               add $(Cx \leq d,\ \alpha^T x \leq \beta')$ and $(Cx \leq d,\ -\alpha^T x \leq -\beta'')$
               to $\mathcal{L}$;
          end
     end
end
if $u = +\infty$ **then** return the message "$S = \emptyset$" **else** return $\bar{x}$ **end**

---

solve LP relaxations, starting with

$$\text{minimize } c^T x \text{ subject to } Ax \leq b$$

and in general

$$\text{minimize } c^T x \text{ subject to } Cx \leq d$$

for some linear system $Cx \leq d$. In a main step of the algorithm, there are three cases depending on the LP solution x^* that we obtain.

Case 1: Suppose the LP bound is less than the value $u = c^T \bar{x}$ of the best point $\bar{x} \in S$ found thus far in the search, and suppose the LP solution is not a point in S. Since the LP relaxation did not provide an optimal solution to the subproblem

$$\text{minimize } c^T x \text{ subject to } Cx \leq d \text{ and } x \in S,$$

we carry out a branching step to continue the search process. To do this, a vector $\alpha \in \mathbf{R}^n$ and scalars β' and β'' are selected such that each member of S satisfies either $\alpha^T x \leq \beta'$ or $\alpha^T x \geq \beta''$. New subproblems are created by imposing the extra constraint $\alpha^T x \leq \beta'$ in one subproblem and $\alpha^T x \geq \beta''$ in the other subproblem.

Case 2: Suppose the LP bound in less than u and suppose x^* is in fact a point in S. Here we update u and $\bar{x}$ by setting $u = c^T x^*$ and $\bar{x} = x^*$.

Case 3: If the LP relaxation is infeasible or if the optimal value is greater than or equal to u, then the subproblem can be discarded (no better point in S satisfies the constraints defining the subproblem).

In the branch-and-cut scheme, this process is augmented by applying the truncated cutting-plane algorithm to each of the subproblems, rather than simply relying on the LP relaxation that is presented. We summarize the scheme in Algorithm 5.4.

Algorithm 5.4 Branch-and-cut method.

---

choose an initial linear system $Ax \leq b$ satisfied by all $x \in S$
set $\mathcal{L} = \{(Ax \leq b)\}$, $u = +\infty$;
while $\mathcal{L} \neq \emptyset$
do remove a system $(Cx \leq d)$ from $\mathcal{L}$;
   if $\{x : Cx \leq d\} = \emptyset$ **then** set $t = +\infty$
   else find x^* that minimizes $c^T x$ subject to $Cx \leq d$
       and set $t = c^T x^*$;
   end
   while $t < u$ and $x^* \notin S$ and FINDCUTS $(S, x^*) \neq \emptyset$
   do   add cuts returned by FINDCUTS(S, x^*) to $Cx \leq d$;
      if $\{x : Cx \leq d\} = \emptyset$ **then** set $t = +\infty$
      else find x^* that minimizes $c^T x$ subject to $Cx \leq d$
          and set $t = c^T x^*$;
      end
   end
   if $t < u$
   then if $x^* \in S$
      then set $u = t, \overline{x} = x^*$;
      else choose a vector α and numbers $\beta' < \beta''$ so that
          each $x \in S$ satisfies either $\alpha^T x \leq \beta'$ or $\alpha^T x \geq \beta''$
          add $(Cx \leq d, \alpha^T x \leq \beta')$ and $(Cx \leq d, -\alpha^T x \leq -\beta'')$
          to $\mathcal{L}$;
      end
   end
end
if $u = +\infty$ **then** return the message "$S = \emptyset$" **else** return $\overline{x}$ **end**

Note that we impose the condition that the separation routine FINDCUTS return cutting planes that are valid for the entire set S rather than for the subproblem solutions $\{x : Cx \leq d\} \cap S$. As we discussed in Section 4.6, this standard practice makes it possible to share the cutting planes that are found with other subproblems by maintaining a list of them in a *cut pool* that can be searched as one of the separation routines.

5.8 HYPERGRAPH INEQUALITIES

It is useful to introduce standard notation for describing the TSP cuts we consider. A *hypergraph* is an ordered pair $(V, \mathcal{F})$ such that $\mathcal{F}$ is a family of (not necessarily distinct) subsets of V; the elements of $\mathcal{F}$ are called the *edges* of the hypergraph. Given a hypergraph $(V, \mathcal{F})$ denoted $\mathcal{H}$, we write

$$\mathcal{H} \circ x = \sum (x(\delta(S)) : S \in \mathcal{F})$$

and we let $\mu(\mathcal{H})$ stand for the minimum of $\mathcal{H} \circ x$ taken over the incidence vectors of tours through V. Thus each tour satisfies the *hypergraph inequality*,

$$\mathcal{H} \circ x \geq t$$

with $t \leq \mu(\mathcal{H})$.

If $\mathcal{F}$ contains only a single set, then $\mu(\mathcal{H}) = 2$ and the hypergraph inequality is a subtour constraint. If $\mathcal{H} = (V, \mathcal{F})$ is a comb with subsets $\mathcal{F} = \{S_0, S_1, \ldots, S_k\}$, then $\mu(\mathcal{H}) = 3k + 1$ and $\mathcal{H} \circ x \geq 3k + 1$ is the corresponding comb inequality.

In these two examples, the natural form of the constraints is as hypergraph inequalities. Although this is not always the case, it is true that any valid TSP inequality can be written in hypergraph form, after scaling and taking a linear combination of degree constraints.

THEOREM 5.9 *Every rational linear inequality satisfied by all incidence vectors of tours through V is a positive scalar multiple of the sum of a linear combination of degree equations and a hypergraph inequality.*

Proof. Suppose $\sum (a_e x_e : e \in E) \geq \beta$ is satisfied by all tours, where a_e is a rational number for each edge e. By subtracting copies of degree constraints if necessary, we may assume $a_e \leq 0$ for all $e \in E$. Furthermore, by scaling, we may also assume that a_e is an even integer for all $e \in E$. Now for any edge $e = (u, v)$ we have

$$-2x_e = x(\delta(\{u, v\})) - x(\delta(u)) - x(\delta(v)).$$

Using this we can write the inequality in the desired form.   □

We take advantage of this result in the Concorde code and by default store all inequalities in hypergraph form.

CUT SKELETONS

It will be convenient to prove here a few simple facts about hypergraph inequalities. Let $\mathcal{H} = (V, \mathcal{F})$ be a hypergraph and for each $e \in E$ let

$$c_e = |S \in \mathcal{F} : e \in \delta(S)|$$

be the edge's coefficient in the expression $\mathcal{H} \circ x$. It is easy to see that $(c_e : e \in E)$ satisfies the *triangle inequality*, that is, $c_{\{u,w\}} + c_{\{w,v\}} \geq c_{\{u,v\}}$ for any triple of vertices u, v, w.

THEOREM 5.10 *Given a hypergraph $\mathcal{H} = (V, \mathcal{F})$, the coefficients of $\mathcal{H} \circ x$ satisfy the triangle inequality.*

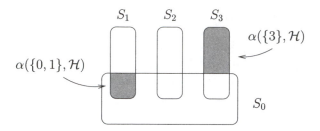

Figure 5.9 The atoms of a hypergraph.

Proof. Consider vertices u, v, w and edge $S \in \mathcal{F}$ such that $\{u, v\} \in \delta(S)$. Replacing S by $V - S$ if necessary, we may assume $u \in S$. If $w \in S$, then $\{w, v\} \in \delta(S)$. If $w \notin S$, then $\{u, w\} \in \delta(S)$.   □

Let $S_1, S_2, \ldots S_m$ be the edges of $\mathcal{H}$. For each subset I of $\{1, 2, \ldots, m\}$, we set

$$\alpha(I, \mathcal{H}) = \bigcap_{i \in I} S_i - \bigcup_{i \notin I} S_i \, ;$$

we refer to each nonempty $\alpha(I, \mathcal{H})$ as an *atom* of $\mathcal{H}$. The atoms are the regions in a Venn diagram representation of $\mathcal{H}$; we indicate two atoms in a representation of a comb in Figure 5.9. The following result shows that $\mu(\mathcal{H})$ is determined by the atoms of $\mathcal{H}$, independent of the number of vertices (≥ 1) that are assigned to each atom.

THEOREM 5.11 *Let $\mathcal{H}$ and $\mathcal{H}'$ be hypergraphs with the same set of vertices and the same number of edges, such that for all $I \subseteq \{1, 2, \ldots, m\}$ we have $\alpha(I, \mathcal{H}) \neq \emptyset$ if and only if $\alpha(I, \mathcal{H}') \neq \emptyset$. Then $\mu(\mathcal{H}) = \mu(\mathcal{H}')$.*

Proof. Let $\bar{x}$ be a tour through V such that $\mathcal{H} \circ \bar{x} = \mu(\mathcal{H})$. Now let $(c_e : e \in E)$ be the coefficients of $\mathcal{H} \circ x$ and let $S \subseteq V$ be an atom of $\mathcal{H}$. Note that $c_e = 0$ for all edges $e \subseteq S$. Making use of the triangle inequality, we can shortcut all but one of the subpaths of $\bar{x}$ that pass through S, so that the tour uses a single 0-cost subpath that visits all vertices in S. Repeating this process, we may assume $\bar{x}(\delta(\alpha(I, \mathcal{H}))) = 2$ for all atoms $\alpha(I, \mathcal{H})$. Visiting the atoms of $\mathcal{H}'$ in the same tour order as $\bar{x}$ yields a tour $\bar{x}'$ with $\mathcal{H}' \circ \bar{x}' = \mu(\mathcal{H})$.   □

Choosing a representative vertex for each atom gives a *skeleton* of the hypergraph that indicates the structure of the corresponding inequality. We record in Concorde a skeleton of each hypergraph inequality used in the LP relaxation, permitting later verification of the right-hand side of the cutting plane.

Related discussions of other standard forms of TSP inequalities can be found in Naddef [417], Naddef and Rinaldi [421], and Queyranne and Wang [467]. These studies include results concerning facets for the convex hull of tours.

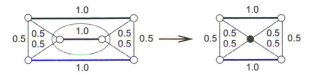

Figure 5.10 Unsafe shrinking.

5.9 SAFE SHRINKING

The intuitive notion of *shrinking* a subset of cities is commonly used as a pre-processing step in the search for cutting planes for the TSP. Formally, shrinking a subset S of V means replacing V with V' defined as $(V - S) \cup \{\sigma\}$ for some new vertex σ (representing the shrunk S) and replacing $\bar{x}$ with $\bar{x}'$ defined on the edges of the complete graph with vertex set V' by

$$\bar{x}'_{\{\sigma, t\}} = \bar{x}(S, \{t\}) \text{ for all } t \in V - S$$

and

$$\bar{x}'_{\{u, v\}} = \bar{x}_{\{u, v\}} \text{ for all } u, v \in V - S.$$

Using Theorem 5.11, any hypergraph inequality that is violated by x' yields a violated cut for $\bar{x}$ by simply unshrinking, that is, by replacing σ by S in all edges of the hypergraph.

Shrinking is used extensively in the computational TSP studies of Grötschel and Holland [232] and Padberg and Rinaldi [446] from the late 1980s, and it was also adopted in earlier work by Land [333] and Crowder and Padberg [147]. While it is permissible to shrink any collection of subsets $S \subseteq V$ in a heuristic separation algorithm, each of the above studies uses only sets with $2 \leq |S| \leq 4$ that are carefully chosen to avoid "hiding" flaws in $\bar{x}$ that could lead to violated cutting planes.

The idea of restricting the sets involved in shrinking operations was formalized in the paper of Padberg and Rinaldi [446]. Given an $\bar{x}$ that cannot be written as a convex combination of tours through V, Padberg and Rinaldi define a set S to be "shrinkable" if $\bar{x}'$ cannot be written as a convex combination of tours through V'. In other words, if there is a violated TSP cut for $\bar{x}$ then there also exists a violated TSP cut for $\bar{x}'$. We say such a set S is *safe* for shrinking.

A natural candidate for safe shrinking is a 2-point set $\{u, v\}$ such that $\bar{x}_{\{u,v\}} = 1$. Even here care must be taken, however, as we indicate in Figure 5.10. In this example, where the nonzero values of $\bar{x}$ are indicated next to the corresponding edges, shrinking the center two vertices is not safe. Indeed, the comb inequality consisting of the handle S_0 and the three teeth S_1, S_2, S_3 indicated in Figure 5.11 is violated by $\bar{x}$, whereas the shrunk vector $\bar{x}'$ is a convex combination of the two tours given in Figure 5.12.

The problem with the above example is the impossibility of extending the tours in Figure 5.12 to cover the edge between the vertices that are shrunk. Padberg and Rinaldi [446] showed, however, that if there exists a vertex t with $\bar{x}(\{t\}, \{u, v\}) = 1$, as in Figure 5.13, then this situation cannot occur.

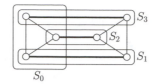

Figure 5.11 Violated comb inequality.

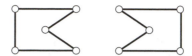

Figure 5.12 Convex combination of tours.

THEOREM 5.12 *If $\bar{x}_{\{u,v\}} = 1$ and there exist a vertex t such that $\bar{x}(\{t\}, \{u, v\}) = 1$, then it is safe to shrink $\{u, v\}$.*

Proof. Suppose $\bar{x}'$ is a convex combination of tours through V', and let T be a tour that is used in this combination. Let σ denote the shrunk vertex corresponding to $\{u, v\}$. Since $\bar{x}'_{\{\sigma,t\}} = 1$, tour T contains the edge $\{\sigma, t\}$. The second edge in T that meets σ corresponds either to an edge (before shrinking) that meets u or an edge that meets v. In the first case, extend T to a tour of V by replacing $\{\sigma, t\}$ by the two edges $\{u, v\}$, $\{v, t\}$. In the second case, extend T to a tour of V by replacing $\{\sigma, t\}$ by the edges $\{u, v\}$, $\{u, t\}$. These two types of replacements are illustrated in Figure 5.14. Carrying this out for all tours in the convex combination yields a representation of $\bar{x}$ as a convex combination of tours through V. □

Repeated application of this rule allows long paths of value 1 edges to be shrunk into a single edge, and it is thus very useful in reducing the size of the support graph.

Theorem 5.12 is a special case of a more general theorem of Padberg and Rinaldi [446]. Another consequence of the Padberg and Rinaldi theorem is the following result on shrinking triples of points.

THEOREM 5.13 *If u, v, w are distinct vertices with $\bar{x}_{\{u,v\}} + \bar{x}_{\{v,w\}} + \bar{x}_{\{w,u\}} = 2$ and there exists a vertex t such that $\bar{x}(\{t\}, \{u, v, w\}) = 1$, then it is safe to shrink $\{u, v, w\}$.* □

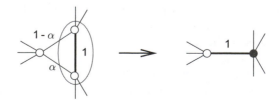

Figure 5.13 Safe shrinking.

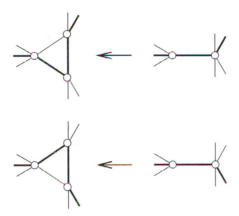

Figure 5.14 Why it is safe to shrink.

TEMPLATE-SAFE SHRINKING

Within the template paradigm, shrinking can often be used to find cuts that match prescribed templates: unshrinking preserves the match in the sense that
- subtour inequalities for $\bar{x}'$ yield subtour inequalities for $\bar{x}$,
- comb inequalities for $\bar{x}'$ yield comb inequalities for $\bar{x}$,

and so on. In this context, however, it is of practical importance to impose an alternative restriction on the types of shrinking operations that are performed. Indeed, if we are only looking for cuts that match a particular template, then shrinking should be safe for that template. Here "safe" means the operation preserves the existence of violated cuts from a specified class of inequalities.

Crowder and Padberg [147] pioneered this concept with their separation routine for subtour inequalities. They observed that if $\bar{x}_{\{u,v\}} = 1$, then it is *subtour-safe* to shrink $\{u, v\}$. Their proof is as follows. Suppose $\bar{x}(\delta(S)) < 2$ for some proper subset of vertices S. By replacing S by $V - S$ if necessary, we may assume $u \in S$. If we also have $v \in S$, then clearly $\bar{x}'(\delta(S - \{u, v\} \cup \{\sigma\})) < 2$, where σ denotes the vertex created in the shrinking step. On the other hand, if $v \notin S$, then we can replace S by $S \cup \{v\}$ and apply the above argument, since $\bar{x}(\delta(S \cup \{v\})) \leq \bar{x}(\delta(S))$. Thus $\bar{x}$ violates a subtour constraint if and only if $\bar{x}'$ also violates a subtour constraint.

Repeated application of the above shrinking rule often greatly reduces the size of the support graph for $\bar{x}$. Note that edges of value 1 can be created in the shrinking step, giving new possibilities for further shrinking. We make use of this process in heuristic and exact separation routines for subtours presented in Chapter 6.

The technique used by Crowder and Padberg is a standard method for proving template-safe shrinking results: given a violated hypergraph inequality $\mathcal{H} \circ x \geq t$ with respect to $\bar{x}$, we manipulate the edges of $\mathcal{H}$ to obtain a violated inequality with respect to the shrunk vector $\bar{x}'$. Further results of this type are discussed in Padberg and Grötschel [450] and in Padberg and Rinaldi [446].

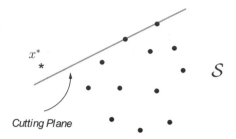

Figure 5.15 Cutting plane from FINDCUTS($\mathcal{S}, x^*$).

5.10 ALTERNATIVE CALLS TO SEPARATION ROUTINES

The branch-and-cut scheme is designed to handle the case where the subroutine
FINDCUTS fails to deliver any cutting planes (or any cutting planes of sufficient
quality to encourage us to continue the process). This does not mean that we are
necessarily happy to terminate the truncated cutting-plane method; a better set of
cuts providing a stronger LP bound is usually preferable. The natural solution
is to employ stronger versions of FINDCUTS, and this is what much of the book
describes. Taking a different tack, in this section we present two techniques that try
to squeeze more out of an existing implementation of FINDCUTS by altering the
input vector x^*.

SEPARATING WITH A CLOSE VECTOR

For many classes of TSP instances, very good heuristic algorithms are available for
finding near-optimal tours. It makes sense to try to take advantage of this when
searching for cutting planes.

In general, given a finite set $\mathcal{S} \subseteq \mathbf{R}^n$ and a cost vector $c \in \mathbf{R}^n$, suppose we have
a vector $\bar{x} \in \mathcal{S}$ that we believe to be a near-optimal solution to

$$\text{minimize } c^T x \text{ subject to } x \in \mathcal{S}.$$

In this situation, it is reasonable to consider augmenting calls to FINDCUTS($\mathcal{S}, x^*$),
where x^* is the optimal LP solution, with calls where x^* is replaced by vectors that
are closer to $\bar{x}$ but still outside of our feasible region P.

We adopt this strategy in the Concorde TSP code, setting

$$x^{**} = \gamma x^* + (1 - \gamma)\bar{x}$$

for some constant γ between 0 and 1, that is, we let x^{**} be a point on the line
segment joining $\bar{x}$ and x^*. Calling FINDCUTS($\mathcal{S}, x^{**}$) can sometimes lead to more
effective cutting planes than working with x^* itself, but, more important, such calls
allow us to reuse our existing heuristic separation routines (a practical consideration when trying to obtain as much improvement as possible from the truncated
cutting-plane method). The process of separating with a close vector is illustrated
in Figures 5.15 and 5.16. This technique is considered in a general context in
Ben-Ameur and Neto [51].

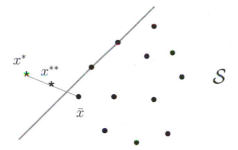

Figure 5.16 Alternative cutting plane from FINDCUTS($\mathcal{S}, x^{**}$).

This "close vector" approach is related to the primal cutting-plane strategy of restricting cuts to inequalities that hold as an equation for $\bar{x}$. Routines that return only cuts of this type are known as *primal separation algorithms*; these cuts remove from the feasible region the entire line segment between $\bar{x}$ and x^*.

With a near-optimal $\bar{x}$, it is clearly attractive to consider primal separation algorithms also in the usual (dual) cutting-plane method. For the TSP, we discuss such an algorithm for subtour inequalities in Section 6.3; primal separation methods for blossoms and combs are treated in Letchford and Lodi [350].

SEPARATING WITH AN INTERIOR VECTOR

In solving combinatorial problems like the TSP, it frequently happens that an LP relaxation does not have a unique optimal solution x^*, that is, the LP problem is *degenerate*. Therefore, cuts returned by FINDCUTS($\mathcal{S}, x^*$) will sometimes not exclude the full set of optimal solutions F^* from the feasible LP set. In this case, the addition of the cuts will make no improvement in the lower bound provided by the relaxation.

An idea for handling such degenerate LP problems is to attempt to call FIND-CUTS with a point x^{**} in the relative interior of the set F^*. Depending on the location of x^{**}, any cuts returned would hopefully exclude a large portion of F^*, rather than just nipping at its corners. This point has been observed in computational studies by Mitchell and Borchers [400] and by Barahona and Ladányi [38], where alternatives to the simplex method are used in the solution of the LP relaxations.

In the Mitchell-Borchers study, building on the earlier work of Mitchell [399], an interior-point LP solver is used in the early stages of a cutting-plane procedure for the linear ordering problem. In the later stages, when the LP bound gets close to the optimal value of the problem, Mitchell and Borchers switch to the simplex method, reporting that the simplex method appears to be faster at this point (owing to its restart capabilities, as we discuss in Chapter 13). Mitchell and Borchers report an improvement in the total number of rounds of cut addition with their approach, when compared to the standard implementation of using only the simplex method. They suggest that this improvement is due to the quality of the cutting planes that is provided early on in the computation, when the interior-point solver provides solutions in the interior of F^*.

The use of interior-point solvers is an interesting option in cutting-plane codes for the TSP, particularly for very large instances having a million cities or more, but several difficulties have limited its application up to now. First, the solutions to LP relaxations for the TSP quickly converge to a small percentage away from an optimal tour, giving less opportunity for the two-headed approach proposed by Mitchell and Borchers. Second, successful implementations of FINDCUTS for the TSP often depend on the sparsity of the support graph G^* of the LP solutions, making it difficult to work with the dense solutions provided by interior-point solvers. It is an important practical research problem to devise techniques based on interior-point solvers that can handle this difficulty with the density of the solution; we discuss this in Chapter 16.

In Barahona and Ladányi [38], an approximate (primal-dual) LP solver of Barahona and Anbil [37] is adopted in branch-and-cut codes for the Steiner tree and max-cut problems in graphs. In their tests, Barahona and Ladányi again report a reduction in the number of rounds of cut addition when compared to an implementation with the simplex algorithm. They write that their solver returns "a point near the center of the optimal face" and cite this feature as the reason for the reduced number of iterations. Again, a difficulty in applying this approach to the TSP is that early in the solution process, the LP relaxations are often quite close in value to the optimal tour, so solutions x^* that are slightly away (even 0.1%) from an optimal LP solution are often too rough an approximation to continue the cutting-plane process. The general framework is attractive, however, and an interesting topic is to study alternative approximation methods that may be suitable for problems like the TSP.

A practical alternative to the use of non-simplex solvers in this context is to coax the simplex method itself into producing a vector x^{**} in the relative interior of F^*. A technique that was studied in early versions of the Concorde code is to produce a family of optimal LP solutions $x^1, \dots, x^k$, using simplex pivot steps chosen at random from those that preserve the optimality of the current LP solution. The target vector x^{**} is then set to the barycenter of the points $x^1, \dots, x^k$. The method is potentially useful in cases where the LP solutions are highly degenerate.

Ben-Ameur and Neto [52] have put this type of approach into a more general context, where separation routines are presented with a set X of points not in P, rather than the single point x^*. In this case, Ben-Ameur and Neto ask for cutting planes that are violated by all points in X and study implementations where X consists of the set of optimal solution x^* to the k previous LP relaxations (with k ranging from 1 up to 50). They report results for a problem on survivable network design; we do not know of any tests of this approach for the TSP.

Chapter Six

Subtour Cuts and PQ-Trees

In the application of the cutting-plane method to the TSP, first and foremost among possible cuts are the subtour inequalities. Cuts that match the subtour template are the means to force the LP solution x^* to be a single connected piece, an obvious requirement for x^* to approximate a tour vector.

Fortunately, the subtour separation problem is very well understood, including a range of efficient exact separation methods. In this chapter we discuss fast subtour heuristics, as well as the exact method of Padberg and Rinaldi [445] that is used in Concorde. We also discuss an effective mechanism for building a repository of sets S satisfying $x^*(\delta(S)) = 2$, using the PQ-tree data structure of Booth and Lueker [79]. This repository is adopted in comb-finding heuristics described in later chapters.

6.1 PARAMETRIC CONNECTIVITY

In Figure 6.1 we display the support graph, G^*, for an optimal solution to the degree LP for a 1,000-city random Euclidean problem, where the cities are points with integer coordinates drawn uniformly from the 1,000,000 by 1,000,000 grid; the travel costs in this example are the Euclidean distances between the points, rounded to the nearest integer. A striking feature of the graph is the large number of small islands of points spread throughout the drawing.

In a support graph such as the one in Figure 6.1, subtour inequalities violated by x^* are readily available: the vertex set S of any connected component of G^* satisfies $x^*(\delta(S)) = 0$. Thus, as a first subtour heuristic we can simply list the connected components of G^*. This is a strategy that goes back to the earliest studies of the cutting-plane method for the TSP; it can be carried out in time linear in the number of edges of G^* using a simple depth-first search algorithm.

For the 1,000-city example, by repeatedly applying the heuristic until G^* is connected, the final LP bound is within 0.972% of the length of an optimal tour. Moreover, the LP value is only 0.409% below the *subtour bound* for this instance, that is, the optimal value for the full subtour relaxation. Similar results hold on a larger example, having 100,000 cities drawn uniformly from the 1,000,000 by 1,000,000 grid. In this case, the LP bound is within 1.048% of the length of a tour found by Keld Helsgaun in 2005 with a variant of his LKH code [260], and it is only 0.394% below the subtour bound, a respectable result for such a large problem.

A drawing of G^* for the LP solution obtained after applying the connectivity heuristic to the 1,000-city example is given in Figure 6.2. This new drawing is big

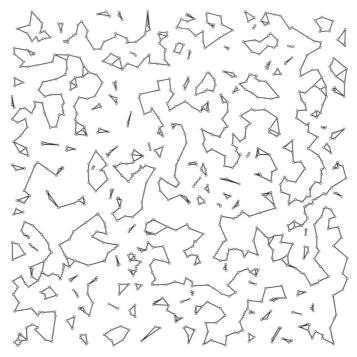

Figure 6.1 Support graph for solution to the degree LP for a 1,000-city problem.

improvement over the initial graph, but it is still possible to spot violated subtours inequalities by eye, focusing on points whose removal would disconnect the graph. It is possible to make further progress by searching for such cut points using an algorithm for 2-connected components. Notice, however, that looking for subtour inequalities violated by x^* simply by listing connected or 2-connected components of G^* means throwing away much information about x^*: all nonzero x_e^*, regardless of their actual values, are treated the same. Such lack of discrimination can have its repercussions. For example, consider an x^* whose G^* is disconnected and let $S_1, \ldots, S_k$ denote the vertex sets of the connected components of G^*: slight perturbations of the components of x^* can make G^* connected while maintaining the conditions

$$0 \le x_e \le 1 \ \text{ for all edges } e,$$
$$\sum(x_e : v \in e) = 2 \ \text{ for all cities } v,$$
$$x^*(\delta(S_i)) < 2 \ \text{ for all } i.$$

In this case, we could have spotted the sets $S_1, S_2, \ldots, S_k$ as the vertex sets of connected components of the graph with edges e such that $x_e^* > \varepsilon$ for some fixed positive ε. Pursuing this idea further, we make ε a parameter ranging from 1 down to 0 and arrive at Algorithm 6.1.

With m standing for the number of edges in the graph G^*, each individual test for $x^*(\delta(S)) < 2$ in Algorithm 6.1 may take time in $\Theta(m)$, which puts the total

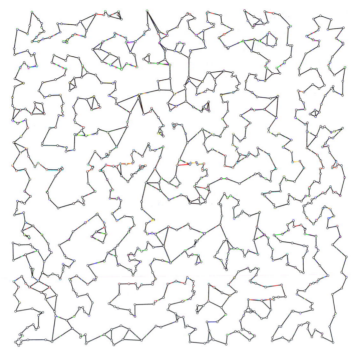

Figure 6.2 Support graph after connectivity heuristic.

Algorithm 6.1 Testing connected components in a parametric family.

initialize an empty list $\mathcal{L}$ of sets;
$F =$ the graph with the vertex set of G^* and with no edges;
for   all edges e of G^* in a nonincreasing order of x_e^*
do   **if**   the two endpoints of e
                belong to distinct connected components of F
        then add edge e to F;
                $S =$ vertex set of the connected component of F
                    that contains e;
                if $x^*(\delta(S)) < 2$ **then** add S to $\mathcal{L}$ **end**
                if F consists of two connected components
                then return $\mathcal{L}$
                end
        end
end
return $\mathcal{L}$;

running time of Algorithm 6.1 in $\Theta(mn)$. (Here $\Theta(f(n))$, for some function $f(n)$, is standard notation to indicate that the worst-case running time of an algorithm is at least $K_1 f(n)$ and at most $K_2 f(n)$ for some constants K_1 and K_2 and for large enough values of n.) It is quicker to first collect all the relevant S and then evaluate all the corresponding values of $x^*(\delta(S))$. All the relevant S may be recorded in a *decomposition forest* whose leaves are the n vertices of G^* and whose interior nodes are in a one-to-one correspondence with sets S for which Algorithm 6.1 tests the inequality $x^*(\delta(S)) < 2$; each interior node w of the decomposition forest corresponds to the set S_w of all leaves of the decomposition forest that are descendants of w. One way of constructing the decomposition forest is Algorithm 6.2; there, roots are nodes equal to their own parents and $\text{ROOT}(w)$ is the root of the tree that contains w.

Algorithm 6.2 Constructing a decomposition forest.

```
for all cities w do parent(w) = w end
counter = n;
for all edges e of G* in a nonincreasing order of x*_e
do   u, v = the two endpoints of e;
     u* = ROOT(u), v* = ROOT(v);
     if   u* ≠ v*
     then get a new node w;
          parent(u*) = w, parent(v*) = w, parent(w) = w;
          counter = counter−1;
          if counter = 2 then return array parent end
     end
end
return array parent;
```

Except for the evaluations of ROOT, straightforward implementations of Algorithm 6.2 take time $\Theta(n + m \log n)$, with the bottleneck $\Theta(m \log n)$ taken up by sorting these edges.

To implement the evaluations of ROOT, we may use the following triple of operations that maintain a changing family $\mathcal{F}$ of disjoint sets, with each set in $\mathcal{F}$ named by one of its elements:

MAKESET(w), with w in no set in $\mathcal{F}$, adds $\{w\}$ to $\mathcal{F}$;
FIND(u),   with u in some set in $\mathcal{F}$, returns the name of this set;
LINK(u, v), with $u \neq v$, deletes the sets named u and v from $\mathcal{F}$
              and adds their union to $\mathcal{F}$.

In our application, members of $\mathcal{F}$ are the sets S_w such that parent(w) = w; if, for each S_w in $\mathcal{F}$, we maintain a pointer root from the name of S_w to the root w of S_w, then we can evaluate ROOT(u) simply as root(FIND(u)). This policy is used in Algorithm 6.3.

Algorithm 6.3 An implementation of Algorithm 6.2.

---

for all cities w
do $\text{parent}(w) = w, \text{root}(w) = w, \text{MAKESET}(w)$;
end
$\text{counter} = n$;
for all edges e of G^* in a nonincreasing order of x_e^*
do $u, v =$ the two endpoints of e;
    $u^* = \text{root}(\text{FIND}(u)), v^* = \text{root}(\text{FIND}(v))$;
    if $u^* \neq v^*$
    then get a new node w;
        $\text{parent}(u^*) = w, \text{parent}(v^*) = w, \text{parent}(w) = w$;
        $\text{LINK}(u^*, v^*), \text{root}(\text{FIND}(u)) = w$;
        $\text{counter} = \text{counter} - 1$;
        if $\text{counter} = 2$ **then** return array $\texttt{parent}$ **end**
    end
end
return array $\texttt{parent}$;

A celebrated result of Tarjan [519] (see also Tarjan and van Leeuwen [521] and Chapter 2 of Tarjan [520]) asserts that a simple and practical implementation of these three operations runs very fast: the time it requires to execute any sequence of k operations is in $O(k\alpha(k))$, with α the very slowly growing function commonly referred to as "the inverse of the Ackermann function." Hence Algorithm 6.3 can be implemented so that the total time spent on calls of MAKESET, FIND, and LINK is in $O(m\alpha(m))$.

The final step in the parametric connectivity procedure is to evaluate $x^*(\delta(S_w))$ for all nodes w of the decomposition tree; we describe this in Algorithm 6.4, making use of the identity $x^*(\delta(S_w)) = 2|S_w| - 2x(\{e : e \subseteq S_w\})$.

Harel and Tarjan [249] and Schieber and Vishkin [492] designed implementations of the first **for** loop in Algorithm 6.4 that run in time in $O(m)$; the second **for** loop runs in time $O(n)$; the third **for** loop runs in time $O(m)$; a straightforward recursive implementation of the fourth **for** loop runs in time $O(n)$; the fifth **for** loop runs in time $O(n)$.

To illustrate its effectiveness on the 1,000-city instance described earlier in this section, we repeatedly apply Algorithm 6.1 until it returns without finding any cuts. The resulting LP bound is within 0.005% of the subtour bound in this case. In the 100,000-city example, the LP bound is within 0.029% of the subtour bound, a nice improvement over the bounds obtained by working only with the connected components of G^*.

Algorithm 6.4 Computing all the values of $x^*(\delta(S_w))$.

***

for all edges e of G^*
do $w(e) =$ the lowest common ancestor of the ends of e;
end
for all nodes w of the decomposition forest
do $\Delta_w = 2$ if w is a leaf and $\Delta_w = 0$ otherwise;
end
for all edges e of G^*
do $\Delta_{w(e)} = \Delta_{w(e)} - 2x_e^*$;
end
for all nodes w in a post-order transversal of the decomposition forest
do $\Delta_w = \Delta_w + \sum(\Delta_v : v$ is child of w);
end
for all nodes w of the decomposition forest
do $x^*(\delta(S_w)) = \Delta_w$;
end

***

6.2 SHRINKING HEURISTIC

Crowder and Padberg [147] and Land [333] developed a heuristic for subtour in-equalities that is based on the shrinking process described in Section 5.9.

The heuristic proceeds by examining the components of the solution vector x^* and shrinking the ends $\{u, v\}$ of any edge satisfying $x_{\{u,v\}}^* = 1$. If this process creates an edge e satisfying $x_e^* > 1$, then the set of original vertices S corresponding to the ends of e gives a violated subtour inequality; we record S and continue by shrinking the ends of e. We repeat this procedure until all edges e in the remaining graph satisfy $x_e^* < 1$. This process is summarized in Algorithm 6.5.

The shrinking heuristic is a very effective technique for finding violated subtour inequalities. Combining the shrinking cuts with the parametric connectivity heuristic, we obtain the an LP bound within 0.00001% of the subtour bound for the 1,000-city instance and within 0.0009% of the subtour bound for our 100,000-city instance. The lower bound produced in this way is thus very close to the optimal value over all subtour inequalities.

6.3 SUBTOUR CUTS FROM TOUR INTERVALS

In this section we present another fast heuristic separation algorithm for subtour inequalities, allowing us to take advantage of any approximation to an optimal tour that we might have obtained by running a tour-finding heuristic. Our motivation for the design of this algorithm comes from the following argument. Since the optimal solution x^* of the current LP relaxation of our TSP instance approximates an optimal tour and since our best heuristically generated tour $\hat{x}$ approximates an

---

Algorithm 6.5 A shrinking heuristic for subtour inequalities.

---

initialize an empty list $\mathcal{L}$ of sets;
while there is an edge e with $x_e^* \geq 1$
do   choose an edge $e = \{u, v\}$ with $x_e^* \geq 1$
      if $x_e^* > 1$
      then S = the set of original vertices corresponding to u and v;
            add S to $\mathcal{L}$;
      end
      shrink $\{u, v\}$;
end
return $\mathcal{L}$;

---

optimal tour, the two vectors x^* and $\hat{x}$ are likely to approximate each other at least in the sense that $x^*(\delta(S)) \approx \hat{x}(\delta(S))$ for most subsets S of V. In particular, sets S that minimize $x^*(\delta(S))$ subject to $S \subset V, S \neq V, S \neq \emptyset$ are likely to be found among sets S that minimize $\hat{x}(\delta(S))$ subject to the same constraints.

This argument may be not entirely convincing, but its conclusion was confirmed by our experience: in examples we have experimented with, many of the sets S such that $x^*(\delta(S)) < 2$ and $S \neq V, S \neq \emptyset$ satisfied $\hat{x}(\delta(S)) = 2$.

Sets S with $\hat{x}(\delta(S)) = 2$ are characterized trivially: if $v_0 v_1 \ldots v_{n-1} v_0$ is the cyclic order on V defined by the tour $\hat{x}$, then $\hat{x}(\delta(S)) = 2$ if and only if S or $V - S$ is one of the *intervals* I_{it} ($1 \leq i \leq t \leq n - 1$) defined by

$$I_{it} = \{v_k : i \leq k \leq t\}.$$

Since $x^*(\delta(V - S)) = x^*(\delta(S))$ for all subsets S of V, we are led to search for intervals I such that $x^*(\delta(I)) < 2$. We might set our goal at finding just one such interval or we might set it at finding all of them. The objective accomplished by our algorithm comes between these two extremes: for each $i = 1, 2, \ldots, n - 2$, we

$$\text{find a } t \text{ that minimizes } x^*(\delta(I_{it})) \text{ subject to } i \leq t \leq n - 1 \qquad (6.1)$$

and, in case $x^*(\delta(I_{it})) < 2$, we record the subtour inequality violated by x^*.

We describe an algorithm that solves the sequence of problems (6.1) in time that, with m standing again for the number of positive components of x^*, is in $\Theta(m \log n)$.

We reduce each of the problems (6.1) to a *minimum prefix-sum problem*,

given a sequence $s_1, s_2, \ldots, s_N$ of numbers,
find a t that minimizes $\sum_{k=1}^{t} s_k$ subject to $1 \leq t \leq N$.

To elaborate, let us write

$$s(i, k) = \begin{cases} 0 & \text{if } 1 \leq k \leq i \leq n - 1, \\ 1 - \sum_{i \leq j < k} x^*(\{v_j, v_k\}) & \text{if } 1 \leq i < k \leq n - 1. \end{cases}$$

If $1 \leq t \leq i$, then $\sum_{k=1}^{t} s(i, k) = 0$; if $i \leq t \leq n - 1$, then

$$\sum_{k=1}^{t} s(i, k) = \sum_{k=i+1}^{t} s(i, k) = (t - i) - \sum_{i \leq j < k \leq t} x^*(\{v_j, v_k\});$$

since

$$x^*(\delta(I_{it})) = 2|I_{it}| - 2 \sum_{i \le j < k \le t} x^*(\{v_j, v_k\}),$$

it follows that

$$\sum_{k=1}^{t} s(i, k) = \begin{cases} 0 & \text{if } t \le i, \\ (x^*(\delta(I_{it}))/2) - 1 & \text{if } t \ge i. \end{cases}$$

Hence problem (6.1) reduces to the problem

$$\text{find a } t \text{ that minimizes } \sum_{k=1}^{t} s(i, k) \text{ subject to } 1 \le t \le n - 1. \qquad (6.2)$$

We solve the sequence of minimum prefix-sum problems (6.2) for $i = n - 2$, $n - 3, \ldots, 1$ in this order; after each decrement of i, we use the formula

$$s(i, k) = \begin{cases} s(i + 1, k) & \text{if } k \le i, \\ 1 - x^*(\{v_i, v_k\}) & \text{if } k = i + 1, \\ s(i + 1, k) - x^*(\{v_i, v_k\}) & \text{if } k > i + 1 \end{cases}$$

to update the input of (6.2). The resulting scheme is Algorithm 6.6.

Algorithm 6.6 Finding intervals I such that $x^*(\delta(I)) < 2$.

---

    initialize an empty list $\mathcal{L}$ of intervals;
    for $k = 1, 2, \ldots, n - 1$ **do** $s_k = 0$ **end**
    for $i = n - 2, n - 3 \ldots, 1$
    do $s_{i+1} = 1$;
        for all edges $\{v_i, v_k\}$ such that $x^*(\{v_i, v_k\}) > 0$ and $i < k$
        do $s_k = s_k - x^*(\{v_i, v_k\})$;
        end
        $t = $ a subscript that minimizes $\sum_{k=1}^{t} s_k$ subject to $1 \le t \le n - 1$;
        if $\sum_{k=1}^{t} s_k < 0$ **then** add I_{it} to $\mathcal{L}$ **end**
    end
    return $\mathcal{L}$;

Each of the minimum prefix-sum problems

$$t = \text{a subscript that minimizes } \sum_{k=1}^{t} s_k \text{ subject to } 1 \le t \le n - 1$$

in Algorithm 6.6 can be solved trivially in time that is in $\Theta(n)$; the total running time of the resulting implementation of Algorithm 6.6 is in $\Theta(n^2)$. Our implementation reduces this total to $\Theta(m \log n)$ by making use of the fact that each of the minimum prefix-sum problems that has to be solved is related to the minimum prefix-sum problem solved in the previous iteration.

    Let us set this implementation in the more general framework of the following three operations:

INITIALIZE(N)   sets $s_1 = s_2 = \ldots = s_N = 0$,
RESET(k, value)   sets $s_k =$ value,
MIN-PREFIX   returns a t that
                     minimizes $\sum_{k=1}^{t} s_k$ subject to $1 \le t \le N$.

We are going to describe a data structure that supports these three operations in such a way that

each INITIALIZE   takes time in $\Theta(N)$,
each RESET   takes time in $\Theta(\log N)$,
each MIN-PREFIX   takes time in $\Theta(\log N)$.

This data structure is a full binary tree T (meaning, as usual, any binary tree in which each node other than a leaf has both a left child and a right child) with leaves $u_1, u_2, \ldots, u_N$ in the left-to-right order and such that each node u of T holds a pair of numbers $s(u), p(u)$ defined recursively by

- $s(u_k) = p(u_k) = s_k$
  whenever u_k is a leaf,

- $s(u) = s(v) + s(w), p(u) = \min\{p(v), s(v) + p(w)\}$
  whenever u is a node with left child v and right child w.

For each node u of T, there are subscripts $a(u)$ and $b(u)$ such that a leaf u_k is a descendant of u if and only if $a(u) \le k \le b(u)$; it is easy to see that

$$s(u) = \sum_{k=a(u)}^{b(u)} s_k \quad \text{and} \quad p(u) = \min\left\{ \sum_{k=a(u)}^{t} s_k : a(u) \le t \le b(u) \right\};$$

in particular, $p(\text{root}) = \min\{\sum_{k=1}^{t} s_k : 1 \le t \le N\}$. These observations suggest the implementations of INITIALIZE, RESET, and MIN-PREFIX that are spelled out in Algorithm 6.7.

To keep the running time of RESET and MIN-PREFIX in $\Theta(\log N)$, it is imperative to choose a T in INITIALIZE so that the depth of T is in $\Theta(\log N)$. Our choice is the *heap structure* with nodes $1, 2, \ldots, 2N - 1$. There, every node i with $i < N$ has left child $2i$ and right child $2i + 1$; nodes $N, N + 1, \ldots, 2N - 1$ are leaves in the left-to-right order; the depth of this tree is $\lfloor \lg(2N - 1) \rfloor$.

The resulting algorithm is presented as Algorithm 6.8.

Adding Algorithm 6.6 to the separation routines actually produces the optimal value over the full subtour relaxation for the 1,000-city instance. Considering the 100,000-city example, taking the algorithm as the only source of cutting planes, the result (3.175% gap to the subtour bound) is worse than that obtained using just the connected components of G. This is not too surprising, given the restricted form of subtour cuts that are produced by Algorithm 6.6 (the argument that $x^*(\delta(S)) \approx \hat{x}(\delta(S))$ does not hold well for the x^* vectors that appear after the addition of only subtour inequalities). If, however, we combine this algorithm with the heuristics presented earlier in this section, the LP bound is within 0.0008% of the subtour bound, a slight improvement over our previous bounds. Although this improvement

---

Algorithm 6.7 Three operations for solving a sequence of minimum prefix-sum problems.

INITIALIZE(N):
$T = $ a full binary tree of depth in $\Theta(\log N)$
      and with leaves $u_1, u_2, \ldots, u_N$ in the left-to-right order;
for each node u of T **do** $s(u) = 0, p(u) = 0$ **end**

RESET(k, value):
$s(u_k) = $ value, $p(u_k) = $ value;
$u = u_k$;
while u is not the root
do $u = $ parent of u;
      $v = $ left child of u, $w = $ right child of u;
      $s(u) = s(v) + s(w), p(u) = \min\{p(v), s(v) + p(w)\}$;
end

MIN-PREFIX:
$u = $ the root;
while u is not a leaf
do $v = $ left child of u, $w = $ right child of u;
      if $p(u) = p(v)$ **then** $u = v$ **else** $u = w$ **end**
end
return the subscript t for which $u = u_t$;

Algorithm 6.8 An efficient implementation of Algorithm 6.6.

---

initialize an empty list $\mathcal{L}$ of intervals;
INITIALIZE($n - 1$);
for $i = n - 2, n - 3, \ldots, 1$
do RESET($i + 1, 1$);
      for all edges $\{v_i, v_k\}$ such that $x^*(\{v_i, v_k\}) > 0$ and $i < k$
      do RESET($k, s(u_k) - x^*(\{v_i, v_k\})$);
      end
      if $p(\text{root}) < 0$
      then $t = $MIN-PREFIX;
            add I_{it} to $\mathcal{L}$;
      end
end
return $\mathcal{L}$;

---

is rather small, the cuts generated by this procedure are particularly useful in the Concorde code, where we store inequalities based on their representation as the union of intervals from the heuristic tour $\hat{x}$ (see Section 12.2).

6.3.1 Enumerating intervals I such that $\eta(I, x^*) = 0$

Recall that $\eta(I, x^*)$ is defined as $x^*(\delta(I)) - 2$. In Chapter 8, we use the list of all intervals I such that $\eta(I, x^*) = 0$, $|I| \geq 2$, and such that both endpoints of I belong to a prescribed subset of $\{v_1, v_2, \ldots, v_{n-1}\}$; the purpose of the present section is to describe an efficient algorithm for producing this list. More precisely, we are going to describe an algorithm that,

> given a number η_0 and subsets LEFT, RIGHT of $\{1, 2, \ldots, n-1\}$, returns the list $\mathcal{L}$ of all intervals I_{it} such that $\eta(I_{it}, x^*) \leq \eta_0$ and $i < t, i \in$ LEFT, $t \in$ RIGHT,

and whose running time is in $\Theta((m + |\mathcal{L}|) \log n)$; in the special case of our interest, where $\eta_0 = 0$ and LEFT = RIGHT, we simply disregard all the intervals I in $\mathcal{L}$ that have $\eta(I_{it}, x^*) < 0$.

The algorithm is a modified version of Algorithm 6.8. For each i such that $i \in$ LEFT, we enumerate all intervals I_{it} such that $\eta(I_{it}, x^*) \leq \eta_0$ and $t > i, t \in$ RIGHT. To do this, we set again

$$s_k = \begin{cases} 0 & \text{if } 1 \leq k \leq i, \\ 1 - \sum_{i \leq j < k} x^*(\{v_j, v_k\}) & \text{if } i < k \leq n-1, \end{cases}$$

so that

$$\sum_{k=1}^{t} s_k = \begin{cases} 0 & \text{if } t \leq i, \\ \eta(I_{it}, x^*)/2 & \text{if } t > i. \end{cases}$$

Again, we use a full binary tree T with leaves $u_1, u_2, \ldots, u_{n-1}$ in the left-to-right order and, for each node u of T, we let $a(u), b(u)$ be the subscripts such that a leaf u_k is a descendant of u if and only if $a(u) \leq k \leq b(u)$; again, each node u of T holds a number $s(u)$ such that

$$s(u) = \sum_{k=a(u)}^{b(u)} s_k.$$

However, the recursive definition of $p(u)$ is changed: we set

- $p(u_t) = \begin{cases} +\infty & \text{if } t > i \text{ and } u_t \notin \text{RIGHT}, \\ s_t & \text{otherwise} \end{cases}$
  whenever u_t is a leaf

and

- $p(u) = \min\{p(v), s(v) + p(w)\}$
  whenever u is a node with left child v and right child w,

so that

$$p(u) = \min \left\{ \sum_{k=a(u)}^{t} s_k : a(u) \le t \le b(u),\, t \in \texttt{RIGHT} \right\} \text{ whenever } a(u) > i.$$

This last identity guides a recursive function

LIST(u, p_0, i) that, given a node u of T, a number p_0, and a subscript i,
returns the list of all intervals I_{it} such that
$$a(u) \le t \le b(u), t > i, t \in \texttt{RIGHT}, \text{ and } \textstyle\sum_{k=a(u)}^{t} s_k \le p_0;$$

let us note that LIST$(\text{root}, \eta_0/2, i)$ returns the list of all intervals I_{it} such that $t > i$, $t \in \texttt{RIGHT}$, and $\eta(I_{it}, x^*) \le \eta_0$.

The resulting algorithm is presented as Algorithm 6.9, where function INITIAL-IZE is as in Algorithm 6.7.

6.4 PADBERG-RINALDI EXACT SEPARATION PROCEDURE

In an exact separation algorithm for subtour inequalities, we must determine for a given vector x^* whether or not there exists a proper subset of vertices S such that $x^*(\delta(S)) < 2$. This can be solved as a special case of the *global minimum-cut problem*. In this more general setting, we are given a vector w of nonnegative edge weights and we must find a proper subset of vertices $S \subseteq V$ such that $w(\delta(S))$ is minimized. To separate subtour inequalities, we let $w_e = x_e^*$ for each edge e, find a global minimum cut S, and check if $x^*(\delta(S)) < 2$. This approach was adopted in the 1972 paper by Hong [270], and it is a common ingredient in implementations of the cutting-plane method for the TSP.

There are a wide range of methods available for efficiently computing a global minimum cut. Hong's technique was to solve the problem with a sequence of $n - 1$ max-flow/min-cut computations. The strategy is straightforward. Since $w(\delta(S)) = w(\delta(V - S))$ for any proper subset S, we may choose some vertex s and consider only sets with $s \in S$. Now for each other vertex t, we compute an (s, t)-*minimum cut* S_{st}, that is, a set $S_{st} \subseteq V$ with $s \in S_{st}$ and $t \notin S_{st}$, minimizing $w(\delta(S_{st}))$. A global minimum cut $w(\delta(S))$ is obtained by letting S be the set S_{st} that minimizes $w(\delta(S_{st}))$ over all choices of t.

Following the safe shrinking discussion in Section 5.9, when applying Hong's procedure to the TSP we can preprocess the support graph G^* by shrinking all pairs of vertices $\{u, v\}$ satisfying $x_{\{u,v\}}^* \ge 1$, and recording the violated subtour inequality if $x_{\{u,v\}}^* > 1$. Note that after each (s, t)-minimum-cut computation we may also shrink the pair of vertices $\{s, t\}$, since any better cut must necessarily have both $s \in S$ and $t \in S$. This latter step again reduces the size of the graph and may create more opportunities to apply the safe-shrinking rule.

Significant further improvements to this process were made by Padberg and Rinaldi [445], who developed a set of rules for permitting additional shrinking of two-point sets of V, together with a strategy for iteratively applying the rules after

---

Algorithm 6.9 Enumerating all intervals I_{it} such that $\eta(I_{it}, x^*) \le \eta_0$ and $i < t$, $i \in$ LEFT, $t \in$ RIGHT

---

initialize an empty list $\mathcal{L}$ of intervals;
INITIALIZE$(n - 1)$;
for $i = n - 2, n - 3, \ldots, 1$
do RESET$^\star(i + 1, 1, \text{RIGHT})$;
$\quad\quad$ **for** all edges $\{v_i, v_k\}$ such that $x^*(\{v_i, v_k\}) > 0$ and $k > i$
$\quad\quad$ **do** RESET$^\star(k, s(u_k) - x^*(\{v_i, v_k\}, \text{RIGHT})$;
$\quad\quad$ **end**
$\quad\quad$ **if** $i \in$ LEFT
$\quad\quad$ **then** $\mathcal{L} = \mathcal{L} \cup$ LIST$(\text{root}, \eta_0/2, i)$;
$\quad\quad$ **end**
end
return $\mathcal{L}$;

RESET$^\star(k, \text{value}, \text{RIGHT})$:
$s(u_k) = \text{value}$;
if $k \in$ RIGHT **then** $p(u_k) = \text{value}$ **else** $p(u_k) = +\infty$ **end**;
$u = u_k$;
while u is not the root
do $\quad u = \text{parent of } u$;
$\quad\quad$ $v = \text{left child of } u, w = \text{right child of } u$;
$\quad\quad$ $s(u) = s(v) + s(w), p(u) = \min\{p(v), s(v) + p(w)\}$;
end

LIST(u, p_0, i):
if $p(u) \le p_0$ and $b(u) > i$
then if u is a leaf
$\quad\quad$ **then** $t = $ the common value of $a(u)$ and $b(u)$;
$\quad\quad\quad\quad$ return $\{I_{it}\}$;
$\quad\quad$ **else** $v = \text{left child of } u, w = \text{right child of } u$;
$\quad\quad\quad\quad$ return LIST$(v, p_0, i) \cup$ LIST$(w, p_0 - s(v), i)$;
$\quad\quad$ **end**
else return $\emptyset$;
end

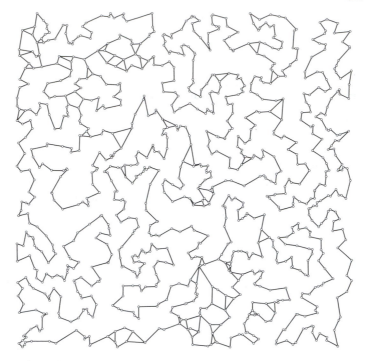

Figure 6.3 Support graph for optimal solution of subtour relaxation.

shrinking steps. For details of this procedure we refer to excellent computational studies by Chekuri et al. [114] and Jünger et al. [299].

We adopt the Padberg-Rinaldi approach in Concorde, using a version of the push-relabel algorithm of Goldberg [210] and Goldberg and Tarjan [211] to solve the (s, t)-minimum-cut problems that arise. Details on implementing push-relabel can be found in Derigs and Meier [155], Anderson and Setubal [12], and Cherkassky and Goldberg [116].

The effectiveness of the Padberg-Rinaldi shrinking rules, together with the good practical performance of Goldberg's algorithm, allows us to apply the separation routine on very large problems. Indeed, we have tested the code by computing the subtour bound on TSP instances having up to 3,000,000 cities.

For the modest 1,000-city example, the exact separation algorithm verifies that our previous x^* is an optimal solution to the subtour relaxation, giving a gap of 0.5609% to the length of an optimal tour; a drawing of the corresponding G^* is given in Figure 6.3. For the 100,000-city instance, the exact subtour separation algorithm produced a subtour bound giving a gap of at most 0.651% to an optimal tour.

It is important to note that a single run of the Padberg-Rinaldi algorithm can produce a large collection of violated subtour inequalities, rather than just the single inequality determined by the global minimum cut. This is crucial for large-scale instances where subtour heuristics usually fail before the subtour bound is reached.

This point is discussed by Levine [353] in his study combining Concorde with Karger and Stein's [305] random-contraction algorithm for global minimum cuts. An interesting direction here is the work of Fleischer [176] describing a fast algorithm for building a cactus representation of all global minimum cuts; a practical implementation of her method is described in Wenger [552].

6.5 STORING TIGHT SETS IN PQ-TREES

We call a subset S of V *tight* if $\eta(S, x^*) = 0$. In its search for comb inequalities violated by x^*, our computer code uses a repository of tight sets in two different ways. The technique described in Chapter 8 attempts to find a comb inequality

$$\eta(H, x) + \eta(T_1, x) + \eta(T_2, x) + \eta(T_3, x) \geq 2$$

violated by x^* by drawing H, T_1, T_2, T_3 from a family of sets S with small values of $\eta(S, x^*)$; this family includes all the tight sets from the repository. The technique described in Chapter 9 attempts to find a comb inequality

$$\eta(H, x) + \sum_{i=1}^{k} \eta(T_i, x) \geq k - 1$$

with all of $T_1, T_2, \ldots, T_k$ drawn from the repository. Construction of the repository is the subject of the present section.

6.5.1 PQ-trees compatible with x^*

Every pair S_1, S_2 of tight sets may point out other tight sets: if the four sets

$$S_1 - S_2, \; S_1 \cap S_2, \; S_2 - S_1, \; V - (S_1 \cup S_2)$$

are nonempty, then all four of them are tight or else x^* violates at least one of the four subtour inequalities defined by these sets. In the former case, the six sets

$$S_1 - S_2, \; S_1 \cap S_2, \; S_2 - S_1, \; S_1, \; S_2, \; S_1 \cup S_2 \qquad (6.3)$$

are tight (and so are their set-theoretic complements); the family of these six sets is like a family of intervals in the sense of having the form

$$V_1, \; V_2, \; V_3, \; V_1 \cup V_2, \; V_2 \cup V_3, \; V_1 \cup V_2 \cup V_3.$$

Repeated applications of this observation led to the organization of our repository of tight sets.

Following Booth and Lueker [79], we call a rooted ordered tree a *PQ-tree* if each of its internal nodes has at least two children and is labeled either a *P-node* or a *Q-node*. For each node u of a PQ-tree T, we let $D(u)$ denote the set of all leaves of T that are descendants of u; we let $\mathcal{I}(T)$ denote the family of

- all sets $D(u)$ such that u is a node of T

and

- all sets $D(u_i) \cup D(u_{i+1}) \cup \ldots \cup D(u_j)$ such that
  $u_1, u_2, \ldots, u_k$ in this order are the children of some Q-node of T
  and $1 \leq i < j \leq k$.

With the leaves of T enumerated in their left-to-right order as $v_1, \ldots, v_{n-1}$, each member of $\mathcal{I}(T)$ has the form $\{v_j : i \leq j \leq k\}$ with $i \leq k$; our notation $\mathcal{I}(T)$ is a mnemonic for "intervals." If $S_1, S_2 \in \mathcal{I}(T)$ and if the three sets $S_1 - S_2$, $S_1 \cap S_2$, $S_2 - S_1$ are nonempty, then all six sets (6.3) belong to $\mathcal{I}(T)$. For the definition of $\mathcal{I}(T)$, it is irrelevant whether an internal node with precisely two children is labeled a P-node or a Q-node; Booth and Lueker label all of them P-nodes; we shall follow their convention in the present section.

Having selected once and for all a city denoted ext for "exterior," we say that a PQ-tree T is *compatible with* x^* if the set of leaves of T is $V - \text{ext}$ and all members of $\mathcal{I}(T)$ are tight. Our choice of PQ-trees compatible with x^* as repositories of tight sets has been motivated by the following theorem.

THEOREM 6.1 *If, in addition to constraints*

$$0 \leq x_e^* \leq 1 \quad \text{for all edges } e,$$
$$\sum(x_e^* : v \in e) = 2 \quad \text{for all cities } v,$$

x^* *satisfies all subtour inequalities, then there is a PQ-tree T whose set of leaves is $V - \text{ext}$ and such that*

$$\mathcal{I}(T) = \{S \subseteq V - \text{ext} : \eta(S, x^*) = 0\}.$$

To see the necessity of the assumption that x^* satisfies all subtour inequalities, consider the incidence vector x^* of a disconnected graph whose components are cycles.

Given an optimal solution x^* of the current LP relaxation, our code first collects a family $\mathcal{F}$ of tight subsets of $V - \text{ext}$ and then it constructs a PQ-tree T compatible with x^*, aiming to include many, if not all, members of $\mathcal{F}$ in $\mathcal{I}(T)$. We will describe first—in Section 6.5.2—our way of constructing T from $\mathcal{F}$ and then—in Section 6.5.3—our way of generating $\mathcal{F}$. Correctness of our construction of T does not depend on the validity of Theorem 6.1; we relegate its proof to Section 6.5.4.

6.5.2 From $\mathcal{F}$ to T

All our PQ-trees have $V - \text{ext}$ for their set of leaves; the *trivial PQ-tree* has no other nodes except for its root, which is a P-node.

Given a family $\mathcal{F}$ of tight subsets of $V - \text{ext}$, we construct a PQ-tree T compatible with x^* and aim to

$$\text{include many, if not all, members of } \mathcal{F} \text{ in } \mathcal{I}(T). \tag{6.4}$$

Objective (6.4) is stated vaguely; at the very least, we want it to mean that $\mathcal{F} \subseteq \mathcal{I}(T)$ whenever

$$\text{some PQ-tree } T' \text{ compatible with } x^* \text{ has } \mathcal{F} \subseteq \mathcal{I}(T'). \tag{6.5}$$

If (6.5) holds, then T may be constructed by Algorithm 6.10. The function INSERT featured in this algorithm has been designed by Booth and Lueker [79]; given a PQ-tree T and a set S such that some PQ-tree T' satisfies

$$\mathcal{I}(T') \supseteq \mathcal{I}(T) \cup \{S\}, \tag{6.6}$$

it returns a PQ-tree T^* such that
- T^* satisfies (6.6) in place of T',
- every PQ-tree T' with property (6.6) has $\mathcal{I}(T') \supseteq \mathcal{I}(T^*)$.

Algorithm 6.10 Prototype of a construction of a PQ-tree

T = the trivial PQ-tree;
for all S in $\mathcal{F}$ **do** $T = $ INSERT(T, S) **end**
return T;

Algorithm 6.10 maintains the invariant $\mathcal{I}(T) \subseteq \mathcal{I}(T')$ and returns a PQ-tree T such that $\mathcal{F} \subseteq \mathcal{I}(T)$; since T' is compatible with x^*, T is compatible with x^*.

There are two reasons why Algorithm 6.10 is unsuitable for our purpose:

(i) If Algorithm 6.10 calls INSERT(T, S) with T and S such that no PQ-tree T' satisfies (6.6), then the Booth-Lueker INSERT destroys T beyond all recovery and, as a failure message, returns the trivial PQ-tree. Such regress is inconsistent with objective (6.4).

(ii) If all calls of INSERT(T, S) are successful, and so Algorithm 6.10 returns a PQ-tree T such that $\mathcal{F} \subseteq \mathcal{I}(T)$, then this PQ-tree is not compatible with x^* when (6.5) fails. PQ-trees incompatible with x^* are unacceptable.

To get around the first drawback, one might make a spare copy of T before each call of INSERT(T, S). Instead, we have designed a variation on Booth-Lueker INSERT. This variation, TEST, given a PQ-tree T and a set S, returns PASS if some PQ-tree T' satisfies (6.6) and returns FAIL otherwise; in either case, it leaves T unchanged. In Algorithm 6.10, we replace each instruction

$$T = \text{INSERT}(T, S)$$

by

if TEST$(T, S) =$PASS **then** $T = $ INSERT(T, S) **end**

The Booth-Lueker INSERT is remarkably efficient: the running time of Algorithm 6.10 is in $O(n + \sum(|S| : S \in \mathcal{F}))$ when all calls of INSERT are successful. Our calls of TEST add to this total a time in $O(kn + \sum(|S| : S \in \mathcal{F}))$ with k at most, and typically far less than, the number of times that TEST returns FAIL.

To get around the second drawback, we have designed a function REBUILD that, given a PQ-tree T, constructs a PQ-tree T^* compatible with x^* and aims to include

---

Algorithm 6.11 Construction of a PQ-tree compatible with x^*

$T = $ the trivial PQ-tree;
for all S in $\mathcal{F}$
do **if** $\text{TEST}(T, S) = $PASS **then** $T = \text{INSERT}(T, S)$ **end**
end
$T = \text{REBUILD}(T, x^*)$;
return T;

---

many, if not all, tight members of $\mathcal{I}(T)$ in $\mathcal{I}(T^*)$; when the modified version of Algorithm 6.10 has returned a PQ-tree T, we may call $\text{REBUILD}(T, x^*)$. (Our computer code uses a slight modification of Algorithm 6.11, which will be presented later as Algorithm 6.14.)

$\text{REBUILD}(T, x^*)$ has the form

$T = \text{CLEAN-UP}(T, x^*)$;
$T = \text{REFINE}(T, x^*)$;
return T;

$\text{CLEAN-UP}(T, x^*)$, given any PQ-tree T, deletes one by one all nodes u such that $D(u)$ is not tight, turning each child of u into a child of the parent of u whenever u is deleted; it labels all the remaining internal nodes as P-nodes. We implement it by Algorithm 6.10 with

$$\mathcal{F} = \{D(u) : \ u \text{ is a node of } T \text{ and } D(u) \text{ is tight}\}; \qquad (6.7)$$

in place of $\mathcal{F}$; since this family is nested in the sense that every two of its members S_1, S_2 satisfy $S_1 \subseteq S_2$ or $S_2 \subseteq S_1$ or $S_1 \cap S_2 = \emptyset$, the resulting PQ-tree T has $\mathcal{I}(T)$ equal to (6.7).

$\text{REFINE}(T, x^*)$ is defined in Algorithm 6.12; given a PQ-tree T compatible with x^*, it returns a PQ-tree T^* compatible with x^* and such that $\mathcal{I}(T^*) \supseteq \mathcal{I}(T)$.

To see the effect of the sequence of instructions

$$\left. \begin{array}{l} S = V - (S_1 \cup S_2 \cup \ldots \cup S_t); \\ \textbf{for } i = j+1, j+2, \ldots, t-1 \textbf{ do } T = \text{INSERT}(S_i \cup S_{i+1}, T) \textbf{ end} \\ T = \text{INSERT}(S_t \cup S, T); \\ T = \text{INSERT}(S \cup S_1, T); \\ \textbf{for } i = 1, 2, \ldots, j-2 \textbf{ do } T = \text{INSERT}(S_i \cup S_{i+1}, T) \textbf{ end} \end{array} \right\} \qquad (6.8)$$

in Algorithm 6.12 on the PQ-tree T, write

$$\begin{array}{l} S'_0 = V - D(u), \\ S'_1 = S_{j+1}, \ S'_2 = S_{j+2}, \ \ldots, \ S'_{t-j} = S_t, \\ S'_{t-j+1} = S, \\ S'_{t-j+2} = S_1, \ S'_{t-j+3} = S_2, \ \ldots, \ S'_t = S_{j-1}; \end{array}$$

Algorithm 6.12 REFINE(T, x^*)

for all P-nodes u of T that have at least three children
do enumerate the children of u as $u_1, \ldots, u_k$;
    $F =$ the graph with vertices $V - D(u), D(u_1), \ldots, D(u_k)$
        and such that A, B are adjacent if and only if $x^*(A, B) = 1$;
    for all components Q of F that have at least two vertices
    do enumerate the vertices of Q as $S_1, S_2, \ldots, S_t$
        in such a way that $x^*(S_i, S_{i+1}) = 1$ for all $i = 1, 2, \ldots, t-1$;
        if $t \geq 3$ and $x^*(S_1, S_t) > 0$ **then** $t = t - 1$ **end**
        if   some S_j is $V - D(u)$
        then $S = V - (S_1 \cup S_2 \cup \ldots \cup S_t)$;
            for $i = j+1, j+2, \ldots, t-1$
            do $T = $ INSERT$(S_i \cup S_{i+1}, T)$
            end
            $T = $ INSERT$(S_t \cup S, T)$;
            $T = $ INSERT$(S \cup S_1, T)$;
            for $i = 1, 2, \ldots, j-2$
            do $T = $ INSERT$(S_i \cup S_{i+1}, T)$
            end
        else **for** $i = 1, 2, \ldots, t-1$
            do $T = $ INSERT$(S_i \cup S_{i+1}, T)$
            end
    end
end
end
return T;

---

observe that $S'_0, S'_1, S'_2, \ldots S'_t$ are pairwise disjoint tight sets whose union is V and that

$$x^*(S'_0, S'_1) = x^*(S'_1, S'_2) = \ldots = x^*(S'_{t-1}, S'_t) = x^*(S'_t, S'_0) = 1.$$

There are children $v_1, v_2, \ldots, v_{t-j}$ and $v_{t-j+2}, v_{t-j+3}, \ldots, v_t$ of u such that $S'_i = D(v_i)$ whenever $1 \leq i \leq t - j$ or $t - j + 2 \leq i \leq t$; enumerate the remaining children of u as $w_1, w_2, \ldots w_{k-t+1}$ and observe that

$$S = D(w_1) \cup D(w_2) \cup \ldots \cup D(w_{k-t+1}).$$

The effect of (6.8) on T is this: the new children of u are $v_1, \ldots, v_t$ in this order (or its reverse), where v_{t-j+1} is a new P-node with children $w_1, w_2, \ldots w_{k-t+1}$ in case $k > t$ and $v_{t-j+1} = w_1$ in case $k = t$; if $t \geq 3$, then u becomes a Q-node; if $t = 2$, then u remains a P-node.

Similarly, let us note the effect of the instruction

$$\textbf{for } i = 1, 2, \ldots, t-1 \textbf{ do } T = \text{INSERT}(S_i \cup S_{i+1}, T) \textbf{ end} \qquad (6.9)$$

in Algorithm 6.12 on the PQ-tree T. There are children $v_1, v_2, \ldots, v_t$ of u such that $S_i = D(v_i)$ whenever $1 \leq i \leq t$. If $v_1, v_2, \ldots, v_t$ are all the children of u, then u becomes a Q-node, with children $v_1, v_2, \ldots, v_t$ in this order (or its reverse); else a new node, with children $v_1, v_2, \ldots, v_t$ in this order (or its reverse), replaces $v_1, v_2, \ldots, v_t$ in the list of children of u; this new node is a Q-node if $t \geq 3$ and a P-node if $t = 2$.

Given an arbitrary PQ-tree T_0, let us write
$$T_1 = \text{CLEAN-UP}(T_0, x^*),$$
and
$$T_2 = \text{REFINE}(T_1, x^*),$$
so that
$$T_2 = \text{REBUILD}(T_0, x^*);$$
let us point out that

$$\text{if } T_0 \text{ is compatible with } x^*, \text{ then } \mathcal{I}(T_2) \supseteq \mathcal{I}(T_0). \qquad (6.10)$$

Since T_0 is compatible with x^*, the passage from T_0 to T_1 amounts to relabeling each internal Q-node as a P-node; the subsequent passage from T_1 to T_2 restores these changed labels to their original form. (In addition, when $\text{REFINE}(T_1, x^*)$ processes P-nodes of T_1 that were labeled as P-nodes in T_0, it may discover and insert into $\mathcal{I}(T_2)$ tight sets that were not included in $\mathcal{I}(T_0)$, and so the set inclusion in (6.10) may hold as strict.)

AN EFFICIENT IMPLEMENTATION OF REBUILD(T, x^*)

Let G denote the graph with vertex set V and with edge set $\{e : x_e^* > 0\}$; let m denote the number of edges of G. Since

$$0 \leq x_e^* \leq 1 \qquad \text{for all edges } e,$$
$$\sum (x_e^* : v \in e) = 2 \qquad \text{for all cities } v,$$

m is at least n; since x^* is a basic optimal solution of the current LP relaxation of our TSP instance, m is at most the number of constraints in this relaxation, including the upper bounds $x_e \leq 1$ and excluding the nonnegativity constraints $x_e \geq 0$. Typically, m is around $2n$; sometimes it rises to $4n$ or more.

To make CLEAN-UP(T, x^*) run fast, we need a quick way of finding out which internal nodes u of T satisfy $\eta(D(u), x^*) = 0$.

Let us say that an edge of G is *internal* if neither of its two endpoints is ext; for each internal node u of T, let us set

$E_u = \{e : \quad e \text{ is an internal edge of } G \text{ and } u \text{ is}$
$\qquad\qquad \text{the lowest common ancestor in } T \text{ of the two endpoints of } e\};$

in this notation, each internal node u of T satisfies

$$\sum (x^*(\delta(D(u)))) = \sum (x^*(\delta(D(v))) : v \text{ is a child of } u)$$
$$-2 \sum (x_e^* : e \in E_u). \qquad (6.11)$$

We first partition the set of all internal edges of G into the sets E_u; then we compute all the numbers $\sum(x_e^* : e \in E_u)$; finally, we use identity (6.11) to compute all the numbers $x^*(\delta(D(u)))$ in a single recursive sweep through the tree.

To implement the first step, we can again use the algorithms of Harel and Tarjan [249] and Schieber and Vishkin [492] that run in time in $O(m)$; the time required to execute the second step and the third step is in $O(m)$.

Our implementation of REFINE(T, x^*) evolves from a few observations concerning an arbitrary internal node u of T and all its children, enumerated as $u_1, \ldots, u_k$.

First, let G_u denote the edge-weighted graph with vertices $V - D(u)$, $D(u_1)$, $\ldots, D(u_k)$, such that A, B are adjacent if and only if $x^*(A, B) > 0$, and such that the weight of an edge with endpoints A, B is $x^*(A, B)$; let n_u denote the number of vertices of G_u and let m_u denote the number of edges of G_u. Observe that

- if adjacency lists representing G_u are available, then, except for the calls of INSERT, the corresponding iteration of the outer **for** loop in Algorithm 6.12 can be executed in time proportional to $n_u + m_u$.

Next, let G_u' denote the edge-weighted graph obtained from G_u by removing vertex $V - D(u)$ and all the edges incident with it; let n_u' denote the number of vertices of G_u' and let m_u' denote the number of edges of G_u'. Note that

$$x^*(D(i), V - D(u)) = 2 - \sum(x^*(D(i), D(j)) : 1 \leq j \leq k, j \neq i),$$

and so

- adjacency lists representing G_u can be reconstructed from adjacency lists representing G_u' in time proportional to $n_u' + m_u'$.

Finally, consider an arbitrary internal edge e of G such that u is the lowest common ancestor in T of the two endpoints of e: in the notation introduced earlier, $e \in E_u$. There are children u_i, u_j of u such that one endpoint of e belongs to $D(u_i)$ and the other endpoint of e belongs to $D(u_j)$; let us refer to u_i, u_j as the *lifted endpoints of e*. Let E_u'' denote the multiset obtained from E_u when the two endpoints of each e in E_u are replaced by the two lifted endpoints of e; let m_u'' denote the number of edges in E_u''. Observe that

- adjacency lists representing G_u' can be reconstructed from E_u'' in time proportional to $n_u' + m_u''$.

Trivially, $n_u' = n_u - 1$ for all u; $\sum n_u < 3n$; $\sum m_u'' < m$; $m_u' \leq m_u''$ for all u; $\sum(m_u - m_u') < n$. We conclude that, except for the calls of INSERT, the running time of Algorithm 6.12 is in $O(t_0 + m)$, with t_0 standing for the time required to compute all the sets E_u''.

As noted earlier, all the sets E_u can be computed in time $O(m)$ by the algorithms of Harel and Tarjan [249] and Schieber and Vishkin [492]. We transform all these sets into sets E_u'' in an order where each node precedes its parent; postorder is a natural choice of this order. During this process, we keep updating an equivalence relation on the set of leaves of T in such a way that, when we are about to compute E''_u, each $D(u_i)$ with u_i a child of u is an equivalence class.

To maintain a family $\mathcal{E}$ of disjoint sets with each set in $\mathcal{E}$ permanently "named" by one of its elements, we can make use of the standard MAKESET, FIND, and LINK operations described in Section 6.1. In our application, each member of $\mathcal{E}$ has the form $D(u)$ for some node u of T; given any member of $\mathcal{E}$, we need to know which node of T defines it; for this purpose, we maintain links $\mathrm{up}(\cdot)$ pointing from the leaf that names $D(u)$ to u. To be able to merge sets $D(u_i)$ and $D(u_j)$ by a call of LINK, we need to know the names of these sets; for this purpose, we maintain links $\mathrm{down}(\cdot)$ pointing from the u to an element of $D(u)$.

Algorithm 6.13 Lifting the endpoints of all internal edges of G.

---

for all nodes u of T in postorder
do **if** u is a leaf
    then MAKESET(u), $\mathrm{up}(u) = u$, $\mathrm{down}(u) = u$;
    else **for** all edges e in E_u
        do v, w = endpoints of e;
        lifted endpoints of e = $\mathrm{up}(\mathrm{FIND}(v)), \mathrm{up}(\mathrm{FIND}(w))$;
        end
        enumerate the children of u as $u_1, u_2, \ldots, u_k$;
        for $i = 1, 2, \ldots, k - 1$
        do LINK($\mathrm{FIND}(\mathrm{down}(u_i))$,$\mathrm{FIND}(\mathrm{down}(u_{i+1}))$);
        end
        $\mathrm{down}(u) = \mathrm{down}(u_1)$;
        $\mathrm{up}(\mathrm{FIND}(\mathrm{down}(u))) = u$;
    end
end

Using the implementation of MAKESET, FIND, and LINK by Tarjan [519], the running time of Algorithm 6.13 is in $O(m\alpha(m))$, where α is the inverse Ackermann function.

6.5.3 Generating $\mathcal{F}$

We use four sources to generate a family of tight sets.

(α) We use all the two-point tight sets: these are simply the edges e such that $x_e^* = 1$.

(β) Enumerate all the cities as $v_0, v_1, \ldots, v_{n-1}$ in an order defined by the best tour through V that we know of and so that $v_0 = \mathrm{ext}$; with subscript arithmetic mod n, define

$$W = \{t : 1 \leq t \leq n - 1 \text{ and } x^*(v_{t-1}v_t) = x^*(v_t v_{t+1}) = 1\}.$$

We use Algorithm 6.9 of Section 6.3.1 to generate all tight sets of the form

$$\{v_j : i \leq j \leq k\} \tag{6.12}$$

with $1 \leq i < k \leq n - 1$ and $i \notin W, k \notin W$.

The restriction $i \notin W, k \notin W$ reduces the ensuing number of calls of INSERT without seriously diminishing $\mathcal{I}(T)$: if $\mathcal{I}(T)$ includes all the tight sets from sources (α) and (β), then it includes all the tight sets of the form (6.12) with $1 \leq i < k \leq n - 1$. To see this, observe that

> if $R, S \in \mathcal{I}(T)$ and if $R - S$, $R \cap S$, $S - R$ are nonempty,
> then $R - S$, $R \cap S$, $S - R$, $R \cup S \in \mathcal{I}(T)$.

(γ) Our cut pool is a set of hypergraph inequalities, $\mathcal{H}_i \circ x \geq t_i$; each $\mathcal{H}_i$ is represented by pointers to the union $\mathcal{A}$ of the edge sets of $\mathcal{H}_1, \mathcal{H}_2, \mathcal{H}_3, \ldots$. We evaluate $\eta(S, x^*)$ for all sets S in $\mathcal{A}$; whenever we come across a set with $\eta(S, x^*) = 0$, we output either S (in case ext $\notin S$) or $V - S$ (in case ext $\in S$).

(δ) Having found an optimal solution x_{old}^* of the preceding LP relaxation of our TSP instance, we constructed a PQ-tree T_{old} compatible with x_{old}^*; this tree has been kept around till now. Typically, x^* does not differ all that much from x_{old}^*, and so it is plausible that quite a few of the sets S in $\mathcal{I}(T_{\text{old}})$, which satisfy $\eta(S, x_{\text{old}}^*) = 0$ by definition, will satisfy $\eta(S, x^*) = 0$.

We treat this source (δ) of tight sets differently from the preceding three sources. Rather than calling the function INSERT(T, S) for each tight member of $\mathcal{I}(T_{\text{old}})$, we call REBUILD$(T_{\text{old}}, x^*)$ before turning to sources (α), (β), (γ) for additional tight sets. This process is described in Algorithm 6.14.

Algorithm 6.14 Our construction of a PQ-tree compatible with x^*.

---

if T_{old} is available
then $T = T_{\text{old}}$;
        $T = $ REBUILD(T, x^*);
else $T = $ the trivial PQ-tree
end
while supply of tights sets S from sources (α), (β), (γ) lasts
do   **if** TEST$(T, S) = $ PASS **then** $T = $ INSERT(T, S) **end**
end
$T = $ REBUILD(T, x^*);
return T;

6.5.4 Proof of Theorem 6.1

For clarity, we split the proof into two steps.

LEMMA 6.2 *Let $\mathcal{F}$ denote the family of all tight subsets of $V - $ ext. If, in addition to constraints*

$$0 \leq x_e^* \leq 1 \quad \text{for all edges } e, \tag{6.13}$$

$$\sum(x_e^* : v \in e) = 2 \quad \text{for all cities } v, \tag{6.14}$$

x^* satisfies all subtour inequalities, then $\mathcal{F}$ has the following properties:

(P1) if $S_1, S_2 \in \mathcal{F}$ and $S_1 \not\subseteq S_2$, $S_2 \not\subseteq S_1$, $S_1 \cap S_2 \neq \emptyset$,
   then $S_1 \cap S_2$, $S_1 \cup S_2$, $S_1 - S_2$, $S_2 - S_1 \in \mathcal{F}$,

(P2) $\mathcal{F}$ includes $V - \mathsf{ext}$ and all the one-point subsets of $V - \mathsf{ext}$,

(P3) if S_1, S_2, S_3 are pairwise disjoint nonempty subsets of $V - \mathsf{ext}$,
   then at least one of $S_1 \cup S_2$, $S_2 \cup S_3$, $S_3 \cup S_1$ does not belong to $\mathcal{F}$.

Proof. Let us write simply $\eta(S)$ for $\eta(S, x^*)$.

Property (P1): Trivially,

$$\eta(S_1 \cap S_2) + \eta(S_1 \cup S_2) = \eta(S_1) + \eta(S_2) - 2x^*(S_1 - S_2, S_2 - S_1),$$
$$\eta(S_1 - S_2) + \eta(S_2 - S_1) = \eta(S_1) + \eta(S_2) - 2x^*(S_1 \cap S_2, V - (S_1 \cup S_2)),$$

and so assumptions $\eta(S_1) = \eta(S_2) = 0$ and (6.13) guarantee that

$$\eta(S_1 \cap S_2) + \eta(S_1 \cup S_2) \leq 0 \quad \text{and} \quad \eta(S_1 - S_2) + \eta(S_2 - S_1) \leq 0;$$

since $S_1 \cap S_2$, $S_1 \cup S_2$, $S_1 - S_2$, $S_2 - S_1$ are nonempty subsets of $V - \mathsf{ext}$ and since x^* satisfies all subtour inequalities, we have

$$\eta(S_1 \cap S_2), \, \eta(S_1 \cup S_2), \, \eta(S_1 - S_2), \, \eta(S_2 - S_1) \geq 0.$$

Property (P2) is guaranteed by (6.14).

Property (P3): Trivially,

$$\eta(S_1 \cup S_2) + \eta(S_2 \cup S_3) + \eta(S_3 \cup S_1) = \eta(S_1) + \eta(S_2) + \eta(S_3) + \eta(S_1 \cup S_2 \cup S_3) + 2;$$

since S_1, S_2, S_3, $S_1 \cup S_2 \cup S_3$ are nonempty subsets of $V - \mathsf{ext}$ and since x^* satisfies all subtour inequalities, we have

$$\eta(S_1), \, \eta(S_2), \, \eta(S_3), \, \eta(S_1 \cup S_2 \cup S_3) \geq 0.$$

$\square$

LEMMA 6.3 *If a family $\mathcal{F}$ of nonempty subsets of a set $V - \mathsf{ext}$ has properties (P1), (P2), (P3), then there is a PQ-tree T with $\mathcal{I}(T) = \mathcal{F}$.*

Proof. We shall proceed by induction on $|V - \mathsf{ext}|$. Let $H(\mathcal{F})$ denote the graph with vertex set $V - \mathsf{ext}$, whose edges are all the two-point members of $\mathcal{F}$.

CASE 1: $\mathcal{F}$ *includes no set S with $1 < |S| < |V - \mathsf{ext}|$.*

Let T be the tree whose root is a P-node with children the elements of $V - \mathsf{ext}$. Property (P2) guarantees that $\mathcal{I}(T) = \mathcal{F}$.

CASE 2: $H(\mathcal{F})$ *is connected.*

To begin, let us show that (i) $H(\mathcal{F})$ has no vertex of degree greater than two and that (ii) $H(\mathcal{F})$ has no cycle.

To prove (i), assume the contrary: $H(\mathcal{F})$ includes a vertex v with three distinct neighbors, v_1, v_2, v_3. Repeated applications of (P1) show that all nonempty subsets of $\{v, v_1, v_2, v_3\}$ belong to $\mathcal{F}$; but then (P3) is contradicted, for instance, by $S_1 = \{v_1\}$, $S_2 = \{v_2\}$, $S_3 = \{v_3\}$.

To prove (ii), assume the contrary: $H(\mathcal{F})$ contains a cycle $v_1 v_2 \ldots v_k v_1$. Repeated applications of (P1) guarantee that

$$\{v_2, v_3, \ldots, v_k\} \in \mathcal{F}$$

and that

$$\{v_3, \ldots, v_k, v_1\} \in \mathcal{F};$$

but then (P3) is contradicted by $S_1 = \{v_1\}$, $S_2 = \{v_2\}$, $S_3 = \{v_3, \ldots, v_k\}$.

From (i) and (ii), it follows that $H(\mathcal{F})$ is a path through $V - \text{ext}$. Let T be the tree whose root is a Q-node with children the elements of $V - \text{ext}$ in the order defined by the path $H(\mathcal{F})$. Repeated applications of (P1) guarantee that $\mathcal{I}(T) \subseteq \mathcal{F}$; in turn, it follows routinely from (P1) and (P3) that $\mathcal{F} \subseteq \mathcal{I}(T)$.

CASE 3: *$\mathcal{F}$ includes a set S with $1 < |S| < |V - \text{ext}|$ and $H(\mathcal{F})$ is disconnected.*
To begin, let us show that there is a set $S^\star$ in $\mathcal{F}$ such that
$(\star)$ $1 < |S^\star| < |V - \text{ext}|$ and
       every S in $\mathcal{F}$ has $S^\star \subseteq S$ or $S \subseteq S^\star$ or $S^\star \cap S = \emptyset$.
If $H(\mathcal{F})$ has no edges, then let $S^\star$ be any minimal set such that $S^\star \in \mathcal{F}$ and $1 < |S^\star| < |V - \text{ext}|$; by assumption, $|S^\star| \geq 3$; now property (P1) and minimality of $S^\star$ guarantee that $S^\star$ has property $(\star)$. If $H(\mathcal{F})$ has at least one edge, then consider an arbitrary connected component of $H(\mathcal{F})$ that has at least two vertices and let A denote the vertex set of this component. If A satisfies $(\star)$ in place of $S^\star$, then we are done; else there is a set B in $\mathcal{F}$ such that $A \not\subseteq B$, $B \not\subseteq A$, $A \cap B \neq \emptyset$; we may assume that B is the smallest set in $\mathcal{F}$ with this property. Since A induces a connected subgraph of $H(\mathcal{F})$, there are a u in $A \cap B$ and a v in $A - B$ such that $\{u, v\} \in \mathcal{F}$, and so $B - \{u\} \in \mathcal{F}$ by property (P1); now minimality of B implies that $A \cap B = \{u\}$; let us write $C = B - \{u\}$. Since A is the vertex set of a connected component of $H(\mathcal{F})$, we must have $|C| \geq 2$; we claim that C satisfies $(\star)$ in place of $S^\star$. To justify this claim, let us assume the contrary: some D in $\mathcal{F}$ has $C \not\subseteq D$, $D \not\subseteq C$, $C \cap D \neq \emptyset$. If $u \in D$, then minimality of B is contradicted by $B \cap D$; if $u \notin D$, then minimality of B is contradicted by $B - D$.

With $S^\star$ satisfying $(\star)$, let us write

$$\mathcal{F}_1 = \{S : S \in \mathcal{F}, S \subseteq S^\star\};$$

having chosen a point σ in $S^\star$, let us write

$$\mathcal{F}_2 = \{S : S \in \mathcal{F}, S \cap S^\star = \emptyset\} \cup \{(S - S^\star) \cup \{\sigma\} : S \in \mathcal{F}, S \supseteq S^\star\}.$$

By the induction hypothesis, there are PQ-trees T_1, T_2 such that $\mathcal{I}(T_1) = \mathcal{F}_1$, $\mathcal{I}(T_2) = \mathcal{F}_2$. Identifying the root of T_1 with leaf σ of T_2, we obtain a PQ-tree T such that $\mathcal{I}(T) = \mathcal{F}$. □

Chapter Seven

Cuts from Blossoms and Blocks

Comb inequalities have a long history in the TSP, going back to the by-hand computations of Dantzig, Fulkerson, and Johnson [151]. The general comb template is now a workhorse in TSP codes. An important open question, however, is to determine the complexity of the exact separation of comb inequalities, given a vector x^* satisfying all subtour constraints. No polynomial-time algorithm is known for the problem, but neither is it known to be $\mathcal{NP}$-hard.

A practical method for the exact separation of combs would likely have a dramatic impact on our ability to solve large-scale instances of the TSP. The most important subclass for which such an algorithm is known are the blossom inequalities, consisting of combs with all teeth having exactly two vertices. This result on blossom separation is due to Padberg and Rao [453], and their algorithm has been adopted in a number of TSP codes, including Concorde. Carr [109] has shown that also combs with a fixed number of teeth can be separated in polynomial time, but the method is not suitable for practical implementation, even in the case of just three teeth, due to its inherently high running time.

In this chapter we discuss the use of the Padberg-Rao algorithm, together with heuristic separation methods for general combs based on the block structure of the graph obtained by deleting edges of value $x_e^* = 1$ from the support graph G^*. The block-based algorithms are due to Padberg and Hong [451] and Padberg and Rinaldi [446], with improvements by Naddef and Thienel [423]. We also discuss shrinking heuristics proposed by Padberg and Grötschel [450], Padberg and Rinaldi [446], and Grötschel and Holland [232], permitting repeated use of the blossom and comb algorithms.

7.1 FAST BLOSSOMS

Recall that a comb is specified by a handle $H \subseteq V$ and an odd number of disjoint teeth $S_1, \ldots, S_k \subseteq V$, such that each tooth has a least one vertex in H and at least one vertex not in H. If a tooth has exactly two vertices, then it can be identified by an edge $e \in E$. Therefore, a blossom can be specified by a handle H and a set of teeth $T \subseteq \delta(H)$. The corresponding blossom inequality is

$$x(\delta(H)) + \sum(x(\delta(\{u_e, v_e\}))) : e = \{u_e, v_e\} \in T) \geq 3|T| + 1. \qquad (7.1)$$

Given an LP solution x^*, let $G_{1/2}^*$ denote the graph having vertex set V and edge set $\{e \in E : 0 < x_e^* < 1\}$. Padberg and Hong [451] propose a blossom separation heuristic that examines the vertex sets $V_1, \ldots, V_q$ of the connected components of

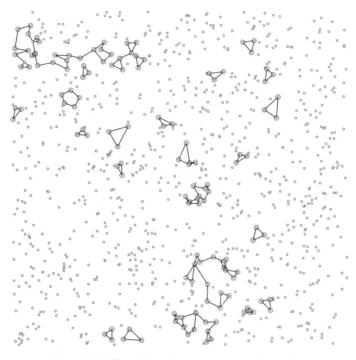

Figure 7.1 $G_{1/2}^*$ with odd vertices marked by circles.

$G_{1/2}^*$. If for some i in $\{1, \ldots, q\}$ the set of edges

$$T = \{e : e \in \delta(V_i), x_e^* = 1\}$$

has odd cardinality, then either a violated blossom inequality or a violated subtour inequality can be constructed as follows. If two of the edges in T intersect outside V_i, then these two edges are removed from the collection and their intersection is added to V_i. Eventually, the set T consists of an odd number of disjoint edges; if there are at least three, then they form the teeth of a violated blossom inequality with V_i as the handle; if there is just one, then V_i alone yields a violated subtour inequality.

We refer to the above technique for finding blossoms as the *odd-component heuristic*. Variants of this method can be found in Hong [270] and in Land [333].

In Figure 7.1, we display the graph $G_{1/2}^*$ for the random Euclidean 1,000-city TSP we considered in the previous chapter. The corresponding vector x^* is an optimal solution to the subtour relaxation. In the drawing, the vertices meeting an odd number of edges with $x_e^* = 1$ are indicated by small circles. It is easy to spot many components having an odd number of these *odd vertices*, each of which gives rise to a violated blossom inequality.

Running the odd-component heuristic, together with the subtour separation routines, produces an LP bound that is 0.274% below the optimal tour length for the 1,000-city TSP. On a 100,000-city example, also from the previous chapter, the LP bound produced is again at most 0.274% below the value of an optimal tour. To obtain these results, we ran the separation algorithms until they returned without any violated cuts. In both examples, the addition of blossom inequalities to the mix of cuts closed over half of the gap between the subtour bound and the optimal TSP value.

GRÖTSCHEL-HOLLAND HEURISTIC

The odd-component heuristic for blossoms suffers from the same problem as the connected-component heuristic for subtours, namely, small perturbations in x^* can hide the odd components that make up the handles of the blossoms. We do not have an analogue of the parametric connectivity procedure in this case, but Grötschel and Holland [231] proposed a method for handling a fixed perturbation ε in the heuristic. Their idea is to consider as possible handles the vertex sets of the components of the graph G_ε^* having vertices V and edges $\{e \in E : \varepsilon \leq x_e^* \leq 1 - \varepsilon\}$. Let V_i denote the vertex set of such a component, and let $e_1, \ldots, e_t$ be the edges in the set

$$\{e \in \delta(V_i) : x_e^* > 1 - \varepsilon\}$$

in nonincreasing order of x_e^*; if t is even, then let e_{t+1} be the edge in

$$\{e \in \delta(V_i) : x_e^* < \varepsilon\}$$

with the greatest x_e^* and increment t by 1, so t is now odd. For each j from 1 up to t, let u_{e_j}, v_{e_j} denote the ends of the edge e_j. For each odd k from 1 up to t such that

$$x^*(\delta(V_i)) + \sum_{j=1}^{k} x^*(\delta(\{u_{e_j}, v_{e_j}\})) < 3k + 1, \tag{7.2}$$

Grötschel and Holland find a subtour inequality or a blossom inequality violated by x^*, using the method described above to deal with edges that intersect. In this case, e_i and e_j may also intersect inside V_i, which is handled by removing the two edges from the collection of potential teeth and deleting the intersection from V_i.

In the Concorde code, we implement a variation of the Grötschel-Holland heuristic, where we consider only $k = t$ or $k = t - 2$. In choosing the value of ε, we follow the recommendation of Grötschel and Holland and set $\varepsilon = 0.3$. Using this algorithm, we obtain an LP bound of 0.239% below the optimal tour for the 1,000-city test instance, and an LP bound of at most 0.246% below the optimal value for the 100,000-city instance. Here, we combine the Grötschel-Holland heuristic, the odd-component heuristic, and the subtour separation routines, running the cutting-plane procedures until no further cuts are produced.

7.2 BLOCKS OF $G_{1/2}^*$

The inclusion of blossom inequalities in the LP relaxation typically has a dramatic impact on the structure of $G_{1/2}^*$. A drawing of this graph for the 1,000-city example is given in Figure 7.2, where the corresponding LP solution x^* is obtained by applying subtour constraints and the odd-component and Grötschel-Holland heuristics for blossoms. A quick inspection shows that although the graph now consists

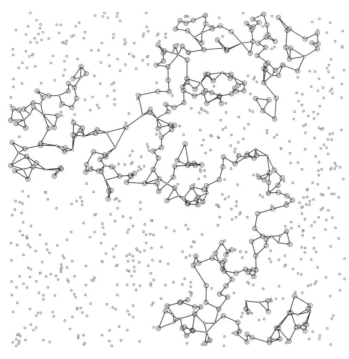

Figure 7.2 $G^*_{1/2}$ after blossom heuristics.

of only two nontrivial components, it does have a rich structure when we consider
its blocks, or 2-connected components.

Padberg and Hong [451] were the first to consider the blocks of $G^*_{1/2}$ as a potential source of handles for blossom inequalities. Letting V_i be the vertex set of a
block, Padberg and Hong search for a set of teeth among the edges $\delta(V_i)$ satisfying
$x^*_e > 0$. To carry this out, note that a potential tooth $e = \{u_e, v_e\} \in \delta(V_i)$ will
contribute to the violation of a blossom inequality if and only if

$$x^*(\delta(\{u_e, v_e\})) < 3. \tag{7.3}$$

Assuming x^* satisfies the degree constraints, inequality (7.3) holds if and only if
$x^*_e > 0.5$. So we would like to include as teeth all edges in the set

$$\{e \in \delta(V_i) : x^*_e > 0.5\}. \tag{7.4}$$

To this end, let $e_1, \dots, e_t$ be the edges of (7.4) in nonincreasing order of x^*_e, as in
the Grötschel-Holland process outlined above. If t is even, let e_{t+1} be the edge in
$\{e \in \delta(V_i) : x^*_e \le 0.5\}$ having greatest x^*_e, and increment t by 1. Now for any odd
$k \in \{1, \dots, t\}$, if (7.2) holds we obtain a violated subtour inequality or a violated
blossom inequality.

COMB INEQUALITIES

Padberg and Rinaldi [446] augment the Padberg-Hong block heuristic by considering as potential teeth certain sets of cardinality greater than two. The idea can

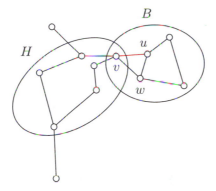

Figure 7.3 Neighborhood of a block of $G^*_{1/2}$.

be motivated by examining the neighborhood of a block of the graph $G^*_{1/2}$ from the 1,000-city example displayed in Figure 7.2; we give a closer view of one such neighborhood in Figure 7.3. When considering the block labeled H as a potential handle of a comb, a natural candidate for a tooth, together with the edges in $\delta(H)$, is the neighboring block labeled B. In their heuristic algorithm, Padberg and Rinaldi consider B as well as the set $\{u, v, w\}$, consisting of the cut vertex v and its G^*-neighbors in B.

In general, let H be a block of $G^*_{1/2}$ and let v be a cut vertex contained in H. We denote by A_v the set having minimum value $x^*(\delta(A_v))$ among

$$B \text{ or } N = \{v\} \cup \{u \in B : x^*_{\{u,v\}} > 0\}$$

for all blocks $B \neq H$ containing v, such that A_v has at least two vertices. If v meets no edge having $x^*_e = 1$, then Padberg and Rinaldi consider A_v as a potential tooth. To construct a comb, the sets A_v with $x^*(\delta(A_v)) < 3$, together with the sets of ends of edges $\{e \in \delta(H) : x^*_e = 1\}$, are placed in nondecreasing order $S_1, \ldots, S_t$ of $x^*(\delta(S))$. We proceed as in the Padberg-Hong blossom heuristic to search for an odd k with

$$x^*(\delta(H)) + \sum_{j=1}^{k} x^*(\delta(S_i)) < 3k + 1. \tag{7.5}$$

If (7.5) holds, then a violated subtour inequality or a violated comb inequality can be identified, after cleaning up any intersecting sets S_i and S_j as described below.

Suppose we have $S_1, \ldots, S_k$ such that (7.5) holds. If two sets S_i and S_j intersect, then one of them, say S_j, must have cardinality two, since the sets A_v are pairwise disjoint. Thus S_i and S_j intersect in a single point outside H, as indicated in Figure 7.4. This intersection can be removed by replacing H by $H \cup S_i$ and deleting S_i and S_j from the collection of potential teeth. To show that (7.5) holds after these changes, we write the comb inequality

$$x^*(\delta(H)) + \sum_{j=1}^{k} x^*(\delta(S_i)) \geq 3k + 1 \tag{7.6}$$

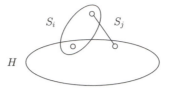

Figure 7.4 Intersecting teeth.

in an alternative form as a $\leq$ constraint. To this end, note that the identity

$$x(\delta(S)) = \sum_{v \in S} x(\delta(v)) - 2x(\{e \in E : e \subseteq S\}) \qquad (7.7)$$

holds for any set $S \subseteq V$. If x satisfies the degree constraints, then (7.7) can be simplified to

$$x(\delta(S)) = 2|S| - 2x(\{e \in E : e \subseteq S\}).$$

Using this expression for the handle and teeth, the comb inequality (7.6) can be written as

$$x(\{e \in E : e \subseteq H\}) + \sum_{j=1}^{k} x(\{e \in E : e \subseteq S_j\}) \leq |H| + \sum_{j=1}^{k} |S_j| - \frac{3k+1}{2}.$$

In this form, it follows directly that the changes to H and $S_1, \ldots, S_k$ do not alter either the left-hand side or the right-hand side of the inequality.

It is possible to apply this heuristic to sets H other than just the blocks of $G^*_{1/2}$. Indeed, Padberg and Rinaldi [446] show by example that restricting H to 2-connected sets can cause the algorithm to miss violated combs. To deal with this possibility, Padberg and Rinaldi propose to consider as a handle any set H such that the subgraph of $G^*_{1/2}$ induced by H is connected and H can be written as the union of blocks, each having at least three vertices. Adopting this proposal for large graphs can be time-consuming, however, since the number possibilities for H can grow quadratically with the number of vertices.

NADDEF AND THIENEL

In a general study of greedy-type separation algorithms for the TSP, Naddef and Thienel [423] propose to extend the Padberg-Rinaldi heuristic by allowing the set of potential teeth to be determined by an algorithm that grows a set starting with a cut vertex v. The goal of the growing process is to produce a set $T \subseteq V$ such that $x^*(\delta(T))$ is small, while satisfying the properties of a tooth, namely T must intersect H, T must contain at least one vertex not in H, and T should not intersect other teeth. Using network-flow techniques, it is possible to find a T that minimizes $x^*(\delta(T))$ subject to these conditions, but such a computation is costly in terms of time and may have the undesirable effect of producing sets T of very large cardinality. Instead, Naddef and Thienel consider two procedures for growing T step by step, in a greedy fashion.

In one case, Naddef and Thienel augment T by searching for a path P in G^* that starts and ends in the set T, adopting the idea from the study of *ear decompositions* in matching theory (see Lovász and Plummer [361]). The vertices of the path are added to T, so the goal in selecting P is to keep the value of $x^*(\delta(T \cup P))$ small. Naddef and Thienel [423] state that their heuristic method for finding such a path is based on breadth-first search, but do not provide further details.

In the second case, the set T is augmented one vertex at a time. Among the vertices that can be added to T, Naddef and Thienel choose v that maximizes

$$\sum_{u \in T} x^*_{\{u,v\}}. \tag{7.8}$$

Naddef and Thienel [423] call (7.8) the *max-back value* of the vertex v and provide a set of tie-breaking rules for choosing among vertices of equal max-back value. We discuss the further application of this greedy approach in Section 10.3.

CONCORDE IMPLEMENTATION

In the Concorde code, we adopt the Naddef-Thienel variant of the Padberg-Rinaldi heuristic. We consider as potential handles H all blocks of $G^*_{1/2}$ having at least three vertices, and all sets that can be written as the union of two such blocks sharing a common cut vertex v. For teeth, we combine the Padberg-Rinaldi sets with those we obtain via a growing process starting at the cut vertices. In our procedure, we add flexibility to the max-back algorithm by allowing vertices also to be removed from T, rather than implementing a strict greedy process. This procedure is described in detail in a more general setting in Chapter 10.

Applying the block comb heuristic improves the lower bound on the 1,000-city Euclidean instance to within 0.137% of the optimal tour value, while on the 100,000-city Euclidean instance the new lower bound leaves a gap of at most 0.145%. To obtain these results, we ran the heuristic, together with the subtour separation methods and the blossom heuristics, stopping the process only when no further cuts were produced. The inclusion of general combs closed over 40% of the remaining gap between the LP lower bound and the optimal tour value in both examples.

A drawing of $G^*_{1/2}$ for the 1,000-city Euclidean problem is given in Figure 7.5, where the corresponding x^* is the optimal solution obtained after the inclusion of block combs in the LP relaxation. Note that the graph is now dominated by one large 2-connected component.

7.3 EXACT SEPARATION OF BLOSSOMS

A *2-factor*, or *simple-perfect 2-matching*, in a graph $G = (V, E)$ is a set of edges $M \subseteq E$ such that every vertex of G is contained in exactly two members of M. In other words, a 2-factor is a set of edges that form disjoint circuits covering all vertices. The incidence vector x of any 2-factor satisfies the linear system

$$x(\delta(v)) = 2 \quad \text{for all vertices } v$$
$$0 \le x_e \le 1 \quad \text{for all edges } e$$

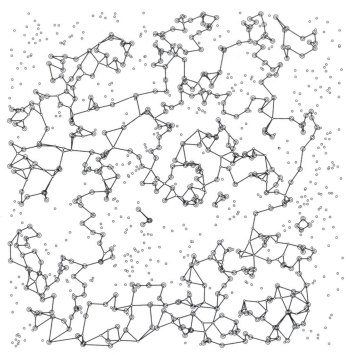

Figure 7.5 $G^*_{1/2}$ after block combs.

consisting of the degree constraints and edge bounds, and every integer solution of
the system is a 2-factor. The *2-factor problem* is to find a minimum-weight 2-factor
in a graph with edge weights $(w_e : e \in E)$.

One of the many consequences of the matching theory of Edmonds [163] is that
incidence vectors of 2-factors satisfy all blossom constraints. Indeed, this result
was the basis for the later introduction of the comb template by Chvátal [125] and
Grötschel and Padberg [236]. The validity of blossom constraints for 2-factors can
be established by following the proof of Theorem 5.8, but it can also be easily seen
by considering an alternative form of the constraints, as we observe below.

Let $H \subseteq V$ be a handle and let $T \subseteq \delta(H)$ be the teeth of a blossom. The form of
the corresponding inequality (7.1) can be modified when we restrict our attention
to vectors x that satisfy the degree constraints, as in the 2-factor problem and also
in the TSP. Note that by summing $x(\delta(u_e)) = 2$ and $x(\delta(v_e)) = 2$ we obtain the
equation

$$x(\delta(\{u_e, v_e\})) = 4 - 2x_e$$

for any edge $e = \{u_e, v_e\}$. Using this, the blossom inequality can be written as

$$x(\delta(H)) + 4|T| - 2\sum(x_e : e \in T) \geq 3|T| + 1$$

or

$$x(\delta(H) - T) - x(T) \geq 1 - |T|. \tag{7.9}$$

Now consider the incidence vector x of a 2-factor. If $x(T) \leq |T| - 1$, then x clearly satisfies (7.9). On the other hand, suppose $x(T) = |T|$. Since T is odd and $x(\delta(v)) = 2$ for all $v \in H$, this implies $x(\delta(H) - T) \geq 1$, and (7.9) is again satisfied.

This argument shows that every 2-factor of G satisfies the linear system

$$x(\delta(v)) = 2 \quad \text{for all vertices } v$$
$$0 \leq x_e \leq 1 \quad \text{for all edges } e$$
$$x(\delta(H) - T) - x(T) \geq 1 - |T| \quad \text{for all blossoms } (H, T)$$

where we write (H, T) to denote a blossom with handle $H \subseteq V$ and teeth $T' \subseteq \delta(H)$. In fact, a much deeper result of Edmonds [163] states that the convex hull of 2-matchings of G is precisely the solution set of the above linear system. As a consequence of this polyhedral theorem, Edmonds was able to obtain a polynomial-time algorithm for solving the 2-factor problem.

By the general theory of Grötschel et al. [233] on the polynomial-time equivalence of optimization and separation, Edmonds' 2-factor algorithm and 2-factor polyhedron theorem together imply that the exact separation problem for blossom inequalities can be solved in polynomial time (see also Padberg and Rao [452]). The algorithm obtained in this way is, however, not a practical way to solve the problem, due to its reliance on repeated application of the ellipsoid method for linear programming (Khachiyan [315], Shor [499], and Yudin and Nemirovskii [561]). Fortunately, a direct combinatorial algorithm for the separation problem was proposed by Padberg and Rao [453], and this polynomial-time method has proved to be effective in practice.

THE PADBERG-RAO ALGORITHM

The exact separation algorithm of Padberg and Rao [453] is an important result in combinatorial optimization. Variants and implementations of the algorithm are given in Grötschel and Holland [231], Padberg and Rinaldi [446], and Letchford et al. [351]. The best-known running-time bound is $O(n^2 m log(n^2/m))$ by Letchford et al., where $n = |V|$ and $m = |E^*|$.

To outline the algorithm, it is useful to rearrange the terms of (7.9) to write the blossom inequality as

$$x(\delta(H) - T) + (|T| - x(T)) \geq 1 \qquad (7.10)$$

for a handle $H \subseteq V$ and teeth $T \subseteq \delta(H)$. The left-hand side of (7.10) is similar to that of a subtour inequality, but with each edge $e \in T$ valued as $1 - x_e$ rather than as simply x_e. To allow for these alternative values, Padberg and Rao replace each edge $e = \{u, v\} \in E$ by a pair of edges $e' = \{u, \sigma_e\}, e'' = \{\sigma_e, v\}$, where σ_e is a new vertex, setting $\bar{x}_{e'} = x_e^*$ and $\bar{x}_{e''} = 1 - x_e^*$. Denote by $\bar{V}$ the vertex set of the graph obtained by this construction and let F denote the set of new edges assigned $1 - x_e^*$ values.

A vertex $v \in \bar{V}$ is called *odd* if it meets an odd number of edges in F, and a set $S \subseteq \bar{V}$ is called odd if it contains an odd number of odd vertices. With a small computation, it can be seen that G^* has a violated blossom inequality (7.10) if and only if there exists an odd set $S \subseteq \bar{V}$ with $\bar{x}(\delta(S)) < 1$. This observation places the exact separation of blossoms in the more general context of finding a *minimum-weight odd cut*, that is, given a graph $G' = (V', E')$ with edge weights $(w_e : e \in E')$ and with an even number of vertices labeled as odd, find an odd set $S \subseteq V'$ that minimizes $w(\delta(S))$. An odd cut in a G' is often referred to as a T-*cut*, where T denotes the set of odd vertices.

A nice treatment of the study of T-cuts can be found in the book by Schrijver [495], including a description of the polynomial-time algorithm proposed by Padberg and Rao [452]. The cornerstone of the algorithm is a result stating that a minimum-weight odd cut can be found among the fundamental cuts of a Gomory-Hu cut-tree for G'. A *fundamental cut* in a tree H is determined by deleting an edge of H and letting S be the vertex set of one of the connected components that remain. A *Gomory-Hu cut-tree* for G' is a tree $H = (V', K)$ with edge weights $(c_e : e \in K)$ such that for any pair of vertices $s, t \in V$, an (s, t)-minimum cut is given by the fundamental cut corresponding to the edge e of minimum weight c_e in the unique path from s to t in H. Gomory and Hu [219] showed that such trees can be constructed in polynomial time; descriptions of the Gomory-Hu algorithm can be found in texts by Hu [274], Cook et al. [138], and Schrijver [495].

On the 1,000-city Euclidean TSP, the addition of this exact separation method for blossoms brings the LP bound to within 0.084% of the optimal tour length, while on the 100,000-city problem the new LP bound is at most 0.125% greater than the optimal tour length. This is again a significant improvement in the bounds, but we should note that the total running time for the repeated calls to the exact blossom algorithm is considerably longer than that for the separation methods described in the previous sections.

7.4 SHRINKING

Padberg and Grötschel [450], Grötschel and Holland [232], and Padberg and Rinaldi [446] propose comb heuristics based on shrinking subsets of vertices, followed by the application of the Padberg-Rao algorithm, or the application of the block comb methods. The shrinking operations adopted in these papers may not be safe when restricted to the comb template, but they have proven to be effective in practice.

Combining the Padberg-Rao blossom algorithm with shrinking is a natural way to extend the reach of the algorithm, since a blossom in a shrunk graph can translate to a general comb in the original graph, when vertices in the teeth of the blossom are expanded. Padberg and Grötschel [450] discuss a direct approach to this idea, starting their algorithm by shrinking each path of 1-edges to a single edge, that is, if P is a path from s to t and each edge $e \in P$ has $x_e^* = 1$, then P is replaced by a single edge $\{s, t\}$ with $x_{\{s,t\}}^* = 1$. Following this step, Padberg and Grötschel propose to search for nested sets $S_1 \subset S_2$ with $x^*(\delta(S_1)) = x^*(\delta(S_2)) = 2$, and shrink S_1 and $S_2 - S_1$, as illustrated in Figure 7.6. With the added condition $x^*(\delta(S_2 - S_1)) = 2$, this configuration is a good candidate for a tooth of a blossom

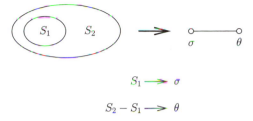

$$S_1 \longrightarrow \sigma$$

$$S_2 - S_1 \longrightarrow \theta$$

Figure 7.6 Shrinking nested tight sets.

Figure 7.7 Shrinking 1-edge in tight triangle.

in the shrunk graph, and Padberg and Grötschel search for such a blossom with the Padberg-Rao algorithm. If the search is unsuccessful, Padberg and Grötschel find a nested pair of tight sets in the shrunk graph and repeat the process. The algorithm terminates when either a violated blossom inequality is discovered or no pair of nested tight sets can be located.

Shrinking paths of 1-edges is a special case of the tight-triangle shrinking rule illustrated in Figure 7.7, where the ends of an edge $e = \{u, v\}$ are shrunk if $x_e^* = 1$ and there exists a vertex w with $x^*(\{u, v\}, \{w\}) = 1$. This operation was shown to be a safe-shrinking rule by Padberg and Rinaldi [446]; a proof and discussion are given in Section 5.9.

GRÖTSCHEL-HOLLAND RULES

Replacing paths of 1-edges and shrinking the 1-edge in a tight triangle are two of five shrinking operations employed by Grötschel and Holland [232]. The full list of Grötschel-Holland rules is the following.

1. Given a path P of 1-edges from vertex s to vertex t, replace P by a single edge $\{s, t\}$ with $x_{\{s,t\}}^* = 1$.

2. Given $\{u, v, w\} \subseteq V$ with $x_{\{u,v\}}^* = 1$ and $x^*(\{u, v\}, \{w\}) = 1$, shrink $\{u, v\}$.

3. Given $S \subseteq V$ with $|S| = 4$ and $x^*(\delta(S)) = 2$, shrink S to a single vertex.

4. Given $\{s, t, u, v\} \subseteq V$ with $x_{\{s,t\}}^* = x_{\{u,v\}}^* = 1$ and $x^*(\{s, t\}, \{u, v\}) = 1$, shrink $\{s, t\}$ and $\{u, v\}$. This rule shrinks the 1-edges in a tight square, as illustrated in Figure 7.8.

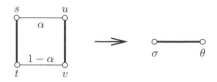

Figure 7.8 Shrinking 1-edges in tight square.

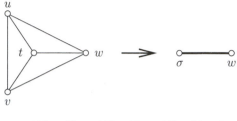

$$x^*(\{u, v\}) + x^*(\{u, t\}) + x^*(\{v, t\}) = 2$$
$$x^*(\{u, w\}) + x^*(\{t, w\}) + x^*(\{v, w\}) = 1$$

Figure 7.9 Shrinking a triple of vertices.

5. Given $\{u, v, w\} \subseteq V$ with $x^*_{\{u,v\}} = 1$ and $x^*(\{u, v\}, \{w\}) \geq 0.5$, shrink $\{u, v\}$. This rule is a weakened version of the tight-triangle rule 2.

Grötschel and Holland adopt these rules in a general shrinking approach for comb inequalities. Their method is to apply a subset of the rules, one after another, in a round of shrinking, followed by a run of the Padberg-Rao algorithm on the shrunk graph. Any set of vertices that match the rules are shrunk, but no newly created vertex is shrunk again during the given round. (In the case of rule 1, neither of the ends of the new edge are shrunk again during the round.)

Based on their computational experiments, Grötschel and Holland [232] package their shrinking rules into four comb heuristics,

$$(1, 2, 3), (1, 4, 3), (1, 5, 3), (1, 4, 5),$$

where (i, j, k) denotes the strategy of first applying rule i, then rule j, and finally rule k, in each round.

PADBERG-RINALDI SHRINKING

Padberg and Rinaldi [446] adopt a less aggressive shrinking process, relying on operations that were proven to be safe-shrinking rules. The engine of their approach is the repeated application of two such operations, the first of which is the tight-triangle rule 2 that was also used by Grötschel and Holland. The second operation takes a set $\{t, u, v, w\} \subseteq V$ with $x^*(\delta(\{t, u, v\})) = 2$ and $x^*(\{t, u, v\}, \{w\}) = 1$, and shrinks $\{t, u, v\}$ to a single vertex; this rule is illustrated in Figure 7.9. Unlike Grötschel and Holland, Padberg and Rinaldi allow newly created vertices to take part in additional shrinking steps. When the shrinking phase terminates, Padberg and Rinaldi apply their block comb heuristic algorithm on the resulting shrunk graph.

This shrinking-based procedure is used within a larger framework, involving the tight-square rule 4 adopted in the Grötschel-Holland heuristic. Padberg and Rinaldi show that if a tight square satisfies certain side constraints, then rule 4 is also a safe-shrinking operation. The overall process is then to search for such a safe tight square in G^*, apply rule 4 to obtain the shrunk graph $\bar{G}^*$, and call the above heuristic algorithm. After recording any violated comb inequalities, $\bar{G}^*$ is searched for another safe tight square, and the process is repeated. The algorithm terminates when no further safe tight squares are discovered.

Padberg and Rinaldi complement this safe-shrinking heuristic with one that involves the use of more aggressive rules, similar to those adopted by Grötschel and Holland. The three rules they employ are the following.

1. Given $\{u, v, w\} \subseteq V$ with $x^*_{\{u,v\}} = 1$ and $x^*(\{u, v\}, \{w\}) > 0.6$, shrink $\{u, v\}$.

2. Given $\{u, v, w\} \subseteq V$ with $x^*(\delta(\{u, v, w\})) = 2$, shrink $\{u, v, w\}$.

3. Given $\{s, t, u, v\} \subseteq V$ with $x^*_{\{s,t\}} = x^*_{\{u,v\}} = 1$ and $x^*(\{s, t\}, \{u, v\}) > 0.6$, shrink $\{s, t\}$ and $\{u, v\}$.

Padberg and Rinaldi use these additional rules when their main procedure does not identify any violated comb inequalities.

CONCORDE IMPLEMENTATION

In the Concorde code, we combine the ideas proposed by Grötschel and Holland and by Padberg and Rinaldi.

The four Grötschel-Holland heuristics are implemented as described above, but with the Padberg-Rao blossom heuristic replaced by the block comb heuristic described in the previous section. This switch is based on the computing time required for the exact blossom algorithm on large-scale graphs, and on results from our computational tests.

We also implement an algorithm with five rounds of shrinking, following each round with a call to Concorde's block comb heuristic. The rounds of shrinking are the following.

1. The tight-triangle rule, as in Figure 7.7, is applied repeatedly.

2. The safe-shrinking rule for triples of vertices, illustrated in Figure 7.9, is applied repeatedly.

3. The tight-square rule, illustrated in Figure 7.8, is applied repeatedly.

4. Any triple $\{u, v, w\}$ having $x^*(\delta(\{u, v, w\})) = 2$ is shrunk.

5. Any quadruple of vertices $\{t, u, v, w\}$ having $x^*(\delta(\{t, u, v, w\})) = 2$ is shrunk.

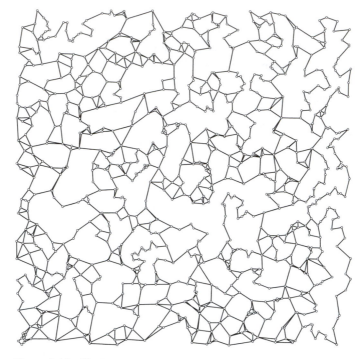

Figure 7.10 G^* after the application of the shrinking-based heuristics.

In moving from one round to the next, we continue to process the shrunk graph that resulted from the previous round, using the additional operations to further shrink the graph.

Applying these heuristics to the 1,000-city Euclidean problem reduced the gap between the LP bound and the optimal tour to 0.081%. On the 100,000-city TSP, the LP bound was improved to at most 0.117% below the value of an optimal tour. The support graph G^* corresponding to the new LP solution for the 1,000-city problem is given in Figure 7.10.

Chapter Eight

Combs from Consecutive Ones

In this chapter we present a heuristic separation algorithm for 3-toothed comb inequalities, exploiting the relationship between these inequalities and the following notion:

> A family $\mathcal{F}$ of subsets of a set W is said to have the consecutive ones property if W can be endowed with a linear order $\prec$ so that each S in $\mathcal{F}$ has the form $\{y \in W : a \preceq y \preceq b\}$ with $a, b \in W$.

To describe this relationship, let us select once and for all a city denoted ext for "exterior" and let us write V^{int} for $V - \{\text{ext}\}$.

If $\mathcal{F}$ is a family of subsets of V^{int} that does not have the consecutive ones property, then every tour x through V satisfies

$$\sum(\eta(S, x) : S \in \mathcal{F}) \geq 2, \tag{8.1}$$

where we use the η notation introduced in Chapter 5, that is, $\eta(S, x) = x(\delta(S)) - 2$. To see that (8.1) holds, note that the tour induces a Hamiltonian path through V^{int}; this path defines a linear order $\prec$, unique up to reversal, on V^{int}; since $\mathcal{F}$ does not have the consecutive ones property, at least one S in $\mathcal{F}$ does not have the form $\{y : a \preceq y \preceq b\}$, and so $\eta(S, x) \geq 2$.

Each comb inequality

$$\eta(H, x) + \eta(T_1, x) + \eta(T_2, x) + \eta(T_3, x) \geq 2. \tag{8.2}$$

with handle H and teeth T_1, T_2, T_3 is an instance of of (8.1). To see this, write (8.2) as

$$\eta(S_0, x) + \eta(S_1, x) + \eta(S_2, x) + \eta(S_3, x) \geq 2$$

with

$$S_0 = \begin{cases} H & \text{if ext} \notin H, \\ V - H & \text{if ext} \in H, \end{cases}$$

and

$$S_i = \begin{cases} T_i & \text{if ext} \notin T_i, \\ V - T_i & \text{if ext} \in T_i \end{cases}$$

for $i = 1, 2, 3$. The resulting subsets S_0, S_1, S_2, S_3 of V^{int} satisfy either

$$\left. \begin{array}{l} S_1 \cap S_2 = \emptyset, \ S_1 \cap S_3 = \emptyset, \ S_2 \cap S_3 = \emptyset, \\ S_0 \cap S_1 \neq \emptyset, \ S_0 \cap S_2 \neq \emptyset, \ S_0 \cap S_3 \neq \emptyset, \\ S_0 \not\supseteq S_1, \ S_0 \not\supseteq S_2, \ S_0 \not\supseteq S_3 \end{array} \right\} \tag{8.3}$$

or, after permuting S_1, S_2, S_3 if necessary,

$$\left. \begin{array}{l} S_1 \cap S_2 = \emptyset, \ S_1 \cup S_2 \subseteq S_3, \\ S_0 \cap S_1 \neq \emptyset, \ S_0 \cap S_2 \neq \emptyset, \\ S_0 \not\subseteq S_3, \ S_0 \not\supseteq S_1, \ S_0 \not\supseteq S_2; \end{array} \right\} \tag{8.4}$$

neither families $\{S_0, S_1, S_2, S_3\}$ with property (8.3) nor families $\{S_0, S_1, S_2, S_3\}$ with property (8.4) have the consecutive ones property.

Among all inequalities (8.1), comb inequalities (8.2) are distinguished by the following fact.

THEOREM 8.1 *Let $\mathcal{F}$ be a family of subsets of V^{int} such that $\mathcal{F}$ does not have the consecutive ones property. Then the inequality*

$$\sum(\eta(S, x) : S \in \mathcal{F}) \geq 2 \tag{8.5}$$

is either a nonnegative linear combination of inequalities

$$x_e \geq 0 \ \text{ for all edges } e,$$

$$\eta(S, x) \geq 0 \ \text{ for all nonempty subsets } S \text{ of } V^{\mathrm{int}},$$

or a nonnegative linear combination of these inequalities and a single comb inequality

$$\eta(H, x) + \eta(T_1, x) + \eta(T_2, x) + \eta(T_3, x) \geq 2 \tag{8.6}$$

such that $H, T_1, T_2 \in \mathcal{F}$ and $T_3 \in \mathcal{F} \cup \{V - S : S \in \mathcal{F}\}$.

This theorem guarantees that, for each family $\mathcal{F}$ of subsets of V^{int} such that $\mathcal{F}$ does not have the consecutive ones property and such that (8.5) is violated by an optimal solution x^* of an LP relaxation of our TSP instance, there is a subtour inequality violated by x^* or a comb inequality (8.6) violated by x^*; the postulate that subtour cuts are always easy to find by direct methods leads us to conclude that either the cut (8.5) is not worth bothering about, or else there is a comb inequality (8.6) whose addition to the LP relaxation would do at least as much good as the addition of constraint (8.5).

Theorem 8.1 motivated our design of an algorithm that is the main subject of the present chapter; however, correctness of the algorithm does not depend on validity of the theorem; we relegate its proof to Section 8.2. The algorithm consists of two phases.

PHASE 1: We collect a suitable family $\mathcal{F}^*$ of subsets of V^{int}
such that $\eta(S, x^*) < 2$ whenever $S \in \mathcal{F}^*$.

PHASE 2: We search for subfamilies $\mathcal{F}$ of $\mathcal{F}^*$ such that
- $\mathcal{F}$ lacks the consecutive ones property
  but each of its proper subfamilies has it,
- $\sum(\eta(S, x^*) : S \in \mathcal{F}) < 2$.

If x^* happens to satisfy all subtour inequalities, then Theorem 8.1 guarantees that each cut $\sum(\eta(S, x) : S \in \mathcal{F}) \geq 2$ found in Phase 2 is a comb inequality. In general, cuts $\sum(\eta(S,x) : S \in \mathcal{F}) \geq 2$ found in Phase 2 may or may not be comb inequalities; we discard those that are not.

In our implementation of Phase 1 we access the repository of tight sets, having $\eta(S, x^*) = 0$, described in Chapter 6, as well as sets that are hypergraph edges in the cut pool that is grown during the execution of the cutting-plane method.

What provoked our interest in this scheme, and led us to prefer it to other ways of searching for families $\mathcal{F}$ of $\mathcal{F}^*$ such that

- $\mathcal{F}$ is a comb with handle H and teeth T_1, T_2, T_3,
- $H, T_1, T_2 \in \mathcal{F}$ and $T_3 \in \mathcal{F} \cup \{V - S : S \in \mathcal{F}\}$,
- $\sum(\eta(S, x^*) : S \in \mathcal{F}) < 2$,

is availability of a remarkably efficient algorithm for recognizing families of sets with the consecutive ones property. Ingredients of this algorithm have been referred to in Section 6.5; let us repeat the pertinent facts here.

A *PQ-tree* is a rooted ordered tree with each of its internal nodes having at least two children labeled either a *P-node* or a *Q-node*. For each node u of a PQ-tree T, we let $D(u)$ denote the set of all leaves of T that are descendants of u; we let $\mathcal{I}(T)$ denote the family of

all sets $D(u)$ such that u is a node of T

and

all sets $D(u_i) \cup D(u_{i+1}) \cup \ldots \cup D(u_j)$ such that

$u_1, u_2, \ldots, u_k$ in this order are the children of some Q-node of T

and $1 \leq i < j \leq k$.

Booth and Lueker [79] designed a function INSERT that, given a PQ-tree T and a set S such that some PQ-tree T' satisfies

$$\mathcal{I}(T') \supseteq \mathcal{I}(T) \cup \{S\}, \tag{8.7}$$

returns a PQ-tree $T^\star$ such that

- $T^\star$ satisfies (8.7) in place of T',
- every PQ-tree T' with property (8.7) has $\mathcal{I}(T') \supseteq \mathcal{I}(T^\star)$;

if no PQ-tree T' satisfies (8.7), then INSERT(T, S) destroys T beyond all recovery and, as a failure message, returns the *trivial PQ-tree* with no internal nodes except for its root, which is a P-node.

PQ-trees are closely related to the consecutive ones property:

A family $\mathcal{F}$ of sets has the consecutive ones property
if and only if there is a PQ-tree T with $\mathcal{F} \subseteq \mathcal{I}(T)$.

(If there is a PQ-tree T with $\mathcal{F} \subseteq \mathcal{I}(T)$, then the left-to-right order of the leaves of T certifies that $\mathcal{F}$ has the consecutive ones property. Conversely, if a linear order $\prec$ on a set W certifies that $\mathcal{F}$ has the consecutive ones property, then $\mathcal{F} \subseteq \mathcal{I}(T)$ is satisfied by the PQ-tree T that consists of the root, labeled a Q-node, and the set W of leaves ordered by $\prec$.)

A primary application of the Booth-Lueker INSERT is an incremental algorithm that tests a prescribed family $\{S_1, S_2, \ldots, S_k\}$ of sets for the consecutive ones property: having constructed a PQ-tree T_i such that

- $S_1, S_2, \ldots, S_i \in \mathcal{I}(T_i)$,

- every PQ-tree T' with $S_1, S_2, \ldots, S_i \in \mathcal{I}(T')$ has $\mathcal{I}(T') \supseteq \mathcal{I}(T_i)$,

we call $\text{INSERT}(T_i, S_{i+1})$. If $\{S_1, S_2, \ldots, S_{i+1}\}$ has the consecutive ones property, then $\text{INSERT}(T_i, S_{i+1})$ returns a PQ-tree T_{i+1} such that

- $S_1, S_2, \ldots, S_{i+1} \in \mathcal{I}(T_{i+1})$,
- every PQ-tree T' with $S_1, S_2, \ldots, S_{i+1} \in \mathcal{I}(T')$ has $\mathcal{I}(T') \supseteq \mathcal{I}(T_{i+1})$;

else $\text{INSERT}(T_i, S_{i+1})$ returns the trivial PQ-tree.

Algorithm 8.1 Testing a family $\mathcal{F}$ of subsets W for the consecutive ones property.

$T =$ the trivial PQ-tree with W the set of leaves;
for all S in $\mathcal{F}$
do **if** $1 < |S| < |W|$
    then $T = \text{INSERT}(T, S)$;
        if $T =$ the trivial PQ-tree **then** return FAIL **end**
    end
end
return PASS;

The running time of Algorithm 8.1 is in $O(n + \sum(|S| : S \in \mathcal{F}))$; this algorithm and its variants are our principal tools in Phase 2.

8.1 IMPLEMENTATION OF PHASE 2

Given a PQ-tree T compatible with x^* along with a family $\mathcal{F}_1^*$ of subsets of V^{int}, we search for subfamilies $\mathcal{F}$ of $\mathcal{I}_0(T) \cup \mathcal{F}_1^*$ such that

the members of $\mathcal{F}$ can be enumerated as S_0, S_1, S_2, S_3 so that
$$S_1 \cap S_2 = \emptyset, \quad S_1 \cap S_3 = \emptyset, \quad S_2 \cap S_3 = \emptyset,$$
$$S_0 \cap S_1 \neq \emptyset, \quad S_0 \cap S_2 \neq \emptyset, \quad S_0 \cap S_3 \neq \emptyset,$$
$$S_0 \not\supseteq S_1, \quad S_0 \not\supseteq S_2, \quad S_0 \not\supseteq S_3$$
or
$$S_1 \cap S_2 = \emptyset, \quad S_1 \cup S_2 \subseteq S_3,$$
$$S_0 \cap S_1 \neq \emptyset, \quad S_0 \cap S_2 \neq \emptyset, \quad S_0 \not\subseteq S_3,$$
$$S_0 \not\supseteq S_1, \quad S_0 \not\supseteq S_2$$

$$(8.8)$$

and such that $\sum(\eta(S, x^*) : S \in \mathcal{F}) < 2$. Let us specify this task in more general terms.

An *independence system* is an ordered pair $(X, \mathcal{I})$ such that X is a finite set and $\mathcal{I}$ is a nonempty collection of subsets of X that is closed under taking subsets (that is to say, if $R \subset S$ and $S \in \mathcal{I}$ then $R \in \mathcal{I}$). Members of $\mathcal{I}$ are referred to as *independent* and the remaining subsets of X are referred to as *dependent*; minimal (with respect to set inclusion) dependent sets are called *circuits*. Our job in Phase

2 generalizes as follows:

$$
\left.
\begin{array}{ll}
\text{given} & \text{an independence system } (X_0 \cup X_1, \mathcal{I}), \\
& \text{a family } \mathcal{C}^\star \text{ of some of its circuits,} \\
& \text{a function } w : X_0 \cup X_1 \rightarrow [0, +\infty), \\
& \text{and a positive number } t \text{ such that} \\
& \quad X_0 \text{ is independent,} \\
& \quad w(s) = 0 \text{ whenever } s \in X_0, \\
& \quad w(s) < t \text{ whenever } s \in X_1, \\
\text{find} & \text{a reasonably long list } \mathcal{L} \text{ of circuits } C \\
& \text{such that } C \in \mathcal{C}^\star \text{ and } \sum(w(s) : s \in C) < t.
\end{array}
\right\}
\quad (8.9)
$$

In our particular instance of (8.9), we have $X_0 = \mathcal{I}_0(T)$, $X_1 = \mathcal{F}_1^\star$, a subset of $X_0 \cup X_1$ is independent if and only if it has the consecutive ones property, family $\mathcal{C}^\star$ consists of all $\{S_0, S_1, S_2, S_3\}$ that satisfy (8.8), function w is defined by $w(S) = \eta(S, x^\star)$ for all S, and we have $t = 2$. Before restricting ourselves to this particular instance, we shall discuss our implementation of Phase 2 in the fully general framework of problem (8.9).

If, in our particular instance of (8.9), $x^\star$ happens to satisfy all subtour inequalities, then Theorem 8.1 guarantees that

$$
\text{every circuit } C \text{ with } \sum(w(s) : s \in C) < t \text{ belongs to } \mathcal{C}^\star. \quad (8.10)
$$

In general, property (8.10) simplifies the statement of problem (8.9):

$$
\left.
\begin{array}{ll}
\text{given} & \text{an independence system } (X_0 \cup X_1, \mathcal{I}), \\
& \text{a function } w : X_0 \cup X_1 \rightarrow [0, +\infty), \\
& \text{and a positive number } t \text{ such that} \\
& \quad X_0 \text{ is independent,} \\
& \quad w(s) = 0 \text{ whenever } s \in X_0, \\
& \quad w(s) < t \text{ whenever } s \in X_1, \\
\text{find} & \text{a reasonably long list } \mathcal{L} \text{ of circuits } C \\
& \text{such that } \sum(w(s) : s \in C) < t.
\end{array}
\right\}
\quad (8.11)
$$

Regardless of condition (8.10), restricting a discussion of problem (8.9) to its special case (8.11) is harmless in the sense that, having found a reasonably long list $\mathcal{L}$ of circuits C such that $\sum(w(s) : s \in C) < t$, one can simply discard those circuits in $\mathcal{L}$ that do not belong to $\mathcal{C}^\star$.

8.1.1 The general case of problem (8.11)

Let us write

$$
w(C) = \sum(w(s) : s \in C)
$$

for every subset C of $X_0 \cup X_1$.

One way of tackling problem (8.11) is to examine all the sets $X_0 \cup \{s\}$ with $s \in X_1$ and, whenever $X_0 \cup \{s\}$ turns out to be dependent, find a circuit C contained in this set: by assumption, $w(C) = w(s) < t$, and so C can be added to the list $\mathcal{L}$. In its formal presentation, Algorithm 8.2 features a function CIRCUIT that, given a sequence $s'_1, s'_2, \ldots, s'_k, s$ of points of $X_0 \cup X_1$ such that

- $\{s'_1, s'_2, \ldots, s'_k\}$ is independent,
- $\{s'_1, s'_2, \ldots, s'_k, s\}$ is dependent, and
- $w(s'_1) \le w(s'_2) \le \ldots \le w(s'_k)$,

always returns a circuit C such that $C \subseteq \{s'_1, s'_2, \ldots, s'_k, s\}$ and, subject to this constraint, attempts to minimize $w(C)$.

Algorithm 8.2 A fast heuristic for tackling problem (8.11).

---

initialize an empty list $\mathcal{L}$ of circuits;
enumerate the elements of X_0 as $s'_1, s'_2, \ldots, s'_k$;
for   all s in X_1
do   **if** $\{s'_1, s'_2, \ldots, s'_k, s\}$ is dependent
         then $C = \text{CIRCUIT}(s'_1, s'_2, \ldots, s'_k, s)$;
                add C to $\mathcal{L}$;
       end
end
return $\mathcal{L}$;

A straightforward implementation of $\text{CIRCUIT}(s'_1, s'_2, \ldots, s'_k, s)$ starts out with the set $\{s'_1, s'_2, \ldots, s'_k, s\}$, dependent by assumption, and keeps removing its elements as long as the resulting set remains dependent; in an attempt to minimize $w(C)$, the various elements s'_i are proposed for removal in an order of nonincreasing $w(s'_i)$. This greedy heuristic is presented as Algorithm 8.3.

Algorithm 8.3 An implementation of $\text{CIRCUIT}(s'_1, s'_2, \ldots, s'_k, s)$.

$C = \{s'_1, s'_2, \ldots, s'_k, s\}$;
for $i = k, k-1, \ldots, 1$
do   **if** $C - \{s'_i\}$ is dependent **then** $C = C - \{s'_i\}$ **end**
end
return C;

The trouble with Algorithm 8.2 is that it may miss quite a few candidates for membership in $\mathcal{L}$: whenever $X_0 \cup \{s\}$ turns out to be independent, which may be the case for quite a few elements s of X_1, we discard s, and with it all the circuits that contain s.

One remedy is to run Algorithm 8.2 repeatedly, substituting a new independent set S' for X_0 in each new iteration: having initialized S' as X_0, we keep adding to it elements s of X_1 as long as $S' \cup \{s\}$ remains independent. In order to increase the chance that the various calls of CIRCUIT come up with $w(C) < t$, we attempt to keep $w(S')$ small as long as possible; for this reason, we propose the elements s

of X_1 for addition to S' in a nondecreasing order of $w(s)$. Algorithm 8.4 presents an implementation of this idea.

Algorithm 8.4 A variation on Algorithm 8.2.

---

initialize an empty list $\mathcal{L}$ of circuits;
enumerate the elements of X_0 as $s'_1, s'_2, \ldots, s'_k$;
$X = X_1$;
repeat for   all s in X
       do   **if** $\{s'_1, s'_2, \ldots, s'_k, s\}$ is dependent
               then $C = \text{Circuit}(s'_1, s'_2, \ldots, s'_k, s)$;
                    if $w(C) < t$ **then** add C to $\mathcal{L}$ **end**
                    remove s from X;
               end
       end
       if $X = \emptyset$
       then   return $\mathcal{L}$;
       else $s = $ an element of X with the smallest $w(s)$;
               $s'_{k+1} = s$;
               $k = k + 1$;
               remove s from X;
       end
end

The trouble with Algorithm 8.4 is that it is too slow for our purpose: the number of times that it tests subsets of $X_0 \cup X_1$ for independence may be quadratic in M. We are going to remove this flaw.

Let $s'_1, s'_2, s'_3, \ldots$ be as in Algorithm 8.4, let $k_{\max}$ denote the largest value of k such that Algorithm 8.4 defines s'_k, and let us write

$$S'_{\max} = \{s'_k : 1 \le k \le k_{\max}\};$$

for each element s of $X_1 - S'_{\max}$, let $a(s)$ be the subscript such that

$$\{s'_1, s'_2, \ldots, s'_k, s\} \text{ is dependent if and only if } a(s) \le k \le k_{\max}.$$

With our implementation of Circuit, which is (equivalent to) Algorithm 8.3,

$$\left. \begin{array}{l} \text{the return value of } \text{Circuit}(s'_1, s'_2, \ldots, s'_k, s) \text{ is} \\ \text{independent of } k \text{ in the range } a(s) \le k \le k_{\max} \text{:} \end{array} \right\} \quad (8.12)$$

to see this, recall that Algorithm 8.3 begins by removing all s'_i such that $i > a(s)$ from $\{s'_1, s'_2, \ldots, s'_k, s\}$. Property (8.12) of Circuit guarantees that Algorithm 8.4 and Algorithm 8.5 return the same $\mathcal{L}$.

Many of the calls of Circuit in Algorithm 8.5 may be useless in the sense of returning a C such that $w(C) \ge t$. In an attempt to avoid them, we may rely on the fact that

$$\text{Circuit}(s'_1, s'_2, \ldots, s'_k, s) \text{ returns a } C \text{ such that}$$
$$\{s'_{a(s)}, s\} \subseteq C \subseteq \{s'_1, s'_2, \ldots, s'_{a(s)}, s\},$$

Algorithm 8.5 An efficient implementation of Algorithm 8.4.

initialize an empty list $\mathcal{L}$ of circuits;
enumerate the elements of X_0 as $s'_1, s'_2, \ldots, s'_k$;
sort X_1 as $s_1, s_2, \ldots, s_M$ so that $w(s_1) \le w(s_2) \le \ldots \le w(s_M)$;
$i = 1$;
repeat **while** $i \le M$ and $\{s'_1, s'_2, \ldots, s'_k, s_i\}$ is dependent
     do $C = \text{CIRCUIT}(s'_1, s'_2, \ldots, s'_k, s_i)$;
            if $w(C) < t$ **then** add C to $\mathcal{L}$ **end**
            $i = i + 1$;
     end
     if $i > M$
     then return $\mathcal{L}$;
     else $s'_{k+1} = s_i$;
            $k = k + 1$;
            $i = i + 1$;
     end
end

---

which is an immediate corollary of (8.12). On the one hand, if

$$w(s'_{a(s)}) + w(s) \ge t, \tag{8.13}$$

then $\text{CIRCUIT}(s'_1, s'_2, \ldots, s'_k, s)$ will return a C such that $w(C) \ge t$; on the other hand, if

$$w(s'_1) + w(s'_2) + \ldots + w(s'_{a(s)}) + w(s) < t, \tag{8.14}$$

then $\text{CIRCUIT}(s'_1, s'_2, \ldots, s'_k, s)$ will return a C such that $w(C) < t$.

These observations suggest modifying Algorithm 8.5 by choosing once and for all an easily computable function σ such that

$$w_a + w \le \sigma(w_1, w_2, \ldots, w_a, w) \le w_1 + w_2 + \ldots + w_a + w$$
$$\text{whenever } 0 \le w_1 \le w_2 \le \ldots \le w_a \le w$$

and skipping each prescribed call of $\text{CIRCUIT}(s'_1, s'_2, \ldots, s'_k, s)$ if and only if

$$\sigma(w(s'_1), \ldots, w(s'_{a(s)}), w(s)) \ge t :$$

with this proviso,

- all the calls of $\text{CIRCUIT}(s'_1, s'_2, \ldots, s'_k, s)$ that are bound to be useless by virtue of (8.13) will be skipped

and

- all the calls of $\text{CIRCUIT}(s'_1, s'_2, \ldots, s'_k, s)$ that are bound to be useful by virtue of (8.14) will be made.

For clarity, let us record this variation on Algorithm 8.5 as Algorithm 8.6.

Algorithm 8.6 A modified version of Algorithm 8.5.

---

initialize an empty list $\mathcal{L}$ of circuits;
enumerate the elements of X_0 as $s'_1, s'_2, \ldots, s'_k$;
sort X_1 as $s_1, s_2, \ldots, s_M$ so that $w(s_1) \le w(s_2) \le \ldots \le w(s_M)$;
$i = 1$;
repeat   **while** $i \le M$ and $\{s'_1, s'_2, \ldots, s'_k, s_i\}$ is dependent
          do   **if** $\sigma(w(s'_1), \ldots, w(s'_a(s)), w(s)) < t$
                 then $C = \text{CIRCUIT}(s'_1, s'_2, \ldots, s'_k, s)$;
                      if $w(C) < t$ **then** add C to $\mathcal{L}$ **end**
              end
              $i = i + 1$;
          end
          if $i > M$
          then   return $\mathcal{L}$;
          else $s'_{k+1} = s_i$;
                $i = i + 1$;
                $k = k + 1$;
          end
end

As long as σ behaves reasonably in the sense that

$$\sigma(w(s'_1), \ldots, w(s'_k), w(s)) \le \sigma(w(s'_1), \ldots, w(s'_k), w(s'_{k+1}), w(s))$$
$$\text{whenever } w(s'_1) \le \ldots \le w(s'_k) \le w(s'_{k+1}),$$

there is a $b(s)$ such that condition

(B) $\sigma(w(s'_1), \ldots, w(s'_k), w(s)) < t$

holds if and only if $k \le b(s)$.

For each element s of $X_1 - S'_{\max}$, Algorithm 8.5 calls $\text{CIRCUIT}(s'_1, \ldots, s'_k, s)$ if and only if $a \le k \le b$. Its inefficiency comes from the fact that it carries out the expensive test of (A) many times and the cheap test of (B) only once: its policy is to

(α)   test (A) with each increment of k until $k = a(s)$, at which point
      $\text{CIRCUIT}(s'_1, s'_2, \ldots, s'_k, s)$ is called if and only if $k \le b$
      and s is removed from X.

To turn this policy around, so as to carry out the expensive test of (A) only once, we may

(β)   test (B) with each increment of k until $k = b(s)$, at which point
      $\text{CIRCUIT}(s'_1, s'_2, \ldots, s'_k, s)$ is called if and only if $a \le k$
      and s is removed from X.

We summarize this class of heuristics in Algorithm 8.7. With our implementation of CIRCUIT, which is Algorithm 8.3, policies (α) and (β) yield the same set of candidates C for membership in $\mathcal{L}$. To see this, recall that Algorithm 8.3 begins by removing all s'_i with $i > a(s)$ from $\{s'_1, s'_2, \ldots, s'_k, s\}$: it follows that

as long as $k \geq k_A$,
the return value of $\text{CIRCUIT}(s'_1, s'_2, \ldots, s'_k, s)$ is independent of k.

Algorithm 8.7 A class of heuristics for tackling problem (8.11).

---

initialize an empty list $\mathcal{L}$ of circuits;
enumerate the elements of X_0 as $s'_1, s'_2, \ldots, s'_k$;
sort X_1 as $s_1, s_2, \ldots, s_M$ so that $w(s_1) \leq w(s_2) \leq \ldots \leq w(s_M)$;
$i = 1, j = M$;
repeat   **while** $i \leq j$ and $\{s'_1, s'_2, \ldots, s'_k, s_i\}$ is dependent
         do $C = \text{CIRCUIT}(s'_1, s'_2, \ldots, s'_k, s_i)$;
                 if $w(C) < t$ **then** add C to $\mathcal{L}$ **end**
                 $i = i + 1$;
         end
         if $i > j$
         then   return $\mathcal{L}$;
         else $s'_{k+1} = s_i$;
                 $i = i + 1$;
                 while $j \geq i$ and $\sigma(w(s'_1), w(s'_2), \ldots, w(s'_{k+1}), w(s_j)) \geq t$
                 do   **if** $\{s'_1, s'_2, \ldots, s'_k, s_j\}$ is dependent
                        then $C = \text{CIRCUIT}(s'_1, s'_2, \ldots, s'_k, s_j)$;
                                if $w(C) < t$ **then** add C to $\mathcal{L}$ **end**
                        end
                        $j = j - 1$;
                 end
                 $k = k + 1$;
         end
end

Algorithm 8.8 describes the same process in different terms: if

     C' denotes the value of C in Algorithm 8.3 and
     C'' denotes the value of C in Algorithm 8.8,

then we have, after each execution of the **for** loop, $C' \supset \{s'_1, s'_2, \ldots, s'_{i-1}\}$, $C'' \cap \{s'_1, s'_2, \ldots, s'_{i-1}\} = \emptyset$, and $C' \cap \{s'_i, s'_{i+1}, \ldots, s'_k\} = C'' \cap \{s'_i, s'_{i+1}, \ldots, s'_k\}$.

8.1.2 Our special case of problem (8.9)

Instead of CIRCUIT, Concorde code uses a variation on Algorithm 8.9; the reasons are the special nature of our oracle for testing independence, which is Algorithm 8.1, and the special nature of our $C^\star$. The special nature of our oracle is that executing the instruction

Algorithm 8.8 An equivalent description of Algorithm 8.3.

$C = \{s\}$;
for $i = k, k - 1, \ldots, 1$
do   **if** $C \cup \{s'_1, s'_2, \ldots, s'_{i-1}\}$ is independent **then** $C = C \cup \{s'_i\}$ **end**
end
return C;

Algorithm 8.9 Another description of Algorithm 8.8.

$C = \{s\}$;
while C is independent
do $i =$ smallest subscript such that $C \cup \{s'_1, s'_2, \ldots, s'_i\}$ is dependent;
     $C = C \cup \{s'_i\}$;
end
return C;

$i =$ smallest subscript such that $C \cup \{s'_1, s'_2, \ldots, s'_i\}$ is dependent

takes no more time than executing the instruction

test $C \cup \{s'_1, s'_2, \ldots, s'_i\}$ for dependence.

This peculiarity makes Algorithm 8.9 run faster than Algorithms 8.3 and 8.8. The special nature of our C^* is that all of its members have size four. This peculiarity allows us to replace the condition

C is independent

in the control of the **while** loop of Algorithm 8.9 by the condition

$|C| < 4$ and C is independent:

once C has reached size four and yet it remains independent, Algorithm 8.9 is bound to return a C such that $C \notin C^*$, which is just as useless as the current value of C.

The resulting function, which we call COMB, is presented in Algorithm 8.10.

Since the running time of Algorithm 8.1 is in $O(n + \sum(|S| : S \in \mathcal{F}))$, the running time of Algorithm 8.10 is in $O(n + \sum_{i=1}^{k}(|S'_i|))$.

We have designed a variation on Booth-Lueker INSERT. This variation, TEST, given a PQ-tree T and a set S, returns PASS if some PQ-tree T' satisfies (8.7) and returns FAIL otherwise; in either case, it leaves T unchanged. The time required to execute the sequence of instructions

$T =$ the trivial PQ-tree;
for   all S in $\mathcal{F}$
do   **if** TEST(T, S) =PASS **then** $T =$ INSERT(T, S) **end**
end
return T;

Algorithm 8.10 Function $\text{COMB}(S'_1, S'_2, \ldots, S'_k, S)$.

$\mathcal{F} = \{S\}$;
while $|\mathcal{F}| < 4$
do $T =$ the trivial PQ-tree;
      for   all $\overline{S}$ in $\mathcal{F}$
      do $T = \text{INSERT}(T, \overline{S})$;
            if $T =$ the trivial PQ-tree **then** return $\mathcal{F}$ **end**
      end
      $i = 0$;
      while $T \neq$ the trivial PQ-tree
      do $i = i + 1$;
            $T = \text{INSERT}(T, S'_i)$;
      end
      add S'_i to $\mathcal{F}$;
end
return $\mathcal{F}$;

---

is in $O(kn + \sum(|S| : S \in \mathcal{F}))$ with k at most, and typically far less than, the number of times that TEST returns FAIL.

A summary of our implementation of Phase 2 is presented in Algorithm 8.11. To give an indication of the effectiveness of these ideas, we applied Concorde's implementation of the heuristic to the 1,000-city Euclidean problem we discussed in the previous two chapters. Using the subtour, blossom, and comb algorithms described earlier, we obtain a 0.081% gap between the LP bound and the optimal tour for this problem; adding consecutive ones combs to the mix lowered the gap to 0.062%. Similarly, for the 100,000-city Euclidean instance the gap of 0.117% was improved to 0.083% with the addition of the consecutive ones heuristic.

8.2 PROOF OF THE CONSECUTIVE ONES THEOREM

To prove Theorem 8.1, we first enumerate all minimal families without the consecutive ones property; to do that, we rely on the following theorem of Tucker [526].

THEOREM 8.2 *A family $\mathcal{F}$ of sets does not have the consecutive ones property if and only if it has at least one of the following five properties:*

(I) There are sets $S_1, S_2, \ldots, S_m$ in $\mathcal{F}$ and a set R consisting of distinct elements $a_1, a_2, \ldots, a_m$ such that $m \geq 3$ and, with subscript arithmetic modulo m, $S_i \cap R = \{a_{i-1}, a_i\}$ for all i.

(II) There are sets $S_1, S_2, \ldots, S_m, T_1, T_2$ in $\mathcal{F}$ and a set R consisting of distinct elements $a_0, a_1, a_2, \ldots, a_m, b$ such that $m \geq 2$, $S_i \cap R = \{a_{i-1}, a_i\}$ for all i, and $T_1 \cap R = \{a_0, a_1, \ldots, a_{m-1}, b\}$, $T_2 \cap R = \{a_1, a_2, \ldots, a_m, b\}$.

Algorithm 8.11 An implementation of Phase 2.

initialize an empty list $\mathcal{L}$ of families of sets;
enumerate the sets in $\mathcal{I}_0(T)$ as $S_1', S_2', \ldots, S_k'$;
sort $\mathcal{F}_1^*$ as $S_1, S_2, \ldots, S_M$ so that $\eta(S_1, x^*) \le \eta(S_2, x^*) \le \ldots \le \eta(S_M, x^*)$;
$i = 1, j = M$;
while $i \le j$
do   **if** $3\eta(S_i, x^*) + \eta(S_j, x^*) \ge 2$
       then   **if**   TEST(T, S_j) =FAIL
           then $\mathcal{F} = $ COMB$(S_1', S_2', \ldots, S_k', S_j)$;
                if $\mathcal{F}$ has property (8.8) **then** add $\mathcal{F}$ to $\mathcal{L}$ **end**
           end
           $j = j - 1$;
       else   **if**   TEST(T, S_i) =FAIL
           then $\mathcal{F} = $ COMB$(S_1', S_2', \ldots, S_k', S_i)$;
                if $\mathcal{F}$ has property (8.8) **then** add $\mathcal{F}$ to $\mathcal{L}$ **end**
           else $k = k + 1$;
                $S_k' = S_i$;
                $T = $ INSERT(T, S_k');
           end
           $i = i + 1$;
       end
end
return $\mathcal{L}$;

(III) *There are sets* $S_1, S_2, \ldots, S_m$, T *in* $\mathcal{F}$ *and a set* R *consisting of distinct elements* $a_0, a_1, a_2, \ldots, a_m, b$ *such that* $m \ge 2$, $S_i \cap R = \{a_{i-1}, a_i\}$ *for all* i, *and* $T \cap R = \{a_1, a_2, \ldots, a_{m-1}, b\}$.

(IV) *There are sets* H, T_1, T_2, T_3 *in* $\mathcal{F}$ *and a set* R *consisting of distinct elements* $a_1, a_2, a_3, b_1, b_2, b_3$ *such that* $H \cap R = \{a_1, a_2, a_3\}$ *and* $T_i \cap R = \{a_i, b_i\}$ *for all* i.

(V) *There are sets* H, T_1, T_2, T_3 *in* $\mathcal{F}$ *and a set* R *consisting of distinct elements* a_1, a_2, a_3, b_1, b_2 *such that* $H \cap R = \{a_1, a_2, a_3\}$ *and* $T_1 \cap R = \{a_1, b_1\}$, $T_2 \cap R = \{a_2, b_2\}$, $T_3 \cap R = \{a_1, a_2, b_1, b_2\}$.

Let us say that

- $\{S_1, S_2, \ldots, S_m\}$ is a *family of type I* if $m \ge 3$ and there are pairwise disjoint sets $C_1, A_1, C_2, A_2, \ldots, C_m, A_m$ such that $A_1, A_2, \ldots, A_m$ are nonempty and, with subscript arithmetic modulo m,
$$S_i = A_{i-1} \cup C_i \cup A_i \text{ for all } i = 1, 2, \ldots, m;$$

- $\{S_1, S_2, \ldots, S_m, T_1, T_2\}$ is a *family of type II* if $m \ge 2$ and there are pairwise disjoint sets $A_0, C_1, A_1, C_2, A_2, \ldots, A_{m-1}, C_m, A_m, B, D_1, D_2$ such

that $A_0, A_1, A_2, \ldots, A_m, B$ are nonempty and
$$S_i = A_{i-1} \cup C_i \cup A_i \text{ for all } i = 0, 1, 2, \ldots, m,$$
$$T_1 = A_0 \cup C_1 \cup A_1 \cup C_2 \cup A_2 \cup \ldots \cup A_{m-1} \cup C_m \cup B \cup D_1,$$
$$T_2 = C_1 \cup A_1 \cup C_2 \cup A_2 \cup \ldots \cup A_{m-1} \cup C_m \cup A_m \cup B \cup D_2;$$

- $\{S_1, S_2, \ldots, S_m, T\}$ is a *family of type III* if $m \geq 2$ and there are pairwise disjoint sets
$$A_0, C_1, A_1, C_2, A_2, \ldots, A_{m-1}, C_m, A_m, B$$
such that $A_0, A_1, A_2, \ldots, A_m, B$ are nonempty and
$$S_i = A_{i-1} \cup C_i \cup A_i \text{ for all } i = 0, 1, 2, \ldots, m,$$
$$T = C_1 \cup A_1 \cup C_2 \cup A_2 \cup \ldots \cup A_{m-1} \cup C_m \cup B;$$

- $\{H, T_1, T_2, T_3\}$ is a *family of type IV* if T_1, T_2, T_3 are pairwise disjoint and if H meets each of T_1, T_2, T_3 but does not contain any;

- $\{H, T_1, T_2, T_3\}$ is a *family of type V* if $T_1 \cap T_2 = \emptyset$, $T_1 \cup T_2 \subseteq T_3$, $H \cap T_1 \neq \emptyset$, $H \cap T_2 \neq \emptyset$, $H \not\supseteq T_1$, $H \not\supseteq T_2$, $H \not\subseteq T_3$;

- $\{S_1, S_2, S_3, S_4\}$ is a *family of type VI* if there are pairwise disjoint nonempty sets A_1, A_2, A_3, A_4, B such that
$$S_1 = A_1 \cup A_2 \cup B, \quad S_2 = A_2 \cup A_3 \cup B,$$
$$S_3 = A_3 \cup A_4 \cup B, \quad S_4 = A_4 \cup A_1 \cup B;$$

- $\{S_1, S_2, S_3\}$ is a *family of type VII* if all four of
$$(S_1 \cap S_2) - S_3, (S_1 \cap S_3) - S_2, (S_2 \cap S_3) - S_1, S_1 \cap S_2 \cap S_3$$
are nonempty.

THEOREM 8.3 *If a family $\mathcal{F}$ of sets lacks the consecutive ones property and if every proper subfamily of $\mathcal{F}$ has it, then $\mathcal{F}$ is a family of type I, II, III, IV, V, VI, or VII.*

Proof. Since $\mathcal{F}$ lacks the consecutive ones property, Theorem 8.2 allows us to distinguish between five cases.

CASE I: *$\mathcal{F}$ satisfies condition (I) of Theorem 8.2.*
We are going to show that $\mathcal{F}$ is a family of type I, VI, or VII.
By assumption of this case, there are sets $S_1, S_2, \ldots, S_m$ in $\mathcal{F}$ and a set R consisting of distinct elements $a_1, a_2, \ldots, a_m$ such that $m \geq 3$ and, with subscript arithmetic modulo m, $S_i \cap R = \{a_{i-1}, a_i\}$ for all i; minimality of $\mathcal{F}$ guarantees that $\mathcal{F} = \{S_1, S_2, \ldots, S_m\}$.
With
$$\mathcal{F}(u) = \{S \in \mathcal{F} : u \in S\} \text{ and } Q = \{u : \mathcal{F}(u) \neq \mathcal{F}(v) \text{ whenever } v \in R\}, \quad (8.15)$$
we claim that
$$\text{every } u \text{ in } Q \text{ satisfies } |\mathcal{F}(u)| \leq 1 \text{ or } \mathcal{F}(u) = \mathcal{F}. \quad (8.16)$$
To justify this claim, assume the contrary: some u in Q has $1 < |\mathcal{F}(u)| < |\mathcal{F}|$. Symmetry allows us to assume that $S_1 \in \mathcal{F}(u)$, $S_2 \notin \mathcal{F}(u)$. Since $|\mathcal{F}(u)| > 1$,

there is a subscript j such that $3 \le j \le m$ and $S_j \in \mathcal{F}(u)$; consider the smallest j with these properties. Note that $\mathcal{F}(a_m) = \{S_1, S_m\}$; since $u \in Q$, it follows that $j < m$. But then condition (I) of Theorem 8.2 is satisfied by j in place of m and by u in place of a_j, contradicting minimality of $\mathcal{F}$; this contradiction proves (8.16).

SUBCASE I.1: *No v in Q has $\mathcal{F}(v) = \mathcal{F}$.*

$\mathcal{F}$ is of type I: (8.16) and the assumption of this subcase guarantee that every u in Q has $\mathcal{F}(u) = \{S_i\}$ for some i or $\mathcal{F}(u) = \emptyset$.

SUBCASE I.2: *$m \ge 4$ and some v in Q has $\mathcal{F}(v) = \mathcal{F}$.*

We are going to show that $\mathcal{F}$ is of type VI by showing that $m = 4$ and that every u in Q has $\mathcal{F}(u) = \mathcal{F}$ or $\mathcal{F}(u) = \emptyset$.

If $m \ge 5$, then condition (III) of Theorem 8.2, with $m = 2$, would be satisfied by S_4 in place of T and by a_m, v, a_2, a_3 in place of a_0, a_1, a_2, b, contradicting minimality of $\mathcal{F}$.

If there were a u in Q with $\mathcal{F}(u) \ne \mathcal{F}$ and $\mathcal{F}(u) \ne \emptyset$, then (8.16) would guarantee $|\mathcal{F}(u)| = 1$ and symmetry would allow us to assume that $\mathcal{F}(u) = \{S_1\}$; but then condition (III) of Theorem 8.2, with $m = 2$, would be satisfied by S_4 in place of T and by u, v, a_2, a_3 in place of a_0, a_1, a_2, b, contradicting minimality of $\mathcal{F}$.

SUBCASE I.1: *$m = 3$ and some v in Q has $\mathcal{F}(v) = \mathcal{F}$.*

By assumption of this subcase, $\mathcal{F}$ is of type VII.

CASE II: *$\mathcal{F}$ satisfies condition (II) of Theorem 8.2.*

We are going to show that $\mathcal{F}$ is a family of type II or VI.

By assumption of this case, there are sets $S_1, S_2, \ldots, S_m, T_1, T_2$ in $\mathcal{F}$ and a set R consisting of distinct elements $a_0, a_1, a_2, \ldots, a_m, b$ such that $m \ge 2$, $S_i \cap R = \{a_{i-1}, a_i\}$ for all i, and $T_1 \cap R = \{a_0, a_1, \ldots, a_{m-1}, b\}$, $T_2 \cap R = \{a_1, a_2, \ldots, a_m, b\}$; minimality of $\mathcal{F}$ guarantees $\mathcal{F} = \{S_1, S_2, \ldots, S_m, T_1, T_2\}$.

With $\mathcal{F}(u)$ and Q defined by (8.15), we claim that

$$\text{for every } u \text{ in } Q, \text{ there are subscripts } i \text{ and } k, \\ \text{with } 1 \le i \le m, 1 \le k \le m \text{ (and possibly } i > k), \qquad (8.17) \\ \text{such that } S_j \in \mathcal{F}(u) \text{ if and only if } i \le j \le k.$$

To justify this claim, assume the contrary: there are a u in Q and subscripts i and k such that $k \ge i + 2$, $S_i, S_k \in \mathcal{F}(u)$ and $S_j \notin \mathcal{F}(u)$ whenever $i < j < k$. But then condition (I) of Theorem 8.2 is satisfied by $S_i, S_{i+1}, \ldots, S_k$ with $R = \{a_i, a_{i+1}, \ldots, a_{k-1}, u\}$, contradicting minimality of $\mathcal{F}$; this contradiction proves (8.17).

SUBCASE II.1: *No u in Q has $T_1 \notin \mathcal{F}(u)$, $T_2 \notin \mathcal{F}(u)$, and $\mathcal{F}(u) \ne \emptyset$.*

We are going to show that $\mathcal{F}$ is of type II by showing that every u in Q has $\mathcal{F}(u) = \{S_i, T_1, T_2\}$ for some i or $\mathcal{F}(u) = \{T_1\}$ or $\mathcal{F}(u) = \{T_2\}$ or $\mathcal{F}(u) = \emptyset$. For this purpose, consider an arbitrary u in Q and let i and k be as in (8.17).

If $T_1 \in \mathcal{F}(u)$ and $T_2 \in \mathcal{F}(u)$, then $\mathcal{F}(u) = \{S_i, T_1, T_2\}$. To see this, note first that $\mathcal{F}(b) = \{T_1, T_2\}$ and that $\mathcal{F}(a_j) = \{S_j, S_{j+1}, T_1, T_2\}$ whenever $1 \le j \le m - 1$; since $u \in Q$, it follows that $i \le k$ and that $i \ne k - 1$. If $i \le k - 2$, then condition (II) of Theorem 8.2 is satisfied when all S_j with $i < j < k$ are removed and u is substituted for the sequence $a_i, a_{i+1}, \ldots, a_{k-1}$; this contradicts minimality of $\mathcal{F}$. Hence $i = k$.

If $T_1 \in \mathcal{F}(u)$ and $T_2 \notin \mathcal{F}(u)$, then $\mathcal{F}(u) = \{T_1\}$. To see this, note first that $\mathcal{F}(a_0) = \{S_1, T_1\}$; since $u \in Q$, it follows that we cannot have $i = k = 1$. If $i \le k$ and $1 < k < m$, then condition (II) of Theorem 8.2 is satisfied when all S_j with $1 \le j < k$ are removed and u is substituted for the sequence $a_0, a_1, \ldots, a_{k-1}$; this contradicts minimality of $\mathcal{F}$. If $i \le k = m$, then condition (I) of Theorem 8.2, with $m = 3$, is satisfied by T_1, T_2, S_m in place of S_1, S_2, S_3 and by b, a_m, u in place of a_1, a_2, a_3; this contradicts minimality of $\mathcal{F}$. Hence $i > k$.

If $T_1 \notin \mathcal{F}(u)$ and $T_2 \in \mathcal{F}(u)$, then $\mathcal{F}(u) = \{T_2\}$; the argument is a mirror image of the argument we have just given.

If $T_1 \notin \mathcal{F}(u)$ and $T_2 \notin \mathcal{F}(u)$, then $\mathcal{F}(u) = \emptyset$ by assumption of this subcase.

SUBCASE II.2: *Some u in Q has $T_1 \notin \mathcal{F}(u)$, $T_2 \notin \mathcal{F}(u)$, and $\mathcal{F}(u) \ne \emptyset$.*

We are going to show that $\mathcal{F}$ is of type VI.

First, we claim that $\mathcal{F}(u) \subseteq \{S_1, S_m\}$. To justify this claim, assume the contrary: there is a subscript j such that $S_j \in \mathcal{F}(u)$ and $1 < j < m$. But then condition (III) of Theorem 8.2, with $m = 2$, is satisfied by T_1, T_2, S_j in place of S_1, S_2, T and by a_0, a_j, a_m, u in place of a_0, a_1, a_2, b, contradicting minimality of $\mathcal{F}$.

Next, we claim that $\mathcal{F}(u) = \{S_1, S_m\}$. To justify this claim, assume the contrary: $\mathcal{F}(u) = \{S_1\}$ or $\mathcal{F}(u) = \{S_m\}$. Symmetry allows us to restrict ourselves to $\mathcal{F}(u) = \{S_m\}$. But then condition (III) of Theorem 8.2 is satisfied by substituting u for a_m and T_2 for T; this contradicts minimality of $\mathcal{F}$.

Now (8.17) implies that $m = 2$. But then the assumptions of Subcase I.2 are satisfied with S_1, S_2, T_2, T_1 in place of S_1, S_2, S_3, S_4, with u, a_2, b, a_0 in place of a_1, a_2, a_3, a_4, and with a_1 in place of v.

CASE III: *$\mathcal{F}$ satisfies condition (III) of Theorem 8.2.*

We are going to show that $\mathcal{F}$ is a family of type III or VII.

By assumption of this case, there are sets $S_1, S_2, \ldots, S_m, T$ in $\mathcal{F}$ and a set R consisting of distinct elements $a_0, a_1, a_2, \ldots, a_m, b$ such that $m \ge 2$, $S_i \cap R = \{a_{i-1}, a_i\}$ for all i, and $T \cap R = \{a_1, a_2, \ldots, a_{m-1}, b\}$; minimality of $\mathcal{F}$ guarantees that $\mathcal{F} = \{S_1, S_2, \ldots, S_m, T\}$.

SUBCASE III.1: $m \ge 3$.

We are going to show that $\mathcal{F}$ is of type III by showing that, with $\mathcal{F}(u)$ and Q defined by (8.15), every u in Q has $\mathcal{F}(u) = \{S_i, T\}$ for some i or $\mathcal{F}(u) = \emptyset$. For this purpose, note first that (8.17) can be justified in Case III by the argument that we used to justify it in Case II; then consider an arbitrary u in Q and let i, k be as in (8.17).

If $T \in \mathcal{F}(u)$, then $\mathcal{F}(u) = \{S_i, T\}$. To see this, note first that $\mathcal{F}(b) = \{T\}$ and that $\mathcal{F}(a_j) = \{S_j, S_{j+1}, T\}$ whenever $1 \le j \le m - 1$; since $u \in Q$, it follows that $i \le k$ and that $i \ne k - 1$. If $i \le k - 2$, then condition (III) of Theorem 8.2 is satisfied when all S_j with $i < j < k$ are removed and u is substituted for the sequence $a_i, a_{i+1}, \ldots, a_{k-1}$; this contradicts minimality of $\mathcal{F}$. Hence $i = k$.

If $T \notin \mathcal{F}(u)$, then $\mathcal{F}(u) = \emptyset$. To see this, note first that $\mathcal{F}(a_0) = \{S_1\}$ and that $\mathcal{F}(a_m) = \{S_m\}$; since $u \in Q$, it follows that we cannot have $i = k = 1$ or $i = k = m$. If $i \le k$ and $1 < k < m$, then condition (III) of Theorem 8.2 is satisfied when all S_j with $1 \le j < k$ are removed and u is substituted for the sequence $a_0, a_1, \ldots, a_{k-1}$; this contradicts minimality of $\mathcal{F}$. If $k = m$ and $i = m - 1$, then condition (III) of Theorem 8.2 is satisfied when S_m is removed and a_{m-1}, a_m are replaced with u; this contradicts minimality of $\mathcal{F}$. If $k = m$ and $i \leq m - 2$, then condition (I) of Theorem 8.2, with $m = 3$, is satisfied by S_{m-2}, S_m, T in place of S_1, S_2, S_3 and by u, a_{m-1}, a_{m-2} in place of a_1, a_2, a_3; this contradicts minimality of $\mathcal{F}$. Hence $i > k$.

SUBCASE III.2: $m = 2$.

If all three of $(S_1 \cap S_2) - T$, $(S_1 \cap T) - S_2$, $(S_2 \cap T) - S_1$ are nonempty, then $\mathcal{F}$ is of type VII; else $\mathcal{F}$ is of type III.

CASE IV: *$\mathcal{F}$ satisfies condition (IV) of Theorem 8.2.*

We are going to show that $\mathcal{F}$ is a family of type IV.

By assumption of this case, there are sets H, T_1, T_2, T_3 in $\mathcal{F}$ and a set R consisting of distinct elements $a_1, a_2, a_3, b_1, b_2, b_3$ such that $H \cap R = \{a_1, a_2, a_3\}$ and $T_i \cap R = \{a_i, b_i\}$ for all i; minimality of $\mathcal{F}$ guarantees $\mathcal{F} = \{H, T_1, T_2, T_3\}$. Showing that $\mathcal{F}$ is of type IV amounts to showing that T_1, T_2, T_3 are pairwise disjoint.

If two of T_1, T_2, T_3 intersect, then symmetry allows us to assume that T_1 and T_2 intersect; let u be a point in $T_1 \cap T_2$. If $u \in H$, then condition (III) of Theorem 8.2, with $m = 2$, is satisfied by T_1, T_2, H in place of S_1, S_2, T and by b_1, u, b_2, a_3 in place of a_0, a_1, a_2, b; if $u \notin H$, then condition (I) of Theorem 8.2, with $m = 3$, is satisfied by T_1, T_2, H in place of S_1, S_2, S_3 and by u, a_2, a_1 in place of a_1, a_2, a_3; in either case, minimality of $\mathcal{F}$ is contradicted.

CASE V: *$\mathcal{F}$ satisfies condition (V) of Theorem 8.2.*

We are going to show that $\mathcal{F}$ is a family of type V.

By assumption of this case, there are sets H, T_1, T_2, T_3 in $\mathcal{F}$ and a set R consisting of distinct elements a_1, a_2, a_3, b_1, b_2 such that $H \cap R = \{a_1, a_2, a_3\}$ and $T_1 \cap R = \{a_1, b_1\}$, $T_2 \cap R = \{a_2, b_2\}$, $T_3 \cap R = \{a_1, a_2, b_1, b_2\}$; minimality of $\mathcal{F}$ guarantees that $\mathcal{F} = \{H, T_1, T_2, T_3\}$. Showing that $\mathcal{F}$ is of type V amounts to showing that $T_1 \cap T_2 = \emptyset$ and that $T_1 \cup T_2 \subseteq T_3$; the argument used in Case IV shows that $T_1 \cap T_2 = \emptyset$.

If $T_1 \cup T_2 \not\subseteq T_3$, then symmetry allows us to assume that $T_1 \not\subseteq T_3$; let u be a point in $T_1 - T_3$. If $u \in H$, then condition (I) of Theorem 8.2, with $m = 3$, is satisfied by T_1, T_3, H in place of S_1, S_2, S_3 and by b_1, a_2, u in place of a_1, a_2, a_3; if $u \notin H$, then condition (III) of Theorem 8.2, with $m = 2$, is satisfied by T_1, T_3, H in place of S_1, S_2, T and by u, a_1, b_2, a_3 in place of a_0, a_1, a_2, b; in either case, minimality of $\mathcal{F}$ is contradicted. $\square$

The converse of Theorem 8.3,

> *if*   *$\mathcal{F}$ is a family of type I, II, III, IV, V, VI, or VII,*
> *then*   *$\mathcal{F}$ lacks the consecutive ones property*
> *and every proper subfamily of $\mathcal{F}$ has it,*

is easy to prove and irrelevant to our purpose.

Proof of Theorem 8.1. Our task is to prove that, for every family $\mathcal{F}$ of subsets of V^{int} such that $\mathcal{F}$ does not have the consecutive ones property, inequality $\sum(\eta(S,x) : S \in \mathcal{F}) \geq 2$ is either a linear combination of constraints

$$\eta(S,x) \geq 0 \quad \text{for all nonempty subsets } S \text{ of } V^{\text{int}},$$
$$x_e \geq 0 \quad \text{for all edges } e,$$

or a linear combination of these constraints and a single comb inequality

$$\eta(H,x) + \eta(T_1,x) + \eta(T_2,x) + \eta(T_3,x) \geq 2.$$

If $\mathcal{F}_0$ is a subfamily of $\mathcal{F}$ and if $\mathcal{F}_0$ does not have the consecutive ones property, then inequality $\sum(\eta(S,x) : S \in \mathcal{F}) \geq 2$ is the sum of constraints

$$\sum(\eta(S,x) : S \in \mathcal{F}_0) \geq 2$$
$$\eta(S,x) \geq 0 \text{ for all } S \text{ in } \mathcal{F} - \mathcal{F}_0.$$

This observation allows us to assume that every proper subfamily of $\mathcal{F}$ has the consecutive ones property; in turn, Theorem 8.3 allows us to distinguish between seven cases.

CASE I: $\mathcal{F}$ *is a family of type I.*
By assumption of this case, $\mathcal{F} = \{S_1, S_2, \ldots, S_m\}$ with $m \geq 4$ and there are pairwise disjoint sets $C_1, A_1, C_2, A_2, \ldots, C_m, A_m$ such that $A_1, \ldots, A_m$ are nonempty and, with subscript arithmetic modulo m,
$$S_i = A_{i-1} \cup C_i \cup A_i \text{ for all } i = 1, 2, \ldots, m.$$
Let us write
$$W = C_1 \cup A_1 \cup C_2 \cup A_2 \cup \ldots \cup C_m \cup A_m.$$
Inequality

$$\eta(A_m \cup C_1 \cup A_1, x) + \sum_{i=2}^{m} \eta(A_{i-1} \cup C_i \cup A_i, x) \geq 2$$

is the sum of constraints

$$\eta(A_i, x) \geq 0 \quad (1 \leq i \leq m),$$
$$\eta(W, x) \geq 0$$

and

$$\begin{array}{ll}
2x(A_1, A_j) \geq 0 & (3 \leq j \leq m-1), \\
2x(A_i, A_j) \geq 0 & (2 \leq i \leq m,\ i+2 \leq j \leq m), \\
2x(A_i, C_j) \geq 0 & (1 \leq i \leq m-1,\ 1 \leq j \leq i-1), \\
2x(A_i, C_j) \geq 0 & (1 \leq i \leq m-1,\ i+2 \leq j \leq m), \\
2x(A_m, C_j) \geq 0 & (2 \leq j \leq m-1), \\
2x(C_i, C_j) \geq 0 & (1 \leq i < j \leq m).
\end{array}$$

CASE II: $\mathcal{F}$ *is a family of type II.*
By assumption of this case, $\mathcal{F} = \{S_1, S_2, \ldots, S_m, T_1, T_2\}$ with $m \geq 2$ and there are pairwise disjoint sets $A_0, C_1, A_1, C_2, A_2, \ldots, A_{m-1}, C_m, A_m, B, D_1, D_2$ such that $A_0, A_1, A_2, \ldots, A_m, B$ are nonempty and
$$S_i = A_{i-1} \cup C_i \cup A_i \text{ for all } i = 0, 1, 2, \ldots, m,$$

$$T_1 = A_0 \cup C_1 \cup A_1 \cup C_2 \cup A_2 \cup \ldots \cup A_{m-1} \cup C_m \cup B \cup D_1,$$
$$T_2 = C_1 \cup A_1 \cup C_2 \cup A_2 \cup \ldots \cup A_{m-1} \cup C_m \cup A_m \cup B \cup D_2.$$

Let us write

$$W = A_0 \cup C_1 \cup A_1 \cup C_2 \cup A_2 \cup \ldots \cup A_{m-1} \cup C_m \cup A_m \cup B \cup D_1 \cup D_2.$$

Inequality

$$\sum_{i=1}^{m} \eta(A_{i-1} \cup C_i \cup A_i, x)$$
$$+ \eta(A_0 \cup C_1 \cup A_1 \cup C_2 \cup A_2 \cup \ldots \cup A_{m-1} \cup C_m \cup B \cup D_1, x)$$
$$+ \eta(C_1 \cup A_1 \cup C_2 \cup A_2 \cup \ldots \cup A_{m-1} \cup C_m \cup A_m \cup B \cup D_2, x) \geq 2$$

is the sum of constraints

$$\eta(A_i, x) \geq 0 \qquad (0 \leq i \leq m),$$
$$\eta(B, x) \geq 0,$$
$$\eta(W, x) \geq 0$$

and

$$2x((V - W) \cup D_1 \cup D_2, A_j) \geq 0 \qquad (1 \leq j \leq m - 1),$$
$$2x((V - W) \cup D_1 \cup D_2, C_j) \geq 0 \qquad (1 \leq j \leq m),$$
$$2x(A_i, A_j) \geq 0 \qquad (0 \leq i \leq m,\ i + 2 \leq j \leq m),$$
$$2x(A_i, C_j) \geq 0 \qquad (0 \leq i \leq m,\ 1 \leq j \leq i - 1),$$
$$2x(A_i, C_j) \geq 0 \qquad (0 \leq i \leq m,\ i + 2 \leq j \leq m),$$
$$2x(C_i, C_j) \geq 0 \qquad (1 \leq i < j \leq m),$$
$$2x(A_0, D_2) \geq 0,$$
$$2x(D_1, A_m) \geq 0,$$
$$2x(D_1, D_2) \geq 0.$$

CASE III: $\mathcal{F}$ *is a family of type III.*

By assumption of this case, $\mathcal{F} = \{S_1, S_2, \ldots, S_m, T\}$ with $m \geq 3$ and there are pairwise disjoint sets

$$A_0, C_1, A_1, C_2, A_2, \ldots, A_{m-1}, C_m, A_m, B$$

such that $A_0, A_1, A_2, \ldots, A_m, B$ are nonempty and

$$S_i = A_{i-1} \cup C_i \cup A_i \text{ for all } i = 0, 1, 2, \ldots, m,$$
$$T = C_1 \cup A_1 \cup C_2 \cup A_2 \cup \ldots \cup A_{m-1} \cup C_m \cup B.$$

Let us write

$$W = A_0 \cup C_1 \cup A_1 \cup C_2 \cup A_2 \cup \ldots \cup A_{m-1} \cup C_m \cup A_m \cup B.$$

Inequality

$$\sum_{i=1}^{m} \eta(A_{i-1} \cup C_i \cup A_i, x)$$
$$+ \eta(C_1 \cup A_1 \cup C_2 \cup A_2 \cup \ldots \cup A_{m-1} \cup C_m \cup B, x) \geq 2$$

is the sum of constraints

$$\eta(A_i, x) \geq 0 \qquad (0 \leq i \leq m),$$
$$\eta(B, x) \geq 0$$

and

$$2x(V - W, A_j) \geq 0 \qquad (1 \leq j \leq m - 1),$$
$$2x(V - W, C_j) \geq 0 \qquad (1 \leq j \leq m),$$
$$2x(A_0, A_j) \geq 0 \qquad (2 \leq j \leq m - 1),$$
$$2x(A_i, A_j) \geq 0 \qquad (1 \leq i \leq m,\ i + 2 \leq j \leq m),$$
$$2x(A_i, C_j) \geq 0 \qquad (0 \leq i \leq m,\ 1 \leq j \leq i - 1),$$
$$2x(A_i, C_j) \geq 0 \qquad (0 \leq i \leq m,\ i + 2 \leq j \leq m),$$
$$2x(C_i, C_j) \geq 0 \qquad (1 \leq i < j \leq m).$$

CASE IV: $\mathcal{F}$ *is a family of type IV.*

By assumption of this case, $\mathcal{F} = \{H, T_1, T_2, T_3\}$ and T_1, T_2, T_3 are pairwise disjoint and H meets each of T_1, T_2, T_3 but does not contain any.

Inequality $\eta(H, x) + \eta(T_1, x) + \eta(T_2, x) + \eta(T_3, x) \geq 2$ is a comb inequality.

CASE V: $\mathcal{F}$ *is a family of type V.*

By assumption of this case, $\mathcal{F} = \{H, T_1, T_2, T_3\}$ and we have $T_1 \cap T_2 = \emptyset$, $T_1 \cup T_2 \subseteq T_3$, $H \cap T_1 \neq \emptyset$, $H \cap T_2 \neq \emptyset$, $H \not\supseteq T_1$, $H \not\supseteq T_2$, $H \not\subseteq T_3$.

Inequality $\eta(H, x) + \eta(T_1, x) + \eta(T_2, x) + \eta(T_3, x) \geq 2$ is identical with the comb inequality $\eta(H, x) + \eta(T_1, x) + \eta(T_2, x) + \eta(V - T_3, x) \geq 2$.

CASE VI: $\mathcal{F}$ *is a family of type VI.*

By assumption of this case, $\mathcal{F} = \{S_1, S_2, S_3, S_4\}$ and there are pairwise disjoint nonempty sets A_1, A_2, A_3, A_4, B such that

$$S_1 = A_1 \cup A_2 \cup B, \quad S_2 = A_2 \cup A_3 \cup B,$$
$$S_3 = A_3 \cup A_4 \cup B, \quad S_4 = A_4 \cup A_1 \cup B.$$

Inequality

$$\eta(A_1 \cup A_2 \cup B, x)$$
$$+ \eta(A_2 \cup A_3 \cup B, x)$$
$$+ \eta(A_3 \cup A_4 \cup B, x)$$
$$+ \eta(A_4 \cup A_1 \cup B, x) \geq 2$$

is the sum of constraints

$$\eta(A_1, x) \geq 0,$$
$$\eta(A_2, x) \geq 0,$$
$$\eta(A_3, x) \geq 0,$$
$$\eta(A_4, x) \geq 0,$$
$$\eta(B, x) \geq 0$$

and

$$2x(A_1, A_3) \geq 0,$$
$$2x(A_2, A_4) \geq 0,$$
$$x(V - W, A_1) \geq 0,$$
$$x(V - W, A_2) \geq 0,$$
$$x(V - W, A_3) \geq 0,$$
$$x(V - W, A_4) \geq 0,$$
$$3x(V - W, B) \geq 0.$$

CASE VII: $\mathcal{F}$ *is a family of type VII.*

By assumption of this case, $\mathcal{F} = \{S_1, S_2, S_3\}$ and all four of $S_1 \cap S_2 \cap S_3$, $(S_1 \cap S_2) - S_3$, $(S_1 \cap S_3) - S_2$, $(S_2 \cap S_3) - S_1$ are nonempty.

Inequality $\eta(S_1, x) + \eta(S_2, x) + \eta(S_3, x) \geq 2$ is the sum of constraints

$$\eta((S_1 \cap S_2) - S_3, x) \geq 0,$$
$$\eta((S_1 \cap S_3) - S_2, x) \geq 0,$$
$$\eta((S_2 \cap S_3) - S_1, x) \geq 0,$$
$$\eta(S_1 \cup S_2 \cup S_3, x) \geq 0$$

and

$$2x(S_1 \cap S_2 \cap S_3, S_1 - (S_2 \cup S_3)) \geq 0,$$
$$2x(S_1 \cap S_2 \cap S_3, S_2 - (S_1 \cup S_3)) \geq 0,$$
$$2x(S_1 \cap S_2 \cap S_3, S_3 - (S_1 \cup S_2)) \geq 0,$$
$$2x(S_1 - (S_2 \cup S_3), S_2 - (S_1 \cup S_3)) \geq 0,$$
$$2x(S_1 - (S_2 \cup S_3), S_3 - (S_1 \cup S_2)) \geq 0,$$
$$2x(S_2 - (S_1 \cup S_3), S_3 - (S_1 \cup S_2)) \geq 0,$$
$$2x(S_1 \cap S_2 \cap S_3, V - (S_1 \cup S_2 \cup S_3)) \geq 0,$$
$$2x((S_1 \cap S_2) - S_3, S_3 - (S_1 \cup S_2)) \geq 0,$$
$$2x((S_1 \cap S_3) - S_2, S_2 - (S_1 \cup S_3)) \geq 0,$$
$$2x((S_2 \cap S_3) - S_1, S_1 - (S_2 \cup S_3)) \geq 0.$$

□

Chapter Nine

Combs from Dominoes

The subject of this chapter is a heuristic separation algorithm for comb inequalities,

$$\eta(H, x) + \sum_{j=1}^{t} \eta(T_j, x) \geq t - 1. \tag{9.1}$$

Earlier we gave a combinatorial proof that (9.1) is satisfied by all tours x through V. The algorithm of the present chapter was motivated by an algebraic proof of the same fact.

The algebraic proof proceeds in two stages in the spirit of the framework for describing Gomory cuts that has been propounded by Chvátal [125]. In the first stage, we show that every nonnegative x which satisfies all subtour inequalities must satisfy the inequality

$$\eta(H, x) + \sum_{j=1}^{t} \eta(T_j, x) \geq t - 2; \tag{9.2}$$

the second stage amounts to observing that every tour x through V gives the left-hand side of (9.2) an even integer value and that $t - 2$ is odd by definition. To complete the first stage, note that every x satisfies

$$\eta(T_j \cap H, x) + \eta(T_j - H, x) = \eta(T_j, x) + 2x(T_j \cap H, T_j - H) - 2,$$

and so it satisfies

$$\sum_{j=1}^{t} x(T_j \cap H, T_j - H) + \sum_{j=1}^{t} \eta(T_j, x) =$$

$$\frac{1}{2} \sum_{j=1}^{t} (\eta(T_j \cap H, x) + \eta(T_j - H, x) + \eta(T_j, x)) + t;$$

if x is nonnegative, then $\eta(H, x) \geq \sum x(T_j \cap H, T_j - H) - 2$; if x satisfies all subtour inequalities, then $\sum(\eta(T_j \cap H, x) + \eta(T_j - H, x) + \eta(T_j, x)) \geq 0$.

The argument used in the first stage has the following corollary. If an optimal solution x^* of an LP relaxation of our TSP instance satisfies all subtour inequalities, then x^* cannot violate any comb inequality (9.1) by more than 1.0, and it violates (9.1) by this much if and only if

(P1) each edge e with $x_e^* > 0$ and precisely one endpoint in H
     has both endpoints in some T_j,

(P2) $\eta(T_j, x^*) = 0$ and $x^*(T_j \cap H, T_j - H) = 1$ for all $j = 1, 2, \ldots, t$.

Conversely, any x^* having properties (P1), (P2) violates (9.1) by precisely 1.0, whether it satisfies all subtour inequalities or not. This observation motivated us to search for combs $\{H, T_1, \ldots, T_t\}$ with properties (P1), (P2); our actual computer code uses a variation on this theme, where we settle for smaller violations in exchange for a chance of finding additional comb inequalities violated by x^*.

By a *domino*, we mean a pair (A, B) of subsets of V such that

$$A \cap B = \emptyset \quad \text{and } A \cup B \neq V;$$

if in addition

$$\eta(A, x^*) = \eta(B, x^*) = \eta(A \cup B, x^*) = 0,$$

then we call (A, B) a *tight domino*. We make various choices of sets F of edges with a small $\sum(x_e^* : e \in F)$ and search for combs $\{H, A_1 \cup B_1, \ldots, A_t \cup B_t\}$ such that

(P3) each (A_j, B_j) with $j = 1, 2, \ldots, t$ is a tight domino,

(P4) $\{e : x_e^* > 0, e \notin F, e \in \delta(H)\}$ is the union of
$\{e : x_e^* > 0, e \notin F, e \cap A_j \neq \emptyset, e \cap B_j \neq \emptyset\}$ with $j = 1, \ldots, t$.

If this search is successful, then (9.1) with $T_j = A_j \cup B_j$ for all j may or may be not violated by x^*. Property (P3) guarantees that

$$\sum(x_e^* : e \cap A_j \neq \emptyset, e \cap B_j \neq \emptyset) = 1 \text{ for all } j;$$

in turn, property (P4) guarantees that

$$\eta(H, x^*) \leq t + \sum(x_e^* : e \in F, e \in \delta(H)) - 2;$$

in particular, (9.1) is violated by x^* whenever $\sum(x_e^* : e \in F) < 1$.

The search consists of two phases.

PHASE 1: We collect a suitable family $\mathcal{D}$ of tight dominoes.
PHASE 2: For each of various choices of sets F of edges with a small
$\sum(x_e^* : e \in F)$, we search for combs $\{H, A_1 \cup B_1, \ldots, A_t \cup B_t\}$
with $(A_1, B_1), \ldots, (A_t, B_t) \in \mathcal{D}$ and property (P4).

Every PQ-tree T compatible with x^* represents a family $\mathcal{D}(T)$ of tight dominoes; this family is the union of families $\mathcal{D}(T, u)$, one for each Q-node u of T; if u is a Q-node of T and if $u_1, u_2, \ldots, u_k$ in this order are the children of u, then each choice of r, s, t with $0 \leq r < s < t \leq k$ yields three tight dominoes in $\mathcal{D}(T, u)$; these three dominoes are (C_1, C_2), (C_2, C_3), and (C_3, C_1) with

$C_1 = D(u_{r+1}) \cup D(u_{r+2}) \cup \ldots \cup D(u_s),$
$C_2 = D(u_{s+1}) \cup D(u_{s+2}) \cup \ldots \cup D(u_t),$
$C_3 = V - (D(u_{r+1}) \cup D(u_{r+2}) \cup \ldots \cup D(u_t)).$

In the definition of $\mathcal{I}(T)$, it is immaterial whether a node with precisely two children is labeled a P-node or a Q-node; Booth and Lueker always label it a P-node; we always label it a Q-node in order to get a larger $\mathcal{D}(T)$.

9.1 PULLING TEETH FROM PQ-TREES

The preceding section describes a way of collecting a family $\mathcal{D}$ of tight dominoes; the subject of the present section is a way of making choices of sets F of edges with a small $\sum(x_e^* : e \in F)$ and searching for combs with handle H and teeth $A_1 \cup B_1, \ldots, A_t \cup B_t$ such that

- $(A_1, B_1), \ldots, (A_t, B_t) \in \mathcal{D}$,
- $\{e : x_e^* > 0, e \notin F, e \cap H \neq \emptyset, e - H \neq \emptyset\}$ is the union of $\{e : x_e^* > 0, e \notin F, e \cap A_j \neq \emptyset, e \cap B_j \neq \emptyset\}$ with $j = 1, \ldots, t$.

The simplest F with a small $\sum(x_e^* : e \in F)$ is the empty set. For clarity, we begin our exposition with $F = \emptyset$; our way of making additional choices of F will be described in Subsection 9.1.3.

Let us view each domino (A, B) as a bipartite graph with vertex set $A \cup B$ and edge set $\{e : e \cap A \neq \emptyset, e \cap B \neq \emptyset, x_e^* > 0\}$; the vertex set of a domino D will be denoted $V(D)$ and its edge set will be denoted $E(D)$. A cut in a graph is a set of edges $\delta(H)$ for some set H of vertices. As usual, let G^* denote the graph with vertex set V and with edge set $\{e : x_e^* > 0\}$. In this notation and terminology, our task with $F = \emptyset$ can be stated as follows: search for subfamilies $\{D_1, D_2, \ldots, D_t\}$ of $\mathcal{D}$ such that

$$V(D_1), V(D_2), \ldots, V(D_t) \text{ are pairwise disjoint,} \qquad (9.3)$$

$$E(D_1) \cup E(D_2) \cup \ldots \cup E(D_t) \text{ is a cut in } G^*, \qquad (9.4)$$

$$t \text{ is odd and at least three.} \qquad (9.5)$$

Let us change this objective by substituting

$$t \text{ is odd} \qquad (9.6)$$

for (9.5). This change is innocuous: each subfamily $\{D_1, D_2, \ldots, D_t\}$ of $\mathcal{D}$ with properties (9.3), (9.4), (9.6) points out either (in case $t \geq 3$) a comb inequality that x^* violates by 1.0 or (in case $t = 1$) a subtour inequality that x^* violates by 1.0. (Our computer code, searching for subfamilies $\{D_1, D_2, \ldots, D_t\}$ of $\mathcal{D}$ with properties (9.3), (9.4), (9.6), dismisses all the subfamilies with $t = 1$ that it finds. This policy is based on the postulate that the algorithms we use for finding subtour cuts directly are adequate.)

We search for subfamilies $\{D_1, D_2, \ldots, D_t\}$ of $\mathcal{D}$ with properties (9.3), (9.4), (9.6) only if G^* is connected. (When G^* is disconnected, each of its connected components points out a subtour inequality violated by 2.0; searching for minor blemishes in the presence of such major faults seems pointless.) Our search strategy evolves from the observation that, as long as (9.3) is satisfied, requirement (9.4) amounts to $m - n + 1$ linear congruences mod 2, with m standing for the number of edges of G^*. Let us explain.

With a spanning tree T of G^* selected once and for all, each edge e of $G^* - T$ defines a unique circuit contained in $T + e$; let C_e denote this circuit. It is well known and easy to prove that

a set of edges of G^* is a cut in G^* if and only if
it meets each C_e with $e \in G^* - T$ in an even number of edges.

Given an edge e of $G^* - T$ and a domino D, let us write

$$a(e, D) = \begin{cases} 0 & \text{if } |C_e \cap E(D)| \text{ is even,} \\ 1 & \text{if } |C_e \cap E(D)| \text{ is odd.} \end{cases}$$

For families $\{D_1, D_2, \ldots, D_t\}$ of tight dominoes that satisfy (9.3), requirement (9.4) amounts to the $m - n + 1$ linear congruences

$$\sum_{i=1}^{t} a(e, D_i) \equiv 0 \bmod 2 \ (e \in G^* - T).$$

This observation suggests searching for subfamilies $\{D_1, D_2, \ldots, D_t\}$ of $\mathcal{D}$ with properties (9.3), (9.4), (9.6) by examining solutions ξ of the system of $m - n + 2$ linear congruences

$$\left. \begin{aligned} \sum(a(e, D)\xi_D : D \in \mathcal{D}) &\equiv 0 \bmod 2 \ (e \in G^* - T), \\ \sum(\xi_D : D \in \mathcal{D}) &\equiv 1 \bmod 2. \end{aligned} \right\} \tag{9.7}$$

On the one hand, the incidence vector ξ of any subfamily $\{D_1, D_2, \ldots, D_t\}$ of $\mathcal{D}$ with properties (9.3), (9.4), (9.6) satisfies (9.7); on the other hand, each solution ξ of (9.7) defines a subfamily $\{D_1, D_2, \ldots, D_t\}$ of $\mathcal{D}$, that satisfies (9.3), (9.4), (9.6) if and only if it satisfies (9.3).

This strategy may be inefficient: (9.7) may have an overwhelming number of solutions and only a small fraction of this number may define subfamilies

$$\{D_1, D_2, \ldots, D_t\}$$

of $\mathcal{D}$ with property (9.3). Our modification attempts to reduce the number of fruitless probes into the solution space of (9.3) by relying on the peculiar structure of our $\mathcal{D}$.

Our $\mathcal{D}$ is represented by a PQ-tree T^* compatible with x^*; it is the union of families $\mathcal{D}(u)$, one for each Q-node u of T^*; if u is a Q-node of T^* and if $u_1, u_2, \ldots, u_k$ in this order are the children of u, then each choice of r, s, t with $0 \le r < s < t \le k$ yields three tight dominoes in $\mathcal{D}(u)$; these three dominoes are (C_1, C_2), (C_2, C_3), and (C_3, C_1) with

$$C_1 = D(u_{r+1}) \cup D(u_{r+2}) \cup \ldots \cup D(u_s),$$
$$C_2 = D(u_{s+1}) \cup D(u_{s+2}) \cup \ldots \cup D(u_t),$$
$$C_3 = V - (D(u_{r+1}) \cup D(u_{r+2}) \cup \ldots \cup D(u_t)).$$

Each Q-node u of T^*, with children $u_1, u_2, \ldots, u_k$ in this order, defines a partition of V into tight sets $S_0, S_1, \ldots, S_k$ by

$$S_0 = V - D(u), \ S_1 = D(u_1), \ S_2 = D(u_2), \ \ldots, \ S_k = D(u_k);$$

these tight sets are like the beads of a necklace in the sense that

$$x^*(S_0, S_1) = x^*(S_1, S_2) = \ldots = x^*(S_{k-1}, S_k) = x^*(S_k, S_0) = 1;$$

the edge set of each tight domino in $\mathcal{D}(u)$ is $\{e : x_e^* > 0, e \cap S_i \ne \emptyset, e \cap S_{i+1} \ne \emptyset\}$ for some subscript i. In this perspective, it is easy to see that distinct tight dominoes arising from the same necklace are indistinguishable in system (9.7):

if u is a Q-node of $T^\star$ and if D', D'' are tight dominoes in $\mathcal{D}(u)$, then $|C \cap E(D')| \equiv |C \cap E(D'')|$ for all circuits C of $G^\star$.

The partition of our $\mathcal{D}$ into subfamilies $\mathcal{D}(u)$ induces an equivalence relation on the set of columns in the left-hand side of (9.7): the equivalence classes are in a one-to-one correspondence with the Q-nodes of $T^\star$ and every two columns in the same equivalence class are identical except for the two names of their variables. We reduce (9.7) by keeping only one column from each equivalence class. To put it differently, let $\mathcal{Q}$ denote the set of all Q-nodes of $T^\star$ and, for each u in $\mathcal{Q}$, let $D_u^\star$ denote an arbitrary tight domino in $\mathcal{D}(u)$. The reduced system is

$$\left. \begin{aligned} \sum(a(e, D_u^\star)\xi_u : u \in \mathcal{Q}) &\equiv 0 \bmod 2 \quad (e \in G^\star - T), \\ \sum(\xi_u : u \in \mathcal{Q}) &\equiv 1 \bmod 2. \end{aligned} \right\} \tag{9.8}$$

We say that a solution ξ of (9.8) is *representable* if there is a family of pairwise vertex-disjoint tight dominoes that consists of one tight domino from each $\mathcal{D}(u)$ such that $\xi_u = 1$. Rather than generating solutions of (9.7) to see which of them define families of pairwise vertex-disjoint tight dominoes, we generate solutions of (9.8) and test them for representability.

9.1.1 Generating solutions of (9.8)

To compute the coefficients $a(e, D_u^\star)$ in (9.8), our computer code relies on the following observations. Let us write

$$b(e, D_u^\star) = \begin{cases} 1 & \text{if } e \in D_u^\star \\ 0 & \text{if } e \notin D_u^\star \end{cases}$$

and, having declared one of the nodes of T as its root, let $c(v, D_u^\star)$ denote the sum mod 2 of all $b(e, D_u^\star)$ with e on the path from the root to v. In this notation,

$$a(vw, D_u^\star) \equiv b(vw, D_u^\star) + c(v, D_u^\star) + c(w, D_u^\star) \pmod{2}.$$

For each Q-node u of $T^\star$, with children $u_1, u_2, \ldots u_k$ in this order, we set

$$D_u^\star = (D(u_1), D(u_2)).$$

Now each e incident with `ext` has $b(e, D_u^\star) = 0$ whenever $u \in \mathcal{Q}$; for each remaining e, there is at most one u in $\mathcal{Q}$ such that $b(e, D_u^\star) = 1$; Algorithm 6.13 provides an efficient way of evaluating all $b(e, D_u^\star)$.

Having transformed (9.8) into an echelon form, we may find that its solution space has a large dimension, which makes enumerating all solutions out of the question. If that is the case, then we resort to generating its solutions one by one, and testing each of them for representability, until our patience runs out. Here, the chances of coming across a representable solution can be improved by a simple trick. Let us say that a solution ξ of (9.8) *majorizes* another solution ξ' if $\xi_u = 1$ whenever $\xi'_u = 1$. If a representable solution ξ of (9.8) majorizes another solution ξ', then ξ' is also representable. In particular, if (9.8) has any representable solution at all, then it has a representable solution which is *minimal* in the sense that it majorizes no other solution of (9.8). This observation leads us to replace each new solution ξ that we generate with a minimal solution ξ' majorized by ξ and then test ξ' rather than ξ for representability.

In our computer code, the dimension of the solution space of (9.8) is considered large if it is at least six; our patience runs out when we have generated a solution ξ of (9.8), reduced ξ to a minimal solution ξ', and tested ξ' for representability 50 times; both of these thresholds have been chosen arbitrarily at least to some extent.

Led by the belief that random samples from the uniform distribution over all minimal solutions are better than biased samples, we use randomization in generating each ξ as well as in the subsequent reduction of ξ into a ξ'. To generate ξ, we assign zero-one values to the free variables at random and solve for the remaining variables. To reduce ξ to ξ', we remove all the variables whose current values are zero, transform the resulting system into an echelon form again if necessary, and repeat the whole process (assign zero-one values to the free variables at random, solve for the remaining variables, remove all the variables whose current values are zero, and transform the resulting system into an echelon form again if necessary) as long as any free variables remain.

9.1.2 Testing solutions of (9.8) for representability

For each Q-node u of T^*, with children $u_1, u_2, \ldots, u_k$, we say that the $k+1$ tight sets

$$V - D(u), \; D(u_1), \; D(u_2), \; \ldots, \; D(u_k)$$

are *the beads of the necklace defined by* u.

LEMMA 9.1 *Let* $V_1, \ldots, V_r$ *be the beads of the necklace defined by a Q-node* v *of* T^* *and let* $W_1, \ldots, W_s$ *be the beads of the necklace defined by another Q-node* w *of* T^*. *Then*

(i) *there are unique subscripts* i *and* j *such that* $V_i \cup W_j = V$,

(ii) *if* $D_v \in \mathcal{D}(v)$ *and* $D_w \in \mathcal{D}(w)$, *then*

$$V(D_v) \cap V(D_w) = \emptyset \quad \text{if and only if} \quad V(D_v) \cap V_i = \emptyset, \; V(D_w) \cap W_j = \emptyset.$$

Proof. (i): If neither v nor w is an ancestor of the other, then

$$(V - D(v)) \cup (V - D(w)) = V.$$

If v is an ancestor of w, then some child v_i of v is an ancestor of w (possibly $v_i = w$) and

$$D(v_i) \cup (V - D(w)) = V.$$

This establishes the existence of i and j. Their uniqueness follows easily from the fact that the beads of each necklace are at least three, pairwise disjoint, and nonempty; here, no references to the PQ-tree are necessary.

(ii): If $V(D_v) \cap V_i = \emptyset$ and $V(D_w) \cap W_j = \emptyset$ then, as $V_i \cup W_j = V$, we have $V(D_v) \subseteq W_j - V_i$, $V(D_w) \subseteq V_i - W_j$, and so $V(D_v) \cap V(D_w) = \emptyset$. Conversely, if $V(D_v) \cap V(D_w) = \emptyset$, then $V(D_v)$ is disjoint from at least two of the beads $W_1, \ldots, W_s$; since all of these beads except W_j are contained in V_i, it follows that $V_i \not\subseteq V(D_v)$, and so $V(D_v) \cap V_i = \emptyset$; the same argument with v and w switched shows that $V(D_w) \subseteq V_i - W_j$. $\qquad\square$

Lemma 9.1 suggests a polynomial-time algorithm for recognizing representable solutions ξ of (9.8) that goes like this: For each pair v, w of Q-nodes of $T^\star$ such that $\xi_v = 1, \xi_w = 1$, let beads V_i, W_j be as in the lemma, and mark both of these beads as taboo. The lemma guarantees that ξ is representable if and only if, for each Q-node u of $T^\star$ such that $\xi_u = 1$, some two consecutive beads in the necklace defined by u remain unmarked.

Let us describe the implementation of this idea that is used in our computer code. To begin, consider an arbitrary Q-node u, with children $u_1, \ldots, u_k$; in the necklace defined by u, associate each bead $D(u_i)$ with the corresponding child u_i and refer to the remaining bead $V - D(u)$ as the *outer* bead. In these terms, the taboo rules can be restated as follows:

- a child v_i of a node v with $\xi_v = 1$ is taboo if and only if some node w with $\xi_w = 1$ is a descendant of v_i.

- the outer bead of a node w with $\xi_w = 1$ is taboo if and only if some node v with $\xi_v = 1$ is not a descendant of v.

(Here, as usual, every node is considered to be its own descendant.)

For each node u of $T^\star$, let $d(u)$ denote the number of descendants v of u such that v is a Q-node with $\xi_v = 1$; let t denote the number of all Q-nodes v with $\xi_v = 1$. We first compute all these numbers in a single recursive sweep through $T^\star$. Then we scan the list of of all Q-nodes v with $\xi_v = 1$; given each item v on this list, we either select a tight domino D_v in $\mathcal{D}(v)$ or return a failure message indicating that ξ is not representable; if we complete the scan without returning the failure message, then all the tight dominoes that we have selected are pairwise vertex disjoint.

To elaborate on the selection of D_v, let $v_1, v_2, \ldots, v_k$ denote the children of v in the order defined by $T^\star$. If v has two consecutive children, v_j and v_{j+1}, with $d(v_j) = d(v_{j+1}) = 0$, then we set $D_v = (D(v_j), D(v_{j+1}))$. If v has no such two consecutive children, then ξ is not representable unless $d(v) = t$; in the exceptional case, we may also have the option of setting $D_v = (V - D(v), D(v_j))$ with $j = 1$ or $j = k$; this option is available if and only if $d(v_j) = 0$.

9.1.3 Choices of F

The subject of Section 9.1 is a way of making choices of sets F of edges with small $\sum(x_e^* : e \in F)$ and, for each of these sets F, searching for pairwise vertex-disjoint tight dominoes $D_1, D_2, \ldots, D_t$ in $\mathcal{D}$ such that t is odd and

$$(E(D_1) - F) \cup (E(D_2) - F) \cup \ldots \cup (E(D_t) - F)$$

is a cut in $G^* - F$.

So far, our exposition has been restricted to $F = \emptyset$; now this restriction will be removed.

We handle nonempty F by a simple adjustment of the method for handling empty F. Let us recall that, with a spanning tree T of G^* selected once and for all, and for each edge e of $G^* - T$, we denoted C_e the unique circuit contained in $T + e$; in addition, let us recall that

> a set of edges of G^* is a cut in G^* if and only if
> it meets each C_e with $e \in G^* - T$ in an even number of edges.

As long as

$$F \text{ includes no edge of } T, \tag{9.9}$$

T is a spanning tree of $G^* - F$, and so

> a set of edges of $G^* - F$ is a cut in $G^* - F$ if and only if
> it meets each C_e with $e \in (G^* - F) - T$ in an even number of edges.

Our choices of F and T always satisfy (9.9); arguing as in the case of $F = \emptyset$, we are led to generate minimal solutions of the system obtained from (9.8) by removing all congruences

$$\sum(a(e, D_u^\star)\xi_u : u \in \mathcal{Q}) \equiv 0 \bmod 2 \quad (e \in F)$$

and to test these solutions for representability.

We first choose a T once and for all; then we complement this choice by several choices of F with property (9.9); since we want F with small values of $\sum(x_e^* : e \in F)$, we let T maximize $\sum(x_e^* : e \in T)$. Now on the one hand, choosing larger sets F may improve our chances of finding the desired sets of vertex-disjoint tight dominoes; on the other hand, larger sets F tend to have larger $\sum(x_e^* : e \in F)$, making it less likely that the comb inequalities we find will be violated by x^*. One idea is to sample a little of both by drawing F from a sequence $F_1, F_2, \ldots, F_t$ such that

$$F_1 \supset F_2 \supset \ldots \supset F_t$$

and

$$x_e^* \geq x_f^* \text{ whenever } e \in F_i - F_j, f \in F_j.$$

The Concorde code implements this idea as follows. We construct a sequence of subsystems of (9.8) by starting with

$$\sum(\xi_u : u \in \mathcal{Q}) \equiv 1 \bmod 2$$

and bringing in the remaining congruences,

$$\sum(a(e, D_u^\star)\xi_u : u \in \mathcal{Q}) \equiv 0 \bmod 2 \quad (e \in G^* - T),$$

one by one in a nonincreasing order of x_e^*. As soon as a new congruence arrives, the current system is transformed into an echelon form; if the newcomer made the system unsolvable, then it is deleted at once. During this process, we monitor the number k of free variables in the current system and compare it to some threshold, initialized as $\frac{2}{3}|Q|$. Whenever k drops below this threshold, we generate an appropriate number of minimal solutions of the current system as in Section 9.1.1, test them for representability as in Section 9.1.2, and see if their representations yield comb inequalities violated by x^*; then we reset the threshold to $\frac{2}{3}k$.

Applying this implementation to our 1,000-city Euclidean problem, the gap between the LP relaxation and the optimal tour is lowered to 0.040%. This is a nice improvement over the gap of 0.062% obtained with the earlier separation routines. In the 100,000-city example, the gap is improved to 0.067% by adding domino combs to the mix of cutting planes, down from a gap of 0.083%. It should be noted that for the larger problem we terminated the cutting-plane process early, due to the large number of rounds of cuts that were generated. This is a common feature of cutting-plane implementations, where a termination rule based on the relative improvement in the LP lower bound is typically used to avoid a very long search. We return to this issue in our report on computational results in Chapter 16, postponing further tests of Concorde until that point.

The idea of solving linear congruences to obtain cutting planes is discussed in a more general context in Caprara et al. [103] and Caprara et al. [104].

9.2 NONREPRESENTABLE SOLUTIONS ALSO YIELD CUTS

The only requirement on the dominoes $D_u^\star$ featured in system (9.8) is that $D_u^\star \in \mathcal{D}(u)$ for all Q-nodes u of T^*; we may always choose $D_u^\star$ so that $V(D_u^\star)$ is disjoint from $V - D(u)$. With this proviso, and with each domino $D_u^\star$ recorded as (A_u, B_u),

($\star$) the sets $\{e : |e \cap A_u| = 1, |e \cap B_u| = 1\}$ ($u \in \mathcal{Q}$) are pairwise disjoint.

Now it follows easily that every solution of (9.8) yields a cut; let us elaborate.

If ($\star$) holds and if ξ is a solution of (9.8), then $\cup(E(D_u^\star) : \xi_u = 1)$ is a cut in G^*; let H denote a set of vertices of G^* such that an edge of G^* belongs to this cut if and only if it has precisely one endpoint in H. Let us enumerate the dominoes $D_u^\star$ with $\xi_u = 1$ as $(A_1, B_1), (A_2, B_2), \ldots, (A_t, B_t)$; for each $i = 1, 2, \ldots, t$, let us set

$$
\begin{aligned}
E_i = &\ \{e : |e \cap A_i| = 1, |e \cap B_i| = 1, e \cap H \neq \emptyset, e - H \neq \emptyset\}, \\
F_i = &\ \{e : |e \cap A_i| = 1, |e \cap B_i| = 1, e \cap H = \emptyset\} \cup \\
&\ \{e : |e \cap A_i| = 1, |e \cap B_i| = 1, e - H = \emptyset\}.
\end{aligned}
$$

Now consider the system of inequalities

$$x_e \geq 0 \ (e \cap H \neq \emptyset, e - H \neq \emptyset, e \notin E_1 \cup E_2 \cup \ldots \cup E_t), \qquad (9.10)$$

$$x_e \geq 0 \ (e \in F_1 \cup F_2 \cup \ldots \cup F_t), \qquad (9.11)$$

$$\tfrac{1}{2}\eta(S, x) \geq 0 \ (S = A_i \text{ or } B_i \text{ or } A_i \cup B_i \text{ with } 1 \leq i \leq t), \qquad (9.12)$$

and let $a^T x - b$ denote the sum of the left-hand sides of all these inequalities. Trivially, $a^T x^* - b = 0$; we claim that

$$a^T x - b \geq 1 \text{ whenever } x \text{ is a tour.}$$

To justify this claim, note first that every tour x through V satisfies all inequalities (9.10), (9.11), (9.12); we are going to derive a contradiction from the assumption that some tour x satisfies all these inequalities as equations. Since x satisfies all inequalities (9.10) as equations, fact $(\star)$ guarantees that $x(H, V - H) = \sum_{i=1}^{t}\sum(x_e : e \in E_i)$; since x satisfies all inequalities (9.11) as equations, we have $\sum(x_e : e \in E_i) = x(A_i, B_i)$ for all $i = 1, 2, \ldots t$; since x satisfies all inequalities (9.11) as equations, we have $x(A_i, B_i) = 1$ for all $i = 1, 2, \ldots t$. It follows that $x(H, V - H) = t$; this is the promised contradiction since $x(H, V - H)$ is even and t is odd.

Explicitly, cut $a^T x \geq b + 1$ may be recorded as

$$\eta(H, x) + \sum_{i=1}^{t}\left(\eta(A_i \cup B_i, x) + 2\sum(x_e : e \in F_i)\right) \geq t - 1. \qquad (9.13)$$

(A combinatorial argument showing that every tour x satisfies (9.13) goes as follows. Write $I = \{i : \eta(A_i \cup B_i, x) + 2\sum(x_e : e \in F_i) = 0\}$. Since $x(H, V - H) \geq |I|$, the left-hand side of (9.13) is at least $(|I| - 2) + 2(t - |I|)$, which is at least $t - 1$ whenever $|I| < t$; if $|I| = t$, then the inequality $x(H, V - H) \geq |I|$ holds as strict since $x(H, V - H)$ is even and t is odd.)

For example, consider $V = \{0, 1, \ldots, 7\}$,

$$x_{01}^* = x_{12}^* = x_{23}^* = x_{34}^* = x_{45}^* = x_{56}^* = x_{67}^* = x_{70}^* = 0.5,$$
$$x_{04}^* = x_{15}^* = x_{26}^* = x_{37}^* = 1,$$

$\mathsf{ext} = 0$, and the PQ-tree T^* with internal nodes $\alpha, \beta, \gamma, \delta, \varepsilon$ such that

α is the root of T and its children are β and 4,
β is a Q-node with children $\gamma, \delta, \varepsilon$ in this order,
γ has children 1 and 5,
δ has children 2 and 6,
ε has children 3 and 7.

Here, the only solution of (9.8) is $\xi_\alpha = \xi_\beta = \xi_\gamma = \xi_\delta = \xi_\varepsilon = 1$ and this solution is not representable. Choosing

$D_\alpha^* = (\{1, 2, 3, 5, 6, 7\}, \{4\}),$
$D_\beta^* = (\{1, 5\}, \{2, 6\}),$
$D_\gamma^* = (\{1\}, \{5\}),$
$D_\delta^* = (\{2\}, \{6\}),$
$D_\varepsilon^* = (\{3\}, \{7\}),$

we obtain $H = \{0, 1, 4, 6, 7\}$; the resulting cut is

$$\left. \begin{aligned} \eta(\{0, 1, 4, 6, 7\}, x) & \\ + (\eta(\{1, 2, 3, 4, 5, 6, 7\}, x) + 2(x_{14} + x_{46} + x_{47})) & \\ + (\eta(\{1, 2, 5, 6\}, x) + 2(x_{16} + x_{25})) & \\ + \eta(\{1, 5\}, x) + \eta(\{2, 6\}, x) + \eta(\{3, 7\}, x) & \geq 4. \end{aligned} \right\} \qquad (9.14)$$

One could do better than that: cut (9.14) is the sum of constraints

$$
\begin{aligned}
-2x_{15} &\geq -2, \\
-2x_{26} &\geq -2, \\
x_{01} + x_{12} + x_{13} + x_{14} + x_{15} + x_{16} + x_{17} &= 2, \\
x_{02} + x_{12} + x_{23} + x_{24} + x_{25} + x_{26} + x_{27} &= 2, \\
x_{05} + x_{15} + x_{25} + x_{35} + x_{45} + x_{56} + x_{57} &= 2, \\
x_{06} + x_{16} + x_{26} + x_{36} + x_{46} + x_{56} + x_{67} &= 2,
\end{aligned}
$$

and inequality

$$
\tilde{a}^T x \geq 12 \tag{9.15}
$$

with

$$
\begin{array}{ccccccc}
\tilde{a}_{01} = 2, & \tilde{a}_{02} = 3, & \tilde{a}_{03} = 3, & \tilde{a}_{04} = 1, & \tilde{a}_{05} = 3, & \tilde{a}_{06} = 2, & \tilde{a}_{07} = 2, \\
& \tilde{a}_{12} = 1, & \tilde{a}_{13} = 3, & \tilde{a}_{14} = 3, & \tilde{a}_{15} = 1, & \tilde{a}_{16} = 2, & \tilde{a}_{17} = 2, \\
& & \tilde{a}_{23} = 2, & \tilde{a}_{24} = 2, & \tilde{a}_{25} = 2, & \tilde{a}_{26} = 1, & \tilde{a}_{27} = 3, \\
& & & \tilde{a}_{34} = 2, & \tilde{a}_{35} = 2, & \tilde{a}_{36} = 3, & \tilde{a}_{37} = 1, \\
& & & & \tilde{a}_{45} = 2, & \tilde{a}_{46} = 3, & \tilde{a}_{47} = 3, \\
& & & & & \tilde{a}_{56} = 1, & \tilde{a}_{57} = 3, \\
& & & & & & \tilde{a}_{67} = 2,
\end{array}
$$

which is satisfied by all tours x through V (in fact, it induces a facet of their set); Naddef and Rinaldi dubbed it the *chain inequality*.

This example illustrates the reason why our computer code dismisses all nonrepresentable solutions of (9.8) that it finds.

On the one hand, the readily available cuts (9.13) may be weak; in addition, these cuts may become quite unwieldy when converted into a hypergraph form. (Our computer code converts each of these cuts into hypergraph form before adding it to the set of constraints of the LP relaxation; for example, (9.14) would be converted into the form

$$
\eta(\{1,2,3,5,6,7\},x) + \eta(\{2,3,4,5\},x) + \eta(\{3,7\},x) + \eta(\{1,2\},x) +
$$
$$
\eta(\{5,6\},x) + \eta(\{1,5\},x) + \eta(\{1,5\},x) + \eta(\{2,6\},x) + \eta(\{2,6\},x) \geq 6.)
$$

On the other hand, searching for stronger cuts (in particular, decomposing cuts (9.13) into nonnegative linear combinations of facet-inducing inequalities and degree equations $x(\delta(u)) = 2$, as we have done in our example) would require an additional effort that we were not willing to invest.

9.3 DOMINO-PARITY INEQUALITIES

The methods described thus far in the chapter are heuristics, in the sense that the algorithms may fail to deliver cutting planes even when there exist comb inequalities violated by 1.0, the maximum possible violation when x^* satisfies all subtour constraints. In an interesting special case, however, Fleischer and Tardos [177] extend the domino techniques to develop an algorithm that is guaranteed to find a comb if one with violation 1.0 is present.

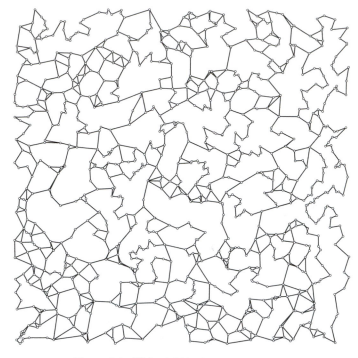

Figure 9.1 G^* for 1,000-city Euclidean TSP.

The case handled by Fleischer and Tardos is when the support graph G^* is planar. Although this assumption is restrictive, when G^* arises in a Euclidean TSP
instance it is often very close to satisfying the planarity condition. In Figure 9.1
we display G^* obtained after the addition of domino combs in our usual 1,000-city
example. As can be seen, the drawing with vertices placed at the coordinates of the
corresponding cities is nearly a planar embedding of the graph.

Fleischer and Tardos pursue the theme of domino-based cuts with a deep analysis
of the properties of necklaces and beads, combined with a separation method based
on planar duality. In our discussion, we assume G^* is given with an embedding
in the plane and that x^* satisfies all subtour constraints. We denote by $\bar{G}^*$ the
planar dual of G^*. The graph $\bar{G}^*$ has a vertex for each face of G^* and an edge $\bar{e}$
corresponding to each $e \in E^*$, with the ends of $\bar{e}$ being the vertices associated with
the faces of G^* incident with e. For any set of edges $F \subseteq E^*$, we denote by $\bar{F}$ the
corresponding edges in $\bar{G}^*$.

A key observation is that tight dominoes have an easily recognized structure in
$\bar{G}^*$. To begin, we first note that if $S \subseteq V$ is a tight set, that is, $x^*(\delta(S)) = 2$,
then the subgraph $G^*[S]$ induced by S must be connected. Indeed, if $G^*[S]$ is
not connected, then S can be partitioned into two nonempty sets S_1 and S_2 with
$x^*(S_1, S_2) = 0$. Since x^* satisfies all subtour constraints, we have

$$x^*(\delta(S)) = x^*(\delta(S_1)) + x^*(\delta(S_2)) \geq 4, \qquad (9.16)$$

contradicting the assumption that S is tight. It follows that $G^*[V - S]$ is also

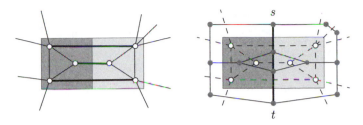

Figure 9.2 A domino in the planar dual.

connected, since $x^*(\delta(S)) = x^*(\delta(V - S))$.

A standard result in graph planarity states that a set of dual edges $\bar{F}$ is a circuit in $\bar{G}^*$ if and only if F is an inclusionwise minimal cut in G^*. In other words, $\bar{F}$ is a circuit if and only if there exists a proper subset S of V such that $G^*[S]$ and $G^*[V - T]$ are connected and $F = \delta_{G^*}(S)$.

Now consider a tight domino (A, B). It follows from the above discussion that each of the graphs

$$G^*[A \cup B], \; G^*[V - (A \cup B)], \; G^*[A], \; G^*[V - A], \; G^*[B], \; G^*[V - B]$$

is connected, and thus $\delta_{G^*}(A \cup B), \delta_{G^*}(A), \delta_{G^*}(B)$ correspond to circuits in $\bar{G}^*$. Let $E^*(A, B)$ denote the edge set of the domino in G^*, that is,

$$E^*(A, B) = \{e \in E^* : e \cap A \neq \emptyset, e \cap B \neq \emptyset\}.$$

The corresponding edges $\overline{E^*(A, B)}$ in $\bar{G}^*$ form a path joining two vertices on the circuit corresponding to $\delta_{G^*}(A \cup B)$. Such a circuit and *domino path* are illustrated in Figure 9.2, where the path meets the circuit in vertices s and t.

Fleischer and Tardos [177] exploit this dual structure of tight dominoes by representing the domino paths as edges in a graph G^d having vertices corresponding to the faces of G^*. The edge representing the domino path $\overline{E^*(A, B)}$ joins the vertices s and t determined by the start and end of the path. The Fleischer-Tardos algorithm then proceeds by searching for a circuit in G^d having an odd number of edges; by planar duality this gives a cut in G^* forming the handle of a violated comb. Here we are skipping many details of the algorithm, and we refer the reader to the original paper [177] for further information.

DOMINO-PARITY INEQUALITIES

The framework of Fleischer and Tardos is adopted in an important study by Letchford [347], providing an exact planar separation algorithm for a generalization of comb inequalities. Letchford uses dominoes as building blocks in his work, both in defining a new class of inequalities and in the main steps of the separation algorithm.

A collection of sets $E_1, \ldots, E_k \subseteq E$ is said to *support* $\delta(S)$ for some $S \subseteq V$ if

$$\delta(S) = \{e \in E : e \text{ is in an odd number of } E_1, \ldots, E_k\}.$$

Consider an odd number of dominoes

$$\mathcal{D} = \{(A_1, B_1), \ldots, (A_r, B_r)\}$$

and a set of edges $F \subseteq E$, such that $F, E(A_1, B_1), \ldots, E(A_r, B_r)$ supports $\delta(H)$ for some $H \subseteq V$. The *domino-parity inequality* corresponding to $\mathcal{D}$ and F is

$$\sum_{j=1}^{r} x(\delta(A_j \cup B_j)) + \sum_{j=1}^{r} x(A_j, B_j) + x(F) \geq 3r + 1; \qquad (9.17)$$

the set H is called the handle of the inequality; we permit the case where $H = \emptyset$. Note that no assumptions are made concerning the intersections of H and the dominoes $\mathcal{D}$. Indeed, for any specified H' we can define F' as the set of all edges such that

$e \in \delta(H')$ and e is contained in an even number of sets $E(A_j, B_j), j = 1, \ldots, r$

or

$e \notin \delta(H')$ and e is contained in an odd number of sets $E(A_j, B_j), j = 1, \ldots, r$

and create a domino-parity constraint with dominoes $\mathcal{D}$ and handle H'.

THEOREM 9.2 *Let* $(A_1, B_1), \ldots, (A_r, B_r)$ *be an odd number of dominoes and let* $F \subseteq E$ *be a set such that*

$$F, E(A_1, B_1), \ldots, E(A_r, B_r) \qquad (9.18)$$

supports $\delta(H)$ *for some* $H \subseteq V$. *Then every tour through* V *satisfies (9.17).*

Proof. We follow the argument given in Letchford [347]. Let x be the incidence vector of a tour and consider a domino (A_j, B_j). Adding the subtour constraints for A_j, B_j, and $A_j \cup B_j$ and dividing by two, we obtain the inequality

$$x(\delta(A_j \cup B_j)) + x(A_j, B_j) \geq 3;$$

combining this with $x(F) \geq 0$, we have

$$\sum_{j=1}^{r} x(\delta(A_j \cup B_j)) + \sum_{j=1}^{r} x(A_j, B_j) + x(F) \geq 3r. \qquad (9.19)$$

Setting μ_e equal to the number of sets (9.18) that contain e for each $e \in E$, inequality (9.19) can be written as

$$\sum_{j=1}^{r} x(\delta(A_j \cup B_j)) + \sum_{e \in E} \mu_e x_e \geq 3r. \qquad (9.20)$$

Rewriting (9.20) as

$$\sum_{j=1}^{r} x(\delta(A_j \cup B_j)) + x(\delta(H)) + \sum_{e \in \delta(H)} (\mu_e - 1) x_e + \sum_{e \notin \delta(H)} \mu_e x_e \geq 3r, \quad (9.21)$$

the theorem follows from the fact that each term in the left-hand side of (9.21) is even, while the right-hand side is odd. $\square$

Given a comb $\{H, T_1, \ldots, T_j\}$, for each j let $A_j = T_j \cap H$ and $B_j = T_j - H$. Setting

$$F = \delta(H) - \bigcup_{j=1}^{r} E(A_j, B_j),$$

we have that (9.17) is the comb inequality for $\{H, T_1, \ldots, T_j\}$. Thus domino-parity inequalities are a natural generalization of combs. Naddef and Wild [425] have shown that, beyond combs, a further broad subclass of domino-parity inequalities also induce facets of the TSP polytope.

LETCHFORD'S ALGORITHM

In our discussion, we again assume x^* satisfies all subtour constraints and that G^* is given with an embedding in the plane.

Consider a domino (A, B). As in the proof of Theorem 9.2, the subtour constraints imply

$$x^*(\delta(A \cup B)) + x^*(A, B) \geq 3.$$

The *weight* of (A, B) is defined as the slack in this inequality, that is,

$$w(A, B) = x^*(\delta(A \cup B)) + x^*(A, B) - 3.$$

In any violated domino-parity inequality, we must have $w(A, B) < 1$ for all dominoes (A, B); call a domino *good* if it satisfies this condition.

A domino (A, B) is called *superconnected* if each of the induced subgraphs

$$G^*[A], G^*[B], G^*[A \cup B], G^*[V - A], G^*[V - B], G^*[V - (A \cup B)]$$

is connected and

$$x^*(A, V - (A \cup B)) > 0 \text{ and } x^*(B, V - (A \cup B)) > 0.$$

Letchford [347] makes use of the following result.

LEMMA 9.3 *Each good domino is superconnected.*

Proof. First note that, for any proper subset S of V with $x^*(\delta(S)) < 4$, the graphs $G^*[S]$ and $G^*[V - S]$ must be connected. This follows from inequality (9.16) and the assumption that x^* satisfies all subtour constraints.

Let (A, B) be a good domino. We have

$$x^*(\delta(A \cup B)) + x^*(A, B) < 4. \tag{9.22}$$

Since $x^*(A, B) \geq 0$, it follows that $G^*[A \cup B]$ and $G^*[V - (A \cup B)]$ are connected. Also, since

$$x^*(\delta(A)) \leq x^*(\delta(A \cup B)) + x^*(A, B) < 4$$

and

$$x^*(\delta(B)) \leq x^*(\delta(A \cup B)) + x^*(A, B) < 4$$

it follows that $G^*[A], G^*[V - A], G^*[B]$, and $G^*[V - B]$ are all connected. Finally,

$$x^*(A, V - (A \cup B)) + x^*(A, B) = x^*(\delta(A)) \geq 2,$$

while (9.22) implies $x^*(A, B) < 2$. We therefore have $x^*(A, V - (A \cup B)) > 0$. A similar argument shows $x^*(B, V - (A \cup B)) > 0$. □

It follows that good dominoes have the dual structure illustrated in Figure 9.2, namely, $\overline{\delta^*(A \cup B)}$ forms a circuit C in $\bar{G}^*$ and $\overline{E^*(A, B)}$ forms a domino path P with ends s and t on C. Here C and P together give three internally disjoint paths joining vertices s and t in $\bar{G}^*$. This observation allows Letchford to construct good dominoes using network-flow techniques, as we describe below.

Given vertices s and t in $\bar{G}^*$, three edge disjoint (s, t)-paths P_1, P_2, P_3 with minimum total weight $x^*(P_1 \cup P_2 \cup P_3)$ can be computed as a minimum-cost flow problem, having supply three at vertex s, demand three at vertex t, and all edge capacities set to one. Using the *successive shortest-path algorithm* described in Ahuja et al. [7], the flow problem can be solved by computing three shortest (s, t)-paths in the residual graph. Letchford [347] notes that since $\bar{G}^*$ is planar, the shortest-path problems can be solved in $O(n)$ time with the algorithm of Henzinger et al. [261], where $n = |V|$. If P_1, P_2, P_3 satisfy

$$x^*(P_1 \cup P_2 \cup P_3) < 4, \tag{9.23}$$

then the paths must be internally disjoint. This follows by noting that if P_1 and P_2 intersect, then $P_1 \cup P_2$ contains two edge disjoint circuits, each having weight at least two, since x^* satisfies the subtour constraints. We can therefore find a minimum-weight good domino having an (s, t)-domino path, or show that no such domino exists.

This technique allows us to find a potentially large supply $\mathcal{D}^*$ of good dominoes, by running the successive shortest-path algorithm for each pair of vertices in $\bar{G}^*$. Letchford's next observation is that if there exists a violated domino-parity inequality, then there exists one of maximum violation that uses only dominoes in $\mathcal{D}^*$. This follows from the fact that, for a specified (s, t), any two dominoes having (s, t)-domino paths are interchangeable, in the sense that we can replace one by another in a domino-parity inequality and obtain another domino-parity inequality. To see this fact, it is useful to have a dual characterization of a domino-parity inequality. A set of edges $K \subseteq E$ is called *Eulerian* if each vertex in the graph (V, K) has even degree; a collection of edges sets $K_1, \ldots, K_t$ is said to *support an Eulerian subgraph* (V, K) if

$$K = \{e \in E : e \text{ is in an odd number of } K_1, \ldots, K_t\}$$

is Eulerian. An odd number of dominoes $(A_1, B_1), \ldots, (A_r, B_r)$ and a set of edges F form a domino-parity inequality if and only if the domino paths $P_1, \ldots, P_r$ and dual edges $\bar{F}$ support an Eulerian subgraph in $\bar{G}^*$. Replacing one (s, t)-path by another (s, t)-path will not change the parity of the degree of any vertex in the subgraph, so the domino-parity conditions are satisfied with either choice of domino. To obtain the maximum violation in an inequality, it therefore suffices to consider the minimum-weight representative for (s, t)-dominoes given in $\mathcal{D}^*$.

In the final step in the separation algorithm, Letchford creates a supergraph M of $\bar{G}^*$ by adding edges $\{s, t\}$ for each domino in $\mathcal{D}^*$ having an (s, t)-domino path; the new edges are assigned the weight of the corresponding domino and each original edge $\bar{e}$ is assigned the weight x_e^*. Using the dual characterization of domino-parity inequalities, it follows that an inequality of maximum violation corresponds to minimum-weight circuit in M having an odd number of edges that correspond to dominoes. Assigning each edge in $\bar{E}^*$ the label *even* and each edge corresponding to a domino the label *odd*, the separation problem is reduced to finding a minimum-weight odd circuit in M. If the odd circuit has weight less than one, then the domino-parity inequality is violated.

This odd-circuit problem can, in turn, be reduced to a series of shortest-path problems in an expanded graph, using an argument described in Barahona and Mahjoub [39] and in Grötschel et al. [233]. The idea is to split each vertex v of

M into two vertices v_1 and v_2. For each odd edge $\{u, v\}$ we create two edges $\{u_1, v_2\}$ and $\{u_2, v_1\}$, and for each even edge $\{u, v\}$ we create two edges $\{u_1, v_1\}$ and $\{u_2, v_2\}$; each edge is given the weight that was assigned to the original edge $\{u, v\}$ in M. Now for each vertex v of M we find a shortest (v_1, v_2)-path in the new graph. Not all of these paths will correspond to circuits in M, since vertices can be repeated, but the path of least cost is a minimum odd circuit in M.

The total running time of Letchford's separation algorithm is $O(n^3)$.

PRACTICAL IMPLEMENTATION

The first computational study of Letchford's algorithm was carried out by Boyd et al. [85], demonstrating the effectiveness of domino-parity inequalities on a selection of TSPLIB examples having up to 1,040 cities. Their tests incorporate the planar separation algorithm into Concorde and measure the improvement in the LP bounds that can be achieved. This was followed by work of Cook et al. [136], implementing the separation algorithm for large-scale instances.

The practical running time of Letchford's algorithm is dominated by the time to generate the collection of dominoes $\mathcal{D}^*$. The Cook et al. study includes a number of techniques for speeding up the minimum-cost flow computations that need to be carried out for each pair of vertices (s, t). One of the main ideas is to limit the computations to a neighborhood around the specified s and t, since vertices beyond a certain distance cannot take part in the solution if the entire set of three paths is to have weight less than four. Details of this technique, including a heuristic setting to further limit the search, are given in [136].

In practical computations, it is important to generate a collection of violated inequalities, rather than a single cutting plane. Fortunately, the odd-circuit algorithm described above can produce a circuit though each vertex of M. Any such circuit of weight less than one gives a violated domino-parity inequality, and Boyd et al. propose to use the entire set as cutting planes. In the Cook et al. study, this idea is extended by adopting a heuristic algorithm to attempt to find additional inequalities. The idea is to sample the odd circuits by performing random walks starting at each vertex, where edges are selected with probability proportional to their weight.

DOMINO-SAFE SHRINKING

Letchford [347] raises the problem of determining safe-shrinking rules for domino-parity inequalities, that is, rules such that the shrunk vector x' is guaranteed to have a violated domino-parity inequality if one exists for the original vector x^*. As usual, safe shrinking is useful as a preprocessing step to reduce the size of the support graph G^*. In the context of the planar separation algorithm, however, it has the added importance of permitting exact separation for some nonplanar support graphs. Indeed, shrinking any connected subgraph maintains graph planarity, that is, if G^* is planar, then the shrunk graph G' is also planar, but the converse is not true, the shrunk graph G' may be planar even though G^* is nonplanar.

One safe-shrinking result is given by Cook et al. [136], who show that it is safe to shrink the ends of an edge $e = \{u, v\}$ if $x_e^* = 1$ and there exists a vertex w with $x^*(\{u, v\}, \{w\}) = 1$; this is the rule of Padberg and Rinaldi [446] discussed in

Section 7.4. In their computational tests, Cook et al. note that repeated application of the rule typically reduces the size of the support graph by a factor of two or more.

FINDING A PLANAR GRAPH

In their initial computational study of domino-parity inequalities, Boyd et al. [85] often encountered planar graphs when processing LP relaxations for their test instances. In the cases where a nonplanar G^* was reached, Boyd et al. halted their computation and carried out a visual inspection of the graph. To proceed further, they carried out, by hand, a sequence of shrinking steps to reduce G^* to a planar graph. The resulting vector x' may not satisfy the degree constraints

$$x(\delta(v)) = 2 \text{ for all } v \in V,$$

but domino-parity inequalities and the planar separation algorithm remain valid in this case, as discussed in Letchford [347].

Cook et al. [136] automate the Boyd et al. process, adopting a straightforward method outlined in Vella [533]. Using a planarity testing code that identifies a K_5 or $K_{3,3}$ subdivision in a nonplanar graph, if G^* is nonplanar we can select two vertices u and v having degree greater than two in a K_5 or $K_{3,3}$ subdivision. The set $\{u, v\}$ is shrunk and the process is repeated until a planar graph is obtained.

As an alternative, Cook et al. also adopt a greedy algorithm that repeatedly deletes edges until G^* is planar. To carry this out, the edges are sorted in nonincreasing value of x_e^* and binary search is used to identify the last e such that the subgraph containing all previous edges is planar; e is deleted and the binary search is repeated. Note that this method will produce a vector x' that does not satisfy the subtour constraints. This implies that the weight of the dominoes found during the domino-generation step may be negative, and this can create negative cycles in M. To avoid this issue, Cook et al. set the weight of all negative dominoes to zero and recompute at the end of the algorithm the actual violation for the inequalities that are produced. This process can cause Letchford's algorithm to overlook potential cutting planes, but in practice Cook et al. found that the total weight of the deleted edges was usually small relative to the maximum violation of the returned inequalities.

In both of the approaches, Cook et al. use J. Boyer's implementation of the planarity testing algorithm of Boyer and Myrvold [90].

Unlike the Boyd et al. [85] study, the larger instances considered by Cook et al. nearly always produced nonplanar support graphs G^*, even after the application of the domino-safe-shrinking rule. The straightforward methods to obtain planar approximations proved to be adequate to produce good cutting planes, but the approximations can certainly be improved, and this remains an important research topic.

SIMPLE DOMINO-PARITY INEQUALITIES

For a general nonplanar support graph, determining the complexity of the exact separation problem for domino-parity inequalities is an open research question.

Like the case of comb inequalities, no polynomial-time algorithm is known for the problem, and it is also not known to be NP-hard.

An interesting result regarding this separation question was obtained by Fleischer et al. [178], continuing a study of Letchford and Lodi [348]. They call a domino-parity inequality *simple* if each domino (A_j, B_j) satisfies $|A_j| = 1$ or $|B_j| = 1$; this class generalizes blossom inequalities and the original version of combs introduced in Chvátal [125], where each tooth intersects the handle in exactly one vertex. For simple domino-parity inequalities, Fleischer et al. show that the exact separation problem can be solved in time $O(n^2 m^2 \log(n^2/m))$, where n is the number of vertices and m is the number of edges.

The Fleischer et al. algorithm is a tour de force, involving a wide range of methods from graph theory, combinatorial optimization, and data structures. To date, no implementation of the method has been studied.

In the case when G^* is planar, Letchford and Pearson [352] have shown that the running time for simple domino-parity separation can be reduced to $O(n^2 \log n)$. The Letchford-Pearson study also includes fast methods for separating subtours and blossoms when the input is restricted to planar graphs.

Chapter Ten

Cut Metamorphoses

Watching a run of the cutting-plane method is like viewing a tug-of-war between the LP solver and the cut finder. When a cutting plane is added to the LP relaxation, the solver often reacts by shifting the defect in x^* prohibited by the cut to an area just beyond the cut's control. This may be one explanation for the footnote in the Dantzig et al. [150] technical report remarking that E. W. Paxson called their procedure the "finger in the dike" method.

One way to deal with the shifting LP solution is to respond to each slight adjustment of x^* with slight adjustments of our cuts, stretching our hand a bit to patch up the next problem in the dike.

Consider the portion of a TSP x^*-vector illustrated in Figure 10.1. The corresponding LP solution arose in a computation on the TSPLIB instance pcb442; each value x_e^* is either 0, 0.5, or 1.0 and, as usual, we only display the edges taking on nonzero values.

A violated 3-tooth comb inequality for the vector in Figure 10.1 is illustrated in Figure 10.2. The handle of the comb is the set H indicated by the large pentagon; one tooth is the complement of the set A of 4 vertices indicated by the rectangle; the other two teeth, B and C, are the indicated 2-vertex sets. We have

$$x^*(\delta(H)) = 3.0 \text{ and } x^*(\delta(A)) = x^*(\delta(B)) = x^*(\delta(C)) = 2.0,$$

so x^* does indeed violate the comb inequality

$$x(\delta(H)) + x(\delta(A)) + x(\delta(B)) + x(\delta(C)) \geq 10. \tag{10.1}$$

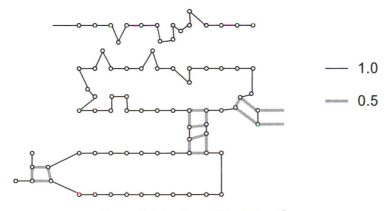

—— 1.0

━ ━ 0.5

Figure 10.1 Portion of LP solution x^*.

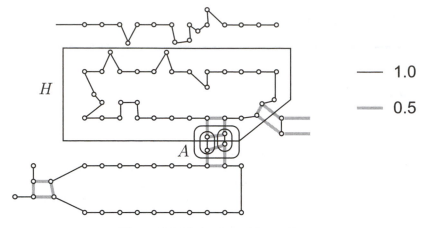

Figure 10.2 Violated comb inequality.

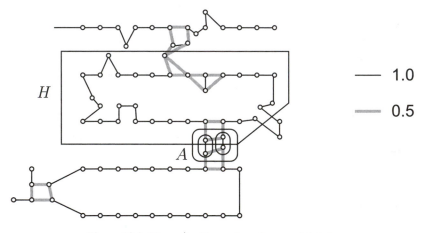

Figure 10.3 New x^* with comb no longer violated.

Adding (10.1) to our LP relaxation and resolving, we obtain the new x^* illus-trated in Figure 10.3. The shift in x^* increases $x^*(\delta(H))$ to 5.0, while keeping each of the teeth at value 2.0. Although the comb inequality is no longer violated (we knew this would be true, since it is part of the LP relaxation), notice that it is easy to alter the comb by enlarging the handle to once again point out a defect in x^*. This new violated comb is illustrated in Figure 10.4.

With the adjusted comb in our LP relaxation, the LP solver returns the x^* indi-cated in Figure 10.5. Again, the shift in the defect in x^* can be tracked down by a modification of our comb inequality—in this case by extending the sets for the teeth out along the paths to reach the edges having value $x_e = 0.5$.

Procedures for automating these types of cut adjustments are the subject of this chapter. We found in our computations that these methods provide a very useful alternative to deriving new cuts from scratch, and they are an essential part of the

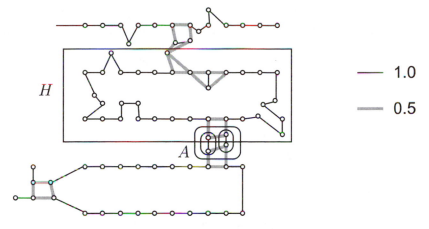

Figure 10.4 Adjusted comb inequality.

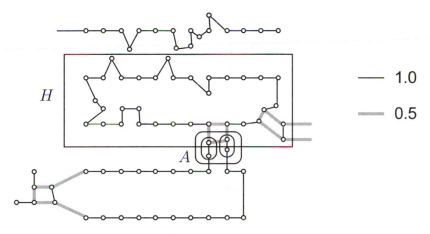

Figure 10.5 New x^* satisfying adjusted comb inequality.

Concorde computer code.

10.1 TIGHTEN

To describe our cut-alteration procedures, it will be convenient to adopt the hypergraph notation we introduced in Section 5.8. Recall that a hypergraph is an ordered pair $(V, \mathcal{F})$ such that $\mathcal{F}$ is a family of subsets of V called the edges of the hypergraph. Given a hypergraph $\mathcal{H} = (V, \mathcal{F})$, a corresponding hypergraph inequality is

$$\mathcal{H} \circ x \geq t$$

with

$$\mathcal{H} \circ x = \sum (x(\delta(S)) : S \in \mathcal{F})$$

and $t \leq \mu(\mathcal{H})$, where $\mu\mathcal{H})$ is the minimum of $\mathcal{H} \circ x$ taken over the incidence vectors of tours through V. All cutting planes in Concorde, other than domino-parity inequalities, are stored as hypergraph inequalities.

Let $E_1, E_2, \ldots E_m$ be the edges of $\mathcal{H}$. For each subset I of $\{1, 2, \ldots, m\}$, we set

$$\alpha(I, \mathcal{H}) = \bigcap_{i \in I} E_i - \bigcup_{i \notin I} E_i$$

as in Section 5.8; we refer to each nonempty $\alpha(I, \mathcal{H})$ as an atom of $\mathcal{H}$. As we noted earlier, the atoms are the regions in a Venn diagram representation of $\mathcal{H}$.

We write $\mathcal{H} \sqsubseteq \mathcal{H}'$ to signify that $\mathcal{H}$ and $\mathcal{H}'$ are hypergraphs with the same set of vertices and the same number of edges such that $\alpha(I, \mathcal{H}') \neq \emptyset$ whenever $\alpha(I, \mathcal{H}) \neq \emptyset$; it is not difficult to see that

$$\mathcal{H} \sqsubseteq \mathcal{H}' \text{ implies } \mu(\mathcal{H}') \geq \mu(\mathcal{H}).$$

By *tightening* a hypergraph $\mathcal{H}_0$, we mean attempting to modify $\mathcal{H}_0$ in such a way that the resulting hypergraph, $\mathcal{H}$, satisfies

$$\mathcal{H}_0 \sqsubseteq \mathcal{H} \quad \text{and} \quad \mathcal{H} \circ x^* < \mathcal{H}_0 \circ x^*. \tag{10.2}$$

Here, "attempting" and "modify" are the operative words: by tightening, we do *not* mean finding a solution $\mathcal{H}$ of (10.2). Rather, we mean a swift and not necessarily exhaustive search for a solution $\mathcal{H}$ of (10.2) such that each edge of $\mathcal{H}$ is either identical with the corresponding edge of $\mathcal{H}_0$, or it differs from it in just a few elements.

When the edges of $\mathcal{H}$ are $E_1, \ldots E_m$ and the edges of $\mathcal{H}'$ are $E'_1, \ldots E'_m$, we write $\mathcal{H}' \approx \mathcal{H}$ to signify that there is a subscript j such that E_i, E'_i are identical whenever $i \neq j$ and differ in precisely one element when $i = j$. Our starting point for tightening a prescribed hypergraph $\mathcal{H}_0$ is the *greedy search* specified in Algorithm 10.1.

Algorithm 10.1 Greedy search.

$\mathcal{H} = \mathcal{H}_0$;
repeat $\mathcal{H}' = $ hypergraph that minimizes $\mathcal{H}' \circ x^*$
      subject to $\mathcal{H}_0 \sqsubseteq \mathcal{H}'$ and $\mathcal{H}' \approx \mathcal{H}$;
      if $\mathcal{H}' \circ x^* < \mathcal{H} \circ x^*$ **then** $\mathcal{H} = \mathcal{H}'$ **else** return $\mathcal{H}$ **end**
end

One trouble with greedy search is that it terminates as soon as it reaches a local minimum, even though a better solution may be just around the corner. One remedy is to continue the search even when $\mathcal{H}' \circ x^* \geq \mathcal{H} \circ x^*$; now the best $\mathcal{H}$ found so far is not necessarily the current $\mathcal{H}$, and so it has to be stored and updated; furthermore, measures that make the search terminate have to be imposed. In our variation on this theme, the search terminates as soon as $\mathcal{H}' \circ x^* > \mathcal{H} \circ x^*$ (we prefer missing a solution of (10.2) to spending an inordinate amount of time in the search and letting

$\mathcal{H}$ stray far away from $\mathcal{H}_0$), but it may go on when $\mathcal{H}' \circ x^* = \mathcal{H} \circ x^*$. We don't have to worry about storing and updating the best $\mathcal{H}$ found so far (this $\mathcal{H}$ is our current $\mathcal{H}$), but we still have to ensure that the modified search terminates.

Consider a vertex v of $\mathcal{H}$ and an edge E_i of $\mathcal{H}$. The *move* (v, i) is the replacement of E_i by $E_i \cup \{v\}$ in case $v \notin E_i$ and by $E_i - \{v\}$ in case $v \in E_i$. In this terminology, $\mathcal{H}' \approx \mathcal{H}$ if and only if $\mathcal{H}'$ can be obtained from $\mathcal{H}$ by a single move; with $\mathcal{H} \oplus (v, i)$ standing for the hypergraph obtained from $\mathcal{H}$ by move (v, i) and with

$$\Delta(v, i) = \begin{cases} x^*(\{v\}, E_i) - 1 & \text{if } v \notin E_i, \\ 1 - x^*(\{v\}, E_i) & \text{if } v \in E_i, \end{cases}$$

we have

$$(\mathcal{H} \oplus (v, i)) \circ x^* = \mathcal{H} \circ x^* - 2\Delta(v, i).$$

In each iteration of our modified greedy search, we find a move (v, i) that maximizes $\Delta(v, i)$ subject to $\mathcal{H}_0 \sqsubseteq \mathcal{H} \oplus (v, i)$. If $\Delta(v, i) > 0$, then we substitute $\mathcal{H} \oplus (v, i)$ for $\mathcal{H}$; if $\Delta(v, i) < 0$, then we return $\mathcal{H}$; if $\Delta(v, i) = 0$, then we either substitute some $\mathcal{H} \oplus (w, j)$ with $\mathcal{H}_0 \sqsubseteq \mathcal{H} \oplus (w, j)$ and $\Delta(w, j) = 0$ for $\mathcal{H}$ or return $\mathcal{H}$. More precisely, we classify all the moves (v, i) with $\Delta(v, i) = 0$ into three categories by assigning to each of them a number $\chi(v, i)$ in $\{0, 1, 2\}$. If

$$\max\{\Delta(v, i) : \mathcal{H}_0 \sqsubseteq \mathcal{H} \oplus (v, i)\} = 0,$$

then we find a move (w, j) that maximizes $\chi(w, j)$ subject to $\mathcal{H}_0 \sqsubseteq \mathcal{H} \oplus (w, j)$ and $\Delta(w, j) = 0$; if $\chi(w, j) > 0$, then we substitute $\mathcal{H} \oplus (w, j)$ for $\mathcal{H}$; if $\chi(w, j) = 0$, then we return $\mathcal{H}$.

To describe this policy in different terms, let us write

$$(\Delta_1, \chi_1) \prec (\Delta_2, \chi_2) \text{ to mean that } \Delta_1 < \Delta_2 \text{ or else } \Delta_1 = \Delta_2 = 0, \chi_1 < \chi_2.$$

(Note that this $\prec$ is a partial order similar to, but not identical with, the lexicographic order on all the pairs (Δ, χ): in $\prec$, the second component of (Δ, χ) is used to break ties only if the first component equals zero.) In each iteration of our modified greedy search, we find a move (v, i) that

$$\text{maximizes } (\Delta(v, i), \chi(v, i)) \text{ subject to } \mathcal{H}_0 \sqsubseteq \mathcal{H} \oplus (v, i)$$

in the sense that no move (w, j) with $\mathcal{H}_0 \sqsubseteq \mathcal{H} \oplus (w, j)$ has $(\Delta(v, i), \chi(v, i)) \prec (\Delta(w, j), \chi(w, j))$. If $(0, 0) \prec (\Delta(v, i), \chi(v, i))$, then we substitute $\mathcal{H} \oplus (v, i)$ for $\mathcal{H}$; if $(\Delta(v, i), \chi(v, i)) \preceq (0, 0)$, then we return $\mathcal{H}$.

The values of $\chi(v, i)$ are defined in Algorithm 10.2.

This definition of $\chi(v, i)$ is motivated by three objectives:

(i)   to make the search terminate,

(ii) to steer the search toward
      returning a hypergraph with relatively small edges,

(iii) to improve the chances of
      finding a hypergraph $\mathcal{H}$ with $\mathcal{H} \circ x^* < \mathcal{H}_0 \circ x^*$.

Algorithm 10.2 Tightening $\mathcal{H}_0$.

---

    for   all vertices v and all $i = 1, 2, \ldots, m$
    do   **if** $v \in E_i$ **then** $\chi(v, i) = 1$ **else** $\chi(v, i) = 2$ **end**
    end
    $\mathcal{H} = \mathcal{H}_0$;
    repeat $(v, i) =$ move that
            maximizes $(\Delta(v, i), \chi(v, i))$ subject to $\mathcal{H}_0 \sqsubseteq \mathcal{H} \oplus (v, i)$;
            if $(\Delta(v, i), \chi(v, i)) \succ (0, 0)$
            then   **if** $\chi(v, i) = 1$
                    then if $\Delta(v, i) = 0$
                          then $\chi(v, i) = 0$;
                          else $\chi(v, i) = 2$;
                    end
                else $\chi(v, i) = 1$;
                end
                $\mathcal{H} = \mathcal{H} \oplus (v, i)$;
            else   return $\mathcal{H}$;
            end
    end

To discuss these three items, let S denote the sequence of moves made by the search. Each move (v, i) in S is either an *add*, $E_i \mapsto E_i \cup \{v\}$, or a *drop*, $E_i \mapsto E_i - \{v\}$; each move (v, i) in S is either *improving*, with $\Delta(v, i) > 0$, or *nonimproving*, with $\Delta(v, i) = 0$.

For an arbitrary but fixed choice of v and i, let S^* denote the subsequence of S that consists of all the terms of S that equal (v, i). Trivially, adds and drops alternate in S^*; hence the definition of $\chi(v, i)$ guarantees that S^* does not contain three consecutive nonimproving terms; since S^* contains only finitely many improving terms, it follows that S^* is finite; in turn, as v and i were chosen arbitrarily, it follows that S is finite.

In S^*, a nonimproving drop cannot be followed by a nonimproving add (which has $\Delta = 0, \chi = 0$), but a nonimproving add can be followed by a nonimproving drop (which has $\Delta = 0, \chi = 1$). This asymmetry pertains to objective (ii): our search returns a hypergraph $\mathcal{H}$ such that, for all choices of v and i with $v \in E_i$, we have $\Delta(v, i) < 0$ or else $\mathcal{H}_0 \not\sqsubseteq \mathcal{H} \oplus (v, i)$.

The algorithm prefers nonimproving adds with $\chi > 0$ (these have $\chi = 2$) to nonimproving drops with $\chi > 0$ (these have $\chi = 1$). This asymmetry pertains to objective (iii): its role is to offset the asymmetry introduced by objective (ii). When no improving move is immediately available, we let edges of $\mathcal{H}$ shift by nonimproving moves in the hope of eventually discovering an improving move. Nonimproving adds that lead nowhere can get undone later by nonimproving drops, after which additional nonimproving drops may lead to a discovery of an improving move. However, nonimproving drops that lead nowhere cannot get undone later by nonimproving adds, and so they may forbid a sequence of nonimproving adds leading to a discovery of an improving move.

The importance of objective (ii) comes from the facts that (a) the LP solver works faster when its constraint matrix gets sparser and (b) cuts defined by hypergraphs with relatively small edges tend to have relatively small numbers of nonzero coefficients. (Part (b) of this argument could be criticized on the grounds that a hypergraph cut is invariant under complementing an edge and so, in problems with n cities, hypergraph edges of size k are just as good as edges of size $n - k$. However, this criticism is just a quibble: in hypergraphs that we use to supply cuts, edges tend to have sizes far smaller than $n/2$.)

To implement the instruction

$$(v, i) = \text{move that}$$
$$\text{maximizes } (\Delta(v, i), \chi(v, i)) \text{ subject to } \mathcal{H}_0 \sqsubseteq \mathcal{H} \oplus (v, i),$$

we maintain

- a priority queue $\mathcal{Q}$ of all pairs (w, j) such that
  $(\Delta(w, j), \chi(w, j)) \succ (0, 0)$ and $\mathcal{H}_0 \sqsubseteq \mathcal{H} \oplus (w, j)$;

to speed up the update $\mathcal{Q}$ after each iteration, we maintain a number of auxiliary objects. Call vertices v and w *neighbors* if $x_{vw}^* > 0$ and write

$$V^*(\mathcal{H}) = \{w : w \text{ belongs to or has a neighbor in an edge of } \mathcal{H}\}.$$

We keep track of

- a superset V^* of $V^*(\mathcal{H})$

initialized as $V^*(\mathcal{H}_0)$ and updated by $V^* = V^* \cup V^*(\mathcal{H})$ after each iteration; for each w in V^*, we maintain

- the values of $(\Delta(w, j), \chi(w, j))$ in an array of length m.

(Note that each w outside V^* has $(\Delta(w, j), \chi(w, j)) = (-1, 2)$ for all j.) In addition, we store

- the family $\mathcal{A}$ of all sets I such that $\alpha(I, \mathcal{H}_0)$ is an atom of $\mathcal{H}_0$;

for each I in $\mathcal{A}$, we maintain

- a doubly linked list $A(I)$ that holds the elements of $\alpha(I, \mathcal{H}) \cap V^*$

(the only I in $\mathcal{A}$ that may have $\alpha(I, \mathcal{H}) \not\subseteq V^*$ is the empty set, which labels the exterior atom of $\mathcal{H}$); for each w in V^*, we define $I(w) = \{j : w \in E_j\}$ and maintain

- a pointer $a(w)$ to $A(I(w))$, with $a(w) = \text{NULL}$ if $I(w) \notin \mathcal{A}$.

Initializing all this information and updating it after each iteration is a routine task; we will discuss its details only for the sake of clarity.

Algorithm 10.3 INSERT(w).

$V^* = V^* \cup \{w\}$;
for $j = 1, 2, \ldots m$ **do** $\Delta(w, j) = -1, \chi(w, j) = 2$ **end**

Inserting a new vertex w into V^*—with $\Delta(w, j)$ and $\chi(w, j)$ set as if w remained outside V^*—is a simple operation; for convenience, we set it apart as function INSERT(w) defined in Algorithm 10.3. The membership of a move (w, j) in $\mathcal{Q}$ may change after each change of $\mathcal{H}$. Algorithm 10.4 defines a function MEMBERSHIP(w, j), that, given a move (w, j) such that $w \in V^*$, inserts (w, j) into $\mathcal{Q}$ or deletes it from $\mathcal{Q}$ as necessary. The initialization defined in Algorithm 10.5 replaces the first four lines of Algorithm 10.2; the update defined in Algorithm 10.6 replaces the line

$$\mathcal{H} = \mathcal{H} \oplus (v, i);$$

of Algorithm 10.2; the current $\mathcal{H}$ is represented by χ since

$$E_j = \{w \in V^* : \chi(w, j) = 1\}.$$

Algorithm 10.4 MEMBERSHIP(w, j).

---

    if $a(w)$ points to a list that includes only one item
    then if $(w, j) \in \mathcal{Q}$ **then** delete (w, j) from $\mathcal{Q}$ **end**
    else **if** $(\Delta(w, j), \chi(w, j)) \succ (0, 0)$
           then if $(w, j) \notin \mathcal{Q}$ **then** insert (w, j) into $\mathcal{Q}$ **end**
           else if $(w, j) \in \mathcal{Q}$ **then** delete (w, j) from $\mathcal{Q}$ **end**
           end
    end

We apply the tightening procedure in our code in two ways. First, we scan the cuts that we currently have in our LP and try tightening each of them in turn. Second, if a scan of the list of inequalities we maintain in a pool (inequalities that have at one time been added to the LP but may no longer be present; see Section 12.1), does not produce sufficiently many cuts, then we try tightening each one that is within some fixed tolerance ε of being violated by the current x^* (we use $\varepsilon = 0.1$).

10.2 TEETHING

The Grötschel-Holland heuristic described in Section 7.1 builds a blossom inequality with a prescribed handle by attaching to this handle a set of 2-vertex teeth selected in a greedy way. Its generalization would replace the set of 2-vertex teeth of any comb inequality by an optimal set of 2-vertex teeth. Unfortunately, such a

Algorithm 10.5 Initialization.

$V^* = \emptyset$;
for $i = 1, 2, \ldots, m$
do   **for**   all v in E_i
       do   **if** $v \notin V^*$ **then** INSERT(v) **end**
           $\chi(v, i) = 1$;
           for   all neighbors w of v
           do   **if** $w \notin V^*$ **then** INSERT(w) **end**
           end
       end
end

$\mathcal{A} = \emptyset$;
for   all v in V^*
do $I = \{i : \chi(v, i) = 1\}$;
    if $I \notin \mathcal{A}$
    then add I to $\mathcal{A}$ and initialize an empty list $A(I)$;
    end
    insert v into $A(I)$ and make $a(v)$ point to $A(I)$;
end

$\mathcal{Q} = \emptyset$;
for   all v in V^*
do   **for**   all neighbors w of v
       do   **if** $w \in V^*$
           then for $i = 1, 2, \ldots m$
               do   **if** $\chi(w, i) \neq \chi(v, i)$
                   then $\Delta(v, i) = \Delta(v, i) + x^*_{vw}$;
               end
           end
       end
       for $i = 1, \ldots, m$ **do** MEMBERSHIP(v, i) **end**
end

---

procedure would have to take care to avoid teeth with nonempty intersection. Fortunately, if a 2-vertex tooth intersects another tooth in a single vertex, then an even stronger inequality can be obtained by adjusting the hypergraph (or discovering a violated subtour inequality); if a 2-vertex tooth intersects another tooth in more than a single vertex, then it must be contained in the other tooth. We can exploit this relationship in an algorithm which we refer to as *teething*.

More precisely, we say that a tooth is *big* if its size is least three, and *small* if its

Algorithm 10.6 Update.

if $a(v) \neq$ NULL
then $A =$ the list that $a(v)$ points to;
         delete v from A;
         if $|A| = 1$
         then $w =$ the unique vertex in A;
                 for $j = 1, \ldots, m$ **do** MEMBERSHIP(w, j) **end**
         end
end

$I = \{j : \chi(v, j) = 1\}$;
if $I \in \mathcal{A}$
then insert v into $A(I)$ and make $a(v)$ point to $A(I)$;
         if $|A(I)| = 2$
         then $w =$ the unique vertex in $A(I)$ other than v;
                 for $j = 1, \ldots, m$ **do** MEMBERSHIP(w, j) **end**
         end
else $a(v) =$ NULL;
end

$\Delta(v, i) = -\Delta(v, i)$;
MEMBERSHIP(v, i);

for   all neighbors w of v
do   **if** $w \notin V^*$
         then INSERT(w);
                 insert w into $A(\emptyset)$ and make $a(w)$ point to $A(\emptyset)$;
         end
         if $\chi(w, i) \equiv \chi(v, i) \bmod 2$
         then $\Delta(w, i) = \Delta(w, i) - x^*_{vw}$;
         else $\Delta(w, i) = \Delta(w, i) + x^*_{vw}$;
         end
         MEMBERSHIP(w, i);
end

---

size is two; given a comb $\mathcal{H}_0$, we set
$$\Delta(\mathcal{H}_0) = \min\{\mathcal{H} \circ x^* - \mu(\mathcal{H}) : \quad \mathcal{H} \text{ is a comb such that}$$
$$\mathcal{H} \text{ and } \mathcal{H}_0 \text{ have the same handle and}$$
$$\text{all big teeth of } \mathcal{H} \text{ are teeth of } \mathcal{H}_0\};$$
teething a comb $\mathcal{H}_0$ means finding either a comb $\mathcal{H}$ such that all big teeth of $\mathcal{H}$ are teeth of $\mathcal{H}_0$ and
$$\text{if } \Delta(\mathcal{H}_0) \leq 0 \text{ then } \mathcal{H} \circ x^* - \mu(\mathcal{H}) \leq \Delta(\mathcal{H}_0)$$

or else a subtour inequality violated by x^*.

As a preliminary step in teething, we test the input $\mathcal{H}_0$ for the property

$$x(\delta(S)) \geq 2 \text{ whenever } S \text{ is an edge of } \mathcal{H}_0; \qquad (10.3)$$

if (10.3) fails, then we have found a subtour inequality violated by x^*. The remainder of the algorithm consists of three parts.

The first part involves the notion of a *pseudocomb*, which is just like a comb except that its small teeth are allowed to intersect—but not to be contained in— other teeth. More rigorously, a pseudocomb is a hypergraph with edge set $\{H\} \cup \mathcal{T}$ such that

- if $T \in \mathcal{T}$, then $T \cap H \neq \emptyset$ and $T - H \neq \emptyset$,
- if $T_1, T_2 \in \mathcal{T}, T_1 \neq T_2$, and $|T_1| \geq 3, |T_2| \geq 3$, then $T_1 \cap T_2 = \emptyset$,
- if $T_1, T_2 \in \mathcal{T}$ and $|T_1| = 2, |T_2| \geq 3$, then $T_1 \not\subset T_2$,
- $|\mathcal{T}|$ is odd.

Given an arbitrary hypergraph $\mathcal{H}$ with edge set $\mathcal{E}$, we write

$$\nu(\mathcal{H}) = \left\{ \begin{array}{ll} 3|\mathcal{E}| - 2 & \text{if } |\mathcal{E}| \text{ is even,} \\ 3|\mathcal{E}| - 3 & \text{if } |\mathcal{E}| \text{ is odd;} \end{array} \right.$$

note that $\nu(\mathcal{H}) = \mu(\mathcal{H})$ whenever $\mathcal{H}$ is a comb. In the first part, we

(i) find a pseudocomb $\mathcal{H}_1$ that minimizes $\mathcal{H}_1 \circ x^* - \nu(\mathcal{H}_1)$
   subject to the constraints that $\mathcal{H}_1$ and $\mathcal{H}_0$ have the same handle
   and that all big teeth of $\mathcal{H}_1$ are teeth of $\mathcal{H}_0$.

Trivially, $\mathcal{H}_1 \circ x^* - \nu(\mathcal{H}_1) \leq \Delta(\mathcal{H}_0)$. If $\mathcal{H}_1 \circ x^* - \nu(\mathcal{H}_1) \geq 0$, then we give up; else we proceed to the second part. This part involves the notion of a *generalized comb*, which is just a comb without some of its teeth. More rigorously, a generalized comb is a hypergraph with edge set $\{H\} \cup \mathcal{T}$ such that

- $H \neq \emptyset$ and $H \neq V$,
- if $T \in \mathcal{T}$, then $T \cap H \neq \emptyset$ and $T - H \neq \emptyset$,
- if $T_1, T_2 \in \mathcal{T}$ and $T_1 \neq T_2$, then $T_1 \cap T_2 = \emptyset$.

In the second part, we

(ii) find either a generalized comb $\mathcal{H}_2$ such that
   all teeth of $\mathcal{H}_2$ are teeth of $\mathcal{H}_1$ and
   $\mathcal{H}_2 \circ x^* - \nu(\mathcal{H}_2) \leq \mathcal{H}_1 \circ x^* - \nu(\mathcal{H}_1)$
   or else a subtour inequality violated by x^*;

in the third part, we

(iii) find either a comb $\mathcal{H}$ such that
   all teeth of $\mathcal{H}$ are teeth of $\mathcal{H}_2$ and
   $\mathcal{H} \circ x^* - \mu(\mathcal{H}) \leq \mathcal{H}_2 \circ x^* - \nu(\mathcal{H}_2)$
   or else a subtour inequality violated by x^*.

To discuss implementations of part (i), let H denote the handle of $\mathcal{H}_0$, set

$$\mathcal{S} = \{T : |T \cap H| = 1, |T - H| = 1\},$$

and let $\mathcal{B}$ denote the set of big teeth of $\mathcal{H}_0$. In this notation, we may state the problem of finding $\mathcal{H}_1$ as

$$
\begin{aligned}
\text{minimize} \quad & \sum_{T \in \mathcal{T}} (x^*(T, V - T) - 3) \\
\text{subject to} \quad & \mathcal{T} \subseteq \mathcal{S} \cup \mathcal{B}, \\
& T_1 \not\subseteq T_2 \text{ whenever } T_1, T_2 \in \mathcal{T}, \\
& |\mathcal{T}| \equiv 1 \bmod 2.
\end{aligned}
\tag{10.4}
$$

Now let us write

$$
\mathcal{S}^* = \{e \in \mathcal{S} : x_e^* > 0\}.
$$

We claim that (10.4) has a solution $\mathcal{T}$ such that $\mathcal{T} \subseteq \mathcal{S}^* \cup \mathcal{B}$, and so the problem of finding $\mathcal{H}_1$ can be stated as

$$
\begin{aligned}
\text{minimize} \quad & \sum_{T \in \mathcal{T}} (x^*(T, V - T) - 3) \\
\text{subject to} \quad & \mathcal{T} \subseteq \mathcal{S}^* \cup \mathcal{B}, \\
& T_1 \not\subseteq T_2 \text{ whenever } T_1, T_2 \in \mathcal{T}, \\
& |\mathcal{T}| \equiv 1 \bmod 2.
\end{aligned}
\tag{10.5}
$$

To justify this claim, note that property (10.3) with $S = H$ guarantees $\mathcal{S}^* \neq \emptyset$ and consider an arbitrary solution $\mathcal{T}$ of (10.4). If some T_1 in $\mathcal{S} - \mathcal{S}^*$ belongs to $\mathcal{T}$, then $x(T_1, V - T_1) = 4 > x(T, V - T)$ for all T in $\mathcal{S}^*$; hence $\mathcal{T}$ must include a set T_2 other than T_1; since property (10.3) with $S = T_2$ guarantees $x(T_2, V - T_2) \geq 2$, the removal of T_1 and T_2 from $\mathcal{T}$ yields another solution of (10.4). Iterating this process, we arrive at the desired conclusion.

Concorde's way of solving problem (10.5) is specified in Algorithm 10.7. The initial part of this algorithm computes sets $R_0(i), R_1(i)$ with $i = 0, 1, 2, \ldots, k$ such that each $R_t(i)$ with $1 \leq i \leq k$

$$
\begin{aligned}
\text{minimizes} \quad & \sum_{T \in \mathcal{T}} (x^*(T, V - T) - 3) \\
\text{subject to} \quad & \mathcal{T} \subseteq \mathcal{S}^*, \\
& T \subset T_i \text{ whenever } T \in \mathcal{T}, \\
& |\mathcal{T}| \equiv t \bmod 2
\end{aligned}
$$

and $R_t(0)$

$$
\begin{aligned}
\text{minimizes} \quad & \sum_{T \in \mathcal{T}} (x^*(T, V - T) - 3) \\
\text{subject to} \quad & \mathcal{T} \subseteq \mathcal{S}^*, \\
& T \not\subseteq T_i \text{ whenever } T \in \mathcal{T} \text{ and } 1 \leq i \leq k, \\
& |\mathcal{T}| \equiv t \bmod 2.
\end{aligned}
$$

The i-th iteration of the last **for** loop computes sets $R_0(0), R_1(0)$ such that $R_t(0)$

$$
\begin{aligned}
\text{minimizes} \quad & \sum_{T \in \mathcal{T}} (x^*(T, V - T) - 3) \\
\text{subject to} \quad & \mathcal{T} \subseteq \mathcal{S}^* \cup \mathcal{B}, \\
& T_1 \not\subseteq T_2 \text{ whenever } T_1, T_2 \in \mathcal{T}, \\
& T \not\subseteq T_j \text{ whenever } T \in \mathcal{T} \text{ and } i < j \leq k, \\
& |\mathcal{T}| \equiv t \bmod 2.
\end{aligned}
$$

Algorithm 10.7 First part of a teething iteration.

H = the handle of $\mathcal{H}_0$;
$T_1, T_2, \ldots, T_k$ = the big teeth of $\mathcal{H}_0$;
for $i = 0, 1, \ldots k$ **do** $\rho_0(i) = 0$, $\rho_1(i) = +\infty$, $R_0(i) = R_1(i) = \emptyset$ **end**
for   all vertices u in H
do   **for**   all vertices v such that $x^*_{uv} > 0$ and $v \notin H$
        do   **if** $u, v \in T_j$ for some j **then** $i = j$ **else** $i = 0$ **end**
            $P_0 = R_0(i)$, $P_1 = R_1(i)$, $\nu_0 = \rho_0(i)$, $\nu_1 = \rho_1(i)$;
            if $\nu_1 + (1 - 2x^*_{uv}) < \nu_0$
            then $\rho_0(i) = \nu_1 + (1 - 2x^*_{uv})$, $R_0(i) = P_1 \cup \{\{u, v\}\}$;
            end
            if $\nu_0 + (1 - 2x^*_{uv}) < \nu_1$
            then $\rho_1(i) = \nu_0 + (1 - 2x^*_{uv})$, $R_1(i) = P_0 \cup \{\{u, v\}\}$;
            end
        end
end

for $i = 1, 2, \ldots, k$
do   **if** $x(T_i, V - T_i) - 3 < \rho_1(i)$
        then $R_1(i) = \{T_i\}$, $\rho_1(i) = x(T_i, V - T_i) - 3$
        end
        $P_0 = R_0(0)$, $P_1 = R_1(0)$, $\nu_0 = \rho_0(0)$, $\nu_1 = \rho_1(0)$;
        if $\nu_1 + \rho_1(i) < \nu_0 + \rho_0(i)$
        then $\rho_0(0) = \nu_1 + \rho_1(i)$, $R_0(0) = P_1 \cup R_1(i)$;
        else $\rho_0(0) = \nu_0 + \rho_0(i)$, $R_0(0) = P_0 \cup R_0(i)$;
        end
        if $\nu_0 + \rho_1(i) < \nu_1 + \rho_0(i)$
        then $\rho_0(0) = \nu_0 + \rho_1(i)$, $R_1(0) = P_0 \cup R_1(i)$;
        else $\rho_0(0) = \nu_1 + \rho_0(i)$, $R_1(0) = P_1 \cup R_0(i)$;
        end
end
$\mathcal{H}_1$ = hypergraph with edge set $\{H\} \cup R_1(0)$;

LEMMA 10.1 *Let A, B and $T_1, \ldots T_s$ be distinct sets such that*

$$T_j \subseteq (A - B) \cup (B - A) \text{ and } |T_j \cap (A - B)| = |T_j \cap (B - A)| = 1$$

whenever $1 \leq j \leq s$. Then

$$x(A \cup B, V - (A \cup B)) + x(A \cap B, V - (A \cap B)) \leq$$
$$x(A, V - A) + x(B, V - B) + \sum_{j=1}^{s} x(T_j, V - T_j) - 4s.$$

Proof. Observe that

$$\sum(x_e^* : e \subseteq A) + \sum(x_e^* : e \subseteq B) + \sum_{j=1}^{s}\sum(x_e^* : e \subseteq T_j) \le$$

$$\sum(x_e^* : e \subseteq A \cup B) + \sum(x_e^* : e \subseteq A \cap B).$$

Substituting $|S| - \frac{1}{2}x^*(S, V - S)$ for each $\sum(x_e^* : e \subseteq S)$ in this inequality yields the desired result. □

Algorithm 10.8 Second part of a teething iteration.

$H = $ the handle of $\mathcal{H}_1$;
$\mathcal{T} = $ the set of all teeth of $\mathcal{H}_1$;
for   all vertices v outside H
do   **if** v belongs to at least two sets in $\mathcal{T}$
        then $T' = $ largest set in $\mathcal{T}$ such that $v \in T'$;
             if $x(T' \cap H, V - (T' \cap H)) \ge 2$
             then $H = H \cup T'$;
                  $\mathcal{T} = \{T \in \mathcal{T} : T \nsubseteq H\}$;
             else return $T' \cap H$;
             end
        end
end
for   all vertices v in H
do   **if** v belongs to at least two sets in $\mathcal{T}$
        then $T' = $ largest set in $\mathcal{T}$ such that $v \in T'$;
             if $x(T' - H, V - (T' - H)) \ge 2$
             then $H = H - T'$;
                  $\mathcal{T} = \{T \in \mathcal{T} : T \cap H \ne \emptyset\}$;
             else return $T' - H$;
             end
        end
end
$\mathcal{H}_2 = $ the hypergraph with edge set $\{H\} \cup \mathcal{T}$;

Let us use Lemma 10.1 to show that Algorithm 10.8 maintains the following invariant:

$$x^*(H, V - H) + \sum_{T \in \mathcal{T}} x^*(T, V - T) \le \qquad\qquad (10.6)$$

$$\begin{cases} \mathcal{H}_1 \circ x^* - \nu(\mathcal{H}_1) + 3|\mathcal{T}| + 1 & \text{if } |\mathcal{T}| \text{ is odd,} \\ \mathcal{H}_1 \circ x^* - \nu(\mathcal{H}_1) + 3|\mathcal{T}| & \text{if } |\mathcal{T}| \text{ is even.} \end{cases}$$

For this purpose, consider first the change of H and $\mathcal{T}$ made by an iteration of the first **for** loop. Lemma 10.1 with $A = H$, $B = T'$, and $T_1, \ldots T_s$ the sets in $\mathcal{T}$ distinct from T' and contained in $H \cup T'$ guarantees that the left-hand side of (10.6) drops by at least $4s + 2$. If s is odd, then the right-hand side of (10.6) drops by $3(s + 1)$; if s is even, then the right-hand side of (10.6) drops by at most $3(s + 1) + 1$; in either case, the right-hand side of (10.6) drops by at most $4s + 2$. Hence all iterations of the first **for** loop maintain invariant (10.6); the same argument with $V - H$ in place of H shows that all iterations of the second **for** loop maintain invariant (10.6).

In particular, if Algorithm 10.8 gets around to constructing $\mathcal{H}_2$, then

$$\mathcal{H}_2 \circ x^* - \nu(\mathcal{H}_2) \leq \mathcal{H}_1 \circ x^* - \nu(\mathcal{H}_1).$$

Note that (10.6) and the assumption $\mathcal{H}_1 \circ x^* - \nu(\mathcal{H}_1) < 0$ guarantee that Algorithm 10.8 maintains the invariant

• $\mathcal{T} \neq \emptyset$;

trivially, it maintains the invariant

• if $T \in \mathcal{T}$, then $T \cap H \neq \emptyset$ and $T - H \neq \emptyset$.

The first **for** loop changes H and $\mathcal{T}$ so that

if $T_1, T_2 \in \mathcal{T}$ and $T_1 \neq T_2$, then $(T_1 \cap T_2) - H = \emptyset$;

the second **for** loop changes H and $\mathcal{T}$ so that

• if $T_1, T_2 \in \mathcal{T}$ and $T_1 \neq T_2$, then $T_1 \cap T_2 = \emptyset$.

To summarize, if the algorithm gets around to constructing $\mathcal{H}_2$, then $\mathcal{H}_2$ is a generalized comb with at least one tooth.

A practical variation on Algorithm 10.8 takes the conditions

$$x^*(\delta(T' \cap H)) \geq 2 \quad \text{and} \quad x^*(\delta(T' - H)) \geq 2$$

for granted: skipping the persistent tests speeds up the computations. If the resulting hypergraph $\mathcal{H}_2$ satisfies

$$\mathcal{H}_2 \circ x^* - \nu(\mathcal{H}_2) \leq \mathcal{H}_1 \circ x^* - \nu(\mathcal{H}_1),$$

then all is well and we proceed to part (iii); else we simply give up. In the latter case, we know that some big tooth T of $\mathcal{H}_1$ satisfies

$$x^*(\delta(T \cap H)) < 2 \quad \text{or} \quad x^*(\delta(T - H)) < 2;$$

the option of finding this T now remains open, even though its appeal may be marred by the fact that violated subtour inequalities can be spotted reasonably fast from scratch.

Part (iii) of teething is trivial. If $\mathcal{H}_2$ has at most two teeth, then

$$\mathcal{H}_2 \circ x^* < 2(1 + |T|),$$

and so at least one edge S of $\mathcal{H}_2$ has $x^*(S, V - S) < 2$. If the number of teeth of $\mathcal{H}_2$ is at least three and odd, then we may set $\mathcal{H} = \mathcal{H}_2$. If the number of teeth of $\mathcal{H}_2$ is at least four and even, then we may let $\mathcal{H}$ be $\mathcal{H}_2$ with an arbitrary tooth deleted: we have $\mu(\mathcal{H}) = \nu(\mathcal{H}_2) - 2$, and (10.3) guarantees that $\mathcal{H} \circ x^* \leq \mathcal{H}_2 \circ x^* - 2$; in this case, the number of distinct cuts we obtain is the number of teeth of $\mathcal{H}_2$.

In our code, as a heuristic separation algorithm, we apply teething to each comb inequality in the current LP.

10.3 NADDEF-THIENEL SEPARATION ALGORITHMS

The study of greedy-type separation algorithms by Naddef and Thienel [423] includes several variations on the theme of modifying an existing inequality in response to a change in the LP solution vector x^*. The main tools in their work are the ear decomposition and "max-back" methods described in Section 7.2. These tools are applied to grow sets S, having small cut-value $x^*(\delta(S))$, that can be used to modify a nonviolated comb to obtain violated inequalities matching other templates for the TSP.

CLIQUE TREES

One of the templates treated by Naddef and Thienel is a generalization of combs introduced by Grötschel and Pulleyblank [239]. The *intersection graph* of a hypergraph $(V, \mathcal{F})$ is the graph with vertex set $\mathcal{F}$ and with two vertices adjacent if and only if these two members of $\mathcal{F}$ intersect. A *clique tree* is any hypergraph $\mathcal{H}$ such that
• the intersection graph of $\mathcal{H}$ is a tree
and such that the edge set of $\mathcal{H}$ can be partitioned into a set of "handles" and a set of "teeth" with the following properties:
• there is at least one handle,
• the handles are pairwise disjoint,
• the teeth are pairwise disjoint,
• the number of teeth that each handle intersects is odd and at least three,
• each tooth includes a point that belongs to no handle.
Grötschel and Pulleyblank [239] proved that, for every clique tree $\mathcal{H}$ with r handles $H_1, \ldots, H_r$ and s teeth $T_1, \ldots, T_s$, the incidence vector x of any tour through V satisfies
$$\mathcal{H} \circ x \geq 2r + 3s - 1,$$
that is,
$$\sum_{i=1}^{r} x(\delta(H_i)) + \sum_{j=1}^{s} x(\delta(T_j)) \geq 2r + 3s - 1.$$
Using the η notation, this inequality can be written as
$$\sum_{i=1}^{r} \eta(H_i, x) + \sum_{j=1}^{s} \eta(T_j, x) \geq s - 1. \tag{10.7}$$
Clique trees with a single handle are combs and (10.7) in this case is a comb inequality. We show the general clique-tree inequality is valid by extending the proof for combs given in Section 5.6.

THEOREM 10.2 *For any clique tree with handles $H_1, \ldots, H_r \subseteq V$ and teeth $T_1, \ldots, T_s \subseteq V$, every tour through V satisfies the clique-tree inequality (10.7).*

Proof. Let t_j denote the number of handles that intersect tooth T_j and let h_i denote the number of teeth that intersect handle H_i; write
$$c_{ij} = \begin{cases} 1 & \text{if the tour includes an edge from } H_i \cap T_j \text{ to } T_j - H_i, \\ 0 & \text{otherwise.} \end{cases}$$

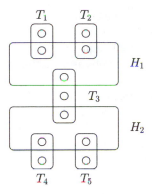

Figure 10.6 Clique tree with two handles.

Since the teeth are pairwise disjoint, we have $\eta(H_i, x) \geq \sum_j c_{ij} - 2$; by definition, we have $\sum_j c_{ij} \leq h_i$; since $\eta(H_i, x)$ is even and h_i is odd, we conclude that

$$\eta(H_i, x) \geq 2 \sum_{j=1}^{s} c_{ij} - h_i - 1. \tag{10.8}$$

The restriction of the tour on a tooth T_j consists of $1 + \eta(T_j, x)/2$ segments; one of these segments passes through a point of T_j that belongs to no handle; since the handles are pairwise disjoint, each i such that $H_i \cap T_j \neq \emptyset$ and $c_{ij} = 0$ adds a new segment; we conclude that

$$\eta(T_j, x) \geq 2(t_j - \sum_{i=1}^{r} c_{ij}). \tag{10.9}$$

From (10.8) and (10.9), we obtain

$$\sum_{i=1}^{r} \eta(H_i, x) + \sum_{j=1}^{s} \eta(T_j, x) \geq 2 \sum_j t_j - \sum_i h_i - r = \sum_j t_j - r;$$

since the intersection graph of $\mathcal{H}$ is a tree, we have $\sum_j t_j = r + s - 1$, and (10.7) follows.   □

An example of a clique tree with two handles is illustrated in Figure 10.6; the atoms containing a vertex (indicated by the small circles) must be nonempty, while the remaining atoms are permitted to contain vertices, but are not required to do so. If the teeth and handles in this example satisfy

$$x^*(\delta(T_1)) = x^*(\delta(T_2)) = x^*(\delta(T_4)) = x^*(\delta(T_5)) = 2$$

and

$$x^*(\delta(T_3)) = x^*(\delta(H_1)) = x^*(\delta(H_2)) = 3,$$

then the clique-tree inequality

$$x(\delta(H_1)) + x(\delta(H_2)) + \sum_{j=1}^{5} x(\delta(T_j)) \geq 18$$

is violated by one. Notice that the combs $\{H_1, T_1, T_2, T_3\}$ and $\{H_2, T_3, T_4, T_5\}$ contained in the clique tree yield inequalities that only hold as equations for such an x^*. This is the type of violated clique tree sought in the Naddef-Thienel search procedure, using one of the two combs as a starting point and growing the remaining handle and teeth.

In their sketch of the clique-tree heuristic, Naddef and Thienel write that they begin with a nonviolated comb and consider growing one or more handles from each tooth T_j such that $x^*(\delta(T_j)) > 2.25$. To accomplish this, they use the ear decomposition or max-back procedure to find a set S intersecting T_j, as in their methods for block combs. A complication, however, is that additional teeth must also be selected to meet the new handles. To manage this, only sets of cardinality two are used as additional teeth. At each step of the handle growing procedure, Naddef and Thienel compute an optimal set of edges J from $\delta(S)$ to serve as potential teeth; the contribution

$$\Delta = 3|J| + 2 - x^*(\delta(S)) - \sum_{e \in J} x^*(\delta(e))$$

to the violation of the inequality is used as a guide to the process.

PATH INEQUALITIES

Consider two combs, $\{H_1, T_1, T_2, T_3\}$ and $\{H_2, T_1, T_2, T_3\}$, with nested handles $H_1 \subset H_2$. Adding the corresponding comb inequalities we obtain

$$x(\delta(H_1)) + x(\delta(H_2)) + 2 \sum_{j=1}^{3} x(\delta(T_j)) \geq 20. \tag{10.10}$$

Fleischmann [181] defines conditions on the handles and teeth that permit inequalities such as (10.10) to be strengthened by reducing the coefficients on the $x(\delta(T_j))$ terms. For example, if each tooth intersects $H_2 - H_1$ and if

$$(H_2 - H_1) - \bigcup_{j=1}^{3} T_j = \emptyset,$$

then

$$x(\delta(H_1)) + x(\delta(H_2)) + \sum_{j=1}^{3} x(\delta(T_j)) \geq 14 \tag{10.11}$$

is valid for all tours. This improves inequality (10.10), since we can obtain (10.10) by adding the three subtour constraints $x(\delta(T_j)) \geq 2$ to (10.11).

In general, suppose we have a hypergraph $\mathcal{H} = (V, \mathcal{F})$ with $\mathcal{F}$ partitioned into handles and teeth; $\mathcal{F}$ is permitted to have multiple copies of each of its members. Let $H_1, \ldots, H_r$ be the distinct handles and let $T_1, \ldots, T_s$ be the distinct teeth; we let α_i denote the multiplicity of handle H_i and we let β_j denote the multiplicity of tooth T_j. Following Fleischmann [181], $\mathcal{H}$ is called a *star hypergraph* if

- $H_1 \subset H_2 \subset \cdots H_r$
- $\{H_i, T_1, T_2, \ldots, T_s\}$ is a comb for each $i = 1, \ldots, r$

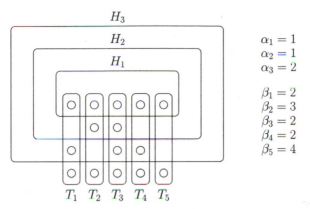

Figure 10.7 Star hypergraph with three handles.

- $(H_{i+1} - H_i) - \bigcup_{j=1}^r T_j = \emptyset$ for each $i = 1, \ldots, r-1$

and the multipliers α_i and β_j satisfy an *interval property*, defined below.

Consider a tooth T_j in $\mathcal{H}$. An inclusionwise maximal set of indices

$$\{l, l+1, \ldots, l+k\} \subseteq \{1, 2, \ldots, r\}$$

with

$$H_l \cap T_j = H_{l+1} \cap T_j = \cdots = H_{l+k} \cap T_j$$

is called a *handle interval* for T_j. The hypergraph $\mathcal{H}$ satisfies the interval property if

$$\sum_{i \in I} \alpha_i \leq \beta_j \tag{10.12}$$

for each j and handle interval I for T_j.

Given a star hypergraph $\mathcal{H}$, the corresponding *star inequality* is

$$\mathcal{H} \circ x \geq (s+1) \sum_{i=1}^r \alpha_i + 2 \sum_{j=1}^s \beta_j,$$

that is,

$$\sum_{i=1}^r \alpha_i x(\delta(H_i)) + \sum_{j=1}^s \beta_j x(\delta(T_j)) \geq (s+1) \sum_{i=1}^r \alpha_i + 2 \sum_{j=1}^s \beta_j. \tag{10.13}$$

Our description of the inequality follows the treatment given in Naddef and Pochet [418], Naddef [417], and Naddef and Rinaldi [422]; a short proof that (10.13) is satisfied by all tours through V can found in Naddef and Pochet [418].

In Figure 10.7 we illustrate a star hypergraph with three handles. Note that a star with a single handle and $\beta_j = 1$ for all teeth is a comb. Thus stars provide another rich generalization of comb inequalities. In the case of clique trees, Grötschel and Pulleyblank [239] proved that each inequality induces a facet of the TSP polytope. This result does not hold for stars. Indeed, if for some j we set β_j larger than the maximum of $\sum_{i \in I} \alpha_i$ taken over all handle intervals for T_j, then (10.13) is the sum of copies of $x(\delta(T_j)) \geq 2$ and a smaller star inequality. Such an inequality would be a poor choice in a cutting-plane algorithm.

Although stars in general can be weak inequalities, Naddef and Rinaldi [422] show that all members of an important subclass do induce facets of the TSP polytope. Call $\mathcal{H}$ a *path hypergraph* if it is a star that satisfies the stronger condition

$$\sum_{i \in I} \alpha_i = \beta_j$$

for each j and each handle interval I for tooth T_j. The corresponding inequalities (10.13) were introduced by Cornuéjols et al. [140] and are known as "path, wheelbarrow, and bicycle" inequalities, or, for short, *path inequalities*. Cornuéjols et al. [140] proved that path inequalities are facet-inducing for the graphical TSP, and Naddef and Rinaldi used this to show they are also facet-inducing for the TSP.

Path inequalities are another target of the greedy approach described in Naddef and Thienel [424]. Starting with a nonviolated comb $\{H, T_1, \ldots, T_r\}$ having one or more teeth T_j with $|T_j| \geq 3$ and $x^*(\delta(T_j)) > 2$, they apply their ear decomposition and max-back algorithms to grow a nested set of handles starting with H. As an alternative, they also consider first partitioning each large tooth T_j, and then building the new handles by assigning the partitions to sets H_i. The main ideas behind these two methods are illustrated on several examples in [424].

A similar approach is also outlined for the *ladder inequalities* studied by Boyd et al. [86], again using a nonviolated comb as the starting point for the search.

Concorde Implementation

Naddef and Thienel [424] report strong computational results with a branch-and-cut code anchored by their greedy algorithms for combs and the separation methods described above. Their implementation was used to solve a collection of instances from the TSPLIB, including the 4,461-city example fnl4461. This is the largest instance reported solved by a code other than Concorde.

The Naddef-Thienel computer code has not been made publicly available, but in Concorde we have implemented variations of several of their ideas. Our algorithms make use of the tighten procedure described earlier in the chapter, rather than the ear decomposition and max-back methods.

The most successful of the Naddef-Thienel-like techniques in Concorde is an implementation of an algorithm for producing path inequalities having two handles, starting from a comb inequality that is in the current LP relaxation. An illustration of a two-handled path hypergraph is given in Figure 10.8; the corresponding path inequality is

$$x(\delta(H_1)) + x(\delta(H_2)) + 2x(\delta(T_1)) + x(\delta(T_2)) + x(\delta(T_3)) \geq 16.$$

To construct such an inequality, consider a comb with handle $H \subseteq V$ and teeth $T_1, \ldots, T_s \subseteq V$. Let $A = \emptyset$. For each j we choose, if possible, a single vertex $v \in T_j - H$ and add v to A; we then let $H_1 = H$ and $H_2 = H \cup A$. Vertex $v \in T_j - H$ is chosen to maximize the value $\sigma_v = x^*(\{v\}, H \cup A)$; if $|T_j - H| = 1$ or if no vertex in $T_j - H$ has $\sigma_v \geq 0.5$, then we do not select a vertex for this tooth.

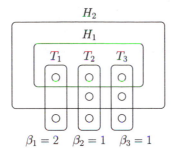

Figure 10.8 Path hypergraph with two handles.

If A has at least one member, then $H_1, H_2, T_1, \ldots, T_s$ form a path hypergraph, where $\beta_j = 1$ if tooth T_j contributed a vertex to A and $\beta_j = 2$ otherwise. The tighten procedure from Section 10.1 is then applied to this configuration, allowing each of the handles and teeth to grow or shrink in an attempt to obtain an inequality violated by x^*.

The above algorithm builds H_2 as a superset of the original handle of the comb. We also run the algorithm to construct H_2 as a subset of the comb's handle. In this case, for each j we choose, if possible, a vertex $v \in T_j \cap H$ and add v to A; we then let $H_1 = H$ and $H_2 = H - A$. Vertex $v \in T_j \cap H$ is chosen to minimize the value $\sigma_v = x^*(\{v\}, (H - A) - \{v\})$; if $|T_j \cap H| = 1$ or if no vertex in $T_j \cap H$ has $2.0 - \sigma_v \geq 0.5$, then we do not select a vertex for this tooth.

To increase the chance of finding a violated path inequality, we run two other variants of the above procedures. In the first, we simply relax the 0.5 tolerance to 0.1, allowing the algorithm to possibly return a hypergraph with more teeth having $\beta_j = 1$. In the second variant, we begin by applying a greedy algorithm that attempts to increase the size of each tooth of the comb before building the new handle. By stretching the teeth in this way, we again give more opportunities to create a hypergraph with teeth having $\beta_j = 1$. The greedy algorithm adopted here is the one described in Section 7.2 and used in our block comb heuristic.

These methods can be directly extended to search for path inequalities with more than two handles, and by default in Concorde we also look for three-handled path hypergraphs. The computational success of the heuristics, together with strong results reported by Naddef and Thienel [424], suggests a potential line of research aimed at developing more sophisticated methods for separating path inequalities.

10.4 GLUING

The purpose of the present section is to describe a way of using cuts from a pool to obtain new cuts, one which amounts to "gluing together" old constraints into a conglomerate cut. Grötschel and Pulleyblank [239] used a notion of gluing in their analysis of clique-tree inequalities. On the one hand, we glue in a way that is slightly more restricted than the Grötschel-Pulleyblank way; on the other hand, we start out with a wider class of elementary building blocks. The resulting recursively defined class of *clique-tree-like inequalities* nearly but not quite subsumes the class of clique-tree inequalities; we postpone discussion of this relationship till the end of this section.

Unlike Grötschel and Pulleyblank, we use gluing not only as an *analytical* tool but also as an *algorithmic* tool; it is the algorithmic use of gluing that is the focal point of this section.

It will be convenient to work with the η notation, and we therefore define the following analogs of $\mathcal{H} \circ x$ and $\mu(\mathcal{H})$ for a specified hypergraph $\mathcal{H} = (V, \mathcal{F})$. We let

$$\mathcal{H} \star x = \sum_{S \in \mathcal{F}} \eta(x, S)$$

and we let $\mu^*(\mathcal{H})$ stand for the minimum of $\mathcal{H} \star x$ taken over the incidence vectors of tours through V.

To define clique-tree-like inequalities, we shall first define *clique-tree-like hypergraphs*, and we shall do that recursively. Two special kinds of hypergraphs serve as elementary blocks in building up clique-tree-like hypergraphs. As usual, a comb is a hypergraph with vertex set V and edge set $\{H, T_0, T_1, \ldots T_{2k}\}$ such that $k \geq 1$ and

- $T_0, T_1, \ldots, T_{2k}$ are pairwise disjoint,
- H meets each of $T_0, T_1, \ldots, T_{2k}$ but does not contain any;

a *flipped comb* is a hypergraph with vertex set V and edge set $\{H, T_0, T_1, \ldots, T_{2k}\}$ such that $k \geq 1$ and

- $H - T_0 \neq \emptyset$, $H \cup T_0 \neq V$, and $T_0 \supseteq T_i$ for all $i = 1, 2, \ldots, 2k$,
- H meets each of $T_1, \ldots, T_{2k}$ but does not contain any,
- $T_1, \ldots, T_{2k}$ are pairwise disjoint.

We say that a hypergraph $\mathcal{H}$ arises from hypergraphs $\mathcal{H}_1, \mathcal{H}_2, \ldots, \mathcal{H}_m$ by *gluing along edge* C if

- $\mathcal{H}_i = (V, \mathcal{F}_i)$ for all i and $\mathcal{H} = (V, \cup_{i=1}^m \mathcal{F}_i)$,
- C belongs to each $\mathcal{F}_i$,
- $A \cap B = \emptyset$ whenever $A \in \mathcal{F}_i - \{C\}$, $B \in \mathcal{F}_j - \{C\}$, and $i \neq j$,
- some vertex in C belongs to no edge of $\mathcal{H}$ other than C,
- some vertex in V belongs to no edge of $\mathcal{H}$.

A *clique-tree-like hypergraph* is any hypergraph $\mathcal{H}$, with edge set partitioned into a set of handles and a set of teeth, such that either

- $\mathcal{H}$ is a comb or a flipped comb,

or else

- $\mathcal{H}$ arises from at least two clique-tree-like hypergraphs by gluing along a tooth.

THEOREM 10.3 *If $\mathcal{H}$ is a clique-tree-like hypergraph with t teeth, then $\mu^*(\mathcal{H}) = t - 1$.*

By a *clique-tree-like inequality*, we mean any inequality

$$\mathcal{H} \star x \geq t - 1$$

such that $\mathcal{H}$ is a clique-tree-like hypergraph with t teeth; Theorem 10.3 guarantees that all clique-tree-like inequalities are satisfied by all incidence vectors x of tours.

Before proving Theorem 10.3, let us describe the way we use it. We sweep through the cut pool; given each new set C from the pool, we search for hypergraphs in our cut pool that can be glued together along their common tooth C into a conglomerate hypergraph $\mathcal{H}$ such that the optimal solution x of our most recent LP relaxation satisfies $\mathcal{H} \star x < \mu^*(\mathcal{H})$. For this purpose, let $\mathcal{H}_1, \mathcal{H}_2, \dots, \mathcal{H}_M$ denote all the hypergraphs in the cut pool that contain C as a tooth; let us write $\mathcal{H}_i = (V, \mathcal{F}_i)$ for all i. We want to find a subset S of $\{1, 2, \dots, M\}$ such that

(i) $A \cap B = \emptyset$ whenever $A \in \mathcal{F}_i - \{C\}$, $B \in \mathcal{F}_j - \{C\}$, and $i \in S$, $j \in S$, $i \neq j$,

and, with $\mathcal{H} = (V, \cup_{i \in S} \mathcal{F}_i)$,

(ii) some vertex in C belongs to no edge of $\mathcal{H}$ other than C,

(iii) some vertex in V belongs to no edge of $\mathcal{H}$,

(iv) $\mathcal{H} \star x < \mu^*(\mathcal{H})$.

The problem of finding such a set S may be cast in the more general setting of independence systems (see Section 8.1 for the standard definition): given an independence system $(X, \mathcal{I})$, given an assignment of nonnegative weights a_i to points i of X, and given a nonnegative number b, find an independent set S that satisfies

(v) $b + \sum_{i \in S}(a_i - b) < 0$

or show that no such S exists. To see that our particular problem fits this general paradigm, write $X = \{1, 2, \dots, M\}$ and set $S \in \mathcal{I}$ if and only if S satisfies (i), (ii), (iii). Trivially, $(X, \mathcal{I})$ is an independence system: $\emptyset \in \mathcal{I}$ and $\mathcal{I}$ is closed under taking subsets. Now write

$$a_i = \mathcal{H}_i \star x - \mu^*(\mathcal{H}_i), \quad b = x(\delta(C)) - 2,$$

and let t_i denote the number of teeth in $\mathcal{H}_i$. Theorem 10.3 guarantees that

$$\mu^*(\mathcal{H}_i) = t_i - 1$$

for all i and that, as long as (i), (ii), (iii) are satisfied,

$$\mu^*(\mathcal{H}) = \sum_{i \in S}(t_i - 1).$$

Since

$$\mathcal{H} \star x = b + \sum_{i \in S}(a_i + \mu^*(\mathcal{H}_i) - b),$$

(iv) and (v) are equivalent whenever S is independent.

Let us describe our search algorithm in the general setting of independence systems. To begin, we may remove from X all points j such that $a_j > b$: removal of all such points transforms each independent set S into an independent set R with $\sum_{i \in R}(a_i - b) \le \sum_{i \in S}(a_i - b)$. After this clean-up, let us write $X = \{1, 2, \dots, M\}$ and, for future reference,

$$c_j = \sum_{i=j+1}^{M} (a_i - b).$$

The algorithm itself is recursive. Its idea may be easiest to grasp in terms of the *enumeration tree*, a complete binary tree of depth M, whose 2^j nodes at level j are in a one-to-one correspondence with the 2^j subsets of $\{1, 2, \dots, j\}$: if a node at level $j - 1$ corresponds to a subset R of $\{1, 2, \dots, j - 1\}$ then its left child corresponds to R and its right child corresponds to $R \cup \{j\}$. This tree gets *pruned* by removing all the descendants of a node whenever it is blatantly obvious that none of these descendants corresponds to an independent set S with property (v). Here, "blatantly obvious" can mean one of two different things. First, if a node corresponds to a set R that is not independent, then no superset of R is independent, and so we may safely disregard all the descendants of this node; second, if a node at level j corresponds to a set R such that $b + \sum_{i \in R}(a_i - b) + c_j \ge 0$, then no superset S of R has property (v), and so we may safely disregard all the descendants of this node. Recursive calls of the algorithm are in a one-to-one correspondence with the nodes of the resulting pruned enumeration tree except for leaves at level M: each of these leaves that persists in the pruned tree corresponds to an independent set S with property (v).

Common sense seems to suggest that our chances of discovering at least one independent set S with property (v) increase as the value of b increases; for this reason, we sweep through the cut pool in a nonincreasing order of $x(\delta(C))$. Independently of this stratagem, the algorithm as described so far may come across a member C of the pool that yields an overwhelmingly large number of independent sets S with property (v); to stop this kind of good luck from turning into a disaster, we terminate the search whenever it has produced 50 new cuts.

Next, let us get started on proving Theorem 10.3. To establish the lower bound, $\mu^*(\mathcal{H}) \ge t - 1$, we prove a more general theorem:

THEOREM 10.4 *If $\mathcal{H}$ arises from $\mathcal{H}_1, \mathcal{H}_2, \dots, \mathcal{H}_m$ by gluing along any common edge, then $\mu^*(\mathcal{H}) \ge \sum_{i=1}^{m} \mu^*(\mathcal{H}_i)$.*

Proof. We may restrict ourselves to the case of $m = 2$. Let C denote the common edge of $\mathcal{H}_1$ and $\mathcal{H}_2$; let us write $\mathcal{H}_i = (V, \mathcal{F}_i)$ and $\mathcal{H}_i' = (V, \mathcal{F}_i - \{C\})$; furthermore, let us write $v \in V_i$ if $v \in V$ and v belongs to at least one edge of $\mathcal{H}_i'$; let us set $V_0 = V - (V_1 \cup V_2)$, $D = V - C$, and $C_i = C \cap V_i$, $D_i = D \cap V_i$ for $i = 0, 1, 2$. By assumption, V_1, V_2 are disjoint and C_0, D_0 are nonempty.

Now consider a tour T through $\mathcal{H}$ whose incidence vector x satisfies $\mathcal{H} \star x = \mu^*(\mathcal{H})$. If this tour has the form $\dots ab \dots cd \dots$ such that
- $a \in C, b \in D, c \in C, d \in D$,

- $a \notin C_1$ or $b \notin D_1$,
- $c \notin C_1$ or $d \notin D_1$,

then replace the pair of edges ab, cd with the pair of edges ac, bd. It is a routine matter to verify that the incidence vector y of the resulting tour satisfies $y(\delta(S)) \leq x(\delta(S))$ for all edges S of $\mathcal{H}_1$. Similarly, if the tour has the form $\ldots ab \ldots cd \ldots$ with

- $a \in D, b \in C, c \in D, d \in C$,
- $a \notin D_1$ or $b \notin C_1$,
- $c \notin D_1$ or $d \notin C_1$,

then replace the pair of edges ab, cd with the pair of edges ac, bd. Again, it is a routine matter to verify that the incidence vector y of the resulting tour satisfies $y(\delta(S)) \leq x(\delta(S))$ for all edges S of $\mathcal{H}_1$. Repeating these transformations as many times as possible, we eventually obtain a tour T_1 through V whose incidence vector x_1 satisfies

$$\mathcal{H}_1' \star x_1 \leq \mathcal{H}_1' \star x \qquad (10.14)$$

and

$$x_1(C, D) \leq x(C_1, D_1) + 2. \qquad (10.15)$$

The same procedure with C_2, D_2 in place of C_1, D_1 yields a tour T_2 through V whose incidence vector x_2 satisfies

$$\mathcal{H}_2' \star x_2 \leq \mathcal{H}_2' \star x \qquad (10.16)$$

and

$$x_2(C, D) \leq x(C_2, D_2) + 2. \qquad (10.17)$$

If $x_1(C, D) + x_2(C, D) \leq x(C, D) + 2$ then (10.14) and (10.16) guarantee that

$$\mathcal{H}_1 \star x_1 + \mathcal{H}_2 \star x_2 = (\mathcal{H}_1' \star x_1 + x_1(C, D) - 2) + (\mathcal{H}_2' \star x_2 + x_2(C, D) - 2)$$
$$\leq \mathcal{H}_1' \star x + \mathcal{H}_2' \star x + x(C, D) - 2 = \mu^*(\mathcal{H}),$$

and the desired conclusion follows. Hence we may assume that

$$x_1(C, D) + x_2(C, D) > x(C, D) + 2.$$

Now, since

$$x_1(C, D) + x_2(C, D) \leq (x(C_1, D_1) + 2) + (x(C_2, D_2) + 2) \leq x(C, D) + 4$$

(the first inequality follows from (10.15) and (10.17); the second inequality is trivial) and since $x_1(C, D), x_2(C, D), x(C, D)$ are all even, we must have

$$x_1(C, D) = x(C_1, D_1) + 2, \qquad (10.18)$$
$$x_2(C, D) = x(C_2, D_2) + 2, \qquad (10.19)$$
$$x(C, D) = x(C_1, D_1) + x(C_2, D_2). \qquad (10.20)$$

Let us say that an edge of T is of *type 1* if one of its endpoints belongs to C_1 and its other endpoint belongs to D_1; let us say that an edge of T is of *type 2* if one of its endpoints belongs to C_2 and its other endpoint belongs to D_2. By (10.20), each edge of T that crosses the boundary of C is either of type 1 or of type 2; by (10.18),

$$\text{the number of edges of type 1 is even;} \qquad (10.21)$$

by (10.19),

$$\text{the number of edges of type 2 is even;} \qquad (10.22)$$

by (10.14), (10.16), (10.18), (10.19), (10.20), we have

$$\mathcal{H}_1 \star x_1 + \mathcal{H}_2 \star x_2 \le \mu^*(\mathcal{H}) + 2. \qquad (10.23)$$

Let us choose once and for all one of the two cyclic orientations of T. We shall say that an edge of type 1 or type 2 is *outward* (with respect to our fixed orientation of T) if its endpoint in C precedes its endpoint in D; otherwise we shall say that the edge is *inward*. Now consider the cyclic sequence S of symbols $+$ and $-$ that arises from the cyclic sequence of edges of type 1 on T by writing a $+$ for each outward edge and writing a $-$ for each inward edge. If

$(*)$ S contains two consecutive identical symbols

then $S = aaS'$ for some S', whose length is even by virtue of (10.21). The block substitution rules

$$++ \mapsto \Lambda, \quad -- \mapsto \Lambda, \quad +-+ \mapsto +, \quad -+- \mapsto -,$$

with Λ standing for the null sequence, transform S' into one of Λ, $+-$, $-+$, and so they transform the cyclic sequence S into Λ. Each of these block substitutions translates into a replacement of a pair of edges ab, cd of type 1 with the pair of edges ac, bd; the entire sequence of these replacements transforms T into a tour T_2' through V, whose incidence vector x_2' satisfies

$$\mathcal{H}_2' \star x_2' \le \mathcal{H}_2' \star x \quad \text{and} \quad x_2'(C, D) = x(C_2, D_2).$$

Now, by virtue of (10.14), (10.18), and (10.20),

$$\mathcal{H}_1 \star x_1 + \mathcal{H}_2 \star x_2' = (\mathcal{H}_1' \star x_1 + x_1(C, D) - 2) + (\mathcal{H}_2' \star x_2' + x_2'(C, D) - 2)$$
$$\le \mathcal{H}_1' \star x + \mathcal{H}_2' \star x + x(C, D) - 2 = \mu^*(\mathcal{H}),$$

and the desired conclusion follows again. Hence we may assume that $(*)$ fails; to put it differently, we may assume that

$$\text{in the cyclic sequence of type 1 on } T, \text{ outward and inward edges alternate.}$$
$$(10.24)$$

Similarly, we may assume that

$$\text{in the cyclic sequence of type 2 on } T, \text{ outward and inward edges alternate.}$$
$$(10.25)$$

(Note that (10.24) is a strengthening of (10.21) and that (10.25) is a strengthening of (10.22).)

In addition, we may also assume that

$$\text{there is an edge of type 2:} \qquad (10.26)$$

else $x(C_2, D_2) = 0$ and $T_1 = T$, contradicting (10.18) and (10.20). Similarly, we may assume that

$$\text{there is an edge of type 1.} \qquad (10.27)$$

Digressing for a while, consider an arbitrary tour T' whose vertex set V is partitioned into pairwise disjoint sets $C_0, C_1, C_2, D_0, D_1, D_2$, and let us write $C = C_0 \cup C_1 \cup C_2$, $D = D_0 \cup D_1 \cup D_2$. Again, let us say that an edge of this tour is of type i if one of its endpoints belongs to C_i and its other endpoint belongs to D_i; let us assume that

(i) each edge of T' with one endpoint in C and the other endpoint in D
     is of type 1 or type 2,

(ii) there are an edge of type 1 and an edge of type 2.

Again, let us say that an edge of type 1 or type 2 is *outward* (with respect to a fixed
orientation of T') if its endpoint in C precedes its endpoint in D and let us say that
the edge is *inward* otherwise; let us assume that

(iii) in the cyclic sequence of edges of type 1 on T', outward and inward edges
       alternate,

(iv) in the cyclic sequence of edges of type 2 on T', outward and inward edges
      alternate.

Removal of all the edges of type 1 from T' breaks T' into a family $\mathcal{P}_1$ of paths;
assumption (iii) guarantees that each of these paths either begins and ends in C_1 or
begins and ends in D_1. Similarly, removal of all the edges of type 2 from T' breaks
T' into a family $\mathcal{P}_2$ of paths; assumption (iv) guarantees that each of these paths
either begins and ends in C_2 or begins and ends in D_2. Let A_1^C denote the union of
vertex sets of all the paths in $\mathcal{P}_1$ that begin and end in C_1; let A_1^D denote the union
of vertex sets of all the paths in $\mathcal{P}_1$ that begin and end in D_1; let A_2^C denote the
union of vertex sets of all the paths in $\mathcal{P}_2$ that begin and end in C_2; let A_2^D denote
the union of vertex sets of all the paths in $\mathcal{P}_2$ that begin and end in D_2. We claim
that

$$A_1^C \cup A_2^C = V \text{ or else } A_1^D \cup A_2^D = V. \tag{10.28}$$

To justify this claim, we shall use induction on the number of edges of type 1 and
type 2. We may assume that some point u of V is outside $A_1^C \cup A_2^C$, and so it
belongs to $A_1^D \cap A_2^D$; let ab be the first edge of type 1 or type 2 that we encounter
proceeding from u along T', and let the b be the successor of a. Since $u \in A_1^D \cap A_2^D$,
we have either $a \in D_1, b \in C_1$ or $a \in D_2, b \in C_2$; symmetry allows us to assume
that $a \in D_1, b \in C_1$. Let cd be the first edge of type 1 or type 2 that we encounter
proceeding from b along T', and let the d be the successor of c. Trivially, we have
$c \in C, d \in D$, and so either $c \in C_1, d \in D_1$ or $c \in C_2, d \in D_2$; the latter case
is excluded since $u \in A_2^D$; hence $c \in C_1, d \in D_1$. If ab and cd are the only two
edges of type 1, then $A_1^D \cup A_2^D = V$ and we are done; else the tour T'' obtained
from T' by transferring all the interior points of the segment $ab \ldots cd$ from C to D
satisfies (i), (ii), (iii), (iv) in place of T'. The transformation of T' into T'' transfers
all the interior points of the segment $ab \ldots cd$ from $A_1^C \cap A_2^D$ to $A_1^D \cap A_2^D$, but
otherwise it leaves $A_1^C, A_2^C, A_1^D, A_2^D$ unchanged; in particular, with respect to T'',
we still have $A_1^C \cup A_2^C \neq V$. Now the induction hypothesis guarantees that, with
respect to T'', we have $A_1^D \cup A_2^D = V$; it follows that, with respect to T', we also
have $A_1^D \cup A_2^D = V$.

By virtue of (10.20), (10.26), (10.27), (10.24), and (10.25), our original tour T
satisfies (i), (ii), (iii) and (iv) in place of T'. Hence (10.28) allows us to distinguish

between two cases.

Case 1: $A_1^C \cup A_2^C = V.$

Consider an arbitrary vertex e in D_0. By assumption of this case, e belongs to at least one A_i^C; symmetry allows us to assume that $i = 1$. It is a routine matter to verify that the sequence of operations transforming T into T_2 maintains the following invariant:

> *the current tour has the form* $\ldots aPd \ldots$ *such that* $a \in D_1$, $d \in D_1$, *and*
>
> P *is a path that begins and ends in* C_1, *contains no edges of type 1, and includes* e.

In particular, T_2 is a concatenation of paths P and Q such that
- P begins and ends in C_1, contains no edges of type 1, and includes e,
- Q begins and ends in D_1, and contains no edges of type 1.

Let b denote the first point of P, let c denote the last point of P, let d denote the first point of Q, let a denote the last point of Q, and let f denote the successor of e on T_2. Now T_2 has the form

$$\ldots ab \ldots ef \ldots cd \ldots;$$

let y denote the incidence vector of the tour obtained from T_2 by replacing the three edges ab, ef, cd with the three edges ae, df, bc. It is a routine matter to verify that $y(C, V - C) = x_2(C, V - C) - 2$ and that $y(S, V - S) \le x_2(S, V - S)$ for all edges S of $\mathcal{H}_2'$ (since none of a, b, c, d, e belong to an edge of $\mathcal{H}_2'$); hence $\mathcal{H}_2 \star y \le \mathcal{H}_2 \star x_2 - 2$, and the desired conclusion follows from (10.23).

Case 2: $A_1^D \cup A_2^D = V.$

Replacing C with D in each of $\mathcal{F}_1$, $\mathcal{F}_2$ and $\mathcal{F}$, we find ourselves in Case 1. □

To establish the upper bound, $\mu^*(\mathcal{H}) \le t - 1$, of Theorem 10.3, we first prove two easy lemmas. By an *optimal* tour through a hypergraph $\mathcal{H}$, we shall mean any tour through $\mathcal{H}$ whose incidence vector x satisfies $\mathcal{H} \star x = \mu^*(\mathcal{H})$.

LEMMA 10.5 *For every hypergraph $\mathcal{H}$ and for every nonempty subset B of an atom of $\mathcal{H}$ there is an optimal tour through $\mathcal{H}$ whose incidence vector x satisfies $x(\delta(B)) = 2$.*

Proof. Let A be the atom of $\mathcal{H}$ that contains B; let P be the concatenation of an arbitrary path through B and an arbitrary path through $A - B$. Let $\mathcal{G}$ be a hypergraph obtained from $\mathcal{H}$ by deleting a set of $|A| - 1$ points in A; consider an optimal tour through $\mathcal{G}$ and let x denote the incidence vector of this tour. Substituting P for the single point of A in this tour, we obtain a tour through $\mathcal{H}$ whose incidence vector y satisfies $\mathcal{H} \star y = \mathcal{G} \star x$; by Theorem 5.11, $\mathcal{G} \star x = \mu^*(\mathcal{H})$. □

By an *end* of a clique-tree-like hypergraph $\mathcal{H}$, we shall mean any hypergraph $\mathcal{H}'$ such that
- $\mathcal{H}'$ is a comb or a flipped comb, and

- $\mathcal{H}$ arises from $\mathcal{H}'$ and some clique-tree-like hypergraph $\mathcal{H}''$ by gluing along their common tooth.

LEMMA 10.6 *Let $\mathcal{H}$ be a clique-tree-like hypergraph. If $\mathcal{H}$ is neither a comb nor a flipped comb, then $\mathcal{H}$ has at least two ends.*

Proof. By induction on the number of edges of $\mathcal{H}$. By definition, $\mathcal{H}$ arises from clique-tree-like hypergraphs $\mathcal{H}_1, \mathcal{H}_2, \ldots, \mathcal{H}_m$ by gluing along their common tooth T. We propose to find hypergraphs $\mathcal{G}_1, \mathcal{G}_2, \ldots, \mathcal{G}_m$ such that each $\mathcal{G}_i$ is an end of $\mathcal{H}$ and all its edges come from $\mathcal{H}_i$. If $\mathcal{H}_i$ is a comb or a flipped comb then we may set $\mathcal{G}_i = \mathcal{H}_i$; else the induction hypothesis guarantees that $\mathcal{H}_i$ has distinct ends, say, $\mathcal{G}_{i1}$ and $\mathcal{G}_{i2}$. If at least one $\mathcal{G}_{ij}$ does not include T in its edge set then this $\mathcal{G}_{ij}$ is an end of $\mathcal{H}$; else both $\mathcal{G}_{i1}$ and $\mathcal{G}_{i2}$ are ends of $\mathcal{H}$. $\qquad\square$

Proof. (*Theorem 10.3*) Let us write $\mathcal{H} = (V, \mathcal{F})$ and proceed by induction on the number of edges of $\mathcal{H}$. We know that that $\mu^*(\mathcal{H}) = t - 1$ whenever $\mathcal{H}$ is a comb with t teeth; since $\mu^*(\mathcal{H})$ is invariant under a substitution of $V - S$ for any S in $\mathcal{F}$, it follows that $\mu^*(\mathcal{H}) = t - 1$ whenever $\mathcal{H}$ is a flipped comb with t teeth; hence Lemma 10.6 allows us to assume that $\mathcal{H}$ arises from a comb or a flipped comb $\mathcal{H}_1$ and a clique-tree-like hypergraph $\mathcal{H}_2$ by gluing along their common tooth T. With t_i standing for the number of teeth of $\mathcal{H}_i$, we have $\mu^*(\mathcal{H}_1) = t_1 - 1$ and, by the induction hypothesis, $\mu^*(\mathcal{H}_2) = t_2 - 1$; trivially, $t = t_1 + t_2 - 1$.

Theorem 10.4 guarantees that $\mu^*(\mathcal{H}) \geq t - 1$.

To show that $\mu^*(\mathcal{H}) \leq t - 1$, let H denote the handle of $\mathcal{H}_1$ and enumerate the teeth of $\mathcal{H}_1$ other than T as $T_1, T_2, \ldots, T_{2k}$; note that $T_1, T_2, \ldots, T_{2k}$ are pairwise disjoint whether $\mathcal{H}_1$ is a comb or a flipped comb; set $B = T_1 \cup T_2 \cup \ldots \cup T_{2k} \cup H$. It is easy to construct a path P through B that

- starts in $T_1 - H$ and ends in $T_{2k} - H$,
- crosses the boundary of T_1 precisely once,
- crosses the boundary of each T_i with $1 < i < 2k$ precisely twice,
- crosses the boundary of T_{2k} precisely once,
- crosses the boundary of H precisely $2k$ times,
- crosses from $H \cap T$ to $H - T$ precisely twice.

Now set $T' = T - H$ if $\mathcal{H}_1$ is a comb and $T' = T \cup H$ if $\mathcal{H}_1$ is a flipped comb; in either case, let $\mathcal{H}'$ denote the hypergraph obtained from $\mathcal{H}_2$ by substituting T' for T. By Lemma 10.5, some optimal tour through $\mathcal{H}'$ is the concatenation of a path Q through $V - B$ and a path R through B. Substituting P for R, we obtain another optimal tour through $\mathcal{H}'$. The incidence vector x of this new tour satisfies $x(\delta(T_i)) = 2$ for all i, $x(\delta(H)) = 2k$, and $x(\delta(T)) = x(\delta(T')) + 2$. Hence $\mathcal{H} \star x = \mu^*(\mathcal{H}') + 2k$ and the desired conclusion follows, since $\mu^*(\mathcal{H}') = \mu^*(\mathcal{H}_2)$ by Theorem 5.11. $\qquad\square$

Let us close this section with a brief discussion of the relationship between clique-tree inequalities and clique-tree-like inequalities.

It is easy to show that a clique-tree-like hypergraph $\mathcal{H}$ is a clique tree if and only if

- each vertex of $\mathcal{H}$ belongs to at most two edges.

In the reverse direction, it is easy to show that a clique tree $\mathcal{H}$ is a clique-tree-like

hypergraph if and only if $\mathcal{H}$ has at least one of the following two properties:
- some vertex of $\mathcal{H}$ belongs to no edge of $\mathcal{H}$,
- $\mathcal{H}$ is a comb.

We know that the inequality

$$\mathcal{H} \star x \geq t - 1$$

for any clique tree $\mathcal{H}$ with t teeth is satisfied by all tours. As we have just observed, this result is not a corollary of our Theorem 10.4. (Nevertheless, it can be derived from Theorem 10.4 with just a little additional effort.)

Chapter Eleven

Local Cuts

The template paradigm of identifying inequalities having a prescribed structure, as outlined in Section 5.6, has been a dominant theme in TSP computation since the work of Grötschel and Padberg [237] on combs. This work has been guided by research into classes of facet-inducing inequalities for the convex hull of tours through the set of cities V. In this chapter we present a separation method that disdains all understanding of the TSP polytope and bashes on regardless of all prescribed templates.

11.1 AN OVERVIEW

The cut-finding technique described in this chapter is an examination of a number of mappings σ from V onto relatively small sets $\overline{V}$ (the action of σ may be thought of as shrinking each subset $\sigma^{-1}(i)$ of V onto the single point i of $\overline{V}$). Each σ induces a linear mapping ϕ that assigns a vector $\phi(x)$ with components indexed by the edges of the complete graph on $\overline{V}$ to every vector x with components indexed by the edges of the complete graph on V: we have $\phi(x) = \overline{x}$ with
$$\overline{x}_{ij} = x(\sigma^{-1}(i), \sigma^{-1}(j)) \quad \text{for all edges } ij.$$
The image $\phi(x)$ of each tour x through V is a vector $\overline{x}$ such that
- each $\overline{x}_e$ is a nonnegative integer,
- the graph with vertex set $\overline{V}$ and edge set $\{e : \overline{x}_e > 0\}$ is connected,
- $\sum(\overline{x}_e : v \in e)$ is even whenever $v \in \overline{V}$.

Let $\mathcal{T}$ denote the set of all the vectors $\overline{x}$ with these properties.

As long as $\overline{V}$ is reasonably small, it is reasonably easy either to find a linear inequality $a^T \overline{x} \geq b$ that is satisfied by all points of $\mathcal{T}$ and violated by $\phi(x^*)$ in place of $\overline{x}$ or else to establish that $\phi(x^*)$ belongs to the convex hull of $\mathcal{T}$. In the former case, the linear inequality $a^T \phi(x) \geq b$ is satisfied by the characteristic vector x of every tour through V and violated by x^* in place of x; in the latter case, we move on to our next choice of σ.

Whenever we manage to separate $\phi(x^*)$ from $\mathcal{T}$, we find the separating linear inequality in a hypergraph form; if $a^T \overline{x} \geq b$ is identical with
$$\sum(\lambda_Q x(Q, \overline{V} - Q) : Q \in \overline{\mathcal{H}}) \geq \mu,$$
then $a^T \phi(x) \geq b$ is identical with
$$\sum(\lambda_Q x(\sigma^{-1}(Q), V - \sigma^{-1}(Q)) : Q \in \overline{\mathcal{H}}) \geq \mu.$$

Algorithm 11.1 A scheme for collecting cuts.

---

initialize an empty list $\mathcal{L}$ of cuts;
for selected small sets $\overline{V}$ and mappings σ from V onto $\overline{V}$
do if   (with ϕ and $\mathcal{T}$ defined by σ)
         $\phi(x^*)$ lies outside the convex hull of $\mathcal{T}$
   then find a hypergraph inequality
$$\sum(\lambda_Q \overline{x}(Q, \overline{V} - Q) : Q \in \overline{\mathcal{H}}) \geq \mu$$
        that is satisfied by all $\overline{x}$ in $\mathcal{T}$
        and violated by $\phi(x^*)$ in place of $\overline{x}$;
        add the cut
$$\sum(\lambda_Q x(\sigma^{-1}(Q), V - \sigma^{-1}(Q)) : Q \in \overline{\mathcal{H}}) \geq \mu$$
        to $\mathcal{L}$;
   end
end
return $\mathcal{L}$;

The notions of σ, ϕ, and $\mathcal{T}$ that are used here are far from original.

Shrinking is a standard preprocessing step in finding TSP cuts that match prescribed templates, as we described in Section 5.9.

Our $\mathcal{T}$ was introduced by Cornuéjols et al. [140], who call the problem of minimizing a prescribed linear function over this $\mathcal{T}$ the *graphical traveling salesman problem*. Inequalities that induce facets of the convex hull of $\mathcal{T}$, and their relationship to inequalities that induce facets of the traveling salesman polytope, have been studied by Cornuéjols et al. [140], by Fleischmann [181], by Naddef and Rinaldi [419], [421], and by Oswald et al. [441].

11.2 MAKING CHOICES OF $\overline{V}$ AND σ

We make each choice of $\overline{V}$ and σ by choosing a partition of V into a small number of nonempty sets $V_0, V_1, \ldots, V_k$; then we set $\overline{V} = \{0, 1, \ldots, k\}$ and

$$\sigma(w) = i \quad \text{if and only if} \quad w \in V_i.$$

Our choices of $V_0, V_1, \ldots, V_k$ are guided by x^* in a way similar to that used by Christof and Reinelt [119] in their algorithm for finding cuts that match templates from a prescribed large catalog. First, we construct once and for all an equivalence relation on V such that each equivalence class V^* of this relation satisfies

$$x^*(V^*, V - V^*) = 2;$$

then we make many different choices of $V_0, V_1, \ldots, V_k$ such that each of $V_1, \ldots, V_k$ is one of these equivalence classes.

The equivalence relation is constructed by iteratively shrinking two-point sets into a single point. At each stage of this process, we have a set W and a mapping

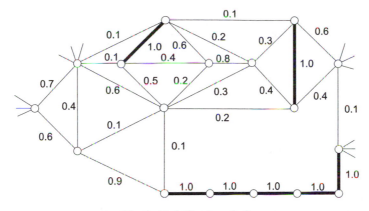

Figure 11.1 Portion of x^*.

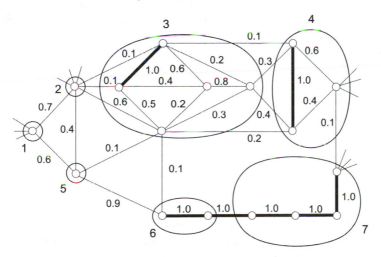

Figure 11.2 Equivalence classes.

$\pi : W \to 2^V$ that defines a partition of V into pairwise disjoint subsets $\pi(w)$ with $w \in W$. If there are distinct elements u, v, w of W such that

$$x^*(\pi(u), \pi(v)) = 1 \quad \text{and} \quad x^*(\pi(u), \pi(w)) + x^*(\pi(v), \pi(w)) = 1, \qquad (11.1)$$

then we replace $\pi(u)$ by $\pi(u) \cup \pi(v)$ and remove v from W; this operation is repeated until there are no u, v, w with property (11.1), in which case we stop. (During this process, we may discover pairs u, v with

$$x^*(\pi(u), \pi(v)) > 1,$$

in which case x^* violates the subtour inequality $x(Q, V - Q) \geq 2$ with $Q = \pi(u) \cup \pi(v)$.) An example of an x^* and the corresponding equivalence classes is given in Figure 11.1 and Figure 11.2.

To make the many different choices of $V_1, \ldots, V_k$, we first set the value of a parameter t that nearly determines the value of k in the sense that $t - 3 \leq k \leq t$.

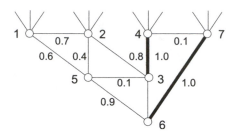

Figure 11.3 Choosing $V_1, V_2, \ldots, V_k$.

Then, for each w in W, we choose a subset C of W such that $w \in C$ and $t - 3 \le |C| \le t$; the corresponding $V_1, \ldots, V_k$ are the $\pi(v)$ with $v \in C$. The choice of C is guided by the graph with vertex set W, where u and v are adjacent if and only if $x^*(\pi(u), \pi(v)) > \varepsilon$ for some prescribed zero tolerance ε: starting at w, we carry out a breadth-first search through this graph, until we collect a set C of $t - 3$ vertices. (The graph for the example in Figure 11.2 is given in Figure 11.3.)

If there are any vertices u outside this C such that $x^*(\pi(u), \pi(v)) = 1$ for some v in C, then we keep adding these vertices u to C as long as $|C| \le t$. (It seems plausible that such a crude way of choosing C can be improved. However, we found its performance satisfactory; none of the alternatives that we tried appeared to work better.)

11.3 REVISIONIST POLICIES

Our computer code deviates from the scheme of Algorithm 11.1 in minor ways. When it comes to including a new cut in $\mathcal{L}$, we are more selective than Algorithm 11.1 suggests. We accept only cuts that have integer coefficients and integer right-hand sides; to produce such cuts, our variation on the theme of Algorithm 11.1 uses integer arithmetic whenever necessary. In addition, cuts that are violated only slightly by $\phi(x^*)$ are of little use to a branch-and-cut algorithm; instead of adding such a weak cut to $\mathcal{L}$, we move on to the next choice of σ as if $\phi(x^*)$ belonged to the convex hull of $\mathcal{T}$.

In certain additional cases, we may also fail to return a cut separating $\phi(x^*)$ from $\mathcal{T}$ even though $\phi(x^*)$ lies well outside the convex hull of $\mathcal{T}$. This happens whenever computations using integer arithmetic are about to create overflow and whenever the number of iterations or recursive calls of some procedure has exceeded a prescribed threshold. In such circumstances, we once again move on to the next choice of σ just as if $\phi(x^*)$ belonged to the convex hull of $\mathcal{T}$. We will take the reader's understanding of this policy for granted and carry on with our exposition in the idealized setting where all cuts are accepted, integers can be arbitrarily large, and we have all the time in the world.

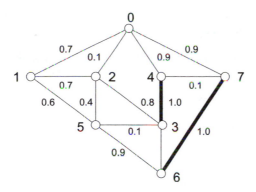

Figure 11.4 EXAMPLE 1.

11.4 DOES $\phi(X^*)$ LIE OUTSIDE THE CONVEX HULL OF $\mathcal{T}$?

11.4.1 EXAMPLE 1

∨ ∨ ∨

We will illustrate the way our code tests the condition

$$\phi(x^*) \text{ lies outside the convex hull of } \mathcal{T}$$

of Algorithm 11.1 on the example where $\overline{V} = \{0, 1, \ldots, 7\}$ and $\phi(x^*) = \overline{x}^*$ with

$$
\begin{aligned}
&\overline{x}_{01}^* = 0.7, \quad \overline{x}_{02}^* = 0.1, \quad \overline{x}_{04}^* = 0.9, \quad \overline{x}_{07}^* = 0.9, \quad \overline{x}_{12}^* = 0.7, \\
&\overline{x}_{15}^* = 0.6, \quad \overline{x}_{23}^* = 0.8, \quad \overline{x}_{25}^* = 0.4, \quad \overline{x}_{34}^* = 1.0, \quad \overline{x}_{35}^* = 0.1, \\
&\overline{x}_{36}^* = 0.1, \quad \overline{x}_{47}^* = 0.1, \quad \overline{x}_{56}^* = 0.9, \quad \overline{x}_{67}^* = 1.0,
\end{aligned}
$$

and $\overline{x}_{ij}^* = 0$ for all the remaining i and j such that $0 \leq i < j \leq 7$. This particular $\overline{x}^*$ is illustrated in Figure 11.4.

∧ ∧ ∧

11.4.2 Rephrasing the question

As in EXAMPLE 1, we will always write

$$\overline{x}^* = \phi(x^*),$$

so that (with $V_0, V_1, \ldots, V_k$ as in Section 11.2),

$$\overline{x}_{ij}^* = x^*(V_i, V_j) \quad \text{whenever } 0 \leq i < j \leq k.$$

∨ ∨ ∨

In EXAMPLE 1, $\overline{x}^*$ is a point in $\mathbf{R}^{28}$ and $\mathcal{T}$ is a subset of the same $\mathbf{R}^{28}$. And yet the problem of separating $\overline{x}^*$ from $\mathcal{T}$ is only 5-dimensional in the following sense: A readily available 5-dimensional affine subspace of the $\mathbf{R}^{28}$ includes $\overline{x}^*$ and intersects $\mathcal{T}$ in a set $\mathcal{T}^*$ such that

$\bar{x}^*$ belongs to the convex hull of $\mathcal{T}$
if and only if it belongs to the convex hull of $\mathcal{T}^*$.

Specifically, this 5-dimensional affine space consists of all the points $\bar{x}$ in $\mathbf{R}^{28}$ that satisfy the 23 equations

$$\bar{x}_{03} = 0, \quad \bar{x}_{05} = 0, \quad \bar{x}_{06} = 0, \quad \bar{x}_{13} = 0, \quad \bar{x}_{14} = 0, \quad \bar{x}_{16} = 0,$$
$$\bar{x}_{17} = 0, \quad \bar{x}_{24} = 0, \quad \bar{x}_{26} = 0, \quad \bar{x}_{27} = 0, \quad \bar{x}_{37} = 0, \quad \bar{x}_{45} = 0,$$
$$\bar{x}_{46} = 0, \quad \bar{x}_{57} = 0, \quad \bar{x}_{34} = 1, \quad \bar{x}_{67} = 1,$$

$$\bar{x}_{36} = 1 - \bar{x}_{56},$$
$$\bar{x}_{15} = 2 - \bar{x}_{25} - \bar{x}_{35} - \bar{x}_{56},$$
$$\bar{x}_{23} = 1 - \bar{x}_{35} - \bar{x}_{36},$$
$$\bar{x}_{01} = 2 - \bar{x}_{12} - \bar{x}_{15},$$
$$\bar{x}_{02} = 2 - \bar{x}_{12} - \bar{x}_{23} - \bar{x}_{25},$$
$$\bar{x}_{04} = 1 - \bar{x}_{47},$$
$$\bar{x}_{07} = 1 - \bar{x}_{47}.$$

(Each point x of this space is determined by the values of its components $\bar{x}_{56}$, $\bar{x}_{25}$, $\bar{x}_{35}$, $\bar{x}_{12}$, $\bar{x}_{47}$.) To see that $\bar{x}^*$ belongs to the convex hull of $\mathcal{T}$ only if it belongs to the convex hull of $\mathcal{T}^*$, note first that each point x of $\mathcal{T}$ satisfies the 14 inequalities

$$\bar{x}_{03} \geq 0, \quad \bar{x}_{05} \geq 0, \quad \bar{x}_{06} \geq 0, \quad \bar{x}_{13} \geq 0, \quad \bar{x}_{14} \geq 0,$$
$$\bar{x}_{16} \geq 0, \quad \bar{x}_{17} \geq 0, \quad \bar{x}_{24} \geq 0, \quad \bar{x}_{26} \geq 0, \quad \bar{x}_{27} \geq 0,$$
$$\bar{x}_{37} \geq 0, \quad \bar{x}_{45} \geq 0, \quad \bar{x}_{46} \geq 0, \quad \bar{x}_{57} \geq 0$$

and the seven inequalities

$$\bar{x}_{01} + \bar{x}_{12} + \bar{x}_{13} + \bar{x}_{14} + \bar{x}_{15} + \bar{x}_{16} + \bar{x}_{17} \geq 2,$$
$$\bar{x}_{02} + \bar{x}_{12} + \bar{x}_{23} + \bar{x}_{24} + \bar{x}_{25} + \bar{x}_{26} + \bar{x}_{27} \geq 2,$$
$$\bar{x}_{03} + \bar{x}_{13} + \bar{x}_{23} + \bar{x}_{34} + \bar{x}_{35} + \bar{x}_{36} + \bar{x}_{37} \geq 2,$$
$$\bar{x}_{04} + \bar{x}_{14} + \bar{x}_{24} + \bar{x}_{34} + \bar{x}_{45} + \bar{x}_{46} + \bar{x}_{47} \geq 2,$$
$$\bar{x}_{05} + \bar{x}_{15} + \bar{x}_{25} + \bar{x}_{35} + \bar{x}_{45} + \bar{x}_{56} + \bar{x}_{57} \geq 2,$$
$$\bar{x}_{06} + \bar{x}_{16} + \bar{x}_{26} + \bar{x}_{36} + \bar{x}_{46} + \bar{x}_{56} + \bar{x}_{67} \geq 2,$$
$$\bar{x}_{07} + \bar{x}_{17} + \bar{x}_{27} + \bar{x}_{37} + \bar{x}_{47} + \bar{x}_{57} + \bar{x}_{67} \geq 2$$

and the two inequalities

$$\begin{aligned} \bar{x}_{03} + \bar{x}_{13} + \bar{x}_{23} + \bar{x}_{35} + \bar{x}_{36} + \bar{x}_{37} + \\ \bar{x}_{04} + \bar{x}_{14} + \bar{x}_{24} + \bar{x}_{45} + \bar{x}_{46} + \bar{x}_{47} \quad &\geq 2, \\ \bar{x}_{06} + \bar{x}_{16} + \bar{x}_{26} + \bar{x}_{36} + \bar{x}_{46} + \bar{x}_{56} + \\ \bar{x}_{07} + \bar{x}_{17} + \bar{x}_{27} + \bar{x}_{37} + \bar{x}_{47} + \bar{x}_{57} \quad &\geq 2 \end{aligned}$$

and then that $\bar{x}^*$ satisfies each of these 23 inequalities as an equation; it follows that $\bar{x}^*$ belongs to the convex hull of $\mathcal{T}$ only if it belongs to the convex hull of those points in $\mathcal{T}$ that satisfy each of the 23 inequalities as an equation.

∧ ∧ ∧

In general (as in Section 11.2), we have $\overline{V} = \{0, 1, \ldots, k\}$ for some k; now $\bar{x}^*$ is a point in $\mathbf{R}^{(k+1)k/2}$ and $\mathcal{T}$ is a subset of the same $\mathbf{R}^{(k+1)k/2}$; a certain affine subspace of this $\mathbf{R}^{(k+1)k/2}$ (typically of a much smaller dimension than $(k+1)k/2$) includes $\bar{x}^*$ and intersects $\mathcal{T}$ in a set $\mathcal{T}^*$ such that

$\overline{x}^*$ belongs to the convex hull of $\mathcal{T}$
if and only if it belongs to the convex hull of $\mathcal{T}^*$.

Specifically, this affine subspace of $\mathbf{R}^{(k+1)k/2}$ consists of all the points x that satisfy the system of equations

$$\overline{x}_e = 0 \quad \text{for all } e \text{ such that } \overline{x}_e^* = 0,$$
$$\overline{x}_e = 1 \quad \text{for all } e \text{ such that } e \subset \{1, 2, \dots, k\} \text{ and } \overline{x}_e^* = 1,$$
$$\sum (\overline{x}_e : u \in e) = 2 \quad \text{for all } u \text{ in } \{1, 2, \dots, k\}.$$

To see that $\overline{x}^*$ belongs to the convex hull of $\mathcal{T}$ only if it belongs to the convex hull of $\mathcal{T}^*$, note first that each point x of $\mathcal{T}$ satisfies the system of inequalities

$$\overline{x}_e \geq 0 \quad \text{for all } e \text{ such that } \overline{x}_e^* = 0,$$
$$\sum (\overline{x}_e : u \in e) \geq 2 \quad \text{for all } u \text{ in } \{1, 2, \dots, k\},$$
$$\overline{x}(e, \overline{V} - e\} \geq 2 \quad \text{for all } e \text{ such that } e \subset \{1, 2, \dots, k\} \text{ and } \overline{x}_e^* = 1$$

and then that $\overline{x}^*$ satisfies each of these inequalities as an equation; it follows that $\overline{x}^*$ belongs to the convex hull of $\mathcal{T}$ only if it belongs to the convex hull of those points $\overline{x}$ in $\mathcal{T}$ that satisfy

$$\overline{x}_e = 0 \quad \text{for all } e \text{ such that } \overline{x}_e^* = 0,$$
$$\sum (\overline{x}_e : u \in e) = 2 \quad \text{for all } u \text{ in } \{1, 2, \dots, k\},$$
$$\overline{x}(e, \overline{V} - e\} = 2 \quad \text{for all } e \text{ such that } e \subset \{1, 2, \dots, k\} \text{ and } \overline{x}_e^* = 1.$$

We replace the condition

$$\overline{x}^* \text{ lies outside the convex hull of } \mathcal{T}$$

in Algorithm 11.1 by the equivalent condition

$$\overline{x}^* \text{ lies outside the convex hull of } \mathcal{T}^*.$$

11.4.3 Rephrasing the question again

Next, let

$\overline{E}$ denote the set of the edges of the complete graph on $\overline{V}$,

let

$\overline{E}_{1/2}$ denote the set of all the edges e such that
$e \subset \{1, 2, \dots, k\}, \overline{x}_e^* \neq 0, \overline{x}_e^* \neq 1$

and, for every vector $\overline{x}$ with components are indexed by elements of $\overline{E}$, let

$\psi(\overline{x})$ denote the restriction of $\overline{x}$
on its components indexed by elements of $\overline{E}_{1/2}$.

∨ ∨ ∨

In EXAMPLE 1, we have

$$\overline{E}_{1/2} = \{12, \ 15, \ 23, \ 25, \ 35, \ 36, \ 47, \ 56\}.$$

and

$$\psi(\overline{x}) = [\overline{x}_{12}, \ \overline{x}_{15}, \ \overline{x}_{23}, \ \overline{x}_{25}, \ \overline{x}_{35}, \ \overline{x}_{36}, \ \overline{x}_{47}, \ \overline{x}_{56}]^T.$$

The edge set $\overline{E}_{1/2}$ is illustrated in Figure 11.5.

∧ ∧ ∧

Let us point out that

$\overline{x}^*$ belongs to the convex hull of $\mathcal{T}^*$

if and only if $\psi(\overline{x}^*)$ belongs to the convex hull of $\psi(\mathcal{T}^*)$.

The "only if" part is trivial; to see the "if" part, observe that there is a linear mapping η such that

$$\eta(\psi(\overline{x})) = \overline{x} \quad \text{whenever } \overline{x} \in \mathcal{T}^* \cup \{\overline{x}^*\}.$$

∨ ∨ ∨

In EXAMPLE 1, this linear mapping sets

$$\overline{x}_{03} = 0, \quad \overline{x}_{05} = 0, \quad \overline{x}_{06} = 0, \quad \overline{x}_{13} = 0, \quad \overline{x}_{14} = 0, \quad \overline{x}_{16} = 0,$$
$$\overline{x}_{17} = 0, \quad \overline{x}_{24} = 0, \quad \overline{x}_{26} = 0, \quad \overline{x}_{27} = 0, \quad \overline{x}_{37} = 0, \quad \overline{x}_{45} = 0,$$
$$\overline{x}_{46} = 0, \quad \overline{x}_{57} = 0, \quad \overline{x}_{34} = 1, \quad \overline{x}_{67} = 1,$$

and

$$\overline{x}_{01} = 2 - \overline{x}_{12} - \overline{x}_{15},$$
$$\overline{x}_{02} = 2 - \overline{x}_{12} - \overline{x}_{23} - \overline{x}_{25},$$
$$\overline{x}_{04} = 1 - \overline{x}_{47},$$
$$\overline{x}_{07} = 1 - \overline{x}_{47}.$$

∧ ∧ ∧

In general, η sets

$$\overline{x}_e = 0 \quad \text{whenever } e \subset \{1, 2, \ldots, k\} \text{ and } \overline{x}_e^* = 0,$$
$$\overline{x}_e = 1 \quad \text{whenever } e \subset \{1, 2, \ldots, k\} \text{ and } \overline{x}_e^* = 1,$$

and

$$\overline{x}_{0i} = 2 - \sum(\overline{x}_{ij} : 1 \le j \le k, \ j \ne i) \quad \text{whenever } i \in \{1, 2, \ldots, k\}.$$

This reduction in the number of variables could be carried a little further. For instance, in EXAMPLE 1 we could set

$$\psi(\overline{x}) = [\overline{x}_{12}, \ \overline{x}_{25}, \ \overline{x}_{35}, \ \overline{x}_{47}, \ \overline{x}_{56}]^T$$

with η recovering $\overline{x}_{36}, \overline{x}_{15}, \overline{x}_{23}$, from the new $\psi(\overline{x})$ by the formulas

$$\overline{x}_{36} = 1 - \overline{x}_{56},$$
$$\overline{x}_{15} = 2 - \overline{x}_{25} - \overline{x}_{35} - \overline{x}_{56},$$
$$\overline{x}_{23} = 1 - \overline{x}_{35} - \overline{x}_{36}.$$

Such variations would certainly make the code more complicated; our experience suggests that they would not make it much faster.

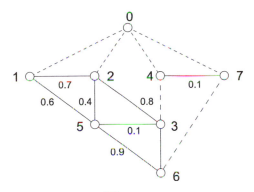

Figure 11.5 $\overline{E}_{1/2}$ for EXAMPLE 1.

11.4.4 A linear programming formulation of the question

The problem of testing a prescribed point for membership in the convex hull of a prescribed set has a number of linear programming formulations. By definition, $\psi(\overline{x}^*)$ belongs to the convex hull of $\psi(\mathcal{T}^*)$ if and only if, with

> A standing for the matrix whose columns are the elements of $\psi(\mathcal{T}^*)$

and with

> e standing for the vector $[1, 1, \ldots, 1]^T$
> whose number of components is determined by the context,

there is a solution λ of the system

$$A\lambda = \psi(\overline{x}^*), \quad e^T\lambda = 1, \quad \lambda \geq 0. \tag{11.2}$$

The duality principle of linear programming guarantees that (11.2) has no solution if and only if there are a vector a and a scalar b such that

$$a^TA - be^T \leq 0, \quad a^T\psi(\overline{x}^*) - b > 0; \tag{11.3}$$

solutions of (11.3) are in a one-to-one correspondence with cuts $a^T\xi \leq b$ that separate $\psi(\overline{x}^*)$ from $\psi(\mathcal{T}^*)$. Rather than settling for any such cut, we want to find a most violated cut; this objective may be interpreted as maximizing $a^T\psi(\overline{x}^*) - b$ subject to a being normalized in some sense; we choose the sense of $\|a\|_1 = 1$. The problem

$$\begin{aligned} \text{maximize} \quad & a^T\psi(\overline{x}^*) - b \\ \text{subject to} \quad & \psi(\mathcal{T}) \subset \{\xi : a^T\xi \leq b\}, \\ & \|a\|_1 = 1 \end{aligned} \tag{11.4}$$

may be formulated as

$$\begin{aligned} \text{maximize} \quad & a^T\psi(\overline{x}^*) - b \\ \text{subject to} \quad & a^TA - be^T \leq 0, \\ & a - u + v = 0, \\ & e^T(u + v) = 1, \\ & u \geq 0, \quad v \geq 0. \end{aligned} \tag{11.5}$$

Folk wisdom has it that the number of simplex iterations required to solve an LP problem tends to grow linearly in the number of constraints other than bounds on individual variables and logarithmically in the number of variables. This premise suggests solving (11.5) by applying the simplex method to its dual,

$$
\begin{aligned}
\text{minimize} \quad & s \\
\text{subject to} \quad A\lambda + w &= \psi(\overline{x}^*), \\
e^T\lambda &= 1, \\
-w + se^T &\geq 0, \\
w + se^T &\geq 0, \\
\lambda &\geq 0:
\end{aligned}
\tag{11.6}
$$

typically, the number N of columns of A is much larger than the number M of its rows; problem (11.5) has $M + N + 1$ constraints other than bounds on individual variables; problem (11.6) has $3M + 1$ constraints other than bounds on individual variables. (By the way, (11.6) formulates the problem of finding a point in the convex hull of $\psi(\mathcal{T}^*)$ that is closest to $\psi(\overline{x}^*)$ in the ℓ_∞ metric.)

Our code tests $\psi(\overline{x}^*)$ for membership in the convex hull of $\psi(\mathcal{T}^*)$ by solving a close relative of (11.6) that has only $M + 1$ constraints other than bounds on individual variables:

$$
\begin{aligned}
\text{maximize} \quad & s \\
\text{subject to} \quad s\psi(\overline{x}^*) - A\lambda + w &= 0, \\
-s + e^T\lambda &= 0, \\
\lambda \geq 0, \quad -e \leq w &\leq e.
\end{aligned}
\tag{11.7}
$$

Since the origin is a feasible solution of (11.7), this problem either has an optimal solution or else is unbounded. In the former case, the simplex method applied to (11.7) finds also an optimal solution of its dual,

$$
\begin{aligned}
\text{minimize} \quad & e^T(u + v) \\
\text{subject to} \quad -a^T A + be^T &\geq 0, \\
a^T\psi(\overline{x}^*) - b &= 1, \\
a + u - v &= 0, \\
u \geq 0, \quad v &\geq 0;
\end{aligned}
\tag{11.8}
$$

in the latter case, (11.8) is infeasible. Problem (11.8) formulates the problem

$$
\begin{aligned}
\text{minimize} \quad & \|a\|_1 \\
\text{subject to} \quad & \psi(\mathcal{T}) \subset \{\xi : a^T\xi \leq b\}, \\
& a^T\psi(\overline{x}^*) = b + 1;
\end{aligned}
\tag{11.9}
$$

problems (11.9) and (11.4) are closely related in the sense that

- if a, b is a feasible solution of (11.9),
  then $a/\|a\|_1, b/\|a\|_1$ is a feasible solution of (11.4);

- if a, b is a feasible solution of (11.4) such that $a^T\psi(\overline{x}^*) > b$,
  then $a/(a^T\psi(\overline{x}^*) - b), b/(a^T\psi(\overline{x}^*) - b)$ is a feasible solution of (11.9).

For every positive t, this relationship establishes a one-to-one correspondence between feasible solutions of (11.9) which the objective function maps to t and feasible solutions of (11.4) which the objective function maps to $1/t$; in particular,

- if a, b is an optimal solution of (11.9),
  then $a/\|a\|_1, b/\|a\|_1$ is an optimal solution of (11.4);

- if (11.4) has no feasible solution,
  then the optimal value of (11.9) is at most zero.

11.4.5 Solving (11.7) by delayed column generation

In general, $\mathcal{T}^*$ may be so large that writing down all the columns of A is out of the question; however, it may be possible to solve (11.7) even in these circumstances. The trick, introduced by Ford and Fulkerson [188] and by Jewell [288], is known as *delayed column generation*. We begin with a restricted version of (11.7), where every column of A comes from $\psi(\mathcal{T}^*)$, but the columns of A may constitute only a subset of $\psi(\mathcal{T}^*)$. Applying the simplex method to (11.7), we either discover that the problem is unbounded or else find its optimal solution. In the former case, $\overline{x}^*$ belongs to the convex hull of $\mathcal{T}$ and we are done. In the latter case, the simplex method has also computed the optimal solution of the dual problem,

$$
\begin{array}{ll}
\text{minimize} & u^T e + v^T e \\
\text{subject to} & -a^T A + b e^T \geq 0, \\
& a^T \psi(\overline{x}^*) - b = 1, \\
& a^T + u^T - v^T = 0, \\
& u^T, v^T \geq 0;
\end{array} \tag{11.10}
$$

If all the elements $\overline{x}$ of $\mathcal{T}^*$ satisfy $a^T \psi(\overline{x}) \leq b$, then $\overline{x}^*$ lies outside the convex hull of $\mathcal{T}$ (to see this, note that $a^T \psi(\overline{x}^*) > b$) and we are done; else some $\overline{x}$ in $\mathcal{T}^*$ satisfies $a^T \psi(\overline{x}) > b$, in which case we add $\psi(\overline{x})$ to A as a new column and iterate.

To generate the columns of A, we use an oracle that,

given   an integer vector p
         with components indexed by elements of $\overline{E}_{1/2}$,
returns   either an $\overline{x}$ that maximizes $p^T \psi(\overline{x})$ subject to $\overline{x} \in \mathcal{T}^*$
         or the message "infeasible" indicating that $\mathcal{T}^*$ is empty.

(An implementation of this oracle will be described in Section 11.8.) Since the simplex method computes a in floating point arithmetic, this oracle cannot be used directly to maximize $a^T \psi(\overline{x})$ over $\mathcal{T}^*$. Instead, we

- compute an integer vector p that approximates closely a multiple of a,
- let r denote the largest component of $p^T A$,

and then we use p, r in place of a, b. More precisely, we call the oracle to

- find an $\overline{x}$ that maximizes $p^T \psi(\overline{x})$ subject to $\overline{x} \in \mathcal{T}^*$

and then we distinguish between two cases.

CASE 1: $p^T \psi(\overline{x}) > r$. We add $\psi(\overline{x})$ to A as a new column and proceed to the next iteration.

CASE 2: $p^T \psi(\overline{x}) = r$. If $p^T \psi(\overline{x}^*) > r + \varepsilon \|p\|_1$ for a prescribed positive ε, then $\overline{x}^*$ lies outside the convex hull of $\mathcal{T}$; else we move on to the next choice of σ.

We may initialize A as the matrix with no columns at all. In this case, (11.7) has an optimal solution; the dual problem (11.10) reads

$$
\begin{aligned}
\text{minimize} \quad & u^T e + v^T e \\
\text{subject to} \quad & a^T \psi(\overline{x}^*) - b && = 1, \\
& a^T + u^T - v^T && = 0, \\
& u^T, v^T \geq 0;
\end{aligned}
$$

and its optimal solution is $a = u = v = 0, b = -1$. Now the first iteration amounts to calling the oracle to maximize $0^T \psi(\overline{x})$ over $\mathcal{T}^*$. If the oracle returns an $\overline{x}$, then we make $\psi(\overline{x})$ the first column of A and proceed to the second iteration; if the oracle returns the message "infeasible," then $\overline{x}^*$ lies outside the convex hull of $\mathcal{T}$.

∨ ∨ ∨

In EXAMPLE 1, we have

$$
\overline{E}_{1/2} = \{12,\ 15,\ 23,\ 25,\ 35,\ 36,\ 47,\ 56\}
$$

and

$$
\psi(\overline{x}) = [\overline{x}_{12},\ \overline{x}_{15},\ \overline{x}_{23},\ \overline{x}_{25},\ \overline{x}_{35},\ \overline{x}_{36},\ \overline{x}_{47},\ \overline{x}_{56}]^T;
$$

in particular,

$$
\psi(\overline{x}^*) = [0.7,\ 0.6,\ 0.8,\ 0.4,\ 0.1,\ 0.1,\ 0.1,\ 0.9]^T.
$$

Let

$$
w_{12},\ w_{15},\ w_{23},\ w_{25},\ w_{35},\ w_{36},\ w_{47},\ w_{56}
$$

denote the eight components of the vector w featured in problem (11.7).

ITERATION 1: We call the oracle to maximize $0^T \psi(\overline{x})$ subject to $\overline{x} \in \mathcal{T}^*$. The oracle returns (say) the element $\overline{x}$ that has

$$
\psi(\overline{x}) = [1,\ 1,\ 1,\ 0,\ 0,\ 0,\ 0,\ 1]^T.
$$

We set up the linear programming problem (11.7) with $\psi(\overline{x})$ being the only column of A.

ITERATION 2: Problem (11.7) has an optimal solution determined by declaring w_{15} to be the unique nonbasic variable and setting its value at one. The simplex method computes

$$
a^T = [0,\ -2.5,\ 0,\ 0,\ 0,\ 0,\ 0,\ 0] \quad \text{and} \quad b = -2.5
$$

as a part of an optimal solution of (11.10); the oracle, called to

$$
\text{maximize } -\overline{x}_{15} \text{ subject to } \overline{x} \in \mathcal{T}^*,
$$

returns (say) the element $\overline{x}$ of $\mathcal{T}^*$ that has

$$
\psi(\overline{x}) = [1,\ 0,\ 0,\ 0,\ 1,\ 0,\ 0,\ 1]^T;
$$

we add $\psi(\overline{x})$ to A as a new column.

ITERATION 3: Problem (11.7) has an optimal solution determined by declaring the variables w_{15}, w_{25} to be nonbasic and setting $w_{15} = 1$, $w_{25} = -1$. The simplex method computes

$$a^T = [0,\ 0,\ 0,\ 2.5,\ 0,\ 0,\ 0,\ 0] \quad \text{and} \quad b = 0$$

as a part of an optimal solution of (11.10); the oracle, called to

$$\text{maximize } \overline{x}_{25} \text{ subject to } \overline{x} \in \mathcal{T}^*,$$

returns (say) the element $\overline{x}$ of $\mathcal{T}^*$ that has

$$\psi(\overline{x}) = [0,\ 1,\ 0,\ 1,\ 0,\ 1,\ 0,\ 0]^T;$$

we add $\psi(\overline{x})$ to A as a new column.

ITERATION 4: Problem (11.7) has an optimal solution determined by declaring the variables w_{15}, w_{23}, w_{25} to be nonbasic and setting $w_{15} = 1$, $w_{23} = -1$, $w_{25} = -1$. The simplex method computes

$$a^T = [0,\ -1.667,\ 1.667,\ 1.667,\ 0,\ 0,\ 0,\ 0] \quad \text{and} \quad b = 0$$

as a part of an optimal solution of (11.10); the oracle, called to

$$\text{maximize } -\overline{x}_{15} + \overline{x}_{23} + \overline{x}_{25} \text{ subject to } \overline{x} \in \mathcal{T}^*,$$

returns the element $\overline{x}$ of $\mathcal{T}^*$ that has

$$\psi(\overline{x}) = [0,\ 0,\ 1,\ 1,\ 0,\ 0,\ 0,\ 1]^T;$$

we add $\psi(\overline{x})$ to A as a new column.

ITERATION 5: Problem (11.7) has an optimal solution determined by declaring the variables $w_{15}, w_{23}, w_{25}, w_{47}$ to be nonbasic and setting $w_{15} = 1$, $w_{23} = -1$, $w_{25} = -1$, $w_{47} = -1$. The simplex method computes

$$a^T = [0,\ 0,\ 0,\ 0,\ 0,\ 0,\ 10,\ 0] \quad \text{and} \quad b = 0$$

as a part of an optimal solution of (11.10); the oracle, called to

$$\text{maximize } \overline{x}_{47} \text{ subject to } \overline{x} \in \mathcal{T}^*,$$

returns the element $\overline{x}$ of $\mathcal{T}^*$ that has

$$\psi(\overline{x}) = [0,\ 1,\ 1,\ 0,\ 0,\ 0,\ 1,\ 1]^T;$$

we add $\psi(\overline{x})$ to A as a new column.

ITERATION 6: Problem (11.7) has an optimal solution determined by declaring the variables $w_{12}, w_{15}, w_{23}, w_{25}, w_{47}$ to be nonbasic and setting $w_{12} = -1$, $w_{15} = 1$, $w_{23} = -1$, $w_{25} = -1$, $w_{47} = -1$. The simplex method computes

$$a^T = [5,\ 0,\ 0,\ 5,\ 0,\ 0,\ 5,\ 0] \quad \text{and} \quad b = 5$$

as a part of an optimal solution of (11.10); the oracle, called to

$$\text{maximize } \overline{x}_{12} + \overline{x}_{25} + \overline{x}_{47} \text{ subject to } \overline{x} \in \mathcal{T}^*,$$

returns an element $\overline{x}$ of $\mathcal{T}^*$ that has $\overline{x}_{12} + \overline{x}_{25} + \overline{x}_{47} = 1$. Since

$$\overline{x}_{12}^* + \overline{x}_{25}^* + \overline{x}_{47}^* = 1.2,$$

we conclude that $\overline{x}^*$ lies outside the convex hull of $\mathcal{T}$.

∧ ∧ ∧

11.4.6 Finding a close integer approximation of a multiple of a

To compute a close integer approximation p of a multiple of a, we first scale a so that the largest of the absolute values of its components equals one: a simple argument involving Cramer's rule (we will give this argument later) suggests that $a/\|a\|_\infty$ stands a relatively good chance of having the form p/q such that p is an integer vector and q is a positive integer smaller than the largest allowed value INT_MAX of an integer variable (normally $2^{31} - 1$).

Now $\|a\|_\infty = 1$ and we examine the components of a one by one in order to make an educated guess of the value of q: we use the continued fraction method to approximate each component a_e of a by a fraction p_e/q_e independently of all the other components of a. If the least common multiple of all the denominators q_e turns out to be at most INT_MAX, then we take it to be our q; else (the case that we never came across in our computations) our code falls back on

$$q = \operatorname{lcm}(1, 2, \dots, 18) = 12,252,240$$

(which is less than INT_MAX and yet makes qa_e an integer whenever a_e is a fraction with a very small denominator). Once the value of q has been set, we let p be the vector qa with each component rounded to the nearest integer.

CRAMER'S RULE AND THE SCALING OF a

In the fourth iteration in EXAMPLE 1, we maximize s subject to

$$
\begin{bmatrix}
0.7 & -1 & -1 & & 1 & & & & & & \\
0.6 & -1 & & & & -1 & 1 & & & & \\
0.8 & -1 & & & & & & 1 & & & \\
0.4 & & & & & -1 & & & 1 & & \\
0.1 & & -1 & & & & & & & 1 & \\
0.1 & & & -1 & & & & & & & 1 \\
0.1 & & & & & & & & & & & 1 \\
0.9 & -1 & -1 & & & & & & & & & & 1 \\
-1 & 1 & 1 & 1 & & & & & & & & &
\end{bmatrix}
\cdot
\begin{bmatrix}
s \\
\lambda_1 \\
\lambda_2 \\
\lambda_3 \\
w_{12} \\
w_{15} \\
w_{23} \\
w_{25} \\
w_{35} \\
w_{36} \\
w_{47} \\
w_{56}
\end{bmatrix}
=
\begin{bmatrix}
0 \\
0 \\
0 \\
0 \\
0 \\
0 \\
0 \\
0 \\
0
\end{bmatrix},
$$

$\lambda_i \geq 0$ for all three i, and $-1 \leq w_e \leq 1$ for all eight e. In the optimal solution, w_{15}, w_{23}, and w_{25} are nonbasic, and so the optimal basis B is the matrix

$$\begin{bmatrix} 0.7 & -1 & -1 & & 1 & & & & \\ 0.6 & -1 & & -1 & & & & & \\ 0.8 & -1 & & & & & & & \\ 0.4 & & & -1 & & & & & \\ 0.1 & & -1 & & 1 & & & & \\ 0.1 & & & -1 & & 1 & & & \\ 0.1 & & & & & & 1 & & \\ 0.9 & -1 & -1 & & & & & 1 & \\ -1 & 1 & 1 & 1 & & & & & \end{bmatrix}.$$

The simplex method computes a and b by solving the system

$$[a_1, a_2, a_3, a_4, a_5, a_6, a_7, a_8, b] \cdot B = [1, 0, 0, 0, 0, 0, 0, 0, 0]. \tag{11.11}$$

The eight equations in this system that correspond to the last eight columns of B form a homogenous system; the set of its solutions is the set of all multiples of

$$[0, -1.667, 1.667, 1.667, 0, 0, 0, 0, 0]; \tag{11.12}$$

we can create a system of equations whose unique solution is a prescribed multiple of (11.12) simply by changing the first column of B in (11.11). The unique solution of the system

$$[\tilde{a}_1, \tilde{a}_2, \tilde{a}_3, \tilde{a}_4, \tilde{a}_5, \tilde{a}_6, \tilde{a}_7, \tilde{a}_8, \tilde{b}] \cdot \tilde{B} = [1, 0, 0, 0, 0, 0, 0, 0, 0] \tag{11.13}$$

with

$$\tilde{B} = \begin{bmatrix} & -1 & -1 & & 1 & & & & \\ -1 & -1 & & -1 & & & & & \\ & -1 & & & & & & & \\ & & & -1 & & & & & \\ & & -1 & & 1 & & & & \\ & & & -1 & & 1 & & & \\ & & & & & & 1 & & \\ & -1 & -1 & & & & & 1 & \\ & 1 & 1 & 1 & & & & & \end{bmatrix}$$

is the vector

$$[0, -1, 1, 1, 0, 0, 0, 0, 0], \tag{11.14}$$

which is the unique multiple of (11.12) with the second component equal to -1; the unique solution of the system (11.13) with

$$\tilde{B} = \begin{bmatrix} & -1 & -1 & & 1 & & & & \\ & -1 & & -1 & & & & & \\ & -1 & & & & & & & \\ 1 & & & -1 & & & & & \\ & & -1 & & 1 & & & & \\ & & & -1 & & 1 & & & \\ & & & & & & 1 & & \\ & -1 & -1 & & & & & 1 & \\ & 1 & 1 & 1 & & & & & \end{bmatrix}$$

is again the vector (11.14), which is the unique multiple of (11.12) with the fourth component equal to 1; and so on.

In general, as s is the unique free variable in (11.7), it is basic, and so $\begin{bmatrix} \psi(\overline{x}^*) \\ -1 \end{bmatrix}$ is the first column of the optimal basis B. The simplex method computes a and b by solving the system

$$[a^T, b] \cdot B = [1, 0, 0, \dots, 0]. \tag{11.15}$$

Removing the equation that corresponds to the first column of B from this system, we get a homogenous system of equations; the set of its solutions is the set of all multiples of the unique solution $[a^T, b]$ of (11.15). Now consider a component of a whose absolute value is the largest; let us say that this component is a_j, the j-th component of a; let $\tilde{B}$ denote the matrix obtained from B by overwriting its first column with zeros everywhere except in the j-th row, where the entry is overwritten with ± 1 and its sign is the sign of a_j. The unique solution $[\tilde{a}^T, \tilde{b}]$ of the system

$$[\tilde{a}^T, \tilde{b}] \cdot \tilde{B} = [1, 0, 0, \dots, 0]$$

is the multiple of $[a^T, b]$ that satisfies $\|\tilde{a}\|_\infty = 1$; by Cramer's rule, the components $\tilde{a}_1, \tilde{a}_2, \dots$ of $\tilde{a}$ satisfy

$$\tilde{a}_i = \frac{\det \tilde{B}_i}{\det \tilde{B}}$$

with $\tilde{B}_i$ the matrix obtained when the i-th row of $\tilde{B}$ is replaced by $[1, 0, 0, \dots, 0]$.

To summarize, $\det \tilde{B}$ is an integer and there is an integer vector p such that

$$\frac{a}{\|a\|_\infty} = \frac{p}{\det \tilde{B}}.$$

Trivially, $|\det \tilde{B}|$ is at most as large as the largest determinant of a zero-one $|E_{1/2}| \times |E_{1/2}|$ matrix. This bound is provably less than $2^{31} - 1$ whenever $|E_{1/2}| \leq 21$, but there are zero-one 23×23 matrices with determinants larger than $2^{31} - 1$. In any case, the upper bound has little relevance to our code: our vectors $a/\|a\|_\infty$ seem to be well approximated by p/q such that p is an integer vector and q is a positive integer less than 2^{31}.

THE CONTINUED FRACTION METHOD

This method, given a real number ξ such that $0 < \xi < 1$, produces a sequence of rational approximations of ξ; its name derives from the sequence of identities

$$\xi = \frac{1}{\xi_1}, \quad \xi = \frac{1}{t_1 + \dfrac{1}{\xi_2}}, \quad \xi = \frac{1}{t_1 + \dfrac{1}{t_2 + \dfrac{1}{\xi_3}}}, \quad \text{and so on,}$$

where each t_i is a positive integer and each ξ_i is a real number greater than one. If an ξ_i happens to be an integer, then (ξ is rational and) the sequence stops at its i-th term; else we construct the next term by replacing ξ_i with

$$t_i + \frac{1}{\xi_{i+1}}$$

such that t_i is a positive integer and ξ_{i+1} is a real number greater than one; the only way to do this is to set

$$t_i = \lfloor \xi_i \rfloor, \quad \xi_{i+1} = \frac{1}{\xi_i - t_i}.$$

The rational numbers r_i obtained when ξ_i is replaced by t_i in the right-hand side of the i-th identity,

$$r_1 = \frac{1}{t_1}, \quad r_2 = \frac{1}{t_1 + \dfrac{1}{t_2}}, \quad r_3 = \frac{1}{t_1 + \dfrac{1}{t_2 + \dfrac{1}{t_3}}}, \quad \text{and so on,}$$

are called *convergents for* ξ. A representation $p_1/q_1, p_2/q_2, p_3/q_3, \dots$ of the sequence $r_1, r_2, r_3, \dots$ such that p_i, q_i are integers may be computed from the recurrence relation

$$p_i = t_i p_{i-1} + p_{i-2},$$
$$q_i = t_i q_{i-1} + q_{i-2},$$

with the initial conditions $p_{-1} = 1$, $q_{-1} = 0$, $p_0 = 0$, $q_0 = 1$: we have

$$r_1 = \frac{1}{t_1}, \quad r_2 = \frac{t_2}{t_2 t_1 + 1}, \quad r_3 = \frac{t_3 t_2 + 1}{t_3(t_2 t_1 + 1) + t_1}, \quad \text{and so on.}$$

For instance, if $\xi = 0.2866817155$, then

$$
\begin{aligned}
\xi_1 &= 3.4881889772980655963738991927\ldots, & t_1 &= 3, \\
\xi_2 &= 2.0483870929135015671321466532\ldots, & t_2 &= 2, \\
\xi_3 &= 20.666668315611240003611926948\ldots, & t_3 &= 20, \\
\xi_4 &= 1.4999962898838866556204881250\ldots, & t_4 &= 1, \\
\xi_5 &= 2.0000148405745738872337833497\ldots, & t_5 &= 2, \\
\xi_6 &= 67382.835820895522388059701492\ldots, & t_6 &= 67382, \\
\xi_7 &= 1.1964285714285714285714285714\ldots, & t_7 &= 1, \\
\xi_8 &= 5.0909090909090909090909090909\ldots, & t_8 &= 5, \\
\xi_9 &= 11, & t_9 &= 11
\end{aligned}
$$

and the convergents for ξ are

$$
\begin{aligned}
p_1/q_1 &= 1/3 & &= 0.3333333333333333333333333333\ldots, \\
p_2/q_2 &= 2/7 & &= 0.2857142857142857142857142857\ldots, \\
p_3/q_3 &= 41/143 & &= 0.2867132867132867132867132867\ldots, \\
p_4/q_4 &= 43/150 & &= 0.2866666666666666666666666666\ldots, \\
p_5/q_5 &= 127/443 & &= 0.2866817155756207674943566659\ldots, \\
p_6/q_6 &= 8557557/29850376 & &= 0.2866817154999999061988364903\ldots, \\
p_7/q_7 &= 8557684/29850819 & &= 0.2866817155000000184249551075\ldots, \\
p_8/q_8 &= 51345977/179104471 & &= 0.2866817154999999997208333230\ldots, \\
p_9/q_9 &= 573363431/2000000000 & &= 0.2866817155.
\end{aligned}
$$

In general, the convergents for ξ have a number of remarkable properties that make them excellent approximations of ξ. In particular, if ξ approximates closely a rational number p/q with a small denominator, then the continued fraction method will recover p and q from ξ. More precisely, a theorem of Legendre (see, for instance, Schrijver [494], Theorem 6.3) states that

if p and q are positive integers and if ξ is a real number
such that $0 < \xi < 1$ and

$$\left| \xi - \frac{p}{q} \right| < \frac{1}{2q^2} \, ,$$

then p/q is a convergent for ξ.

Our use of the continued fraction method is motivated by this theorem.

As approximations of ξ, the convergents $p_1/q_1, p_2/q_2, p_3/q_3, \ldots$ become progressively less desirable in the sense that their denominators increase, and they tend to become progressively more desirable in the sense that

$$\left| \xi - \frac{p_i}{q_i} \right| < \frac{1}{q_i^2} \quad \text{for all } i.$$

This trade-off and the fact that we have no useful information on the precision with which the a and the b computed by the simplex method approximate the exact solution of (11.15) conspire to obscure the choice of a convergent that is best for our purpose. The choice made by our computer code is this:

($\star$) as soon as $\xi_i - t_i$ gets smaller than a prescribed threshold ε
(10^{-5} is one representative value of ε),
we declare p_i/q_i to be our approximation of ξ
and cease computing further convergents for ξ.

Three mutually related heuristic arguments in support of this stopping rule go as follows. If $\xi_i - t_i = 0$, then $\xi = p_i/q_i$ and further convergents for ξ simply do not exist; this fact suggests rule ($\star$) with ε a zero tolerance. If $\xi_i - t_i > 0$, then

$$t_{i+1} = \left\lfloor \frac{1}{\xi_i - t_i} \right\rfloor ,$$

and so $\xi_i - t_i$ is very small if and only if t_{i+1} is very large. We have

$$t_{i+1} = \left\lfloor \frac{|\xi q_{i-1} - p_{i-1}|}{|\xi q_i - p_i|} \right\rfloor$$

and the quantity $|\xi q - p|$ normalizes the error of the approximation $\xi \approx p/q$ with respect to the denominator. In this sense, $\xi_i - t_i$ is very small if and only if the approximation $\xi \approx p_i/q_i$ is very much better than the approximation $\xi \approx p_{i-1}/q_{i-1}$; one may suspect that the first dramatic improvement indicates p_i/q_i hitting its target. Finally, we have

$$q_{i+1} = t_{i+1}q_i + q_{i-1} \text{ and } 1 \leq q_{i-1} < q_i,$$

and so $\xi_i - t_i$ is very small if and only if q_{i+1}/q_i is very large; when q_{i+1}/q_i becomes very large, one could argue that the approximation $\xi \approx p_i/q_i$ was already good enough and that the transition to $\xi \approx p_{i+1}/q_{i+1}$ must have been forced only to account for inaccuracies in the computed value of ξ.

Stopping rule ($\star$) may not apply before overflow occurs; if j is the smallest subscript such that $q_j > \texttt{INT\_MAX}$ and if $\xi_i - t_i \geq \varepsilon$ for all $i = 1, 2, \ldots, j - 1$, then we choose the approximation $\xi \approx p_i/q_i$ that minimizes $\xi_i - t_i$ subject to $1 \leq i < j$.

11.5 SEPARATING $\phi(X^*)$ FROM $\mathcal{T}$: THE THREE PHASES

Whenever we set out to separate $\bar{x}^*$ from $\mathcal{T}$, we know not only that $\bar{x}^*$ lies outside the convex hull of $\mathcal{T}$: we also have an integer vector p and an integer r such that

$$\mathcal{T}^* \subset \{\bar{x} : p^T \psi(\bar{x}) \le r\} \quad \text{and} \quad p^T \psi(\bar{x}^*) > r + \varepsilon||p||_1.$$

To convert the cut $p^T \psi(\bar{x}) \le r$ that separates $\bar{x}^*$ from $\mathcal{T}^*$ into a hypergraph cut that separates $\bar{x}^*$ from $\mathcal{T}$, we proceed in three phases. These three phases correspond to the three set inclusions in the chain

$$\mathcal{T}^* \subseteq \mathcal{T}'' \subseteq \mathcal{T}' \subseteq \mathcal{T}$$

with $\mathcal{T}'$ and $\mathcal{T}''$ defined as follows:

- $\mathcal{T}'$ consists of all $\bar{x}$ in $\mathcal{T}$ that satisfy

$$\sum(\bar{x}_e : w \in e) = 2 \quad \text{for all } w \text{ in } \{1, 2, \dots, k\};$$

- $\mathcal{T}''$ consists of all $\bar{x}$ in $\mathcal{T}'$ that satisfy

$$\bar{x}_e = 0 \quad \text{for all } e \text{ such that } e \subset \{1, 2, \dots, k\} \text{ and } \bar{x}_e^* = 0,$$
$$\bar{x}_e = 1 \quad \text{for all } e \text{ such that } e \subset \{1, 2, \dots, k\} \text{ and } \bar{x}_e^* = 1.$$

In this notation,

- $\mathcal{T}^*$ consists of all $\bar{x}$ in $\mathcal{T}''$ that satisfy

$$\bar{x}_{0v} = 0 \quad \text{for all } v \text{ in } \{1, 2, \dots, k\} \text{ such that } \bar{x}_{0v}^* = 0.$$

The four sets $\mathcal{T}^* \subseteq \mathcal{T}'' \subseteq \mathcal{T}' \subseteq \mathcal{T}$ are illustrated in Figure 11.6.

PHASE 1: We find an integer vector a'' and an integer b'' such that the inequality

$$a''^T \bar{x} \le b''$$

separates $\bar{x}^*$ from $\mathcal{T}''$, induces a facet of the convex hull of $\mathcal{T}''$, and satisfies

$$\{\bar{x} \in \mathcal{T} : p^T \psi(\bar{x}) = r\} \subseteq \{\bar{x} \in \mathcal{T} : a''^T \bar{x} = b''\}.$$

PHASE 2: With the notation

$$\overline{E}_0 = \{e : e \subset \{1, 2, \dots, k\}, \bar{x}_e^* = 0\},$$
$$\overline{E}_1 = \{e : e \subset \{1, 2, \dots, k\}, \bar{x}_e^* = 1\},$$

we find integers a'_e ($e \in \overline{E}_0 \cup \overline{E}_1$) such that the inequality

$$a''^T \bar{x} + \sum(a'_e \bar{x}_e : e \in \overline{E}_0 \cup \overline{E}_1) \le b'' + \sum(a'_e : e \in \overline{E}_1)$$

induces a facet of the convex hull of $\mathcal{T}'$. (Note that this inequality is violated by $\bar{x}^*$.)

PHASE 3: Writing

$$a'^T \bar{x} \le b'$$

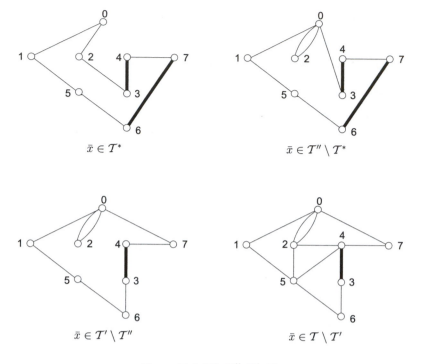

$\bar{x} \in \mathcal{T}^*$ $\bar{x} \in \mathcal{T}'' \setminus \mathcal{T}^*$ $\bar{x} \in \mathcal{T}' \setminus \mathcal{T}''$ $\bar{x} \in \mathcal{T} \setminus \mathcal{T}'$

Figure 11.6 $\mathcal{T}^*, \mathcal{T}'', \mathcal{T}', \mathcal{T}$.

for the inequality, produced in PHASE 2, that separates $\bar{x}^*$ from $\mathcal{T}'$ and induces a facet of the convex hull of $\mathcal{T}'$, we find a hypergraph inequality

$$\sum (\lambda_Q x(Q, \overline{V} - Q) : Q \in \overline{\mathcal{H}}) \geq \mu$$

satisfied by all $\bar{x}$ in $\mathcal{T}$ and such that, for some integers $\pi_1, \pi_2, \dots, \pi_k$,

$$\sum (\lambda_Q x(Q, \overline{V} - Q) : Q \in \overline{\mathcal{H}}) = \sum_{w=1}^{k} \pi_w \sum (\overline{x}_e : w \in e) - 2a'^T \overline{x},$$

$$\mu = \sum_{w=1}^{k} 2\pi_w - 2b'.$$

(Note that this inequality is violated by $\bar{x}^*$.)

In PHASE 1 and PHASE 2, we access $\mathcal{T}'$ by function ORACLE that,

given

integer vectors c, ℓ, u
with components indexed by elements of $\overline{E}$
and a threshold t (either an integer or $-\infty$),
returns either
an $\bar{x}$ that maximizes $c^T \bar{x}$

subject to $\overline{x} \in T'$, $\ell \leq \overline{x} \leq u$, $c^T \overline{x} > t$

or

the message "infeasible," indicating that no $\overline{x}$ in T' satisfies $\ell \leq \overline{x} \leq u$, $c^T \overline{x} > t$.

This function was also used in Section 11.4 to generate columns of A; it will be referred to as ORACLE; its implementation is outlined in Section 11.8.

11.6 PHASE 1: FROM T^* TO T''

11.6.1 Trivial subtour cuts and the dimension of T''

In Section 11.4, we defined

$$\overline{E}_{1/2} = \{e : e \subset \{1, 2, \ldots, k\}, \overline{x}_e^* \neq 0, \overline{x}_e^* \neq 1\};$$

in Section 11.5, we defined

$$\overline{E}_0 = \{e : e \subset \{1, 2, \ldots, k\}, \overline{x}_e^* = 0\},$$
$$\overline{E}_1 = \{e : e \subset \{1, 2, \ldots, k\}, \overline{x}_e^* = 1\}.$$

We may assume that

$$\overline{E}_{1/2} = \{e : e \subset \{1, 2, \ldots, k\}, 0 < \overline{x}_e^* < 1\} :$$

otherwise $\overline{x}_e^* > 1$ for some e such that $e \subset \{1, 2, \ldots, k\}$, and so $\overline{x}^*$ is separated from T by the subtour inequality

$$\overline{x}(e, \overline{V} - e) \geq 2.$$

(As noted by Cornuéjols et al. [140], subtour inequalities induce facets of the convex hull of T.)

Next, consider the graph with vertex set $\{1, 2, \ldots, k\}$ and edge set $\overline{E}_1$ and, for each e in $\overline{E}_{1/2}$, consider the graph with vertex set $\{1, 2, \ldots, k\}$ and edge set $\overline{E}_1 \cup \{e\}$. If any of these $1 + |\overline{E}_{1/2}|$ graphs has a connected component which is a cycle, then the vertex set S of this connected component yields the subtour inequality

$$\overline{x}(S, \overline{V} - S) \geq 2$$

that separates $\overline{x}^*$ from T and we are done; hence we may assume that each of these graphs is a system of paths, and so $\psi(T'')$ includes the $1 + |\overline{E}_{1/2}|$ vertices of the unit simplex,

$$\begin{bmatrix} 0 \\ 0 \\ 0 \\ \vdots \\ 0 \end{bmatrix}, \begin{bmatrix} 1 \\ 0 \\ 0 \\ \vdots \\ 0 \end{bmatrix}, \begin{bmatrix} 0 \\ 1 \\ 0 \\ \vdots \\ 0 \end{bmatrix}, \begin{bmatrix} 0 \\ 0 \\ 1 \\ \vdots \\ 0 \end{bmatrix}, \ldots, \begin{bmatrix} 0 \\ 0 \\ 0 \\ \vdots \\ 1 \end{bmatrix}.$$

Now $\dim \psi(T'') = |\overline{E}_{1/2}|$; since the linear mapping η defined in Section 11.4 satisfies $\eta(\psi(T'')) = T''$, it follows that

$$\dim T'' = |\overline{E}_{1/2}|.$$

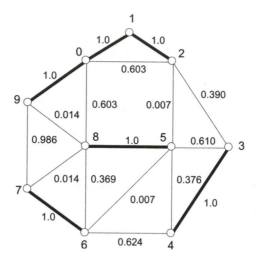

Figure 11.7 EXAMPLE 2.

11.6.2 EXAMPLE 2

∨ ∨ ∨

We will illustrate PHASE 1 on EXAMPLE 1 and also on the example where $\overline{V} = \{0, 1, \ldots, 9\}$ and

$$\begin{aligned}
&\overline{x}_{01}^* = 1.000, \quad \overline{x}_{02}^* = 0.603, \quad \overline{x}_{08}^* = 0.603, \quad \overline{x}_{09}^* = 1.000, \quad \overline{x}_{12}^* = 1.000, \\
&\overline{x}_{23}^* = 0.390, \quad \overline{x}_{25}^* = 0.007, \quad \overline{x}_{34}^* = 1.000, \quad \overline{x}_{35}^* = 0.610, \quad \overline{x}_{45}^* = 0.376, \\
&\overline{x}_{46}^* = 0.624, \quad \overline{x}_{56}^* = 0.007, \quad \overline{x}_{58}^* = 1.000, \quad \overline{x}_{67}^* = 1.000, \quad \overline{x}_{68}^* = 0.369, \\
&\overline{x}_{78}^* = 0.014, \quad \overline{x}_{79}^* = 0.986, \quad \overline{x}_{89}^* = 0.014
\end{aligned}$$

with $\overline{x}_{ij}^* = 0$ for all the remaining i and j such that $0 \le i < j \le 9$. This example $\overline{x}^*$ is illustrated in Figure 11.7.

∧ ∧ ∧

11.6.3 Separating $\overline{x}^*$ from $\mathcal{T}''$

It is easy to convert the cut $p^T \psi(\overline{x}) \le r$ that separates $\overline{x}^*$ from $\mathcal{T}^*$ into a cut that separates $\overline{x}^*$ from $\mathcal{T}''$.

∨ ∨ ∨

In EXAMPLE 1, $p^T \psi(\overline{x}) \le r$ reads

$$\overline{x}_{12} + \overline{x}_{25} + \overline{x}_{47} \le 1$$

and $\mathcal{T}^*$ consists of all $\overline{x}$ in $\mathcal{T}''$ that satisfy

$$\overline{x}_{03} = 0, \quad \overline{x}_{05} = 0, \quad \overline{x}_{06} = 0.$$

If $\overline{x} \in T''$, then $\psi(\overline{x})$ is a zero-one vector, and so

$$\overline{x}_{12} + \overline{x}_{25} + \overline{x}_{47} \leq 3;$$

it follows that the inequality

$$\overline{x}_{12} + \overline{x}_{25} + \overline{x}_{47} - 2(\overline{x}_{03} + \overline{x}_{05} + \overline{x}_{06}) \leq 1$$

is satisfied by all $\overline{x}$ in T'' and violated by $\overline{x}^*$.

In EXAMPLE 2, $p^T \psi(\overline{x}) \leq r$ reads

$$\overline{x}_{89} \leq 0$$

and T^* consists of all $\overline{x}$ in T'' that satisfy

$$\overline{x}_{03} = 0, \ \overline{x}_{04} = 0, \ \overline{x}_{05} = 0, \ \overline{x}_{06} = 0, \ \overline{x}_{07} = 0.$$

If $\overline{x} \in T''$, then $\psi(\overline{x})$ is a zero-one vector, and so

$$\overline{x}_{89} \leq 1;$$

it follows that the inequality

$$\overline{x}_{89} - (\overline{x}_{03} + \overline{x}_{04} + \overline{x}_{05} + \overline{x}_{06} + \overline{x}_{07}) \leq 0$$

is satisfied by all $\overline{x}$ in T'' and violated by $\overline{x}^*$.

∧ ∧ ∧

In general, the inequality

$$p^T \psi(\overline{x}) - (||p||_1 - r) \sum (\overline{x}_{0v} : \overline{x}_{0v}^* = 0) \leq r \qquad (11.16)$$

is satisfied by all $\overline{x}$ in T'' and violated by $\overline{x}^*$ (to see that $||p||_1 - r$ is always a positive integer, note that $r < p^T \psi(\overline{x}^*) \leq ||p||_1$); the bulk of PHASE 1 is taken up by transforming this cut into a cut that induces a facet of the convex hull of T''.

11.6.4 From a cut to a facet-inducing cut: An overview

∨ ∨ ∨

In EXAMPLE 1, the optimal basis of problem (11.7) has five columns $\begin{bmatrix} -\psi(\overline{x}) \\ 1 \end{bmatrix}$ such that $\overline{x} \in T^*$; specifically, with

$$\psi(\overline{x}) = [\, \overline{x}_{12}, \ \overline{x}_{15}, \ \overline{x}_{23}, \ \overline{x}_{25}, \ \overline{x}_{35}, \ \overline{x}_{36}, \ \overline{x}_{47}, \ \overline{x}_{56} \,]^T,$$

the five vectors $\psi(\overline{x})$ are

$$[\, 1, \ 1, \ 1, \ 0, \ 0, \ 0, \ 0, \ 1\,]^T,$$
$$[\, 1, \ 0, \ 0, \ 0, \ 1, \ 0, \ 0, \ 1\,]^T,$$
$$[\, 0, \ 1, \ 0, \ 1, \ 0, \ 1, \ 0, \ 0\,]^T,$$
$$[\, 0, \ 0, \ 1, \ 1, \ 0, \ 0, \ 0, \ 1\,]^T,$$
$$[\, 0, \ 1, \ 1, \ 0, \ 0, \ 0, \ 1, \ 1\,]^T;$$

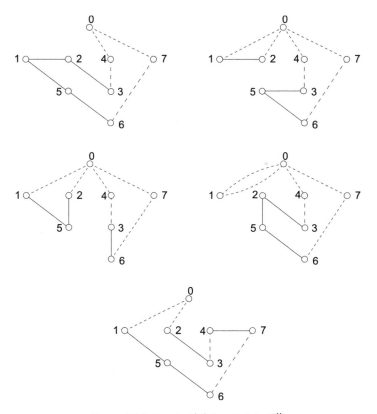

Figure 11.8 $\bar{x}$ and $\psi(\bar{x})$ for all $\bar{x}$ in $\mathcal{T}_0''$.

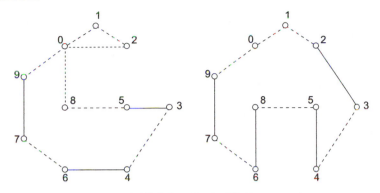

Figure 11.9 $\psi(\overline{x})$ for all $\overline{x}$ in $\mathcal{T}_0''$, EXAMPLE 2.

these vectors are illustrated in Figure 11.8. Since the basis is a nonsingular matrix, these five vectors are affinely independent; let $\mathcal{T}_0''$ denote the set of corresponding five affinely independent elements of $\mathcal{T}^*$. Each of the five points of $\mathcal{T}_0''$ satisfies $\overline{x}_{12} + \overline{x}_{25} + \overline{x}_{47} = 1$, and so (being a point of $\mathcal{T}^*$) it satisfies $\overline{x}_{12} + \overline{x}_{25} + \overline{x}_{47} - 2(\overline{x}_{03} + \overline{x}_{05} + \overline{x}_{06}) = 1$.

In EXAMPLE 2, the optimal basis of problem (11.7) has two columns $\begin{bmatrix} -\psi(\overline{x}) \\ 1 \end{bmatrix}$ such that $\overline{x} \in \mathcal{T}^*$; specifically, with

$$\psi(\overline{x}) = \left[\, \overline{x}_{23},\ \overline{x}_{25},\ \overline{x}_{35},\ \overline{x}_{45},\ \overline{x}_{46},\ \overline{x}_{56},\ \overline{x}_{68},\ \overline{x}_{78},\ \overline{x}_{79},\ \overline{x}_{89} \,\right]^T,$$

the two vectors $\psi(\overline{x})$ are

$$[\, 0,\ 0,\ 1,\ 0,\ 1,\ 0,\ 0,\ 0,\ 1,\ 0 \,]^T,$$
$$[\, 1,\ 0,\ 0,\ 1,\ 0,\ 0,\ 1,\ 0,\ 1,\ 0 \,]^T.$$

The two vectors $\psi(\overline{x})$ are illustrated in Figure 11.9. With $\mathcal{T}_0''$ standing for the set of the corresponding two affinely independent elements of $\mathcal{T}^*$, both points of $\mathcal{T}_0''$ satisfy $\overline{x}_{89} = 0$, and so (being points of $\mathcal{T}^*$) they satisfy $\overline{x}_{89} - (\overline{x}_{03} + \overline{x}_{04} + \overline{x}_{05} + \overline{x}_{06} + \overline{x}_{07}) = 0$. The two vectors $\psi(\overline{x})$ are illustrated in Figure 11.9.

∧ ∧ ∧

In general, the optimal basis of problem (11.7) provides a set $\mathcal{T}_0''$ of affinely independent elements of $\mathcal{T}^*$ that satisfy (11.16) with the sign of equality. (It may turn out that $\mathcal{T}_0'' = \emptyset$; this happens if and only if $\mathcal{T}^* = \emptyset$, in which case (11.16) reads $\sum(\overline{x}_{0v} : \overline{x}_{0v}^* = 0) \leq -1$.) Writing $a^T \overline{x} \leq b$ for (11.16), we have an integer vector a, and integer b, and a (possibly empty) set $\mathcal{T}_0''$ such that

$$\mathcal{T}'' \subset \{\overline{x} : a^T \overline{x} \leq b\}, \tag{11.17}$$
$$\mathcal{T}_0'' \text{ is an affinely independent subset of } \mathcal{T}'', \tag{11.18}$$
$$\mathcal{T}_0'' \subset \{\overline{x} : a^T \overline{x} = b\}, \tag{11.19}$$
$$a^T \overline{x}^* > b. \tag{11.20}$$

Our code maintains these four invariants while adding new elements to T_0'' and adjusting a and b if necessary: when $|T_0''|$ reaches $|\overline{E}_{1/2}|$, the current cut $a^T\overline{x} \le b$ induces a facet of the convex hull of T''. This process consists of two stages.

In the first stage, function STRENGTHEN attempts to maximize $|T_0''|$ by relatively effortless tricks. Its input cut ($= (11.16)$) is a nonnegative linear combination of its output cut and a linear inequality satisfied by all points in T'' as well as by $\overline{x}^*$. If STRENGTHEN fails to raise $|T_0''|$ all the way to $|\overline{E}_{1/2}|$, then we proceed to the more laborious second stage: function FACET is guaranteed to raise $|T_0''|$ to $|\overline{E}_{1/2}|$, and so it produces a cut inducing a facet of the convex hull of T''. The input cut of FACET ($=$ the output cut of STRENGTHEN) is a nonnegative linear combination of its output cut and a linear inequality satisfied by all points in T'' (but not necessarily by $\overline{x}^*$).

To add new elements to T_0'' and to adjust a and b if necessary, STRENGTHEN and FACET rely on a function TILT, which, given a nonzero integer vector a and an integer b, a nonzero integer vector v and an integer w, and an element $\overline{x}^0$ of T'' that satisfy (11.17) as well as the condition

(PRE1) if $T'' \subset \{\overline{x} : v^T\overline{x} \le w\}$, then $a^T\overline{x}^0 < b$ and $v^T\overline{x}^0 = w$,

returns a nonzero integer vector a^+, an integer b^+, and an element $\overline{x}^+$ of T'' such that

(POST1) if $T'' \subset \{\overline{x} : v^T\overline{x} \le w\}$, then $a^+ = v$, $b^+ = w$, and $\overline{x}^+ = \overline{x}^0$,
(POST2) if $T'' \not\subset \{\overline{x} : v^T\overline{x} \le w\}$, then
        • inequality $a^{+T}\overline{x} \le b^+$ is a nonnegative linear combination
         of inequalities $a^T\overline{x} \le b$ and $v^T\overline{x} \le w$,
        • $T'' \subset \{\overline{x} : a^{+T}\overline{x} \le b^+\}$,
        • $a^{+T}\overline{x}^+ = b$ and $v^T\overline{x}^+ > w$.

Whenever STRENGTHEN or FACET calls TILT, it has chosen v and w so that

(PRE2) $T_0'' \subset \{\overline{x} : v^T\overline{x} = w\}$.

From (11.17), (11.18), (11.19), (PRE1), (PRE2), (POST1), and (POST2), it follows that

(POST3) $T'' \subset \{\overline{x} : a^{+T}\overline{x} \le b^+\}$.
(POST4) $\overline{x}^+ \notin T_0''$ and $T_0'' \cup \{\overline{x}^+\}$ is an affinely independent set,
(POST5) $T_0'' \cup \{\overline{x}^+\} \subset \{\overline{x} : a^{+T}\overline{x} = b^+\}$,

which means that invariants (11.17), (11.18), (11.19) are maintained when a, b, and T_0'' are replaced by a^+, b^+, and $T_0'' \cup \{\overline{x}^+\}$.

11.6.5 Implementation of TILT

Condition $T'' \subset \{\overline{x} : v^T\overline{x} \le w\}$ in TILT $(a, b, v, w, \overline{x}^0)$ can be tested by calling the ORACLE to maximize $v^T \overline{x}$ subject to $\overline{x} \in T''$.

In case $T'' \not\subset \{\overline{x} : v^T\overline{x} \leq w\}$, condition (POST2) can be enforced as follows:

(i) find an element $\overline{x}^+$ of T'' that
   minimizes $\dfrac{b - a^T\overline{x}}{v^T\overline{x} - w}$ subject to $v^T\overline{x} > w$,

(ii) set $\lambda = v^T\overline{x}^+ - w$, $\mu = b - a^T\overline{x}^+$, $a^+ = \lambda a + \mu v$, $b^+ = \lambda b + \mu w$.

Problem (i) may be visualized in terms of the mapping $\pi : T'' \to \mathbf{R}^2$ defined by

$$\pi(\overline{x}) = [\pi_1, \pi_2] \text{ with } \pi_1 = a^T\overline{x}, \pi_2 = v^T\overline{x}.$$

In these terms, $\pi(T'')$ is a finite subset of the closed half-plane defined by $\pi_1 \leq b$ and at least one of its points belongs to the open half-plane defined by $\pi_2 > w$. For each point in the upper left quadrant

$$\{[\pi_1, \pi_2] : \pi_1 \leq b, \pi_2 > w\},$$

quantity $(b - \pi_1)/(\pi_2 - w)$ is the tangent of the angle between two half-lines that begin at $[b, w]$: the vertical half-line that points up and the half-line that passes through $[\pi_1, \pi_2]$. For every value of a parameter t, the line defined by

$$t\pi_1 + (1 - t)\pi_2 = tb + (1 - t)w \qquad (11.21)$$

passes through the point $[b, w]$; as t moves from 0 to 1, this line rotates clockwise around the hinge $[b, w]$ from its horizontal position corresponding to $t = 0$ to its vertical position corresponding to $t = 1$. The last point in the open half-plane $\pi_2 > w$ that the line encounters during this circular sweep is $\pi(\overline{x}^+)$.

Unfortunately, we can access $\pi(T'')$ only through ORACLE, which has no notion of fractional objective functions such as $(b-\pi_1)/(\pi_2-w)$; fortunately, the rotation involved in minimizing $(b - \pi_1)/(\pi_2 - w)$ can be divided into incremental steps, each of which requires just one call of ORACLE. Our current line (11.21) has not yet encountered all the points $[\pi_1, \pi_2]$ of $\pi(T'')$ such that $\pi_2 > w$ if and only if some point $[\pi_1, \pi_2]$ of $\pi(T'')$ satisfies

$$t\pi_1 + (1 - t)\pi_2 > tb + (1 - t)w.$$

If ORACLE finds such a point $[\pi_1, \pi_2]$, then we make the line rotate further until it passes through this $[\pi_1, \pi_2]$, after which we try again; else we stop. Since $\pi(T'')$ is finite, this process eventually terminates. The reduction of a single rotation into a sequence of pushes that is used in Algorithm 11.2 is known in the discipline of fractional programming as the *Dinkelbach method*; see, for instance, Section 5.6 of Craven [144] or Section 4.5 of Stancu-Minasian [507]. In turn, the Dinkelbach method is sometimes construed as the variant of the Newton-Raphson method that attempts to find a root of a function $f : \mathbf{R} \to \mathbf{R}$ as the limit of a sequence $t_1, t_2, t_3, \ldots$ constructed iteratively by

$$t_{i+1} = t_i - \frac{f(t_i)}{g(t_i)}$$

with g a *subgradient of f*, meaning any function $g : \mathbf{R} \to \mathbf{R}$ such that

$$f(t') \geq f(t) + g(t)(t' - t) \quad \text{ for all } t' \text{ and } t.$$

The TILT process is illustrated in Figures 11.10 through 11.15, using a total of three pushes.

Algorithm 11.2 TILT $(a, b, v, w, \overline{x}^0)$.

$\overline{x}$ = element of $\mathcal{T}''$ that maximizes $v^T \overline{x}$;
if $v^T \overline{x} > w$
then   **if** $a^T \overline{x} < b$
            then $\lambda = v^T \overline{x} - w, \mu = b - a^T \overline{x}$;
                     $v = \lambda a + \mu v, w = \lambda b + \mu w$;
                     $d =$ the gcd of all the components of $[v^T, w]$;
                     $v = v/d, w = w/d$;
                     return TILT $(a, b, v, w, \overline{x})$;
            else return $(a, b, \overline{x})$;
            end
else   return $(v, w, \overline{x}^0)$;
end

Figure 11.10 Critical points in x^*.

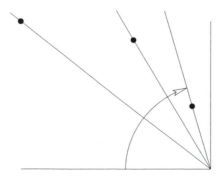

Figure 11.11 Rotational optimization.

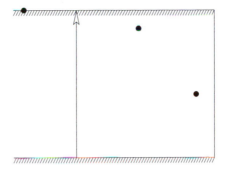

Figure 11.12 First push.

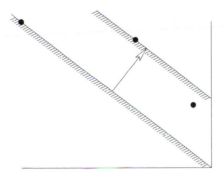

Figure 11.13 Second push.

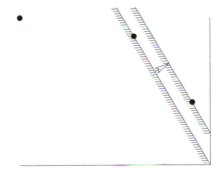

Figure 11.14 Third push.

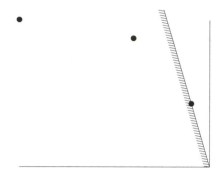

Figure 11.15 Optimal—no push.

11.6.6 Function STRENGTHEN

∨ ∨ ∨

In EXAMPLE 1, we have $\overline{V} = \{0, 1, \ldots, 7\}$ and

$$\overline{x}_{01}^* = 0.7, \quad \overline{x}_{02}^* = 0.1, \quad \overline{x}_{04}^* = 0.9, \quad \overline{x}_{07}^* = 0.9, \quad \overline{x}_{12}^* = 0.7,$$
$$\overline{x}_{15}^* = 0.6, \quad \overline{x}_{23}^* = 0.8, \quad \overline{x}_{25}^* = 0.4, \quad \overline{x}_{34}^* = 1.0, \quad \overline{x}_{35}^* = 0.1,$$
$$\overline{x}_{36}^* = 0.1, \quad \overline{x}_{47}^* = 0.1, \quad \overline{x}_{56}^* = 0.9, \quad \overline{x}_{67}^* = 1.0$$

with $\overline{x}_{ij}^* = 0$ for all the remaining i and j such that $0 \le i < j \le 7$. The inequality satisfied by all $\overline{x}$ in $\mathcal{T}''$ and violated by $\overline{x}^*$ is

$$-2\overline{x}_{03} - 2\overline{x}_{05} - 2\overline{x}_{06} + \overline{x}_{12} + \overline{x}_{25} + \overline{x}_{47} \le 1; \qquad (11.22)$$

in the notation

$$\psi(\overline{x}) = [\overline{x}_{12}, \overline{x}_{15}, \overline{x}_{23}, \overline{x}_{25}, \overline{x}_{35}, \overline{x}_{36}, \overline{x}_{47}, \overline{x}_{56}]^T,$$

the five vectors in $\psi(\mathcal{T}_0'')$ are

$$[\,1,\ 1,\ 1,\ 0,\ 0,\ 0,\ 0,\ 1\,]^T,$$
$$[\,1,\ 0,\ 0,\ 0,\ 1,\ 0,\ 0,\ 1\,]^T,$$
$$[\,0,\ 1,\ 0,\ 1,\ 0,\ 1,\ 0,\ 0\,]^T,$$
$$[\,0,\ 0,\ 1,\ 1,\ 0,\ 0,\ 0,\ 1\,]^T,$$
$$[\,0,\ 1,\ 1,\ 0,\ 0,\ 0,\ 1,\ 1\,]^T.$$

Each of the five points of $\mathcal{T}_0''$, being a point of $\mathcal{T}^*$, satisfies

$$\overline{x}_{03} = 0, \quad \overline{x}_{05} = 0, \quad \overline{x}_{06} = 0;$$

in order to raise $|\mathcal{T}_0''|$ to $|\overline{E}_{1/2}|$, we must admit into $\mathcal{T}_0''$ an element of $\mathcal{T}''$ which has $\overline{x}_{03} > 0$, an element of $\mathcal{T}''$ which has $\overline{x}_{05} > 0$, and an element of $\mathcal{T}''$ which has $\overline{x}_{06} > 0$. (There may be fewer than three of these new elements as one or two of them can play a multiple role.) This observation is the key to STRENGTHEN.

In order to admit into $\mathcal{T}_0''$ an element $\overline{x}^+$ of $\mathcal{T}''$ that has $\overline{x}_{03}^+ > 0$, we call TILT $(a, b, v, w, \overline{x}^0)$ with $a^T \overline{x} \le b$ the inequality (11.22), with $v^T \overline{x} \le w$ the inequality

$\overline{x}_{03} \leq 0$, and with $\overline{x}^0$ an arbitrary point of $\mathcal{T}''$. TILT returns the original inequality (11.22) as $a^{+T}\overline{x} \leq b^+$ and it returns the $\overline{x}^+$ defined by

$$\psi(\overline{x}^+) = [1, 0, 0, 1, 0, 0, 1, 1]^T.$$

We leave $a^T\overline{x} \leq b$ unchanged and add $\overline{x}^+$ to $\mathcal{T}_0''$.

Next, each of the six points of $\mathcal{T}_0''$ still satisfies $\overline{x}_{05} = 0$ and $\overline{x}_{06} = 0$. In order to admit into $\mathcal{T}_0''$ an element $\overline{x}^+$ of $\mathcal{T}''$ that has $\overline{x}_{05}^+ > 0$, we call TILT $(a, b, v, w, \overline{x}^0)$ with $a^T\overline{x} \leq b$ the inequality (11.22), with $v^T\overline{x} \leq w$ the inequality $\overline{x}_{05} \leq 0$, and with $\overline{x}^0$ an arbitrary point of $\mathcal{T}''$. TILT returns the inequality

$$-2\overline{x}_{03} - \overline{x}_{05} - 2\overline{x}_{06} + \overline{x}_{12} + \overline{x}_{25} + \overline{x}_{47} \leq 1 \tag{11.23}$$

as $a^{+T}\overline{x} \leq b^+$ and it returns the $\overline{x}^+$ defined by

$$\psi(\overline{x}^+) = [1, 0, 0, 1, 0, 1, 0, 0]^T.$$

We replace $a^T\overline{x} \leq b$ by (11.23) and add $\overline{x}^+$ to $\mathcal{T}_0''$.

Finally, each of the seven points of $\mathcal{T}_0''$ still satisfies $\overline{x}_{06} = 0$. In order to admit into $\mathcal{T}_0''$ an element $\overline{x}^+$ of $\mathcal{T}''$ that has $\overline{x}_{06}^+ > 0$, we call TILT $(a, b, v, w, \overline{x}^0)$ with $a^T\overline{x} \leq b$ the inequality (11.23), with $v^T\overline{x} \leq w$ the inequality $\overline{x}_{06} \leq 0$, and with $\overline{x}^0$ an arbitrary point of $\mathcal{T}''$. TILT returns inequality (11.23) as $a^{+T}\overline{x} \leq b^+$ and it returns the $\overline{x}^+$ defined by

$$\psi(\overline{x}^+) = [1, 0, 0, 1, 1, 0, 1, 0]^T.$$

We leave $a^T\overline{x} \leq b$ unchanged and add $\overline{x}^+$ to $\mathcal{T}_0''$.

The three points that we have added to $\mathcal{T}_0''$ are illustrated in Figure 11.16.

Now $|\mathcal{T}_0''|$ has reached dim $\mathcal{T}''$, and so (11.23) induces a facet of the convex hull of $\mathcal{T}''$. We have decomposed the input cut (11.22) into a sum of the cut (11.23) and the linear inequality $-\overline{x}_{05} \leq 0$; since $-\overline{x}_{05} \leq 0$ is satisfied by all points in $\mathcal{T}''$ as well as by $\overline{x}^*$, the facet-inducing inequality (11.23) must be a cut.

∧ ∧ ∧

Algorithm 11.3 generalizes our treatment of EXAMPLE 1. As noted earlier in this section, this algorithm maintains invariants (11.17), (11.18), (11.19); let us point out that it maintains (11.20), too. In fact, the algorithm maintains the following stronger invariant:

($\star$) the input cut is a nonnegative linear combination
    of the current inequality $a^T\overline{x} \leq b$
    and a linear inequality satisfied by all points in $\mathcal{T}''$ as well as by $\overline{x}^*$.

In the beginning, when the current inequality $a^T\overline{x} \leq b$ is the input cut, we may take the trivial $0^T x \leq 0$ for the linear inequality satisfied by all points in $\mathcal{T}''$ as well as by $\overline{x}^*$. Every time Algorithm 11.3 replaces $a^T\overline{x} \leq b$ by $a^{+T}\overline{x} \leq b^+$, condition (POST2) guarantees the existence of numbers λ and μ such that $\lambda > 0$, $\mu \geq 0$, and

$$a^{+T} = \lambda a + \mu v, \quad b^+ = \lambda b + \mu w;$$

it follows that $a^T\overline{x} \leq b$ is a nonnegative linear combination of $a^{+T}\overline{x} \leq b^+$ and the linear inequality $-v^T\overline{x} \leq -w$ satisfied by all points in $\mathcal{T}''$ as well as by $\overline{x}^*$.

Our code uses Algorithm 11.3 with $\mathcal{C}$ the union of catalogs $\mathcal{C}_1$ and $\mathcal{C}_2$:

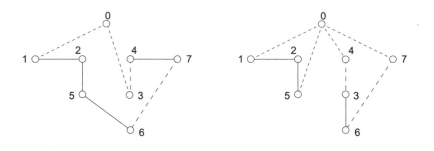

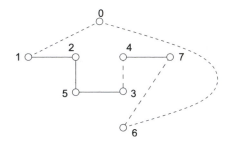

Figure 11.16 STRENGTHEN EXAMPLE 1.

Algorithm 11.3 STRENGTHEN$(a, b; T_0'')$.

$\bar{x}_0$ = an arbitrary element of T'';
$\mathcal{C}$ = a catalog of vectors $[v^T, w]$ such that
$\quad\quad T'' \subset \{\bar{x} : v^T\bar{x} \geq w\}$, $v^T\bar{x}^* \geq w$, $T'' \not\subset \{\bar{x} : v^T\bar{x} = w\}$;
for all $[v^T, w]$ in $\mathcal{C}$
do remove $[v^T, w]$ from $\mathcal{C}$
$\quad\quad$ **if** $\quad T_0'' \subset \{\bar{x} : v^T\bar{x} = w\}$
$\quad\quad$ **then** $(a^+, b^+, \bar{x}^+)$ =TILT $(a, b, v, w, \bar{x}^0)$;
$\quad\quad\quad\quad a = a^+, b = b^+, T_0'' = T_0'' \cup \{\bar{x}^+\}$;
$\quad\quad$ **end**
end
return $(a, b; T_0'')$;

- C_1 consists of all $[v^T, w]$ such that $v^T \overline{x} \geq w$ reads $\overline{x}_e \geq 0$ or $-\overline{x}_e \geq -1$ for some e in $\overline{E}_{1/2}$,

- C_2 consists of all $[v^T, w]$ such that $v^T \overline{x} \geq w$ reads $\overline{x}_{0u} \geq 0$ for some u in $\{1, 2, \ldots, k\}$.

In the **for** loop of Algorithm 11.3, we scan C_1 before scanning C_2.

∨ ∨ ∨

In EXAMPLE 1, the cut produced by STRENGTHEN is

$$-2\overline{x}_{03} - \overline{x}_{05} - 2\overline{x}_{06} + \overline{x}_{12} + \overline{x}_{25} + \overline{x}_{47} \leq 1;$$

we may convert this cut into its equivalent form

$$\overline{x}_{12} + \overline{x}_{15} + 2\overline{x}_{23} + 2\overline{x}_{25} + 3\overline{x}_{35} + 4\overline{x}_{36} + \overline{x}_{47} + 3\overline{x}_{56} \leq 7$$

by substituting for $\overline{x}_{03}, \overline{x}_{05}, \overline{x}_{06}$ from the equations

$$\overline{x}_{03} = 1 - \overline{x}_{23} - \overline{x}_{35} - \overline{x}_{36},$$
$$\overline{x}_{05} = 2 - \overline{x}_{15} - \overline{x}_{25} - \overline{x}_{35} - \overline{x}_{56},$$
$$\overline{x}_{06} = 1 - \overline{x}_{36} - \overline{x}_{56},$$

which hold identically for every $\overline{x}$ in $T'' \cup \{\overline{x}^*\}$.

∧ ∧ ∧

In general, the cut produced by STRENGTHEN may be converted into its equivalent form

$$\sum (a_e \overline{x}_e : e \in E_{1/2}) \leq b$$

by substituting for all $\overline{x}_{0k}$ from the equations

$$\overline{x}_{0k} = \begin{cases} 2 - \sum(\overline{x}_e : e \in E_{1/2}, k \in e) & \text{if } k \text{ is an endpoint of no edge in } E_1, \\ 1 - \sum(\overline{x}_e : e \in E_{1/2}, k \in e) & \text{if } k \text{ is an endpoint of an edge in } E_1, \end{cases}$$

which hold identically for every $\overline{x}$ in $T'' \cup \{\overline{x}^*\}$. Our code records the cut in this form.

∨ ∨ ∨

In EXAMPLE 2, we have $\overline{V} = \{0, 1, \ldots, 9\}$ and

$$\begin{array}{lllll}
\overline{x}_{01}^* = 1.000, & \overline{x}_{02}^* = 0.603, & \overline{x}_{08}^* = 0.603, & \overline{x}_{09}^* = 1.000, & \overline{x}_{12}^* = 1.000, \\
\overline{x}_{23}^* = 0.390, & \overline{x}_{25}^* = 0.007, & \overline{x}_{34}^* = 1.000, & \overline{x}_{35}^* = 0.610, & \overline{x}_{45}^* = 0.376, \\
\overline{x}_{46}^* = 0.624, & \overline{x}_{56}^* = 0.007, & \overline{x}_{58}^* = 1.000, & \overline{x}_{67}^* = 1.000, & \overline{x}_{68}^* = 0.369, \\
\overline{x}_{78}^* = 0.014, & \overline{x}_{79}^* = 0.986, & \overline{x}_{89}^* = 0.014
\end{array}$$

with $\overline{x}_{ij}^* = 0$ for all the remaining i and j such that $0 \leq i < j \leq 9$. The inequality satisfied by all $\overline{x}$ in T'' and violated by $\overline{x}^*$ is

$$-\overline{x}_{03} - \overline{x}_{04} - \overline{x}_{05} - \overline{x}_{06} - \overline{x}_{07} + \overline{x}_{89} \leq 0; \qquad (11.24)$$

in the notation

$$\psi(\overline{x}) = [\,\overline{x}_{23},\ \overline{x}_{25},\ \overline{x}_{35},\ \overline{x}_{45},\ \overline{x}_{46},\ \overline{x}_{56},\ \overline{x}_{68},\ \overline{x}_{78},\ \overline{x}_{79},\ \overline{x}_{89}\,]^T,$$

the two vectors in $\psi(T_0'')$ are

$$[\,0,\ 0,\ 1,\ 0,\ 1,\ 0,\ 0,\ 0,\ 1,\ 0\,]^T,$$
$$[\,1,\ 0,\ 0,\ 1,\ 0,\ 0,\ 1,\ 0,\ 1,\ 0\,]^T.$$

The successive iterations of the **for** loop of Algorithm 11.3, where condition $T_0'' \subset \{\overline{x} : v^T\overline{x} = w\}$ is satisfied go as follows.

ITERATION 1: $v^T\overline{x} \geq w$ is $\overline{x}_{25} \geq 0$.

TILT $(a, b, v, w, \overline{x}^0)$ returns the original inequality (11.24) as $a^{+T}\overline{x} \leq b^+$ and it returns the $\overline{x}^+$ defined by

$$\psi(\overline{x}^+) = [\,0,\ 1,\ 0,\ 0,\ 1,\ 0,\ 0,\ 0,\ 1,\ 1\,]^T.$$

We leave $a^T\overline{x} \leq b$ unchanged and add $\overline{x}^+$ to T_0''.

ITERATION 2: $v^T\overline{x} \geq w$ is $\overline{x}_{56} \geq 0$.

TILT $(a, b, v, w, \overline{x}^0)$ returns inequality

$$-\overline{x}_{03} - \overline{x}_{04} - \overline{x}_{05} - \overline{x}_{06} - \overline{x}_{07} + \overline{x}_{56} + \overline{x}_{89} \leq 0 \qquad (11.25)$$

as $a^{+T}\overline{x} \leq b^+$ and it returns the $\overline{x}^+$ defined by

$$\psi(\overline{x}^+) = [\,1,\ 0,\ 0,\ 0,\ 0,\ 1,\ 0,\ 0,\ 0,\ 1\,]^T.$$

We replace $a^T\overline{x} \leq b$ by (11.25) and add $\overline{x}^+$ to T_0''.

ITERATION 3: $v^T\overline{x} \geq w$ is $\overline{x}_{78} \geq 0$.

TILT $(a, b, v, w, \overline{x}^0)$ returns inequality

$$-\overline{x}_{03} - \overline{x}_{04} - \overline{x}_{05} - \overline{x}_{06} - \overline{x}_{07} + \overline{x}_{56} + \overline{x}_{78} + \overline{x}_{89} \leq 0 \qquad (11.26)$$

as $a^{+T}\overline{x} \leq b^+$ and it returns the $\overline{x}^+$ defined by

$$\psi(\overline{x}^+) = [\,1,\ 0,\ 0,\ 1,\ 0,\ 0,\ 0,\ 1,\ 0,\ 0\,]^T.$$

We replace $a^T\overline{x} \leq b$ by (11.26) and add $\overline{x}^+$ to T_0''.

ITERATION 4: $v^T\overline{x} \geq w$ is $\overline{x}_{05} \geq 0$.

TILT $(a, b, v, w, \overline{x}^0)$ returns (11.26) as $a^{+T}\overline{x} \leq b^+$ and it returns the $\overline{x}^+$ defined by

$$\psi(\overline{x}^+) = [\,1,\ 0,\ 0,\ 0,\ 1,\ 0,\ 0,\ 1,\ 0,\ 0\,]^T$$

We leave $a^T\overline{x} \leq b$ unchanged and add $\overline{x}^+$ to T_0''.

No more vectors $[v^T, w]$ in $\mathcal{C}$ satisfy $T_0'' \subset \{\overline{x} : v^T\overline{x} = w\}$, and so STRENGTHEN terminates; it has decomposed the input cut (11.24) into a sum of the cut (11.26) and the linear inequality $-\overline{x}_{56} - \overline{x}_{78} \leq 0$, which is satisfied by all points in T'' as well as by $\overline{x}^*$. Substituting for $\overline{x}_{03}, \overline{x}_{04}, \overline{x}_{05}, \overline{x}_{06}, \overline{x}_{07}$ from the equations

$$\overline{x}_{03} = 1 - \overline{x}_{23} - \overline{x}_{35},$$
$$\overline{x}_{04} = 1 - \overline{x}_{45} - \overline{x}_{46},$$
$$\overline{x}_{05} = 1 - \overline{x}_{25} - \overline{x}_{35} - \overline{x}_{45} - \overline{x}_{56},$$
$$\overline{x}_{06} = 1 - \overline{x}_{46} - \overline{x}_{56} - \overline{x}_{68},$$
$$\overline{x}_{07} = 1 - \overline{x}_{78} - \overline{x}_{79},$$

which hold identically for every $\overline{x}$ in $T'' \cup \{\overline{x}^*\}$, we convert (11.26) into its equivalent form

$$\overline{x}_{23} + \overline{x}_{25} + 2\overline{x}_{35} + 2\overline{x}_{45} + 2\overline{x}_{46} + 3\overline{x}_{56} + \overline{x}_{68} + 2\overline{x}_{78} + \overline{x}_{79} + \overline{x}_{89} \leq 5.$$

∧ ∧ ∧

11.6.7 A theorem of Minkowski and function FACET

If STRENGTHEN fails to raise $|T_0''|$ all the way to $|\overline{E}_{1/2}|$, then we turn to the following corollary of a classic theorem of Minkowski [398].

THEOREM 11.1 *Let T'' be a finite subset of some $\mathbf{R}^d$, let $\overline{x}^*$ be a point in $\mathbf{R}^d$, and let an inequality $a^T \overline{x} \leq b$ separate T'' from $\overline{x}^*$ in the sense that*

$$T'' \subset \{\overline{x} : a^T \overline{x} \leq b\} \text{ and } a^T \overline{x}^* > b.$$

If $T'' \not\subseteq \{\overline{x} : a^T \overline{x} = b\}$, then there is an inequality $a''^T \overline{x} \leq b''$ that separates T'' from $\overline{x}^*$, induces a facet of the convex hull of T'', and satisfies

$$\{\overline{x} \in T'' : a^T \overline{x} = b\} \subseteq \{\overline{x} \in T'' : a''^T \overline{x} = b''\}.$$

In our application of Theorem 11.1, $a^T \overline{x} \leq b$ is supplied by STRENGTHEN; to construct the corresponding inequality $a''^T \overline{x} \leq b''$, we appeal repeatedly to an algorithmic proof of the following lemma.

LEMMA 11.2 *Let T'' be a finite subset of some $\mathbf{R}^d$, let a be a vector and let b be a number such that*

$$T'' \subset \{\overline{x} : a^T \overline{x} \leq b\},$$

and let T_0'' be an affinely independent subset of T'' such that

$$T_0'' \subset \{\overline{x} : a^T \overline{x} = b\}.$$

If $|T_0''| < \dim T''$, then there are vectors a^-, a^+, numbers b^-, b^+, and elements $\overline{x}^-, \overline{x}^+$ of $T'' - T_0''$ such that

(i) *$T'' \subset \{\overline{x} : a^{-T} \overline{x} \leq b^-\}$, $T_0'' \cup \{\overline{x}^-\}$ is affinely independent, and $T_0'' \cup \{\overline{x}^-\} \subset \{\overline{x} : a^{-T} \overline{x} = b^-\}$,*

(ii) $T'' \subset \{\overline{x} : a^{+T}\overline{x} \leq b^+\}$, $T_0'' \cup \{\overline{x}^+\}$ is affinely independent, and
$T_0'' \cup \{\overline{x}^+\} \subset \{\overline{x} : a^{+T}\overline{x} = b^+\}$,

(iii) $a^T\overline{x} \leq b$ is a nonnegative linear combination of
$a^{-T}\overline{x} \leq b^-$ and $a^{+T}\overline{x} \leq b^+$.

Proof. Assumption $|T_0''| < \dim T''$ implies $\dim T'' > 0$. There are a subset S of $\{1, 2, \ldots, d\}$ and a linear mapping $\eta : \mathbf{R}^S \to \mathbf{R}^d$ such that $|S| = \dim T''$ and,

with $\psi(\overline{x})$ standing for the restriction of $\overline{x}$ on S,
we have $\eta(\psi(\overline{x})) = \overline{x}$ whenever $\overline{x} \in T''$;

to highlight the special case of our interest, we will write $\overline{E}_{1/2}$ for S. This notation is used in Algorithm 11.4; our proof of Lemma 11.2 amounts to an analysis of Algorithm 11.4, illustrated in Figure 11.17. Existence of an element $\overline{x}^0$ of T'' –

Algorithm 11.4 Finding the a^-, a^+, b^-, b^+, and $\overline{x}^-, \overline{x}^+$ of Lemma 11.2.

$\overline{x}^0 =$ an element of $T'' - T_0''$ such that
$\qquad T_0'' \cup \{\overline{x}^0\}$ is affinely independent;
if $\quad a^T\overline{x}^0 = b$
then $a^- = a^+ = a, b^- = b^+ = b, \overline{x}^- = \overline{x}^+ = \overline{x}^0$;
else find a nonzero vector v and a number w such that
$\qquad\qquad v^T\overline{x} - w = 0 \quad$ for all $\overline{x}$ in $T_0'' \cup \{\overline{x}^0\}$,
$\qquad\qquad\qquad v_e = 0 \quad$ for all e outside $\overline{E}_{1/2}$;
$\qquad$ **if** $\quad T'' \subset \{\overline{x} : v^T\overline{x} \geq w\}$
$\qquad$ **then** $\overline{x}^- = \overline{x}^0$;
$\qquad$ **else** $\overline{x}^- =$ an element of T'' that
$\qquad\qquad\qquad$ minimizes $(b - a^T\overline{x})/(w - v^T\overline{x})$ subject to $v^T\overline{x} < w$;
$\qquad$ **end**
$\qquad \lambda^- = w - v^T\overline{x}^-, \ \mu^- = b - a^T\overline{x}^-,$
$\qquad a^- = \lambda^- a - \mu^- v, \ b^- = \lambda^- b - \mu^- w$;
$\qquad$ **if** $\quad T'' \subset \{\overline{x} : v^T\overline{x} \leq w\}$
$\qquad$ **then** $\overline{x}^+ = \overline{x}^0$;
$\qquad$ **else** $\overline{x}^+ =$ an element of T'' that
$\qquad\qquad\qquad$ minimizes $(b - a^T\overline{x})/(v^T\overline{x} - w)$ subject to $v^T\overline{x} > w$;
$\qquad$ **end**
$\qquad \lambda^+ = v^T\overline{x}^+ - w, \ \mu^+ = b - a^T\overline{x}^+,$
$\qquad a^+ = \lambda^+ a + \mu^+ v, \ b^+ = \lambda^+ b + \mu^+ w$;
end
return a^-, a^+, b^-, b^+, and $\overline{x}^-, \overline{x}^+$;

T_0'' such that $T_0'' \cup \{\overline{x}^0\}$ is affinely independent follows from the assumption that $|T_0''| < \dim T''$. The system

$$v^T\overline{x} - w = 0 \text{ for all } \overline{x} \text{ in } T_0'' \cup \{\overline{x}^0\}, \tag{11.27}$$

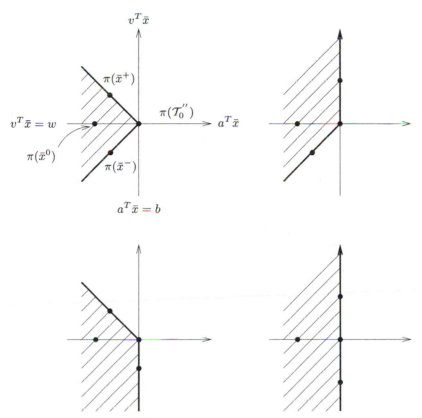

Figure 11.17 Algorithm 11.4 when $a^T \bar{x}^0 < b$. (We may also have $\bar{x}^+ = \bar{x}^0$ or $\bar{x}^- = \bar{x}^0$, but not both.)

$$v_e = 0 \text{ for all } e \text{ outside } \overline{E}_{1/2} \qquad (11.28)$$

has a nonzero solution since the number $d + 1$ of its variables exceeds the number $(|\mathcal{T}_0''| + 1) + (d - |\overline{E}_{1/2}|)$ of its equations. Verifying that a^-, b^-, and $\bar{x}^-$ have property (i) and that a^+, b^+, and $\bar{x}^+$ have property (ii) is a routine matter; we will complete the proof by showing that a^-, b^-, a^+, and b^+ have property (iii).

Since $\psi(v)$ is a nonzero vector and since $\psi(\mathcal{T}'')$ has full dimension in $\mathbf{R}^S$, the function defined on $\psi(\mathcal{T}'')$ by $\xi \mapsto \psi(v)^T \xi$ is not constant; since (11.28) guarantees $v^T \bar{x} = \psi(v)^T \psi(\bar{x})$ for all $\bar{x}$ in $\mathbf{R}^d$, it follows that the function defined on $\mathcal{T}''$ by $\bar{x} \mapsto v^T \bar{x}$ is not constant, and so

at least one of λ^+, λ^- is positive.

Trivially, all of $\lambda^+, \mu^+, \lambda^-, \mu^-$ are nonnegative. If $\mu^+ = 0$, then $\lambda^+ > 0$, and so $a^T \bar{x} \le b$ is a positive multiple of $a^{+T} \bar{x} \le b^+$; if $\mu^- = 0$, then $\lambda^- > 0$, and so $a^T \bar{x} \le b$ is a positive multiple of $a^{-T} \bar{x} \le b^-$; if $\mu^+ > 0$ and $\mu^- > 0$, then $\lambda^+ \mu^- + \lambda^- \mu^+ > 0$ and

$$a = \frac{\mu^+}{\lambda^+ \mu^- + \lambda^- \mu^+} a^- + \frac{\mu^-}{\lambda^+ \mu^- + \lambda^- \mu^+} a^+,$$

$$b = \frac{\mu^+}{\lambda^+ \mu^- + \lambda^- \mu^+} b^- + \frac{\mu^-}{\lambda^+ \mu^- + \lambda^- \mu^+} b^+.$$

$\square$

Our code records the cut produced by STRENGTHEN in the form

$$\sum (a_e \overline{x}_e : e \in E_{1/2}) \le b;$$

let us point out that

$$b > 0. \tag{11.29}$$

For this purpose, assume the contrary: $b \le 0$. Since $\psi(T'')$ includes the $1 + |\overline{E}_{1/2}|$ vertices of the unit simplex, it follows that $b = 0$ and that $a_e \le 0$ for all e in $\overline{E}_{1/2}$; in turn, since $(a_e : e \in \overline{E}_{1/2})$ is not the zero vector, we must have $a_f < 0$ for some f in $\overline{E}_{1/2}$. Since T_0'' has been preprocessed by STRENGTHEN, it includes an element $\overline{x}'$ such that $\overline{x}'_f = 1$. On the one hand, the fact that $\overline{x}' \in T_0''$ guarantees that $\sum(a_e \overline{x}'_e : e \in \overline{E}_{1/2}) = b$; on the other hand, we have $\sum(a_e \overline{x}'_e : e \in \overline{E}_{1/2}) \le a_f \overline{x}'_f < 0 = b$; this contradiction completes the proof of (11.29).

If $\overline{x}^0$ denotes the element of T'' such that $\psi(\overline{x}^0)$ is the vector of all zeros, then (11.29) guarantees that $a^T \overline{x}^0 < b$; in particular, $\overline{x}^0 \notin T_0''$ and $T_0'' \cup \{\overline{x}^0\}$ is affinely independent. Consequently, Algorithm 11.4 can be replaced by Algorithm 11.5 in our special case.

Algorithm 11.5 Algorithm 11.4 streamlined for our special case.

---

find a nonzero integer vector v such that
$$v^T \overline{x} = 0 \quad \text{for all } \overline{x} \text{ in } T_0'',$$
$$v_e = 0 \quad \text{for all } e \text{ outside } \overline{E}_{1/2};$$
$\overline{x}^0 = $ the element of T'' such that $\psi(\overline{x}^0)$ is the vector of all zeros;
$(a^-, b^-, \overline{x}^-) = \text{TILT }(a, b, -v, 0, \overline{x}^0);$
$(a^+, b^+, \overline{x}^+) = \text{TILT }(a, b, v, 0, \overline{x}^0):$
return $a^-, a^+, b^-, b^+,$ and $\overline{x}^-, \overline{x}^+;$

Proof of Theorem 11.1 This proof amounts to straightforward analysis of Algorithm 11.6. The initial value of T_0'' in this algorithm can be any affinely independent subset of T'' such that $T_0'' \subset \{\overline{x} : a^T \overline{x} = b\}$ (a default is $T_0'' = \emptyset$); assumption $T'' \not\subset \{\overline{x} : a^T \overline{x} = b\}$ of the theorem guarantees that the initial $|T_0''|$ is at most $\dim T''$. The **while** loop of this algorithm maintains the four invariants

$$T'' \subset \{\overline{x} : a^T \overline{x} \le b\},$$
$$T_0'' \text{ is an affinely independent subset of } T'',$$
$$T_0'' \subset \{\overline{x} : a^T \overline{x} = b\},$$
$$a^T \overline{x}^* > b;$$

Algorithm 11.6 Finding the inequality $a''^T \overline{x} \le b''$ of Theorem 11.1.

while $|\mathcal{T}_0''| < \dim \mathcal{T}''$

do   find vectors a^-, a^+, numbers b^-, b^+, and elements $\overline{x}^-, \overline{x}^+$
        of $\mathcal{T}'' - \mathcal{T}_0''$ with properties (i), (ii), (iii) of Lemma 11.2;
        if $a^{+T}\overline{x}^* > b^+$
        then replace $(a, b; \mathcal{T}_0'')$ by $(a^+, b^+; \mathcal{T}_0'' \cup \{\overline{x}^+\})$;
        else replace $(a, b; \mathcal{T}_0'')$ by $(a^-, b^-; \mathcal{T}_0'' \cup \{\overline{x}^-\})$;
        end

end
return the inequality $a^T \overline{x} \le b$;

in particular, the fourth of these invariants is maintained since property (iii) of Lemma 11.2 guarantees that at least one of $a^{-T}\overline{x}^* > b^-$ and $a^{+T}\overline{x}^* > b^+$ holds true.   □

We modify the way Algorithm 11.6 chooses between $(a^+, b^+; \mathcal{T}_0'' \cup \{\overline{x}^+\})$ and $(a^-, b^-; \mathcal{T}_0'' \cup \{\overline{x}^-\})$: the point is that cuts violated only slightly by $\overline{x}^*$ are of little use to us and yet, given the choice between inequality $a^{+T}\overline{x} \le b^+$ violated only slightly by $\overline{x}^*$ and inequality $a^{-T}\overline{x} \le b^-$ violated significantly by $\overline{x}^*$, Algorithm 11.6 will choose the former.

This observation raises the issue of measuring the amount by which point $\overline{x}^*$ violates inequality $a^T \overline{x} \le b$. On the one hand, this measure should be invariant under scaling the row vector $[a^T, b]$; this argument suggests measures such as

$$\frac{a^T \overline{x}^* - b}{\|a\|_1} \quad \text{or} \quad \frac{a^T \overline{x}^*}{b}.$$

On the other hand, the inequality returned by Algorithm 11.6 differs from the cut used in solving the TSP instance and the worth of that cut is ultimately measured by its effect, imponderable beforehand, on the overall time required to solve the instance. We measure the violation simply by $a^T \overline{x}^* - b$. In the special case of our

Algorithm 11.7 A modification of Algorithm 11.6.

while $|\mathcal{T}_0''| < \dim \mathcal{T}''$

do   find vectors a^-, a^+, numbers b^-, b^+, and elements $\overline{x}^-, \overline{x}^+$
        of $\mathcal{T}'' - \mathcal{T}_0''$ with properties (i), (ii), (iii) of Lemma 11.2;
        if $a^{+T}\overline{x}^* - b^+ \ge a^{-T}\overline{x}^* - b^-$
        then replace $(a, b; \mathcal{T}_0'')$ by $(a^+, b^+; \mathcal{T}_0'' \cup \{\overline{x}^+\})$;
        else replace $(a, b; \mathcal{T}_0'')$ by $(a^-, b^-; \mathcal{T}_0'' \cup \{\overline{x}^-\})$;
        end

end
return the inequality $a^T \overline{x} \le b$;

interest, we may use Algorithm 11.5 to execute the instructions

find vectors a^-, a^+, numbers b^-, b^+, and elements $\overline{x}^-, \overline{x}^+$ of $\mathcal{T}'' - \mathcal{T}_0''$
with properties (i), (ii), (iii) of Lemma 11.2;

in Algorithm 11.7; for clarity, let us spell out the result as Algorithm 11.8.

Algorithm 11.8 Algorithm 11.7 in our special case.

$\overline{x}^0 =$ the element of $\mathcal{T}''$ such that $\psi(\overline{x}^0)$ is the vector of all zeros;
while $|\mathcal{T}_0''| < \dim \mathcal{T}''$
do   find a nonzero integer vector v such that
$$v^T \overline{x} = 0 \quad \text{for all } \overline{x} \text{ in } \mathcal{T}_0'',$$
$$v_e = 0 \quad \text{for all } e \text{ outside } \overline{E}_{1/2};$$
$(a^-, b^-, \overline{x}^-) = \text{TILT}\,(a, b, -v, 0, \overline{x}^0);$
$(a^+, b^+, \overline{x}^+) = \text{TILT}\,(a, b, v, 0, \overline{x}^0):$
if $a^{+T}\overline{x}^* - b^+ \geq a^{-T}\overline{x}^* - b^-$
then replace $(a, b; \mathcal{T}_0'')$ by $(a^+, b^+; \mathcal{T}_0'' \cup \{\overline{x}^+\});$
else replace $(a, b; \mathcal{T}_0'')$ by $(a^-, b^-; \mathcal{T}_0'' \cup \{\overline{x}^-\});$
end
end
return the inequality $a^T \overline{x} \leq b$;

∨ ∨ ∨

We will illustrate Algorithm 11.8 on EXAMPLE 2. Here, the input inequality $a^T \overline{x} \leq b$
is

$$\overline{x}_{23} + \overline{x}_{25} + 2\overline{x}_{35} + 2\overline{x}_{45} + 2\overline{x}_{46} + 3\overline{x}_{56} + \overline{x}_{68} + 2\overline{x}_{78} + \overline{x}_{79} + \overline{x}_{89} \leq 5;$$

in the notation

$$\psi(\overline{x}) = [\overline{x}_{23}, \overline{x}_{25}, \overline{x}_{35}, \overline{x}_{45}, \overline{x}_{46}, \overline{x}_{56}, \overline{x}_{68}, \overline{x}_{78}, \overline{x}_{79}, \overline{x}_{89}]^T,$$

$\psi(\mathcal{T}_0'')$ consists of the six vectors

$$[0, 0, 1, 0, 1, 0, 0, 0, 1, 0]^T,$$
$$[1, 0, 0, 1, 0, 0, 1, 0, 1, 0]^T,$$
$$[0, 1, 0, 0, 1, 0, 0, 0, 1, 1]^T,$$
$$[1, 0, 0, 0, 0, 1, 0, 0, 0, 1]^T,$$
$$[1, 0, 0, 1, 0, 0, 0, 1, 0, 0]^T,$$
$$[1, 0, 0, 0, 1, 0, 0, 1, 0, 0]^T$$

and $\psi(\overline{x}^*)$ is

$$[0.390, 0.007, 0.610, 0.376, 0.624, 0.007, 0.369, 0.014, 0.986, 0.014]^T.$$

Since $|\mathcal{T}_0''| = 6$ and $|\overline{E}_{1/2}| = 10$, the algorithm will go through four iterations. In
all of them, we let $\overline{x}^0$ stand for the element of $\mathcal{T}''$ such that $\psi(\overline{x}^0)$ is the vector of all
zeros.

ITERATION 1: First, we have to find a nonzero integer vector v such that $v_e = 0$ for all e outside $\overline{E}_{1/2}$ and such that

$$
\begin{array}{lllllll}
& v_{35} & +v_{46} & & +v_{79} & & = 0, \\
v_{23} & & +v_{45} & +v_{68} & +v_{79} & & = 0, \\
& v_{25} & +v_{46} & & +v_{79} & +v_{89} & = 0, \\
v_{23} & & +v_{56} & & & +v_{89} & = 0, \\
v_{23} & +v_{45} & & +v_{78} & & & = 0, \\
v_{23} & +v_{46} & & +v_{78} & & & = 0.
\end{array}
$$

As luck has it, just a reordering of variables converts this system of equations into an upper echelon form,

$$
\begin{array}{lllllll}
v_{35} & & +v_{46} & & +v_{79} & & = 0, \\
& v_{68} & +v_{45} & +v_{23} & +v_{79} & & = 0, \\
& v_{25} & +v_{46} & & +v_{79} & +v_{89} & = 0, \\
& v_{56} & & +v_{23} & & +v_{89} & = 0, \\
& v_{45} & & +v_{23} & +v_{78} & & = 0, \\
& v_{46} & & +v_{23} & +v_{78} & & = 0.
\end{array}
$$

Any choice of values of the dependent variables $v_{23}, v_{78}, v_{79}, v_{89}$ other than $v_{23} = v_{78} = v_{79} = v_{89} = 0$ yields a nonzero solution of this system. Choosing arbitrarily $[v_{23}, v_{78}, v_{79}, v_{89}] = [-1, 0, 0, 0]$, we get

$$
\begin{array}{l}
[\ v_{23},\ v_{25},\ v_{35},\ v_{45},\ v_{46},\ v_{56},\ v_{68},\ v_{78},\ v_{79},\ v_{89}\ ] = \\
[\ -1,\ -1,\ -1,\ 1,\ 1,\ 1,\ 0,\ 0,\ 0,\ 0\ ] .
\end{array}
$$

TILT $(a, b, -v, 0, \overline{x}^0)$ provides

$$
2\overline{x}_{23} + 2\overline{x}_{25} + 3\overline{x}_{35} + \overline{x}_{45} + \overline{x}_{46} + 2\overline{x}_{56} + \overline{x}_{68} + 2\overline{x}_{78} + \overline{x}_{79} + \overline{x}_{89} \leq 5
$$

as $a^{-T}\overline{x} \leq b^{-}$ and it provides the $\overline{x}^{-}$ defined by

$$
\psi(\overline{x}^{-}) = [\ 0,\ 0,\ 1,\ 0,\ 0,\ 0,\ 1,\ 0,\ 1,\ 0]^{T};
$$

TILT $(a, b, v, 0, \overline{x}^0)$ provides

$$
\overline{x}_{35} + 3\overline{x}_{45} + 3\overline{x}_{46} + 4\overline{x}_{56} + \overline{x}_{68} + 2\overline{x}_{78} + \overline{x}_{79} + \overline{x}_{89} \leq 5 \qquad (11.30)
$$

as $a^{+T}\overline{x} \leq b^{+}$ and it provides the $\overline{x}^{+}$ defined by

$$
\psi(\overline{x}^{+}) = [\ 0,\ 0,\ 0,\ 0,\ 0,\ 1,\ 0,\ 0,\ 0,\ 1]^{T}.
$$

We have $a^{-T}\overline{x} - b^{-} = a^{+T}\overline{x} - b^{+} = 0.035$ and the tie is broken in favor of a^{+}, b^{+}, and $\overline{x}^{+}$: we replace $a^T\overline{x} \leq b$ by (11.30) and add $\overline{x}^{+}$ to $\mathcal{T}_0''$.

ITERATION 2: First, we have to find a nonzero integer vector v such that $v_e = 0$ for all e outside $\overline{E}_{1/2}$ and such that

$$
\begin{array}{lllllll}
v_{35} & & +v_{46} & & +v_{79} & & = 0, \\
& v_{68} & +v_{45} & +v_{23} & +v_{79} & & = 0, \\
& v_{25} & +v_{46} & & +v_{79} & +v_{89} & = 0, \\
& v_{56} & & +v_{23} & & +v_{89} & = 0, \\
& v_{45} & & +v_{23} & +v_{78} & & = 0, \\
& v_{46} & & +v_{23} & +v_{78} & & = 0, \\
& v_{56} & & & & +v_{89} & = 0.
\end{array}
$$

Again, we are lucky in being able to convert this system into an upper echelon form just by reordering its variables and its equations:

$$
\begin{aligned}
v_{35} &&&& +v_{46} &&&& +v_{79} && &= 0,\\
& v_{68} && +v_{45} && +v_{23} && +v_{79} && &= 0,\\
&& v_{25} && +v_{46} &&&& +v_{79} &+v_{89} &= 0,\\
&&& v_{45} && +v_{23} && +v_{78} && &= 0,\\
&&&& v_{46} &+v_{23} && +v_{78} && &= 0,\\
&&&&& v_{23} &+v_{56} &&& +v_{89} &= 0,\\
&&&&&& v_{56} &&& +v_{89} &= 0.
\end{aligned}
$$

Choosing arbitrarily $[v_{78}, v_{79}, v_{89}] = [-1, 0, 0]$, we get

$$
\begin{aligned}
[\; v_{23},\; v_{25},\; v_{35},\; v_{45},\; v_{46},\; v_{56},\; v_{68},\; v_{78},\; v_{79},\; v_{89}\;] = \\
[\;\; 0,\; -1,\; -1,\;\; 1,\;\; 1,\;\; 0,\; -1,\; -1,\;\; 0,\;\; 0\;].
\end{aligned}
$$

TILT $(a, b, -v, 0, \overline{x}^0)$ provides $a^T\overline{x} \leq b$ as $a^{-T}\overline{x} \leq b^-$ and it provides the $\overline{x}^-$ defined by

$$\psi(\overline{x}^-) = [\,0, 1, 0, 0, 1, 0, 0, 1, 0, 0\,]^T;$$

TILT $(a, b, v, 0, \overline{x}^0)$ provides $a^T\overline{x} \leq b$ as $a^{+T}\overline{x} \leq b^+$ and it provides the $\overline{x}^+$ defined by

$$\psi(\overline{x}^+) = [\,0, 0, 0, 1, 0, 0, 0, 0, 1, 1\,]^T.$$

We leave $a^T\overline{x} \leq b$ unchanged and add $\overline{x}^+$ to T_0''.

ITERATION 3: First, we have to find a nonzero integer vector v such that $v_e = 0$ for all e outside $\overline{E}_{1/2}$ and such that

$$
\begin{aligned}
v_{35} &&&& +v_{46} &&&& +v_{79} && &= 0,\\
& v_{68} && +v_{45} && +v_{23} && +v_{79} && &= 0,\\
&& v_{25} && +v_{46} &&&& +v_{79} &+v_{89} &= 0,\\
&&& v_{45} && +v_{23} && +v_{78} && &= 0,\\
&&&& v_{46} &+v_{23} && +v_{78} && &= 0,\\
&&&&& v_{23} &+v_{56} &&& +v_{89} &= 0,\\
&&&&&& v_{56} &&& +v_{89} &= 0,\\
&&& v_{45} &&&&& +v_{79} &+v_{89} &= 0.
\end{aligned}
$$

Our luck keeps holding: reordering the variables and the equations, we convert the system into the upper echelon form

$$
\begin{aligned}
v_{35} &&&& +v_{46} &+v_{79} &&&& &= 0,\\
& v_{68} &&&& +v_{79} &+v_{45} &+v_{23} && &= 0,\\
&& v_{25} &+v_{46} && +v_{79} &&&& +v_{89} &= 0,\\
&&& v_{46} &&&& +v_{23} && +v_{78} &= 0,\\
&&&&& v_{79} &+v_{45} &&& +v_{89} &= 0,\\
&&&&&& v_{45} &+v_{23} && +v_{78} &= 0,\\
&&&&&&& v_{23} &+v_{56} &+v_{89} &= 0,\\
&&&&&&&& v_{56} &+v_{89} &= 0.
\end{aligned}
$$

Choosing arbitrarily $[v_{78}, v_{89}] = [-1, 0]$, we get

$$
\begin{aligned}
[\; v_{23},\; v_{25},\; v_{35},\; v_{45},\; v_{46},\; v_{56},\; v_{68},\; v_{78},\; v_{79},\; v_{89}\;] = \\
[\;\; 0,\;\; 0,\;\; 0,\;\; 1,\;\; 1,\;\; 0,\;\; 0,\; -1,\; -1,\;\; 0\;].
\end{aligned}
$$

TILT $(a, b, -v, 0, \overline{x}^0)$ provides $a^T \overline{x} \leq b$ as $a^{-T} \overline{x} \leq b^-$ and it provides the $\overline{x}^-$ defined by

$$\psi(\overline{x}^-) = [\, 0, \ 0, \ 0, \ 0, \ 0, \ 1, \ 0, \ 0, \ 1, \ 0\,]^T;$$

TILT $(a, b, v, 0, \overline{x}^0)$ provides $a^T \overline{x} \leq b$ as $a^{+T} \overline{x} \leq b^+$ and it provides the $\overline{x}^+$ defined by

$$\psi(\overline{x}^+) = [\, 0, \ 0, \ 1, \ 0, \ 1, \ 0, \ 0, \ 0, \ 0, \ 1\,]^T.$$

We leave $a^T \overline{x} \leq b$ unchanged and add $\overline{x}^+$ to $\mathcal{T}_0''$.

ITERATION 4: First, we have to find a nonzero integer vector v such that $v_e = 0$ for all e outside $\overline{E}_{1/2}$ and such that

v_{35}		$+v_{46}$	$+v_{79}$						$= 0,$
	v_{68}		$+v_{79}$	$+v_{45}$	$+v_{23}$				$= 0,$
	v_{25}	$+v_{46}$	$+v_{79}$					$+v_{89}$	$= 0,$
		v_{46}			$+v_{23}$		$+v_{78}$		$= 0,$
			v_{79}	$+v_{45}$				$+v_{89}$	$= 0,$
				v_{45}	$+v_{23}$		$+v_{78}$		$= 0,$
					v_{23}	$+v_{56}$		$+v_{89}$	$= 0,$
						v_{56}		$+v_{89}$	$= 0,$
v_{35}		$+v_{46}$						$+v_{89}$	$= 0.$

Reordering the variables and the equations, we convert the system into the form

v_{68}				$+v_{79}$	$+v_{45}$	$+v_{23}$			$= 0,$
	v_{25}		$+v_{46}$	$+v_{79}$				$+v_{89}$	$= 0,$
		v_{56}						$+v_{89}$	$= 0,$
			v_{35}	$+v_{46}$	$+v_{79}$				$= 0,$
				v_{46}		$+v_{23}$	$+v_{78}$		$= 0,$
					v_{79}	$+v_{45}$		$+v_{89}$	$= 0,$
						v_{45}	$+v_{23}$	$+v_{78}$	$= 0,$
		v_{56}					$+v_{23}$	$+v_{89}$	$= 0,$
			v_{35}	$+v_{46}$				$+v_{89}$	$= 0.$

Then we subtract the third row from the eighth row; subsequent reordering of variables and equations yields the form

v_{68}				$+v_{79}$	$+v_{45}$			$+v_{23}$	$= 0,$
	v_{25}		$+v_{46}$	$+v_{79}$		$+v_{89}$			$= 0,$
		v_{56}				$+v_{89}$			$= 0,$
			v_{35}	$+v_{46}$	$+v_{79}$				$= 0,$
				v_{46}			$+v_{78}$	$+v_{23}$	$= 0,$
					v_{79}	$+v_{45}$	$+v_{89}$		$= 0,$
						v_{45}	$+v_{78}$	$+v_{23}$	$= 0,$
			v_{35}	$+v_{46}$		$+v_{89}$			$= 0,$
								v_{23}	$= 0.$

(In general, when we have to resort to row operations in order to convert the system

$$v^T \overline{x} = 0 \quad \text{for all } \overline{x} \text{ in } \mathcal{T}_0''$$

to an upper echelon form, we choose the pivot elements by the rule of Markowitz [374] that is designed to minimize fill-in.) Next, we subtract the fourth row from the eighth row; another reordering of variables and equations yields the upper echelon form

$$
\begin{aligned}
v_{68} \qquad\qquad\qquad\qquad +v_{45} \;+v_{79} \;+v_{23} \qquad\qquad &= 0, \\
v_{25} \qquad\qquad +v_{46} \qquad\quad +v_{79} \qquad\quad +v_{89} &= 0, \\
v_{56} \qquad\qquad\qquad\qquad\qquad\qquad\qquad +v_{89} &= 0, \\
v_{35} \;+v_{46} \qquad\qquad +v_{79} \qquad\qquad &= 0, \\
v_{46} \;+v_{78} \qquad\qquad\qquad +v_{23} \qquad\quad &= 0, \\
v_{78} \;+v_{45} \qquad\qquad +v_{23} \qquad\quad &= 0, \\
v_{45} \;+v_{79} \qquad\qquad +v_{89} &= 0, \\
-v_{79} \qquad\qquad +v_{89} &= 0, \\
v_{23} \qquad\qquad &= 0.
\end{aligned}
$$

Setting $v_{89} = -1$, we get

$$
\begin{aligned}
[\; v_{23}, \; v_{25}, \; v_{35}, \; v_{45}, \; v_{46}, \; v_{56}, \; v_{68}, \; v_{78}, \; v_{79}, \; v_{89} \;] = \\
[\quad 0, \quad 0, \; -1, \quad 2, \quad 2, \quad 1, \; -1, \; -2, \; -1, \; -1 \;] \; .
\end{aligned}
$$

TILT $(a, b, -v, 0, \overline{x}^0)$ provides

$$
\overline{x}_{35} + \overline{x}_{45} + \overline{x}_{46} + 2\overline{x}_{56} + \overline{x}_{68} + 2\overline{x}_{78} + \overline{x}_{79} + \overline{x}_{89} \leq 3 \tag{11.31}
$$

as $a^{-T}\overline{x} \leq b^-$ and it provides the $\overline{x}^-$ defined by

$$
\psi(\overline{x}^-) = [\, 0, \, 0, \, 1, \, 0, \, 0, \, 0, \, 0, \, 1, \, 0, \, 0]^T;
$$

TILT $(a, b, v, 0, \overline{x}^0)$ provides

$$
\overline{x}_{45} + \overline{x}_{46} + \overline{x}_{56} \leq 1
$$

as $a^{+T}\overline{x} \leq b^+$ and it provides the $\overline{x}^+$ defined by

$$
\psi(\overline{x}^+) = [\, 0, \, 0, \, 0, \, 1, \, 0, \, 0, \, 0, \, 0, \, 0, \, 0]^T.
$$

Since $a^{-T}\overline{x} - b^- = 0.021$ and $a^{+T}\overline{x} - b^+ = 0.007$, we replace $a^T\overline{x} \leq b$ by (11.31) and add $\overline{x}^-$ to $\mathcal{T}_0''$.

Now $|\mathcal{T}_0''|$ has reached $|\overline{E}_{1/2}|$, and so Algorithm 11.8 returns inequality (11.31).

$\wedge \wedge \wedge$

FUNCTION FACET

In actual computations, Algorithm 11.8 may fail to return an inequality that separates $\mathcal{T}''$ from $\overline{x}^*$ and induces a facet of the convex hull of $\mathcal{T}''$. Reasons for such a failure have been mentioned in Section 11.3. One of them may be an imminent threat of overflow in computations using integer arithmetic, such as finding a nonzero integer solution v of the system

$$
\begin{aligned}
v^T\overline{x} = 0 \quad &\text{for all } \overline{x} \text{ in } \mathcal{T}_0'', \\
v_e = 0 \quad &\text{for all } e \text{ outside } \overline{E}_{1/2}
\end{aligned}
$$

or the iterated updates of v and w in TILT (Algorithm 11.2 in Section 11.6.5); another reason may be that the number of recursive calls of ORACLE (this function is used in TILT and its recursive implementation will be described in Section 11.8) has exceeded a prescribed threshold. And even when Algorithm 11.8 does return an inequality that separates $\mathcal{T}''$ from $\bar{x}^*$ and induces a facet of the convex hull of $\mathcal{T}''$, this inequality may be violated by $\bar{x}^*$ so slightly that the success is treated as a failure: with ε a prescribed zero tolerance ($\varepsilon = 0.01$ is one representative value), we treat an inequality $a^T \bar{x} \le b$ (whose left-hand side is normalized in the sense that the coefficients are integers and the greatest common divisor of all of them equals 1) as violated by $\bar{x}^*$ if and only if $a^T \bar{x}^* > b + \varepsilon$.

Rather than accepting such defeat at once, we backtrack to the most recent point in the computations where we had chosen one of two inequalities $a^{-T} \bar{x} \le b^-$ and $a^{+T} \bar{x} \le b^+$ such that $a^{-T} \bar{x}^* > b^-$ and $a^{+T} \bar{x}^* > b^+$; then we switch to the other choice and try again. This policy is formalized in Algorithm 11.9.

Earlier versions of our computer code used a slightly different version of FACET; this version is presented here as Algorithm 11.10.

11.7 PHASE 2: FROM $\mathcal{T}''$ TO $\mathcal{T}'$

Given the linear inequality $a^T \bar{x} \le b$ constructed in PHASE 1, we find integers Δ_e ($e \in \overline{E}_0 \cup \overline{E}_1$) such that the inequality

$$a^T x + \sum(\Delta_e \bar{x}_e : e \in \overline{E}_0 \cup \overline{E}_1) \le b + \sum(\Delta_e : e \in \overline{E}_1)$$

induces a facet of the convex hull of $\mathcal{T}'$. We compute the integers Δ_e one by one; this technique, known as *sequential lifting*, originated in the work of Gomory [218] and was elaborated by Balas [31], Hammer, Johnson, and Peled [247], Padberg [448], [449], Wolsey [557], [558], and others.

In Algorithm 11.11, each of the two **while** loops maintains the invariant

$a^T \bar{x} \le b$ induces a facet of the convex hull of the set of all $\bar{x}$ in $\mathcal{T}'$
such that $\bar{x}_e = 0$ whenever $e \in Q_0$ and $\bar{x}_e = 1$ whenever $e \in Q_1$;

each new μ can be computed by a single call of ORACLE; as we will observe later, $\mu \ge b$ in every iteration of the first **while** loop and $\mu \le b$ in every iteration of the second **while** loop. The process is illustrated in Figure 11.18.

11.7.1 EXAMPLE 3

∨ ∨ ∨

We will illustrate PHASE 2 on on the example where $\overline{V} = \{0, 1, \ldots, 8\}$ and

$$\begin{array}{lllll}
\bar{x}_{05}^* = 0.45, & \bar{x}_{06}^* = 0.55, & \bar{x}_{07}^* = 1.00 & \bar{x}_{08}^* = 1.10, & \bar{x}_{13}^* = 0.45, \\
\bar{x}_{14}^* = 1.00, & \bar{x}_{15}^* = 0.10, & \bar{x}_{16}^* = 0.45, & \bar{x}_{23}^* = 0.55, & \bar{x}_{24}^* = 0.55, \\
\bar{x}_{28}^* = 0.90, & \bar{x}_{37}^* = 1.00, & \bar{x}_{45}^* = 0.45, & \bar{x}_{56}^* = 1.00
\end{array}$$

with $\bar{x}_{ij}^* = 0$ for all the remaining i and j such that $0 \le i < j \le 8$. The cut produced in PHASE 1 reads

$$\bar{x}_{13} + \bar{x}_{15} + \bar{x}_{16} + \bar{x}_{23} + \bar{x}_{24} + \bar{x}_{28} + \bar{x}_{45} \le 3; \tag{11.32}$$

Algorithm 11.9 FACET$(a, b; \mathcal{T}_0'')$.

---

if $|\mathcal{T}_0''| = |\overline{E}_{1/2}|$

then if $a^T \overline{x}^* > b + \varepsilon$

     then return $a^T \overline{x} \leq b$;

     else   return a failure message;

     end

else $\overline{x}^0 = $ the element of $\mathcal{T}''$ such that $\psi(\overline{x}^0)$ is the vector of all zeros;

     find a nonzero integer vector v such that

         $v^T \overline{x} = 0$ for all $\overline{x}$ in $\mathcal{T}_0''$,

           $v_e = 0$ for all e outside $\overline{E}_{1/2}$;

     $(a^+, b^+, \overline{x}^+) = $ TILT $(a, b, v, 0, \overline{x}^0)$;

     $(a^-, b^-, \overline{x}^-) = $ TILT $(a, b, -v, 0, \overline{x}^0)$;

     if $a^{+T} \overline{x}^* - b^+ \geq a^{-T} \overline{x}^* - b^-$

     then if   FACET$(a^+, b^+; \mathcal{T}_0' \cup \{\overline{x}^+\})$ returns an $a''^T \overline{x} \leq b''$

         then return $a''^T \overline{x} \leq b''$;

         else **if** $a^{-T} \overline{x}^* > b^-$ and FACET$(a^-, b^-; \mathcal{T}_0'' \cup \{\overline{x}^-\})$

                  returns an $a''^T \overline{x} \leq b''$

            then return $a''^T \overline{x} \leq b''$;

            else   return a failure message;

            end

         end

     else **if**   FACET$(a^-, b^-; \mathcal{T}_0' \cup \{\overline{x}^-\})$ returns an $a''^T \overline{x} \leq b''$

         then return $a''^T \overline{x} \leq b''$;

         else **if** $a^{+T} \overline{x}^* > b^+$ and FACET$(a^+, b^+; \mathcal{T}_0'' \cup \{\overline{x}^+\})$

                  returns an $a''^T \overline{x} \leq b''$

            then return $a''^T \overline{x} \leq b''$;

            else   return a failure message;

            end

         end

     end

end

Algorithm 11.10 An earlier version of $\text{FACET}(a, b; \mathcal{T}_0'')$.

---

if $|\mathcal{T}_0''| = |\overline{E}_{1/2}|$

then if $a^T \overline{x}^* > b + \varepsilon$

       then return $a^T \overline{x} \leq b$;

       else return a failure message;

       end

else $\overline{x}^0 = $ the element of $\mathcal{T}''$ such that $\psi(\overline{x}^0)$ is the vector of all zeros;

       find a nonzero integer vector v such that

$$v^T \overline{x} = 0 \quad \text{for all } \overline{x} \text{ in } \mathcal{T}_0'',$$
$$v_e = 0 \quad \text{for all } e \text{ outside } \overline{E}_{1/2};$$

       $(a^+, b^+, \overline{x}^+) = \text{TILT}\,(a, b, v, 0, \overline{x}^0)$;

       if $a^T \overline{x}^+ = b$ **then** return $\text{FACET}(a, b; \mathcal{T}_0'' \cup \{\overline{x}^+\})$ **end**

       $(a^-, b^-, \overline{x}^-) = \text{TILT}\,(a, b, -v, 0, \overline{x}^0)$;

       if $a^T \overline{x}^- = b$ **then** return $\text{FACET}(a, b; \mathcal{T}_0'' \cup \{\overline{x}^-\})$ **end**

       if $a^{+T} \overline{x}^* - b^+ \geq a^{-T} \overline{x}^* - b^-$

       then if $\text{FACET}(a^+, b^+; \mathcal{T}_0'' \cup \{\overline{x}^+\})$ returns an $a''^T \overline{x} \leq b''$

              then return $a''^T \overline{x} \leq b''$;

              else if $a^{-T} \overline{x}^* > b^-$ and $\text{FACET}(a^-, b^-; \mathcal{T}_0'' \cup \{\overline{x}^-\})$

                          returns an $a''^T \overline{x} \leq b''$

                then return $a''^T \overline{x} \leq b''$;

                else return a failure message;

                end

           end

       else if $\text{FACET}(a^-, b^-; \mathcal{T}_0' \cup \{\overline{x}^-\})$ returns an $a''^T \overline{x} \leq b''$

           then return $a''^T \overline{x} \leq b''$;

           else if $a^{+T} \overline{x}^* > b^+$ and $\text{FACET}(a^+, b^+; \mathcal{T}_0'' \cup \{\overline{x}^+\})$

                          returns an $a''^T \overline{x} \leq b''$

                then return $a''^T \overline{x} \leq b''$;

                else return a failure message;

                end

           end

       end

end

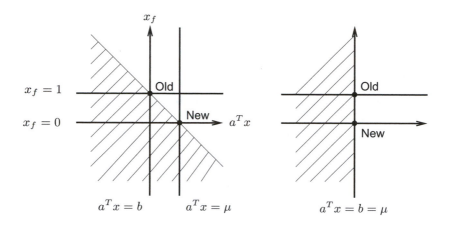

First while loop

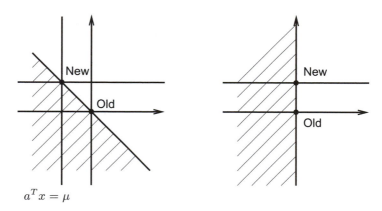

Second while loop

Figure 11.18 Algorithm 11.11, sequential lifting.

Algorithm 11.11 Sequential lifting from $\mathcal{T}''$ to $\mathcal{T}'$.

$Q_0 = \overline{E}_0$, $Q_1 = \overline{E}_1$;
while $Q_1 \neq \emptyset$
do $f =$ an edge in Q_1;
             $\mu = \max\{a^T\overline{x} : \overline{x} \in \mathcal{T}',$
                              $\overline{x}_e = 0$ whenever $e \in Q_0 \cup \{f\}$,
                              $\overline{x}_e = 1$ whenever $e \in Q_1 - \{f\}\}$;
             replace $a^T\overline{x} \leq b$ by $a^T\overline{x} + (\mu - b)\overline{x}_f \leq \mu$;
             delete f from Q_1;
end
while $Q_0 \neq \emptyset$
do $f =$ an edge in Q_0;
             $\mu = \max\{a^T\overline{x} : \overline{x} \in \mathcal{T}',$
                              $\overline{x}_e = 0$ whenever $e \in Q_0 - \{f\}$,
                              $\overline{x}_f = 1\}$;
             replace $a^T\overline{x} \leq b$ by $a^T\overline{x} + (b - \mu)\overline{x}_f \leq b$;
             delete f from Q_0;
end

in the notation

$$\psi(\overline{x}) = [\overline{x}_{13},\ \overline{x}_{15},\ \overline{x}_{16},\ \overline{x}_{23},\ \overline{x}_{24},\ \overline{x}_{28},\ \overline{x}_{45}]^T,$$

$\psi(\mathcal{T}_0'')$ consists of the seven vectors

$$[\,1,\ 0,\ 0,\ 0,\ 1,\ 1,\ 0\,]^T,$$
$$[\,0,\ 0,\ 1,\ 1,\ 1,\ 0,\ 0\,]^T,$$
$$[\,0,\ 0,\ 1,\ 1,\ 0,\ 1,\ 0\,]^T,$$
$$[\,1,\ 0,\ 0,\ 0,\ 0,\ 1,\ 1\,]^T,$$
$$[\,0,\ 1,\ 0,\ 1,\ 1,\ 0,\ 0\,]^T,$$
$$[\,0,\ 0,\ 0,\ 1,\ 0,\ 1,\ 1\,]^T,$$
$$[\,0,\ 1,\ 0,\ 0,\ 1,\ 1,\ 0\,]^T.$$

∧ ∧ ∧

11.7.2 Streamlining Algorithm 11.11

We have defined $\overline{V}$ as $\{0, 1, \ldots, k\}$ and we have defined $\overline{E}$ as the set of all the edges of the complete graph on $\overline{V}$. Now let us define $\overline{V}^{\mathrm{int}}$ as $\{1, 2, \ldots, k\}$, let

$\overline{E}^{\mathrm{int}}$ denote the set of the edges of the complete graph on $\overline{V}^{\mathrm{int}}$,

and let

$\chi(\overline{x})$ denote the restriction of $\overline{x}$
on its components indexed by elements of $\overline{E}^{\mathrm{int}}$.

In this notation,

$$\overline{E}^{\text{int}} = \overline{E}_0 \cup \overline{E}_{1/2} \cup \overline{E}_1$$

and each $\overline{x}$ in T' is uniquely determined by $\chi(\overline{x})$; elements of $\chi(T')$ are precisely the incidence vectors of the edge sets of *path systems*—meaning graphs whose connected components are paths—with vertex set $\overline{V}^{\text{int}}$.

∨ ∨ ∨

In EXAMPLE 3, the seven elements of $\chi(T_0'')$ are

- the path system with components 5-6 and 7-3-1-4-2-8,
- the path system with components 5-6-1-4-2-3-7 and 8,
- the path system with components 5-6-1-4 and 7-3-2-8,
- the path system with components 2-8 and 6-5-4-1-3-7,
- the path system with components 6-5-1-4-2-3-7 and 8,
- the path system with components 6-5-4-1 and 7-3-2-8,
- the path system with components 3-7 and 6-5-1-4-2-8.

∧ ∧ ∧

$\chi(T')$ is monotone in the sense that changing 1 to 0 in any component of any element of $\chi(T')$ yields another element of $\chi(T')$; it follows that $\mu \geq b$ in every iteration of the first **while** loop in Algorithm 11.11 and that

$$\mu \leq b \text{ in every iteration of the second \textbf{while} loop.} \tag{11.33}$$

Our code exploits (11.33): it deletes an edge uv in Q_0 from Q_0 without any calls of ORACLE and without any alterations of our inequality $a^T \overline{x} \leq b$ whenever we have on hand an element $\overline{x}$ of T' such that

(i) $\overline{x}_e = 0$ for all e in Q_0 and $a^T \overline{x} = b$

and such that

(ii) u and v are endpoints of distinct paths in the path system $\chi(\overline{x})$.

This $\overline{x}$ certifies the existence of an $\overline{x}'$ in T' such that

$$\overline{x}'_e = 0 \text{ for all } e \text{ in } Q_0 - \{uv\}, \ \overline{x}'_{uv} = 1, \text{ and } a^T \overline{x}' = b;$$

it follows that, in the iteration of the second **while** loop where $f = uv$, we will have $\mu \geq b$, and so $\mu = b$ by virtue of (11.33).

After the execution of the first **while** loop, we have on hand the $|\overline{E}_{1/2}|$ elements $\overline{x}$ of T_0''; each of them has property (i); for each of these $\overline{x}$, we delete from Q_0 all the edges uv that satisfy (ii). Each subsequent call of ORACLE in PHASE 2 returns an $\overline{x}^{\max}$; again, we delete from Q_0 all the edges uv that satisfy (ii) with $\overline{x}^{\max}$ in place of $\overline{x}$.

∨ ∨ ∨

In EXAMPLE 3,

- the path system with components 5-6 and 7-3-1-4-2-8 eliminates edges 57, 58, 67, 68 from Q_0,

- the path system with components 5-6-1-4-2-3-7 and 8 eliminates edges 58, 78 from Q_0,

- the path system with components 5-6-1-4 and 7-3-2-8 eliminates edges 47, 48, 57, 58 from Q_0,

- the path system with components 2-8 and 6-5-4-1-3-7 eliminates edges 26, 27, 68, 78 from Q_0,

- the path system with components 6-5-1-4-2-3-7 and 8 eliminates edges 68, 78 from Q_0,

- the path system with components 6-5-4-1 and 7-3-2-8 eliminates edges 17, 18, 67, 68 from Q_0,

- the path system with components 3-7 and 6-5-1-4-2-8 eliminates edges 36, 38, 67, 78 from Q_0;

Altogether, these seven elements of $\chi(\mathcal{T}_0'')$ eliminate the 13 edges 17, 18, 26, 27, 36, 38, 47, 48, 57, 58, 67, 68, 78 from Q_0, leaving us with

$$Q_0 = \{12, 25, 34, 35, 46\}.$$

∧ ∧ ∧

Another consequence of monotonicity of $\chi(\mathcal{T}')$ is the fact that, in both **while** loops of Algorithm 11.11, the instruction

$$\mu = \max\{a^T \overline{x} : \overline{x} \in \mathcal{T}', \ldots\}$$

is equivalent to the instruction

$$\mu = \max\{a^T \overline{x} : \overline{x} \in \mathcal{T}', \ldots, \overline{x}_e = 0 \text{ whenever } e \notin Q_0 \cup Q_1 \text{ and } a_e = 0\};$$

our code uses the latter variant in Algorithm 11.12.

∨ ∨ ∨

In EXAMPLE 3, the first **while** loop of Algorithm 11.12—with edges of Q_1 scanned in their lexicographic order—goes as follows.

ITERATION 1: $f = 14$, $Q_1 = \{14, 37, 56\}$, $Q_0 = \{12, 25, 34, 35, 46\}$.

ORACLE, called to maximize the left-hand side $a^T \overline{x}$ of (11.32) subject to $\overline{x} \in \mathcal{T}'$ and

---

Algorithm 11.12 PHASE 2.

$Q_0 = \overline{E}_0$, $Q_1 = \overline{E}_1$;

for   all elements $\overline{x}$ of T_0''

do   delete from Q_0 all the edges uv such that u and v are
                  endpoints of distinct paths in the path system $\chi(\overline{x})$;

end

while $Q_1 \neq \emptyset$

do $f =$ an edge in Q_1;
       find an element $\overline{x}^{\max}$ of T' that maximizes $a^T\overline{x}$ subject to

$$\overline{x}_e = 0 \text{ whenever } e \in Q_0 \cup \{f\},$$
$$\overline{x}_e = 1 \text{ whenever } e \in Q_1 - \{f\},$$
$$\overline{x}_e = 0 \text{ whenever } e \notin Q_0 \cup Q_1 \text{ and } a_e = 0;$$

       replace $a^T\overline{x} \leq b$ by $a^T\overline{x} + (a^T\overline{x}^{\max} - b)\overline{x}_f \leq a^T\overline{x}^{\max}$;
       delete from Q_0 all the edges uv such that u and v are
                 endpoints of distinct paths in the path system $\chi(\overline{x}^{\max})$;
       delete f from Q_1;

end

while $Q_0 \neq \emptyset$

do $f =$ an edge in Q_0;
       find an element $\overline{x}^{\max}$ of T' that maximizes $a^T\overline{x}$ subject to

$$\overline{x}_e = 0 \text{ whenever } e \in Q_0 - \{f\},$$
$$\overline{x}_f = 1,$$
$$\overline{x}_e = 0 \text{ whenever } e \notin Q_0 \text{ and } a_e = 0;$$

       replace $a^T\overline{x} \leq b$ by $a^T\overline{x} + (b - a^T\overline{x}^{\max})\overline{x}_f \leq b$;
       delete from Q_0 all the edges uv such that u and v are
                 endpoints of distinct paths in the path system $\chi(\overline{x}^{\max})$;
       delete f from Q_0;

end

$$\overline{x}_{12} = \overline{x}_{17} = \overline{x}_{18} = \overline{x}_{25} = \overline{x}_{26} = \overline{x}_{27} = \overline{x}_{34} = \overline{x}_{35} = \overline{x}_{36} =$$
$$\overline{x}_{38} = \overline{x}_{46} = \overline{x}_{47} = \overline{x}_{48} = \overline{x}_{57} = \overline{x}_{58} = \overline{x}_{67} = \overline{x}_{68} = \overline{x}_{78} = 0,$$
$$\overline{x}_{37} = \overline{x}_{56} = 1, \text{ and } \overline{x}_{14} = 0,$$

returns the $\overline{x}^{\max}$ represented by

- the path system with the single component 7-3-1-6-5-4-2-8;

since $a^T \overline{x}^{\max} = 5$, we replace (11.32) by

$$\overline{x}_{13} + 2\overline{x}_{14} + \overline{x}_{15} + \overline{x}_{16} + \overline{x}_{23} + \overline{x}_{24} + \overline{x}_{28} + \overline{x}_{45} \le 5. \qquad (11.34)$$

ITERATION 2: $f = 37$, $Q_1 = \{37, 56\}$, $Q_0 = \{12, 25, 34, 35, 46\}$.

ORACLE, called to maximize the left-hand side $a^T \overline{x}$ of (11.34) subject to $\overline{x} \in \mathcal{T}'$ and

$$\overline{x}_{12} = \overline{x}_{17} = \overline{x}_{18} = \overline{x}_{25} = \overline{x}_{26} = \overline{x}_{27} = \overline{x}_{34} = \overline{x}_{35} = \overline{x}_{36} =$$
$$\overline{x}_{38} = \overline{x}_{46} = \overline{x}_{47} = \overline{x}_{48} = \overline{x}_{57} = \overline{x}_{58} = \overline{x}_{67} = \overline{x}_{68} = \overline{x}_{78} = 0,$$
$$\overline{x}_{56} = 1, \text{ and } \overline{x}_{37} = 0,$$

returns the $\overline{x}^{\max}$ represented by

- the path system with components 6-5-4-1-3-2-8 and 7;

since $a^T \overline{x}^{\max} = 6$, we replace (11.34) by

$$\overline{x}_{13} + 2\overline{x}_{14} + \overline{x}_{15} + \overline{x}_{16} + \overline{x}_{23} + \overline{x}_{24} + \overline{x}_{28} + \overline{x}_{37} + \overline{x}_{45} \le 6. \qquad (11.35)$$

ITERATION 3: $f = 56$, $Q_1 = \{56\}$, $Q_0 = \{12, 25, 34, 35, 46\}$.

ORACLE, called to maximize the left-hand side $a^T \overline{x}$ of (11.35) subject to $\overline{x} \in \mathcal{T}'$ and

$$\overline{x}_{12} = \overline{x}_{17} = \overline{x}_{18} = \overline{x}_{25} = \overline{x}_{26} = \overline{x}_{27} = \overline{x}_{34} = \overline{x}_{35} = \overline{x}_{36} =$$
$$\overline{x}_{38} = \overline{x}_{46} = \overline{x}_{47} = \overline{x}_{48} = \overline{x}_{57} = \overline{x}_{58} = \overline{x}_{67} = \overline{x}_{68} = \overline{x}_{78} = 0,$$
$$\text{and } \overline{x}_{56} = 0,$$

returns the $\overline{x}^{\max}$ represented by

- the path system with components 5-4-1-6 and 7-3-2-8;

since $a^T \overline{x}^{\max} = 7$, we replace (11.35) by

$$\overline{x}_{13} + 2\overline{x}_{14} + \overline{x}_{15} + \overline{x}_{16} + \overline{x}_{23} + \overline{x}_{24} + \overline{x}_{28} + \overline{x}_{37} + \overline{x}_{45} + \overline{x}_{56} \le 7. \qquad (11.36)$$

The second **while** loop—with edges of Q_1 scanned in their lexicographic order—goes as follows.

ITERATION 1: $f = 12$, $Q_0 = \{12, 25, 34, 35, 46\}$.

ORACLE, called to maximize the left-hand side $a^T \overline{x}$ of (11.36) subject to $\overline{x} \in \mathcal{T}'$ and

$$\overline{x}_{17} = \overline{x}_{18} = \overline{x}_{25} = \overline{x}_{26} = \overline{x}_{27} = \overline{x}_{34} = \overline{x}_{35} = \overline{x}_{36} =$$
$$\overline{x}_{38} = \overline{x}_{46} = \overline{x}_{47} = \overline{x}_{48} = \overline{x}_{57} = \overline{x}_{58} = \overline{x}_{67} = \overline{x}_{68} = \overline{x}_{78} = 0$$
and $\overline{x}_{12} = 1$,

returns the $\overline{x}^{\max}$ represented by

- the path system with components 6-5-4-1-2-3-7 and 8;

since $a^T \overline{x}^{\max} = 6$, we replace (11.36) by

$$\overline{x}_{12} + \overline{x}_{13} + 2\overline{x}_{14} + \overline{x}_{15} + \overline{x}_{16} + \overline{x}_{23} + \overline{x}_{24} + \overline{x}_{28} + \overline{x}_{37} + \overline{x}_{45} + \overline{x}_{56} \leq 7. \quad (11.37)$$

ITERATION 2: $f = 25$, $Q_0 = \{25, 34, 35, 46\}$.

ORACLE, called to maximize the left-hand side $a^T \overline{x}$ of (11.37) subject to $\overline{x} \in \mathcal{T}'$ and

$$\overline{x}_{17} = \overline{x}_{18} = \overline{x}_{26} = \overline{x}_{27} = \overline{x}_{34} = \overline{x}_{35} = \overline{x}_{36} =$$
$$\overline{x}_{38} = \overline{x}_{46} = \overline{x}_{47} = \overline{x}_{48} = \overline{x}_{57} = \overline{x}_{58} = \overline{x}_{67} = \overline{x}_{68} = \overline{x}_{78} = 0$$
and $\overline{x}_{25} = 1$,

returns the $\overline{x}^{\max}$ represented by

- the path system with components 4-1-3-7 and 6-5-2-8;

since $a^T \overline{x}^{\max} = 6$, we replace (11.37) by

$$\overline{x}_{12} + \overline{x}_{13} + 2\overline{x}_{14} + \overline{x}_{15} + \overline{x}_{16} + \overline{x}_{23} + \overline{x}_{24} + \overline{x}_{25} + \overline{x}_{28} + \overline{x}_{37} + \overline{x}_{45} + \overline{x}_{56} \leq 7; \quad (11.38)$$

since 4 and 6 are endpoints of distinct paths in $\chi(\overline{x}^{\max})$, we delete edge 46 from Q_0.

ITERATION 3: $f = 34$, $Q_0 = \{34, 35\}$.

ORACLE, called to maximize the left-hand side $a^T \overline{x}$ of (11.38) subject to $\overline{x} \in \mathcal{T}'$ and

$$\overline{x}_{17} = \overline{x}_{18} = \overline{x}_{26} = \overline{x}_{27} = \overline{x}_{35} = \overline{x}_{36} =$$
$$\overline{x}_{38} = \overline{x}_{46} = \overline{x}_{47} = \overline{x}_{48} = \overline{x}_{57} = \overline{x}_{58} = \overline{x}_{67} = \overline{x}_{68} = \overline{x}_{78} = 0$$
and $\overline{x}_{34} = 1$,

returns the $\overline{x}^{\max}$ represented by

- the path system with the single component 7-3-4-1-6-5-2-8;

since $a^T \overline{x}^{\max} = 7$, we leave (11.38) unchanged.

ITERATION 4: $f = 35$, $Q_0 = \{35\}$.

ORACLE, called to maximize the left-hand side $a^T \overline{x}$ of (11.38) subject to $\overline{x} \in \mathcal{T}'$ and

$$\overline{x}_{17} = \overline{x}_{18} = \overline{x}_{26} = \overline{x}_{27} = \overline{x}_{34} = \overline{x}_{36} =$$
$$\overline{x}_{38} = \overline{x}_{46} = \overline{x}_{47} = \overline{x}_{48} = \overline{x}_{57} = \overline{x}_{58} = \overline{x}_{67} = \overline{x}_{68} = \overline{x}_{78} = 0$$
and $\overline{x}_{35} = 1$,

returns the $\overline{x}^{\max}$ represented by

- the path system with the single component 7-3-5-6-1-4-2-8;

since $a^T \overline{x}^{\max} = 7$, we leave (11.38) unchanged. This inequality, (11.38), induces a facet of the convex hull of $\mathcal{T}'$ and is violated by $\overline{x}^*$.

The output of Algorithm 11.12 may depend on the order in which the elements of Q_1 and the elements of Q_0 are scanned: in EXAMPLE 3, the second **while** loop with edges of Q_1 scanned in their lexicographic order goes as follows.

ITERATION 1: $f = 46$, $Q_0 = \{46, 35, 34, 25, 12\}$.

ORACLE, called to maximize the left-hand side $a^T \overline{x}$ of (11.36) subject to $\overline{x} \in \mathcal{T}'$ and

$$\overline{x}_{12} = \overline{x}_{17} = \overline{x}_{18} = \overline{x}_{25} = \overline{x}_{26} = \overline{x}_{27} = \overline{x}_{34} = \overline{x}_{35} = \overline{x}_{36} =$$
$$\overline{x}_{38} = \overline{x}_{47} = \overline{x}_{48} = \overline{x}_{57} = \overline{x}_{58} = \overline{x}_{67} = \overline{x}_{68} = \overline{x}_{78} = 0$$
$$\text{and } \overline{x}_{46} = 1,$$

returns the $\overline{x}$ represented by

- the path system with components 5-6-4-1-3-7 and 2-8;

since $a^T \overline{x}^{\max} = 6$, we replace (11.36) by

$$\overline{x}_{13} + 2\overline{x}_{14} + \overline{x}_{15} + \overline{x}_{16} + \overline{x}_{23} + \overline{x}_{24} + \overline{x}_{28} + \overline{x}_{37} + \overline{x}_{45} + \overline{x}_{46} + \overline{x}_{56} \leq 7; \quad (11.39)$$

since 2 and 5 are endpoints of distinct paths in $\chi(\overline{x}^{\max})$, we delete edge 25 from Q_0.

ITERATION 2: $f = 35$, $Q_0 = \{35, 34, 12\}$.

ORACLE, called to maximize the left-hand side $a^T \overline{x}$ of (11.39) subject to $\overline{x} \in \mathcal{T}'$ and

$$\overline{x}_{12} = \overline{x}_{17} = \overline{x}_{18} = \overline{x}_{25} = \overline{x}_{26} = \overline{x}_{27} = \overline{x}_{34} = \overline{x}_{36} =$$
$$\overline{x}_{38} = \overline{x}_{47} = \overline{x}_{48} = \overline{x}_{57} = \overline{x}_{58} = \overline{x}_{67} = \overline{x}_{68} = \overline{x}_{78} = 0$$
$$\text{and } \overline{x}_{35} = 1,$$

returns the $\overline{x}$ represented by

- the path system with the single component 7-3-5-6-1-4-2-8;

since $a^T \overline{x}^{\max} = 7$, we leave (11.39) unchanged.

ITERATION 3: $f = 34$, $Q_0 = \{34, 12\}$.

ORACLE, called to maximize the left-hand side $a^T \overline{x}$ of (11.39) subject to $\overline{x} \in \mathcal{T}'$ and

$$\overline{x}_{12} = \overline{x}_{17} = \overline{x}_{18} = \overline{x}_{25} = \overline{x}_{26} = \overline{x}_{27} = \overline{x}_{35} = \overline{x}_{36} =$$
$$\overline{x}_{38} = \overline{x}_{47} = \overline{x}_{48} = \overline{x}_{57} = \overline{x}_{58} = \overline{x}_{67} = \overline{x}_{68} = \overline{x}_{78} = 0$$
$$\text{and } \overline{x}_{34} = 1,$$

returns the $\overline{x}$ represented by

• the path system with components 6-5-1-4-3-2-8 and 7;

since $a^T \overline{x}^{\max} = 6$, we replace (11.39) by

$$\overline{x}_{13} + 2\overline{x}_{14} + \overline{x}_{15} + \overline{x}_{16} + \overline{x}_{23} + \overline{x}_{24} + \overline{x}_{28} + \overline{x}_{34} + \overline{x}_{37} + \overline{x}_{45} + \overline{x}_{46} + \overline{x}_{56} \leq 7. \quad (11.40)$$

ITERATION 4: $f = 12$, $Q_0 = \{12\}$.

ORACLE, called to maximize the left-hand side $a^T \overline{x}$ of (11.40) subject to $\overline{x} \in T'$ and

$$\overline{x}_{17} = \overline{x}_{18} = \overline{x}_{25} = \overline{x}_{26} = \overline{x}_{27} = \overline{x}_{35} = \overline{x}_{36} =$$
$$\overline{x}_{38} = \overline{x}_{47} = \overline{x}_{48} = \overline{x}_{57} = \overline{x}_{58} = \overline{x}_{67} = \overline{x}_{68} = \overline{x}_{78} = 0$$
and $\overline{x}_{12} = 1$,

returns the $\overline{x}$ represented by

• the path system with components 6-5-4-1-2-3-7 and 8;

since $a^T \overline{x}^{\max} = 6$, we replace (11.40) by

$$\overline{x}_{12} + \overline{x}_{13} + 2\overline{x}_{14} + \overline{x}_{15} + \overline{x}_{16} + \overline{x}_{23}$$
$$+\overline{x}_{24} + \overline{x}_{28} + \overline{x}_{34} + \overline{x}_{37} + \overline{x}_{45} + \overline{x}_{46} + \overline{x}_{56} \leq 7. \quad (11.41)$$

This inequality induces a facet of the convex hull of T' and is violated by $\overline{x}^*$.

∧ ∧ ∧

11.8 IMPLEMENTING ORACLE

ORACLE has to solve the following problem:

Given
> *integer vectors c, ℓ, u*
> *with components indexed by elements of $\overline{E}$*
> *and a threshold t (either an integer or $-\infty$),*
return either
> *an $\overline{x}$ that maximizes $c^T \overline{x}$*
> *subject to $\overline{x} \in T'$, $\ell \leq \overline{x} \leq u$, $c^T \overline{x} > t$*
> *or*
> *the message "infeasible" indicating that*
> *no $\overline{x}$ in T' satisfies $\ell \leq \overline{x} \leq u$, $c^T \overline{x} > t$.*

We are going to describe two recursive algorithms for solving this problem. The more primitive branch-and-bound algorithm is suited to easy instances and the more sophisticated branch-and-cut algorithm is suited to hard instances. In our application, we are faced with a large number of instances and do not know how easy or how hard each of them is until we have solved it. For this reason, we use the two algorithms in tandem: if the branch-and-bound algorithm fails to solve the instance within a prescribed number (4,000 is one representative value) of recursive calls of itself, then we switch to the branch-and-cut algorithm. If the branch-and-cut algorithm fails to solve the instance within a prescribed number (200 is one representative value) of recursive calls of itself, then we we return a failure message.

11.8.1 The branch-and-bound algorithm

Here, we begin with an iterative function that, given ℓ and u, returns

either

integer vectors $\bar{\ell}, \bar{u}$ with components indexed by elements of $\overline{E}$ such that, for each e in $\overline{E}$,

$$\bar{\ell}_e = \ell_e, \bar{u}_e = u_e \text{ or else } \ell_e \le \bar{\ell}_e = \bar{u}_e \le u_e \tag{11.42}$$

and such that

$$\{\bar{x} \in \mathcal{T}' : \bar{\ell} \le \bar{x} \le \bar{u}\} = \{\bar{x} \in \mathcal{T}' : \ell \le \bar{x} \le u\} \tag{11.43}$$

or

the message "infeasible" indicating that no $\bar{x}$ in $\mathcal{T}'$ satisfies $\ell \le \bar{x} \le u$.

Each iteration of this function begins with $\bar{\ell}$ and $\bar{u}$ that satisfy (11.42), (11.43). If
- there is a v in $\overline{V}$ such that $\sum(\bar{u}_e : v \in e) < 2$ or
- there is a v in $\overline{V}^{\text{int}}$ such that $\sum(\bar{\ell}_e : v \in e) > 2$ or
- $\sum(\bar{\ell}_e : 0 \in e)$ equals $\sum(\bar{u}_e : 0 \in e)$ and is odd or
- the graph with vertex set $\overline{V}^{\text{int}}$ and edge set $\{e \in \overline{E}^{\text{int}} : \bar{\ell}_e = 1\}$
  has a connected component that is not a path

(one-vertex graphs are paths, too), then we return "infeasible." Otherwise
- if there is a v in $\overline{V}$ such that $\sum(\bar{\ell}_e : v \in e) < \sum(\bar{u}_e : v \in e) = 2$,
  then we set $\bar{\ell}_e = \bar{u}_e$ for all e such that $v \in e$;
- if there is a v in $\overline{V}^{\text{int}}$ such that $2 = \sum(\bar{\ell}_e : v \in e) < \sum(\bar{u}_e : v \in e)$,
  then we set $\bar{u}_e = \bar{\ell}_e$ for all e such that $v \in e$;
- if there are a v in $\overline{V}$ and an edge e such that $v \in e, \bar{\ell}_e = 0, \bar{u}_e = 2$,
  $\bar{\ell}_f = \bar{u}_f$ for all other edges incident with v, and $\sum(\bar{\ell}_f : v \in f)$ is odd,
  then we set $\bar{\ell}_e = 1, \bar{u}_e = 1$;
- if the graph with vertex set $\overline{V}^{\text{int}}$ and edge set $\{e \in \overline{E}^{\text{int}} : \bar{\ell}_e = 1\}$
  has a connected component that is a path with at least two edges
  and with endpoints v, w such that $\bar{u}_{vw} = 1$, then we set $\bar{u}_{vw} = 0$;

after each of these updates, we proceed to the next iteration; when updates are no longer forced and "infeasible" has not been returned, we return $\bar{\ell}$ and $\bar{u}$.

Next, we compute

$$\lambda_v(c, \bar{\ell}, \bar{u}) = \max\{\sum_{v \in e} c_e \bar{x}_e : \bar{\ell} \le \bar{x} \le \bar{u}, \sum_{v \in e} \bar{x}_e = 2\} \text{ for all } v \text{ in } \overline{V}^{\text{int}},$$

$$\lambda_0(c, \bar{\ell}, \bar{u}) = \max\{\sum_{0 \in e} c_e \bar{x}_e : \bar{\ell} \le \bar{x} \le \bar{u}, \sum_{0 \in e} \bar{x}_e \ge 2, \sum_{0 \in e} \bar{x}_e \text{ even}\},$$

and

$$\lambda(c,\bar{\ell},\bar{u}) = \left\lfloor \frac{1}{2} \sum (\lambda_v(c,\bar{\ell},\bar{u}) : v \in \overline{V}) \right\rfloor.$$

Note that $c^T\bar{x} \leq \lambda(c,\bar{\ell},\bar{u})$ whenever $\bar{x} \in \mathcal{T}'$ and $\bar{\ell} \leq \bar{x} \leq \bar{u}$; if $\lambda(c,\bar{\ell},\bar{u}) \leq t$, then (11.43) allows us to return the message "infeasible".

Now if $\bar{\ell} = \bar{u}$, then we return the common value of $\bar{\ell}$ and $\bar{u}$; otherwise we resort to branching. We branch on an edge e in $\overline{E}^{\text{int}}$ that maximizes $|c_e|$ subject to $\bar{\ell}_e = 0$, $\bar{u}_e = 1$; of the two subproblems, we solve first that which has $\bar{\ell}_e = 1$ in case $c_e > 0$ and $\bar{u}_e = 0$ in case $c_e < 0$.

11.8.2 The branch-and-cut algorithm

This algorithm can access, in addition to c, ℓ, u, and t, a system

$$A'\bar{x} = b', A''\bar{x} \geq b''$$

of linear constraints satisfied by all $\bar{x}$ in $\mathcal{T}'$. System $A'\bar{x} = b'$ consists of all the equations $\bar{x}(\{w\}, \overline{V} - \{w\}) = 2$ with $w \in \overline{V}^{\text{int}}$; system $A''\bar{x} \geq b''$ is initially empty and recursive calls of the branch-and-cut algorithm may add to it inequalities that are returned by a function FINDCUTS.

FINDCUTS, given a ξ such that $A'\xi = b'$, attempts to find inequalities satisfied by all $\bar{x}$ in $\mathcal{T}'$ and violated by ξ. Our current implementation of it uses both heuristic and exact separation algorithms for subtour inequalities as well as heuristic separation algorithms for blossom inequalities; at the root node of the branch-and-cut tree, we also use an exact separation algorithm for blossom inequalities based on the algorithm of Padberg and Rao [453].

If the optimal solution ξ of the problem

$$\begin{aligned} \text{maximize} \quad & c^T\bar{x} \\ \text{subject to} \quad & A'\bar{x} = b', \ A''\bar{x} \geq b'', \ \ell \leq \bar{x} \leq u, \ c^T\bar{x} \geq t + 1 - \varepsilon \end{aligned}$$

with a prescribed small ε (our current setting is $\varepsilon = 10^{-10}$) turns out to be an integer vector, then we return ξ; else we call FINDCUTS(ξ). In the latter case, if FINDCUTS(ξ) returns cuts, then we add these cuts to $A''\bar{x} \geq b''$ and iterate; else we resort to branching. We branch on an edge e in $\overline{E}^{\text{int}}$ that minimizes $|\xi_e - 0.5|$; we solve first the subproblem with $\ell_e = u_e = 0$ and then the subproblem with $\ell_e = u_e = 1$ (and with t increased to $c^T\xi$ in case the first subproblem returned a ξ with $c^T\xi > t$).

With $t_{\max}$ standing for the final value of t, the lower bound ℓ on $\bar{x}$ and the upper bound u on $\bar{x}$ at each leaf in the binary tree of subproblems satisfy

$$\{\bar{x} : A'\bar{x} = b', \ A''\bar{x} \geq b'', \ \ell \leq \bar{x} \leq u, \ c^T\bar{x} \geq t_{\max} + 1\} = \emptyset. \tag{11.44}$$

As usual, we certify (11.44) by exhibiting vectors $\bar{c}, y', y''$ with components indexed by elements of $\overline{E}$ and a number z such that

$$\bar{c}^T = y'^T A' + y''^T A'' + zc^T, \tag{11.45}$$

$$y'' \geq 0, z \geq 0, \tag{11.46}$$

$$\sum(\bar{c}_e u_e : \bar{c}_e > 0) + \sum(\bar{c}_e \ell_e : \bar{c}_e < 0) < y'^T b' + y''^T b'' + z(t_{\max}+1) : \tag{11.47}$$

any element $\bar{x}$ of the left-hand side set in (11.44) would have to satisfy

$$\bar{c}^T \bar{x} \geq y'^T b' + y''^T b'' + z(t_{\max} + 1)$$

because of (11.45), (11.46) and it would have to satisfy

$$\bar{c}^T \bar{x} < y'^T b' + y''^T b'' + z(t_{\max} + 1)$$

because of (11.47). The LP solver has computed $\bar{c}, y', y''$ and z (which equals 0 or 1) that satisfy (11.45), (11.46), and

$$\sum(\bar{c}_e u_e : \bar{c}_e > 0) + \sum(\bar{c}_e \ell_e : \bar{c}_e < 0) < y'^T b' + y''^T b'' + z(t_{\max} + 1 - \varepsilon)$$

in floating point arithmetic and within small zero tolerances. To make sure that the system (11.45), (11.46), (11.47) really is solvable, we use fixed point arithmetic, where each number is represented by 32 bits (including the sign bit) before the decimal point and 32 bits after the decimal point. Addition and subtraction of any two such numbers as well as multiplication of any such number by an integer can be carried out without rounding errors; however, any of these operations may result in overflow. We start out with the best fixed point approximations to the floating point y' and y'' returned by the LP solver (and with each negative component of y'' replaced by zero); then we compute $\bar{c}$ from (11.45) and finally we verify (11.47). If (11.47) fails or an overflow occurs, then we may branch out on any edge e such that $\ell_e = 0$ and $u_e = 1$.

11.9 PHASE 3: FROM $\mathcal{T}'$ TO $\mathcal{T}$

PHASE 2 has produced an inequality $a^T \bar{x} \leq b$ that induces a facet of the convex hull of $\mathcal{T}'$ and is violated by $\bar{x}^*$; the left-hand side coefficients of this inequality are nonnegative integers a_e such that

$$a_{0w} = 0 \text{ for all } w \text{ in } \overline{V}^{\text{int}}.$$

In PHASE 3, we find a hypergraph inequality

$$\sum(\lambda_Q x(Q, \overline{V} - Q) : Q \in \overline{\mathcal{H}}) \geq \beta \tag{11.48}$$

satisfied by all $\bar{x}$ in $\mathcal{T}$ and such that, for some integers $\pi_w (w \in \overline{V}^{\text{int}})$, the left-hand side of (11.48) is identically equal to

$$\sum(\pi_w \sum(\bar{x}_e : w \in e) : w \in \overline{V}^{\text{int}}) - 2a^T \bar{x} \tag{11.49}$$

and the right-hand side of (11.48) equals

$$\sum(2\pi_w : w \in \overline{V}^{\text{int}}) - 2b.$$

(This relationship between (11.48) and $a^T \bar{x} \leq b$ implies that (11.48) induces a facet of the convex hull of $\mathcal{T}'$ and is violated by $\bar{x}^*$.) Algorithm 11.13 accomplishes this objective with a negligible amount of computations.

Algorithm 11.13 PHASE 3.

---

construct a hypergraph $\overline{\mathcal{H}}$ on $\overline{V}^{\text{int}}$ and positive integers $\lambda_Q(Q \in \overline{\mathcal{H}})$
such that the linear form
$$\sum \lambda_Q(\sum(\overline{x}_e : e \subseteq Q) : Q \in \overline{\mathcal{H}})$$
is identically equal to $a^T \overline{x}$;
return the inequality
$$\sum(\lambda_Q x(Q, \overline{V} - Q) : Q \in \overline{\mathcal{H}}) \geq \sum(2\lambda_Q|Q| : Q \in \overline{\mathcal{H}}) - 2b;$$

11.9.1 Correctness of Algorithm 11.13

EXISTENCE OF THE DESIRED INTEGERS $\pi_w(w \in \overline{V}^{\text{int}})$

For every subset Q of $\overline{V}$, the linear form $x(Q, \overline{V} - Q)$ is identically equal to

$$\sum \left(\sum(\overline{x}_e : w \in e) : w \in Q \right) - 2 \sum(\overline{x}_e : e \subseteq Q).$$

It follows that the left-hand side of (11.48) is identically equal to

$$\sum \left(\lambda_Q \sum \left(\sum(\overline{x}_e : w \in e) : w \in Q \right) : Q \in \overline{\mathcal{H}} \right) - 2a^T \overline{x},$$

which equals (11.49) with

$$\pi_w = \sum(\lambda_Q : w \in Q, Q \in \overline{\mathcal{H}}).$$

With this definition of π_w, we have

$$\sum(\pi_w : w \in \overline{V}^{\text{int}}) = \sum(\lambda_Q|Q| : Q \in \overline{\mathcal{H}}).$$

THE OUTPUT INEQUALITY OF ALGORITHM 11.13 IS TRIANGULAR

Following Naddef and Rinaldi [421], let us say that an inequality $\alpha^T \overline{x} \geq \beta$ is
triangular if its left-hand side coefficients α_e satisfy

$$\alpha_{uv} \leq \alpha_{uw} + \alpha_{wv} \quad \text{for all choices of distinct } u, v, w.$$

Every inequality $x(Q, \overline{V}-Q) \geq 0$ is triangular, and so every hypergraph inequality
(11.48) with all coefficients λ_Q positive is triangular.

THE OUTPUT INEQUALITY OF ALGORITHM 11.13 IS SATISFIED BY ALL $\overline{x}$ IN
$\mathcal{T}$.

More generally, there is the following fact.

THEOREM 11.3 *If a triangular inequality is satisfied by all points of $\mathcal{T}'$, then it is
satisfied by all points of $\mathcal{T}$.*

Proof. Consider an arbitrary triangular inequality $\alpha^T \overline{x} \geq \beta$ satisfied by all $\overline{x}$ in $\mathcal{T}'$.
Given any $\overline{x}$ in $\mathcal{T} - \mathcal{T}'$, we will find an $\overline{x}'$ in $\mathcal{T}$ such that $\alpha^T \overline{x}' \leq \alpha^T \overline{x}$ and, with e
standing for the all-ones vector, $e^T \overline{x}' < e^T \overline{x}$. Then it will follow by induction on
$e^T \overline{x}$ that $\alpha^T \overline{x} \geq \beta$ is satisfied by all $\overline{x}$ in $\mathcal{T}$.

To find $\overline{x}'$, we use arguments introduced by Naddef and Rinaldi [421]. To begin, observe that elements of $\mathcal{T}$ can be interpreted as incidence vectors of closed walks through $\overline{V}$, meaning cyclic sequences of elements of $\overline{V}$ that include each element of $\overline{V}$ at least once. Every closed walk, $w_0 w_1 \ldots w_0$, that corresponds to $\overline{x}$ passes through some element w of $\overline{V}^{\text{int}}$ at least twice: $w = w_i$ and $w = w_j$ with $j \neq i$. Removing w_i from the cyclic sequence $w_0 w_1 \ldots w_0$ (which amounts to removing the two edges $w_{i-1} w_i$, $w_i w_{i+1}$ and, in case $w_{i-1} \neq w_{i+1}$, inserting the single edge $w_{i-1} w_{i+1}$) yields a closed walk, whose incidence vector $\overline{x}'$ satisfies $e^T \overline{x}' < e^T \overline{x}$; since $\alpha^T \overline{x} \geq \beta$ is triangular, we have $\alpha^T \overline{x}' \leq \alpha^T \overline{x}$. (To see that $\alpha^T \overline{x}' \leq \alpha^T \overline{x}$ holds even when $w_{i-1} = w_{i+1}$, note that $\alpha_e \geq 0$ for all e: we have $\alpha_{vw} \geq \alpha_{vz} - \alpha_{zw}$ and $\alpha_{vw} \geq \alpha_{zw} - \alpha_{vz}$ whenever v, w, z are distinct points of $\overline{V}$.) □

11.9.2 How we construct the $\overline{\mathcal{H}}$ and the $\lambda_Q (Q \in \overline{\mathcal{H}})$

A straightforward way of implementing the instruction

> construct a hypergraph $\overline{\mathcal{H}}$ on $\overline{V}^{\text{int}}$ and positive integers $\lambda_Q (Q \in \overline{\mathcal{H}})$
> such that the linear form
> $$\sum \lambda_Q (\sum (\overline{x}_e : e \subseteq Q) : Q \in \overline{\mathcal{H}})$$
> is identically equal to $a^T \overline{x}$;

in Algorithm 11.13 is to let $\overline{\mathcal{H}}$ consist of all two-point subsets $\{u, v\}$ such that $a_{uv} > 0$ and to set $\lambda_{\{u,v\}} = a_{uv}$ for all $\{u, v\}$ in $\overline{\mathcal{H}}$. Instead, our code uses Algorithm 11.14.

Algorithm 11.14 How we choose $\overline{\mathcal{H}}$ and $\lambda_Q (Q \in \overline{\mathcal{H}})$ in Algorithm 11.13.

$\overline{\mathcal{H}} = \emptyset$;
while $a \neq 0$
do   choose a maximal (with respect to set inclusion)
            subset Q of $\overline{V}^{\text{int}}$ such that $\min\{a_e : e \subseteq Q\} > 0$;
        $\overline{\mathcal{H}} = \overline{\mathcal{H}} \cup \{Q\}$;
        $\lambda_Q = \min\{a_e : e \subseteq Q\}$;
        for all a_e such that $e \subseteq Q$ **do** $a_e = a_e - \lambda_Q$ **end**
end

∨ ∨ ∨

In EXAMPLE 1, the inequality

$$\overline{x}_{12} + \overline{x}_{15} + 2\overline{x}_{23} + 2\overline{x}_{25} + 3\overline{x}_{35} + 4\overline{x}_{36} + \overline{x}_{47} + 3\overline{x}_{56} \leq 7,$$

constructed in Section 11.6, induces a facet of the convex hull of $\mathcal{T}''$ and is violated by $\overline{x}^*$. Algorithm 11.11 of Section 11.7 converts this inequality into the inequality

$$\overline{x}_{12} + \overline{x}_{13} + \overline{x}_{15} + \overline{x}_{16} + 2\overline{x}_{23} + 2\overline{x}_{25} + 2\overline{x}_{26}$$
$$+4\overline{x}_{34} + 3\overline{x}_{35} + 4\overline{x}_{36} + \overline{x}_{37} + \overline{x}_{46} + \overline{x}_{47} + 3\overline{x}_{56} + 4\overline{x}_{67} \leq 15,$$

which induces a facet of the convex hull of $\mathcal{T}'$ and is violated by $\bar{x}^*$. With $a^T x$ standing for the left-hand side of the latter inequality. we may construct the $\overline{\mathcal{H}}$ and the $\lambda_Q (Q \in \overline{\mathcal{H}})$ in Algorithm 11.13 as follows:

1. We choose $Q = \{1, 2, 3, 5, 6\}$, which yields $\lambda_Q = 1$ and leaves us with

$$a^T x = \bar{x}_{23} + \bar{x}_{25} + \bar{x}_{26} + 4\bar{x}_{34} + 2\bar{x}_{35} + 3\bar{x}_{36} + \bar{x}_{37} + \bar{x}_{46} + \bar{x}_{47} + 2\bar{x}_{56} + 4\bar{x}_{67}.$$

2. We choose $Q = \{2, 3, 5, 6\}$, which yields $\lambda_Q = 1$ and leaves us with

$$a^T x = 4\bar{x}_{34} + \bar{x}_{35} + 2\bar{x}_{36} + \bar{x}_{37} + \bar{x}_{46} + \bar{x}_{47} + \bar{x}_{56} + 4\bar{x}_{67}.$$

3. We choose $Q = \{3, 4, 6, 7\}$, which yields $\lambda_Q = 1$ and leaves us with

$$a^T x = 3\bar{x}_{34} + \bar{x}_{35} + \bar{x}_{36} + \bar{x}_{56} + 3\bar{x}_{67}.$$

4. We choose $Q = \{3, 4\}$, which yields $\lambda_Q = 3$ and leaves us with

$$a^T x = \bar{x}_{35} + \bar{x}_{36} + \bar{x}_{56} + 3\bar{x}_{67}.$$

5. We choose $Q = \{3, 5, 6\}$, which yields $\lambda_Q = 1$ and leaves us with

$$a^T x = 3\bar{x}_{67}.$$

6. We choose $Q = \{6, 7\}$, which yields $\lambda_Q = 3$ and leaves us with

$$a^T x = 0.$$

The resulting hypergraph inequality,

$$\bar{x}(\{1, 2, 3, 5, 6\}, \{0, 4, 7\})$$
$$+ \bar{x}(\{2, 3, 5, 6\}, \{0, 1, 4, 7\})$$
$$+ \bar{x}(\{3, 4, 6, 7\}, \{0, 1, 2, 5\})$$
$$+ 3\bar{x}(\{3, 4\}, \{0, 1, 2, 5, 6, 7\})$$
$$+ \bar{x}(\{3, 5, 6\}, \{0, 1, 2, 4, 7\})$$
$$+ 3\bar{x}(\{6, 7\}, \{0, 1, 2, 3, 4, 5\}) \geq 26,$$

is satisfied by all points of $\mathcal{T}$ and it is violated by $\bar{x}^*$. (By the way, this inequality belongs to the class of *path inequalities* of Cornuéjols et al. [140].)

In EXAMPLE 2, the inequality

$$\bar{x}_{35} + \bar{x}_{45} + \bar{x}_{46} + 2\bar{x}_{56} + \bar{x}_{68} + 2\bar{x}_{78} + \bar{x}_{79} + \bar{x}_{89} \leq 3,$$

constructed in Section 11.6, induces a facet of the convex hull of $\mathcal{T}''$ and is violated by $\bar{x}^*$. Algorithm 11.11 of Section 11.7 converts this inequality into the inequality

$$\bar{x}_{34} + \bar{x}_{35} + \bar{x}_{36} + \bar{x}_{45} + \bar{x}_{46} + 2\bar{x}_{56}$$
$$+ \bar{x}_{57} + 3\bar{x}_{58} + 3\bar{x}_{67} + \bar{x}_{68} + 2\bar{x}_{78} + \bar{x}_{79} + \bar{x}_{89} \leq 10,$$

which induces a facet of the convex hull of $\mathcal{T}'$ and is violated by $\bar{x}^*$. In Phase 3, we convert the latter inequality into the hypergraph inequality

$$\bar{x}(\{3,4,5,6\},\{0,1,2,7,8,9\})$$
$$+\bar{x}(\{5,6,7,8\},\{0,1,2,3,4,9\})$$
$$+2\bar{x}(\{5,8\},\{0,1,2,3,4,6,7,9\})$$
$$+2\bar{x}(\{6,7\},\{0,1,2,3,4,5,8,9\})$$
$$+\bar{x}(\{7,8,9\},\{0,1,2,3,4,5,6\}) \geq 18\,,$$

which is satisfied by all points of $\mathcal{T}$ and violated by $\bar{x}^*$. (Again, this inequality belongs to the class of *path inequalities* of Cornuéjols et al. [140].)

In EXAMPLE 3, the inequality

$$\bar{x}_{13} + \bar{x}_{15} + \bar{x}_{16} + \bar{x}_{23} + \bar{x}_{24} + \bar{x}_{28} + \bar{x}_{45} \leq 3 \qquad (11.50)$$

induces a facet of the convex hull of $\mathcal{T}''$ and is violated by $\bar{x}^*$. In Section 11.7, we pointed out that the output of Algorithm 11.11 depends on the order in which the restrictions $x_f = 1$ ($f \in \bar{E}_1$) and $x_f = 0$ ($f \in \bar{E}_0$) imposed on elements of $\mathcal{T}''$ are lifted: we converted (11.50) first into the inequality

$$\bar{x}_{12} + \bar{x}_{13} + 2\bar{x}_{14} + \bar{x}_{15} + \bar{x}_{16} + \bar{x}_{23} + \bar{x}_{24}$$
$$+\bar{x}_{25} + \bar{x}_{28} + \bar{x}_{37} + \bar{x}_{45} + \bar{x}_{56} \leq 7, \qquad (11.51)$$

which induces a facet of the convex hull of $\mathcal{T}'$ and is violated by $\bar{x}^*$, and then into the inequality

$$\bar{x}_{12} + \bar{x}_{13} + 2\bar{x}_{14} + \bar{x}_{15} + \bar{x}_{16} + \bar{x}_{23} + \bar{x}_{24}$$
$$+\bar{x}_{28} + \bar{x}_{34} + \bar{x}_{37} + \bar{x}_{45} + \bar{x}_{46} + \bar{x}_{56} \leq 7, \qquad (11.52)$$

which also induces a facet of the convex hull of $\mathcal{T}'$ and is violated by $\bar{x}^*$. In Phase 3, we convert (11.51) into the hypergraph inequality

$$\bar{x}(\{1,2,3\},\{0,4,5,6,7,8\})$$
$$+\bar{x}(\{1,4,5\},\{0,2,3,6,7,8\})$$
$$+\bar{x}(\{1,4\},\{0,2,3,5,6,7,8\})$$
$$+\bar{x}(\{1,6\},\{0,2,3,4,5,7,8\})$$
$$+\bar{x}(\{2,4\},\{0,1,3,5,6,7,8\})$$
$$+\bar{x}(\{2,5\},\{0,1,3,4,6,7,8\})$$
$$+\bar{x}(\{2,8\},\{0,1,3,4,5,6,7\})$$
$$+\bar{x}(\{3,7\},\{0,1,2,4,5,6,8\})$$
$$+\bar{x}(\{5,6\},\{0,1,2,3,4,7,8\}) \geq 26\,,$$

which is satisfied by all points of $\mathcal{T}$ and violated by $\bar{x}^*$; we convert (11.52) into the comb inequality

$$\bar{x}(\{1,2,3,4\},\{0,5,6,7,8\}) + \bar{x}(\{1,4,5,6\},\{0,2,3,7,8\})$$
$$+\bar{x}(\{2,8\},\{0,1,3,4,5,6,7\}) + \bar{x}(\{3,7\},\{0,1,2,4,5,6,8\}) \geq 10\,,$$

which is also satisfied by all points of $\mathcal{T}$ and violated by $\bar{x}^*$.

∧ ∧ ∧

11.9.3 Cuts that induce facets of the convex hull of $\mathcal{T}$

The cut produced by Algorithm 11.13 is easily transformed into a cut that induces a facet of the convex hull of $\mathcal{T}$. We will prove at the end of this section that, in case

$$\overline{V}^{\text{int}} \text{ includes two distinct points such that} \tag{11.53}$$
$$\text{no member of } \overline{\mathcal{H}} \text{ contains both of them,}$$

Algorithm 11.15 transforms the cut

$$\sum (\lambda_Q x(Q, \overline{V} - Q) : Q \in \overline{\mathcal{H}}) \geq \beta$$

produced by Algorithm 11.13 into a cut that induces a facet of the convex hull of $\mathcal{T}$. We will also prove at the end of this section that, in case (11.53) fails, the input inequality $a^T \overline{x} \leq b$ of Algorithm 11.13 reads

$$\lambda \sum (\overline{x}_e : e \in \overline{E}^{\text{int}}) \leq \lambda(k - 1)$$

for some positive λ, and so Algorithm 11.13 returns a positive multiple of the subtour inequality

$$\overline{x}(\{0\}, \overline{V}^{\text{int}}) \geq 2;$$

as noted in Section 11.6, Cornuéjols et al. [140] pointed out that subtour inequalities induce facets of the convex hull of $\mathcal{T}$.

Algorithm 11.15 From a hypergraph cut to a facet-inducing hypergraph cut.

---

1. For all choices of distinct points u, v, w of $\overline{V}$, define

$$\tau(u, v, w) = \sum (\lambda_Q : Q \in \overline{\mathcal{H}}, u \in Q, v \in Q, w \notin Q) +$$
$$\sum (\lambda_Q : Q \in \overline{\mathcal{H}}, u \notin Q, v \notin Q, w \in Q).$$

2. For all points w of $\overline{V}^{\text{int}}$, evaluate

$$\Delta_w = \min\{\tau(u, v, w) : uv \in \overline{E}, u \neq w, v \neq w\}.$$

3. Return the inequality

$$\sum (\lambda_Q x(Q, \overline{V} - Q) : Q \in \overline{\mathcal{H}})$$
$$- \sum (\Delta_w \overline{x}(\{w\}, \overline{V} - \{w\}) : w \in \overline{V}^{\text{int}}) \geq$$
$$\beta - 2 \sum (\Delta_w : w \in \overline{V}^{\text{int}}).$$

∨ ∨ ∨

In EXAMPLE 1, $\overline{\mathcal{H}}$ consists of

$$\{1, 2, 3, 5, 6\}, \ \{2, 3, 5, 6\}, \ \{3, 4, 6, 7\}, \ \{3, 4\}, \ \{3, 5, 6\}, \ \{6, 7\};$$

since no member of $\overline{\mathcal{H}}$ contains both 1 and 4, assumption (11.53) is satisfied; since

$$\Delta_1 = \tau(0,3,1) = 0,$$
$$\Delta_2 = \tau(0,3,2) = 0,$$
$$\Delta_3 = \tau(4,5,3) = 0,$$
$$\Delta_4 = \tau(0,3,4) = 0,$$
$$\Delta_5 = \tau(0,3,5) = 0,$$
$$\Delta_6 = \tau(3,7,6) = 0,$$
$$\Delta_7 = \tau(0,6,7) = 0,$$

the output inequality of Algorithm 11.13 induces a facet of the convex hull of $\mathcal{T}$.

In EXAMPLE 2, $\overline{\mathcal{H}}$ consists of

$$\overline{x}(\{3,4,5,6\}, \; \overline{x}(\{5,6,7,8\}, \; \overline{x}(\{5,8\}, \; \overline{x}(\{6,7\}, \; \overline{x}(\{7,8,9\};$$

since no member of $\overline{\mathcal{H}}$ contains both 1 and 2, assumption (11.53) is satisfied; since

$$\Delta_1 = \tau(0,2,1) = 0,$$
$$\Delta_2 = \tau(0,1,2) = 0,$$
$$\Delta_3 = \tau(0,4,3) = 0,$$
$$\Delta_4 = \tau(0,3,4) = 0,$$
$$\Delta_5 = \tau(3,8,5) = 0,$$
$$\Delta_6 = \tau(3,7,6) = 0,$$
$$\Delta_7 = \tau(6,8,7) = 0,$$
$$\Delta_8 = \tau(5,7,8) = 0,$$
$$\Delta_9 = \tau(0,7,9) = 0,$$

the output inequality of Algorithm 11.13 induces a facet of the convex hull of $\mathcal{T}$.

In the first cut in EXAMPLE 3, $\overline{\mathcal{H}}$ consists of

$$\overline{x}(\{1,2,3\}, \; \overline{x}(\{1,4,5\}, \; \overline{x}(\{1,4\}, \; \overline{x}(\{1,6\},$$
$$\overline{x}(\{2,4\}, \; \overline{x}(\{2,5\}, \; \overline{x}(\{2,8\}, \; \overline{x}(\{3,7\}, \; \overline{x}(\{5,6\},$$

and we have $\lambda_Q = 1$ for all Q in $\overline{\mathcal{H}}$; since no member of $\overline{\mathcal{H}}$ contains both 1 and 7, assumption (11.53) is satisfied; we have

$$\Delta_1 = \tau(4,6,1) = 1,$$
$$\Delta_2 = \tau(1,8,2) = 2,$$
$$\Delta_3 = \tau(1,7,3) = 0,$$
$$\Delta_4 = \tau(1,2,4) = 1,$$
$$\Delta_5 = \tau(2,6,5) = 1,$$
$$\Delta_6 = \tau(1,5,6) = 1,$$
$$\Delta_7 = \tau(0,3,7) = 0,$$
$$\Delta_8 = \tau(0,2,8) = 0,$$

and so the output inequality of Algorithm 11.13 gets transformed into the hypergraph inequality

$$\overline{x}(\{1,2,3\}, \{0,4,5,6,7,8\})$$
$$+ \overline{x}(\{1,4,5\}, \{0,2,3,6,7,8\})$$

$$+\bar{x}(\{1,4\},\{0,2,3,5,6,7,8\})$$
$$+\bar{x}(\{1,6\},\{0,2,3,4,5,7,8\})$$
$$+\bar{x}(\{2,4\},\{0,1,3,5,6,7,8\})$$
$$+\bar{x}(\{2,5\},\{0,1,3,4,6,7,8\})$$
$$+\bar{x}(\{2,8\},\{0,1,3,4,5,6,7\})$$
$$+\bar{x}(\{3,7\},\{0,1,2,4,5,6,8\})$$
$$+\bar{x}(\{5,6\},\{0,1,2,3,4,7,8\})$$
$$-\bar{x}(\{1\},\{0,2,3,4,5,6,7,8\})$$
$$-2\bar{x}(\{2\},\{0,1,3,4,5,6,7,8\})$$
$$-\bar{x}(\{4\},\{0,1,2,3,5,6,7,8\})$$
$$-\bar{x}(\{5\},\{0,1,2,3,4,6,7,8\})$$
$$-\bar{x}(\{6\},\{0,1,2,3,4,5,7,8\})\geq 14\,,$$

which induces a facet of the convex hull of $\mathcal{T}$ and is violated by $\bar{x}^*$.

In the second cut in EXAMPLE 3, $\overline{\mathcal{H}}$ consists of

$$\{1,2,3,4\},\ \{1,4,5,6\},\ \{2,8\},\ \{3,7\};$$

since no member of $\overline{\mathcal{H}}$ contains both 1 and 7, assumption (11.53) is satisfied; since

$$\Delta_1 = \tau(2,5,1) = 0,$$
$$\Delta_2 = \tau(1,8,2) = 0,$$
$$\Delta_3 = \tau(1,7,3) = 0,$$
$$\Delta_4 = \tau(2,5,4) = 0,$$
$$\Delta_5 = \tau(0,1,5) = 0,$$
$$\Delta_6 = \tau(0,1,6) = 0,$$
$$\Delta_7 = \tau(0,3,7) = 0,$$
$$\Delta_8 = \tau(0,2,8) = 0,$$

the output inequality of Algorithm 11.13 induces a facet of the convex hull of $\mathcal{T}$.

∧ ∧ ∧

The current version of our code can handle only hypergraph constraints with nonnegative coefficients; it always settles for the cut produced by Algorithm 11.13, even though (like the first cut in EXAMPLE 3) this cut may not induce a facet of the convex hull of $\mathcal{T}$. Algorithm 11.14 aims to mitigate the effects of this carelessness by reducing the number of points w such that $\Delta_w > 0$.

11.9.4 Postponed proofs

WHEN ASSUMPTION (11.53) HOLDS

Naddef and Rinaldi [421] say that a triangular inequality $\alpha^T x \geq \beta$ is *tight triangular* if

for each w in $\overline{V}$ there are u and v such that $\alpha_{uv} = \alpha_{uw} + \alpha_{wv}$.

Consider the relationship between an arbitrary inequality $\alpha^T \overline{x} \geq \beta$ and corresponding inequalities

$$\alpha^T \overline{x} - \sum (d_w \overline{x}(\{w\}, \overline{V} - \{w\}) : w \in \overline{V}) \geq$$
$$\beta - 2 \sum (d_w : w \in \overline{V}) \qquad (11.54)$$

with arbitrary constants $d_w : (w \in \overline{V})$: two pertinent observations are

(i) inequality (11.54) is tight triangular if and only if

$$2d_w = \min\{\alpha_{uw} + \alpha_{wv} - \alpha_{uv} : uv \in \overline{E}, u \neq w, v \neq w\}$$

for all w in $\overline{V}$,

(ii) if $\alpha^T \overline{x} \geq \beta$ induces a facet of the convex hull of T' and if $d_0 = 0$, then inequality (11.54) induces a facet of the convex hull of T'.

In the special case where the linear form $\alpha^T \overline{x}$ is identically equal to

$$\sum (\lambda_Q x(Q, \overline{V} - Q) : Q \in \overline{\mathcal{H}}),$$

we have

$$\alpha_{uw} + \alpha_{wv} - \alpha_{uv} = 2 \sum (\lambda_Q : Q \in \overline{\mathcal{H}}, u \in Q, v \in Q, w \notin Q) +$$
$$2 \sum (\lambda_Q : Q \in \overline{\mathcal{H}}, u \notin Q, v \notin Q, w \in Q).$$

In particular, if $\overline{\mathcal{H}}$ is a hypergraph on $\overline{V}^{\text{int}}$, then

$$\alpha_{u0} + \alpha_{0v} - \alpha_{uv} = 2 \sum (\lambda_Q : Q \in \overline{\mathcal{H}}, u \in Q, v \in Q),$$

and so assumption (11.53) guarantees the existence of distinct points u, v in $\overline{V}^{\text{int}}$ such that $\alpha_{u0} + \alpha_{0v} - \alpha_{uv} = 0$. We conclude that, as long as (11.53) holds, the cut produced by Algorithm 11.13 is tight triangular and induces a facet of the convex hull of T'.

THEOREM 11.4 *If a tight triangular inequality induces a facet of the convex hull of T', then it induces a facet of the convex hull of T.*

Proof. Let $\alpha^T \overline{x} \geq \beta$ be a tight triangular inequality inducing a facet of the convex hull of T'. Theorem 11.3 guarantees that

$$T \subset \{\overline{x} : \alpha^T \overline{x} \geq \beta\};$$

since $\alpha^T \overline{x} \geq \beta$ induces a facet of the convex hull of T', there is a set T_0' of dim T' affinely independent vectors in T' such that

$$T_0' \subset \{\overline{x} : \alpha^T \overline{x} = \beta\}.$$

We propose to extend T_0' into a set T_0 of dim T affinely independent vectors in T such that

$$T_0 \subset \{\overline{x} : \alpha^T \overline{x} = \beta\} :$$

for each w in $\overline{V}^{\text{int}}$, we will find an $\overline{x}$ in $\mathcal{T}$ such that

$$\overline{x}(\{w\}, \overline{V} - \{w\}) = 4,$$

$$\overline{x}(\{z\}, \overline{V} - \{z\}) = 2 \quad \text{whenever } z \in \overline{V}^{\text{int}} \text{ and } z \neq w,$$

$$\alpha^T \overline{x} = \beta.$$

Since $\alpha^T \overline{x} \geq \beta$ is tight triangular, there are u and v such that $\alpha_{uv} = \alpha_{uw} + \alpha_{wv}$; since $\alpha^T \overline{x} \geq \beta$ induces a facet of the convex hull of $\mathcal{T}'$, some $\overline{x}'$ in $\mathcal{T}'$ has $\alpha^T \overline{x}' = \beta$ and $\overline{x}'_{uv} > 0$. In the corresponding closed walk, $w_0 w_1 \ldots w_0$, some edge $w_i w_{i+1}$ equals uv; inserting w between w_i and w_{i+1} in this cyclic sequence (which amounts to replacing the single edge $w_i w_{i+1}$ by the two edges $w_i w$ and $w w_{i+1}$) yields a closed walk, whose incidence vector $\overline{x}$ has the desired properties. $\square$

It is important to note, however, that Oswald et al. [441] have discovered families of facet-inducing inequalities for the graphical TSP in tight-triangular form that are not facet-inducing for the TSP. Nonetheless, Oswald et al. [442] show that under mild conditions, the local-cuts prodedure will produce inequalities that are facet-inducing for the TSP.

WHEN ASSUMPTION (11.53) FAILS

Every input inequality $a^T \overline{x} \leq b$ of Algorithm 11.13 satisfies

$$a_e = 0 \quad \text{whenever } e \notin \overline{E}^{\text{int}};$$

if (11.53) fails, then

$$a_e > 0 \quad \text{whenever } e \in \overline{E}^{\text{int}}.$$

We claim that these assumptions imply

$$\{\overline{x} \in \mathcal{T}' : a^T \overline{x} = b\} \subset \{\overline{x} : \sum(\overline{x}_e : e \in \overline{E}^{\text{int}}) = k - 1\}. \tag{11.55}$$

To justify this claim, consider an arbitrary $\overline{x}$ in $\mathcal{T}'$ such that

$$\sum(\overline{x}_e : e \in \overline{E}^{\text{int}}) < k - 1,$$

and so the path system $\chi(\overline{x})$—as defined in Section 11.7—consists of at least two paths. With u and v standing for two elements of $\overline{V}^{\text{int}}$ that are endpoints of two distinct paths in the path system, set

$$\overline{x}'_{uv} = 1, \ \overline{x}'_{0u} = \overline{x}_{0u} - 1, \ \overline{x}'_{0v} = \overline{x}_{0v} - 1$$

and $\overline{x}'_e = \overline{x}_e$ for all other choices of e. The resulting $\overline{x}'$ belongs to $\mathcal{T}'$, and so

$$a^T \overline{x} = a^T \overline{x}' - a_{uv} \leq b - a_{uv} < b,$$

which completes the argument.

Every input inequality $a^T \overline{x} \leq b$ of Algorithm 11.13 induces a facet of the convex hull of $\mathcal{T}'$, and so it induces a facet of the convex hull of $\chi(\mathcal{T}')$; since $\chi(\mathcal{T}')$ is full-dimensional, relation (11.55) implies the existence of a multiplier λ such that the linear form $a^T \overline{x}$ is identically equal to $\lambda \sum(\overline{x}_e : e \in \overline{E}^{\text{int}})$. Since $a^T \overline{x} \leq b$ induces a facet of the convex hull of $\mathcal{T}'$, we must have $\lambda > 0$ and $b = \lambda(k - 1)$.

11.10 GENERALIZATIONS

The cutting-plane method and its descendants are applicable to any problem

$$\text{minimize } c^T x \text{ subject to } x \in \mathcal{S}$$

such that $\mathcal{S}$ is a finite subset of some $\mathbf{R}^m$ and such that an efficient algorithm to recognize points of $\mathcal{S}$ is available. The corresponding general problem of finding cuts is this:

> given (by means of an efficient membership-testing oracle)
> a finite subset $\mathcal{S}$ of some $\mathbf{R}^m$ and
> given a point x^* in $\mathbf{R}^m$ that lies outside the convex hull of $\mathcal{S}$,
> find a vector a and a scalar b such that
> $\mathcal{S} \subset \{x : a^T x \le b\}$ and $a^T x^* > b$.

Algorithm 11.1, dealing with the special case case where $m = n(n-1)/2$ and $\mathcal{S}$ is the set of the incidence vectors of all the tours through a set of n cities, generalizes to Algorithm 11.16.

Algorithm 11.16 A very general scheme for collecting cuts.

---

initialize an empty list $\mathcal{L}$ of cuts;
for selected small integers d, linear mappings $\phi : \mathbf{R}^m \to \mathbf{R}^d$, and
    finite subsets $\mathcal{T}$ of $\mathbf{R}^d$ such that $\phi(\mathcal{S}) \subseteq \mathcal{T}$
do if $\phi(x^*)$ lies outside the convex hull of $\mathcal{T}$
    then find a vector a and a scalar b such that
            $\mathcal{T} \subset \{\overline{x} : a^T \overline{x} \le b\}$ and $a^T \phi(x^*) > b$;
           add the cut $a^T \phi(x) \le b$ to $\mathcal{L}$;
    end
end
return $\mathcal{L}$;

The trick of trying to separate x^* from $\mathcal{S}$ by separating $\phi(x^*)$ from $\mathcal{T}$ was used previously by Crowder et al. [146] in the context of integer linear programming, where $\mathcal{S}$ consists of all integer solutions of some explicitly recorded system

$$Ax = b, \ \ell \le x \le u \tag{11.56}$$

and x^* satisfies (11.56) in place of x. Crowder et al. consider systems (11.56) such that A is sparse and $\ell = 0, u = e$; for each equation $\alpha^T x = \beta$ in the system $Ax = b$, they consider the set $\mathcal{T}$ of all integer solutions of

$$\alpha^T x = \beta, \ \ 0 \le x \le e$$

restricted on components x_j such that $\alpha_j \ne 0$; with $\phi(x)$ standing for the restriction of x on these components, they try to separate x^* from $\mathcal{S}$ by separating $\phi(x^*)$ from $\mathcal{T}$. In the attempt to separate $\phi(x^*)$ from $\mathcal{T}$, they use exclusively inequalities that match certain prescribed templates and they use special-purpose separation algorithms to find such cuts.

Boyd [83] starts out with the choices of ϕ and $\mathcal{T}$ made by Crowder et al., but then he separates $\phi(x^*)$ from $\mathcal{T}$ by a general-purpose procedure. He solves the problem

$$\begin{aligned}
\text{maximize} \quad & z \\
\text{subject to} \quad & z - a^T\phi(x^*) + a^T\phi(x) \le 0 \text{ for all } x \text{ in } \mathcal{T}, \\
& ||a||_1 \le \gamma, \ ||a||_\infty \le 1
\end{aligned}$$

with a prescribed constant γ by a method that is essentially the simplex method (in particular, the value of z increases with each nondegenerate iteration); to access $\mathcal{T}$, he uses an oracle implemented by a dynamic programming algorithm. If the optimal value z^* turns out to be positive, then he returns the cut

$$a^T\phi(x) \le a^T\phi(x^*) - z^*,$$

which he calls a *Fenchel cutting plane*.

The technique developed in Section 11.4 provides another implementation of the body of the **for** loop in Algorithm 11.16 for a fairly general class of sets $\mathcal{T}$: it requires only an efficient oracle that, given any vector a in $\mathbf{R}^d$, returns either an $\overline{x}$ that maximizes $a^T\overline{x}$ subject to $\overline{x} \in \mathcal{T}$ or the message "infeasible" indicating that $\mathcal{T}$ is empty. This technique is reviewed in Algorithm 11.17: function SEPARATE, given (by means of an efficient maximization oracle) a finite subset $\mathcal{T}$ of some $\mathbf{R}^d$ and given a point $\overline{x}^*$ in $\mathbf{R}^d$, returns either a failure message indicating that $\overline{x}^*$ lies inside the convex hull of $\mathcal{T}$ or else a vector a and a scalar b such that

$$\mathcal{T} \subset \{\overline{x} : a^T\overline{x} \le b\} \text{ and } a^T\overline{x}^* > b.$$

One way of solving the linear programming problem in each iteration of the **repeat** loop in Algorithm 11.17 is to apply the simplex method to its dual with relatively few rows,

$$\begin{aligned}
\text{maximize} \quad & s \\
\text{subject to} \quad & s\overline{x}^* - A\lambda + w & = 0, \\
& -s + e^T\lambda & = 0, \\
& \lambda \ge 0, \quad -e \le w \le e,
\end{aligned} \tag{11.57}$$

just as our code does (with $\mathcal{T}^*$ in place of $\mathcal{T}$) in the special case of the TSP. Some of the modifications of SEPARATE($\phi(x^*), \mathcal{T}$) used by our code in the special case of the TSP may also carry over to other problems. In particular, if we have at our disposal a nonempty family $\mathcal{F}$ of row vectors $[v^T, w]$ such that

$$\mathcal{T} \subset \{\overline{x} : v^T\overline{x} \ge w\} \text{ and } v^T\phi(x^*) = w,$$

then we may test the condition

$\phi(x^*)$ lies outside the convex hull of $\mathcal{T}$

in Algorithm 11.16 by calling SEPARATE($\phi(x^*), \mathcal{T}^*$) with

$$\mathcal{T}^* = \{\overline{x} \in \mathcal{T} : v^T\overline{x} = w \text{ for all } [v^T, w] \text{ in } \mathcal{F}\}.$$

The point of this substitution is that SEPARATE($\phi(x^*), \mathcal{T}^*$) tends to run faster as the dimension of $\{\overline{x} \in \mathbf{R}^d : v^T\overline{x} = w \text{ for all } [v^T, w] \text{ in } \mathcal{F}\}$ decreases; if (as in

Algorithm 11.17 SEPARATE($\overline{x}^*, \mathcal{T}$).

if $\mathcal{T} \neq \emptyset$
then $A =$ any $d \times 1$ matrix whose single column is an element of $\mathcal{T}$;
     repeat if   the linear programming problem
$$\text{minimize } u^T e + v^T e$$
$$\begin{aligned} \text{subject to } \quad -a^T A + b e^T &\geq 0, \\ a^T \overline{x}^* - b &= 1, \\ a^T + u^T - v^T &= 0, \\ u^T, v^T &\geq 0 \end{aligned}$$

        is infeasible
        then return a failure message;
        else find an element $\overline{x}$ of $\mathcal{T}$ that maximizes $a^T \overline{x}$;
        if $a^T \overline{x} > b$
        then add $\overline{x}$ to A as a new column;
        else return a and b;
        end
       end
   end
else return 0 and -1;
end

---

our special case of the TSP) most of our choices of d, ϕ, and $\mathcal{T}$ in Algorithm 11.16 place $\phi(x^*)$ in the convex hull of $\mathcal{T}$, then we may save time even if each successful call of SEPARATE($\phi(x^*), \mathcal{T}^*$) is followed by a call of SEPARATE($\phi(x^*), \mathcal{T}$).

In fact, if SEPARATE($\phi(x^*), \mathcal{T}^*$) returns an inequality $a^{*T}\overline{x} \leq b^*$ that separates $\phi(x^*)$ from $\mathcal{T}^*$, then there is no need to call SEPARATE($\phi(x^*), \mathcal{T}$): for every nonnegative M such that

$$M \geq \frac{a^{*T}\overline{x} - b^*}{\sum(v^T \overline{x} - w : [v^T, w] \in \mathcal{F})} \quad \text{for all } \overline{x} \text{ in } \mathcal{T} - \mathcal{T}^*,$$

the inequality

$$a^{*T}\overline{x} - M \cdot \sum(v^T \overline{x} - w : [v^T, w] \in \mathcal{F}) \leq b^*$$

separates $\phi(x^*)$ from $\mathcal{T}$ and may be converted into a cut inducing a facet of the convex hull of $\mathcal{T}$ by the techniques developed in Section 11.6.

Specifically, there are a subset S of $\{1, 2, \ldots, d\}$ and a linear mapping $\eta : \mathbf{R}^S \to \mathbf{R}^d$ such that $|S| = \dim \mathcal{T}$ and,

    with $\psi(\overline{x})$ standing for the restriction of $\overline{x}$ on S,
    we have $\eta(\psi(\overline{x})) = \overline{x}$ whenever $\overline{x} \in \mathcal{T}$;

the substitution of $\eta(\psi(\overline{x}))$ for $\overline{x}$ converts any cut into an equivalent form $a^T \overline{x} \leq b$ with the property that

$$a_e = 0 \quad \text{for all } e \text{ outside } S;$$

in turn, Algorithm 11.18, given any cut $a^T \overline{x} \le b$ with this property, returns a cut inducing a facet of the convex hull of $\mathcal{T}$. Here, a default $\mathcal{T}_0$ is supplied by the optimal basis of problem (11.57) and a default $\mathcal{C}$ consists of all $[v^T, w]$ in $\mathcal{F}$ such that $\mathcal{T}'' \not\subset \{\overline{x} : v^T \overline{x} = w\}$.

Success of Algorithm 11.16 hinges on the ability to make choices of ϕ and $\mathcal{T}$ (which may be guided by x^* and $\mathcal{S}$) in such a way that
    (i) chances of $\phi(x^*)$ falling outside the convex hull of $\mathcal{T}$ are reasonable
and
    (ii) cuts $a^T \phi(x) \le b$ collected in $\mathcal{L}$ are not too weak.

If one were to use Algorithm 11.16 in the context of integer linear programming, where $\mathcal{S}$ consists of all integer solutions of some explicitly recorded system

$$Ax = b, \; \ell \le x \le u,$$

then one would have to design a way of making choices of ϕ and $\mathcal{T}$. One option is to choose each ϕ and $\mathcal{T}$ by choosing a $d \times m$ integer matrix P, setting $\phi(x) = Px$, and letting $\mathcal{T}$ consist of all integer vectors $\overline{x}$ such that some x satisfies

$$Ax = b, \; \ell \le x \le u, Px = \overline{x}.$$

Maximizing a linear function over $\mathcal{T}$ amounts to solving a mixed-integer linear programming problem in d integer and m noninteger variables; as long as d is small, this problem can be solved quickly. As long as at least one row r^T of P satisfies

$$\lfloor \max\{r^T x : Ax = b, \; \ell \le x \le u\} \rfloor < r^T x^*, \tag{11.58}$$

$\phi(x^*)$ falls outside the convex hull of $\mathcal{T}$; Gomory's methods mentioned in Section 4.3 provide vectors r^T with property (11.58) at insignificant computational cost. Computational experiments with this scheme, and also with other mappings for general MIP problems, have been carried out by Espinoza [168].

Algorithm 11.18 From a cut $a^T \overline{x} \le b$ to a facet-inducing cut.

$\mathcal{T}_0 =$ an affinely independent subset of $\{\overline{x} \in \mathcal{T} : a^T \overline{x} = b\}$;
$\mathcal{C} \;=$ a catalog of vectors $[v^T, w]$ such that
$\qquad \mathcal{T} \subset \{\overline{x} : v^T \overline{x} \ge w\}, \; v^T \overline{x}^* \ge w, \; \mathcal{T} \not\subset \{\overline{x} : v^T \overline{x} = w\}$;
$\overline{x}_0 =$ an arbitrary element of $\mathcal{T}$;
for   all $[v^T, w]$ in $\mathcal{C}$
do   remove $[v^T, w]$ from $\mathcal{C}$
$\qquad$ **if** $\mathcal{T}_0 \subset \{\overline{x} : v^T \overline{x} = w\}$
$\qquad$ **then** $(a^+, b^+, \overline{x}^+) = \text{TILT}\,(a, b, v, w, \overline{x}^0)$;
$\qquad\qquad\; a = a^+, b = b^+, \mathcal{T}_0 = \mathcal{T}_0 \cup \{\overline{x}^+\}$;
$\qquad$ **end**
end
while $|\mathcal{T}_0| < \dim \mathcal{T}$
do $\overline{x}^0 =$ an element of $\mathcal{T} - \mathcal{T}_0$ such that
$\qquad\qquad \mathcal{T}_0 \cup \{\overline{x}^0\}$ is affinely independent;
$\qquad$ **if** $a^T \overline{x}^0 = b$
$\qquad$ **then** replace $(a, b; \mathcal{T}_0)$ by $(a, b; \mathcal{T}_0 \cup \{\overline{x}^0\})$;
$\qquad$ **else** find a nonzero vector v and a number w such that
$\qquad\qquad\qquad v^T \overline{x} - w = 0$ for all $\overline{x}$ in $\mathcal{T}_0 \cup \{\overline{x}^0\}$,
$\qquad\qquad\qquad\qquad v_e = 0$ for all e outside S;
$\qquad\qquad (a^-, b^-, \overline{x}^-) = \text{TILT}\,(a, b, -v, w, \overline{x}^0)$;
$\qquad\qquad (a^+, b^+, \overline{x}^+) = \text{TILT}\,(a, b, v, w, \overline{x}^0)$;
$\qquad$ **if** $a^{+T} \phi(x^*) - b^+ \ge a^{-T} \phi(x^*) - b^-$
$\qquad$ **then** replace $(a, b; \mathcal{T}_0)$ by $(a^+, b^+; \mathcal{T}_0 \cup \{\overline{x}^+\})$;
$\qquad$ **else** replace $(a, b; \mathcal{T}_0)$ by $(a^-, b^-; \mathcal{T}_0 \cup \{\overline{x}^-\})$;
$\qquad$ **end**
$\qquad$ **end**
end
return a and b;

$\text{TILT}\,(a, b, v, w, \overline{x}^0)$:
$\overline{x} =$ element of $\mathcal{T}$ that maximizes $v^T \overline{x}$;
if $v^T \overline{x} > w$
then **if** $a^T \overline{x} < b$
$\qquad$ **then** $\lambda = v^T \overline{x} - w, \mu = b - a^T \overline{x}$;
$\qquad\qquad$ return $\text{TILT}\,(a, b, \lambda a + \mu v, \lambda b + \mu w, \overline{x})$;
$\qquad$ **else** return $(a, b, \overline{x})$;
$\qquad$ **end**
else return $(v, w, \overline{x}^0)$;
end

Chapter Twelve

Managing the Linear Programming Problems

The solution of linear programming problems is a crucial part of the Dantzig-Fulkerson-Johnson approach to the TSP. Although the individual LPs are often not particularly difficult to solve, the great number of them that arise make the time spent in the LP solver a dominant part of a TSP computation.

In this chapter we describe machinery to handle the flow of cuts and edges into and out of the LP relaxations, as well methods used for interacting with an LP solver. The actual solution of the LPs will be described in Chapter 13.

12.1 THE CORE LP

The LPs that need to be solved during a TSP computation can be very large. Right from the start, they are very wide, containing a column for every pair of cities. Moreover, as time goes on and more and more cutting planes are added, the LPs may eventually become long, that is, they may have many rows.

To deal with the width of the LPs, the problems are solved implicitly by solving a *core LP* containing columns corresponding only to a small subset of the complete set of edges. The edges not in the core LP are handled by periodically computing the reduced costs that the corresponding columns would give with respect to the current LP dual solution. Based on these calculations, some of the columns can be added to the core LP. This is a version of delayed column generation; it can be viewed as the dual of cutting-plane generation. Column generation is a standard technique for handling LPs containing many columns, as we discussed in Section 11.4.5. For background material on linear programming and column generation we refer the reader to standard reference works by Chvátal [126], Schrijver [494], Nemhauser and Wolsey [428], Wolsey [559], and Vanderbei [532].

Naturally, column generation can experience the same difficulty that can arise in cutting-plane generation: over time we may add enough columns to the core LP so that it itself becomes wide. The art of column generation, and of cutting-plane generation, is to proceed using the smallest LPs that we can efficiently get away with. This means that we need to be selective in choosing cuts and columns to enter the core LP, beyond the obvious conditions that cuts should be violated and columns should have negative reduced cost. Moreover, rules are needed for deciding whether or not a particular cut or column should be removed from the core. This reflects the fact that a cut or column that appeared to be very important when it was initially added to the core LP, may, after a sequence of further additions, no longer play a role in determining the LP optimal solution. In such a case, we may be better off deleting the particular cut or column. For instance, a cut that is no longer satisfied as an equation by the current LP optimal solution is a candidate for deletion. We discuss these issues in this section.

12.1.1 The initial LP

To start a computation, we build an initial core LP and solve it. The obvious choice for the LP is the formulation without any cutting planes, namely,

$$\text{Minimize } \sum(c_e x_e : e \in E') \tag{12.1}$$

$$\text{subject to}$$

$$x(\delta_{G'}(v)) = 2, \text{ for all } v \in V$$

$$0 \leq x_e \leq 1, \text{ for all } e \in E'$$

where E' is some subset of the edges and $G' = (V, E')$.

For large TSP instances, the bound provided by the solution of this LP can be rather weak. The poor quality of the bound means that considerable work is needed in the cutting-plane computation in order to get a good approximation of the TSP value. This prompted Jünger et al. [296] to include some predetermined set of cuts to jump start the cutting-plane method. They proposed to take advantage of the observation that geometrically isolated sets of cities are likely to correspond to important subtour inequalities. The idea is to precompute a selection of these sets to include in the initial LP.

The goal of Jünger et al. [296] was to obtain good bounds on large problems (in particular, the TSPLIB instance pla33810). They used an initial edge set obtained from the Delaunay triangulation of the point set, and employed column generation to find an optimal solution over the complete set of edges. The LPs that arose were solved with the interior point LP code LOQO of Vanderbei [531].

This jump-starting technique could be extended using subtour inequalities obtained via a short run of the Held-Karp iterative algorithm or using blossom inequalities obtained by solving the 2-matching relaxation using the algorithm of Edmonds [163]. In our discussion below, however, we will focus only on the simple strategy of starting without any predetermined cuts.

12.1.2 Initial edge set

It is obviously impractical to work with the complete edge set of a TSP instance when solving the LP relaxations—the memory requirements would easily overwhelm the LP solver. Fortunately, if we are selective in adding and deleting cutting planes, the number of edges that are assigned nonzero values in optimum basic solutions to the LPs is relatively small. This is illustrated in Table 12.1, where we list the number of nonzero edges used in several computations; the values are given as multiples of the number of cities in the instance.

It is not efficient to try to keep the number of edges in the core LPs close to these minimum values, but slightly larger sets of between $1.5|V|$ and $3|V|$ edges can be used throughout most computations. We try to select our initial edge set to

Table 12.1 Number of nonzero edges in TSP LPs.

Problem	Maximum	Average
fnl4461	1.34	1.27
rl11849	1.39	1.27
usa13509	1.36	1.30
pla33810	1.10	1.08

lie within this range. There are many different possibilities for choosing such an E' and we consider only a small sample of the alternatives in our tests below.

The idea of using a core edge set appears in the work of Padberg and Hong [451]. Although they did not propose a general method, they did use restricted sets of edges in several of their tests. In particular, they needed to use external pricing to handle a 318-city instance, since the 50,403 edges in the problem exceeded the 50,000 limit that was part of their computer code. To get an initial edge set for the instance, they chose an appropriate constant K and selected the 30,939 edges having cost at most K.

The first systematic use of core edge sets is in the work of Land [333]. Her test set of problem instances included a number of 100-city examples (so 4,950 edges), but she restricted the number of core edges to a maximum of 500. The initial edge set she chose consisted of the union of the edges in the best tour found and the 4-nearest-neighbor edge set, that is, the four minimum-cost edges containing each city.

Land's edge set was also used by Grötschel and Holland [232]; they ran tests with the initial set consisting of the union of the best available tour and the k-nearest-neighbor edge set, for $k \in \{0, 2, 5, 10\}$. Jünger et al. [300] carried out further tests, using $k \in \{2, 5, 10, 20\}$. In this latter study, the conclusion was that $k = 20$ produced far too many columns and slowed down their solution procedure, but the smaller values of k all led to reasonable results.

A different approach was adopted by Padberg and Rinaldi [447], taking advantage of the form of their tour-finding heuristic. In their study, Padberg and Rinaldi compute a starting tour by making k independent runs of the heuristic algorithm of Lin and Kernighan [357]; for their initial edge set, they take the union of the edge sets of the k tours. Their recommendation is to use

$$k = \text{Minimum}\{50, \text{Maximum}\{10, 0.1|V|\}\}$$

when solving an instance with $|V|$ cities. Although we do not use repeated runs of Lin-Kernighan to obtain an initial tour, the Padberg-Rinaldi tour-union method remains an attractive idea since it provides an excellent sample of the edges of the complete graph (with respect to the TSP).

In Table 12.2, we compare the Padberg-Rinaldi idea with the k-nearest set on a 100,000-city random Euclidean instance. The tours were obtained with short runs of the Chained Lin-Kernighan heuristic of Martin et al. [379], using the implementation described in Chapter 15; the short runs each used $|V|/100 + 1$ iterations of the heuristic. For the k-nearest sets, we also include the edges in the best starting tour we found for the instance.

In each case, we ran a selection of the cutting-plane separation algorithms we described in earlier chapters, together with column generation over the edges of the complete graph on the 100,000 points. For each of the edge sets we report the initial

Table 12.2 Initial edge set for 100,000-city TSP.

Edge set	$|E|$-Initial	$|E|$-Final	CPU Time (scaled)
2-Nearest	146,481	231,484	1.037
3-Nearest	194,569	240,242	1.721
4-Nearest	246,716	270,096	1.129
5-Nearest	300,060	312,997	1.103
10 Tours	167,679	224,308	1.000
50 Tours	215,978	247,331	1.016

number of edges and the final number of edges in the core LP after the cutting-plane and column-generation routines terminated, as well as the total running time scaled so that the 10-tour time is 1.000.

We chose the union of 10 tours in our implementation—it has the lowest CPU time in the test and it maintains the smallest edge set during the computations.

12.1.3 Initial solution

The solution technique chosen for the initial LP, regardless of its form, must be tied to the technique used to solve the subsequent core LPs. We will in general be working with the dual simplex algorithm, so to start things off we need not only an optimal solution to the initial LP, but also an optimal basis. In this section we consider the problem of finding such a basis for (12.1).

We refer to the LP (12.1) as the *fractional 2-matching problem*. This name is derived from the fact that integral solutions to (12.1) correspond to sets of edges having the property that each vertex in V is met by exactly two edges in the set, that is, a 2-matching.

A variety of methods are available for solving the fractional 2-matching problem. The LP itself is only of modest size, as long as E' is sufficiently sparse and the TSP instance is not extremely large, so one possibility is to solve the problem directly with general LP solution methods such as those found in CPLEX. Alternatively, the problem can be solved by a primal-dual algorithm, similar to the Hungarian algorithm for bipartite matching problems, as we describe below. Other combinatorial approaches can be found in Boyd [87] (including a combinatorial version of the primal simplex algorithm, similar in spirit to the network simplex algorithm for min-cost flow problems) and in Trick [524].

A FRACTIONAL 2-MATCHING ALGORITHM

The primal-dual algorithm can be viewed as a simplified version of Edmonds' b-matching algorithm, described in papers of Edmonds [163] and Edmonds and Johnson [166].

The dual LP of (12.1) is

$$\text{Maximize } \sum(2y_v : v \in V) - \sum(z_e : e \in E') \tag{12.2}$$

$$\text{subject to}$$

$$y_u + y_v - z_e \le c_e \text{ for all } e = \{u, v\} \in E'$$

$$z_e \ge 0, \text{ for all } e \in E'.$$

Given a solution $(\bar{y}, \bar{z})$ to this LP, we call an edge $e = \{u, v\} \in E'$ *tight* if

$$\bar{y}_u + \bar{y}_v - \bar{z}_e = c_e.$$

The complementary slackness conditions, for the pair of LPs (12.1) and (12.2), are that for all edges $e \in E'$

(a) if $\bar{x}_e > 0$, then e is tight

(b) if $\bar{z}_e > 0$, then $\bar{x}_e = 1$.

Using a standard augmenting path approach, the algorithm will construct primal and dual solutions that satisfy these conditions. The solutions $\bar{x}$ that appear in the algorithm will also have the property that $\bar{x}_e$ is either $0, 1$, or $1/2$ for all edges e, and the edges with $\bar{x}_e = 1/2$ will form disjoint odd circuits.

To begin, let $\bar{x}_e = 0$ and $\bar{z}_e = 0$ for all $e \in E'$ and for each vertex $v \in V$ let

$$\bar{y}_v = \frac{1}{2}\min(c_e : e \in (\{v\}, V - \{v\})).$$

At each iteration, we choose a vertex r such that $\bar{x}(\delta(r)) < 2$, and grow a tree T rooted at r having the following properties: each edge in T is tight, no edge e in T has $\bar{z}_e > 0$, and for each vertex v in T, the unique path in T from v to r alternates between matched edges (that is, $\bar{x}_e = 1$) and unmatched edges (that is, $\bar{x}_e = 0$). The vertices of T are labeled "+" and "−" according to the parity of the number of edges in the path back to the root r, that is, vertex r and all vertices of even distance from r are labeled "+" and all vertices of odd distance from r are labeled "−". We grow T by appending matched edges e satisfying $\bar{z}_e = 0$ that join "−" vertices to vertices not yet in T, or by appending tight unmatched edges that join "+" vertices to vertices not yet in T.

Whenever we add a new vertex v to T, we check to see if we can perform an *augmentation*, according to the following rules.

First, suppose v is a "+" vertex that meets an edge f having $\bar{x}_f = 1/2$. Then we can increase $\sum(\bar{x}_e : e \in E')$ by $1/2$ if we perform the changes illustrated in Figure 12.1. This augmentation has two steps. We begin by walking around the odd circuit of $1/2$-valued edges that contains f, starting at f, and alternately setting $\bar{x}_e = 1$ and $\bar{x}_e = 0$, so that the two edges meeting v in the circuit get assigned the value 1. We then walk along the path in T from v to r, replacing $\bar{x}_e$ by $1 - \bar{x}_e$ for edge e in the path.

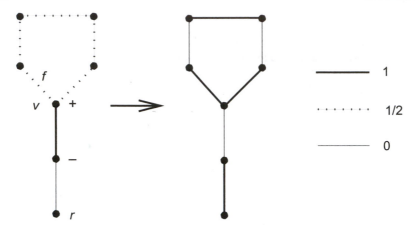

Figure 12.1 A 1/2–augmentation from a "+" vertex.

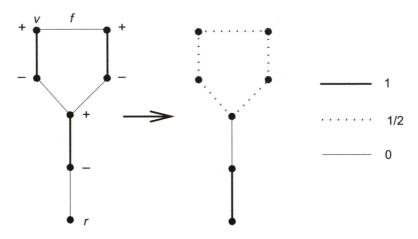

Figure 12.2 A 1/2–cycle augmentation from a "+" vertex.

Now suppose v is a "+" vertex that does not meet any $1/2$-valued edge but does meet a tight, unmatched edge f such that the other end of f is also a "+" vertex. In this case, an augmentation can be carried out that forms an odd circuit of $1/2$'s and increases $\sum(\bar{x}_e : e \in E')$ by $1/2$. This augmentation is illustrated in Figure 12.2. To describe it, let P_1 denote the path in T from v to r and let P_2 denote the path in T from the other end of f to r. The union of P_1, P_2, and f forms an odd circuit C and a path P. The augmentation is to set $\bar{x}_e = 1/2$ for all edges e in the circuit C, and to set $\bar{x}_e$ to $1 - \bar{x}_e$ for all edges e in the path P. (The reason we first check whether v meets a $1/2$-valued edge before we consider this circuit-forming augmentation is to maintain the condition that the $1/2$-valued edges from disjoint odd circuits. This is important in the basis finding routine that we outline below.)

If v is a "−" vertex, then there are three types of augmentations. The first two are analogous to the "+" augmentations described above and are illustrated in Fig-

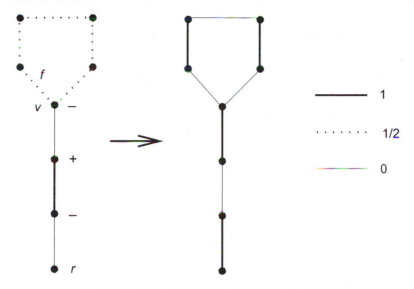

Figure 12.3 A 1/2–augmentation from a "−" vertex.

ure 12.3 and Figure 12.4, respectively.   In the first case v meets a $1/2$-valued edge, and in the second case v does not meet any $1/2$-valued edges but does meet a matched edge f such that $\bar{z}_f = 0$ and the other end of f is also a "−" vertex. The third type of augmentation is when v does not meet a $1/2$-valued edge, but $\bar{x}(\delta(v)) < 2$. In this case, we simply augment along the path in T from v to r, replacing $\bar{x}_e$ by $1 - \bar{x}_e$ for each edge e in the path. Such an augmentation increases $\sum(\bar{x}_e : e \in E')$ by 1.

If we can neither make an augmentation, nor grow the tree T any further, then we must alter the dual solution $(\bar{y}, \bar{z})$ to either create a new tight, unmatched edge e or a new matched edge e having $z_e = 0$, such that e allows us either to grow the tree or to augment $\bar{x}$. The constraints on the dual change are that all edges e in T must remain tight and continue to have $\bar{z}_e = 0$, that $(\bar{y}, \bar{z})$ continues to satisfy the constraints of the LP (12.2), and that $(\bar{y}, \bar{z})$ and $\bar{x}$ continue to satisfy the complementary slackness conditions. These requirements can be met by adding some $\epsilon > 0$ to the $\bar{y}$-value of all "+" vertices and subtracting ϵ from the "−" vertices, and adjusting $\bar{z}_e$ for all matched edges e that are not in T but meet at least one vertex in T. Subject to the conditions listed above, we choose ϵ as large as possible. What stops us from making ϵ arbitrarily large is that (i) $\bar{z}_e$ becomes 0 for some matched edge joining two "−" vertices, or (ii) $\bar{z}_e$ becomes 0 for some matched edge joining a "−" vertex to a vertex not in T, or (iii) some unmatched edge joining two "+" vertices becomes tight, or (iv) some unmatched edge joining a "+" vertex to a vertex not in T becomes tight. If none of these events happens, that is, if ϵ can be made arbitrarily large, then there is no solution to the fractional matching problem. If either (i) or (iii) occurs, then we can augment $\bar{x}$ as we described above. If either (ii) or (iv) occurs, then we can grow the tree T.

The overall algorithm is to repeatedly grow T, alter the dual solution, and grow

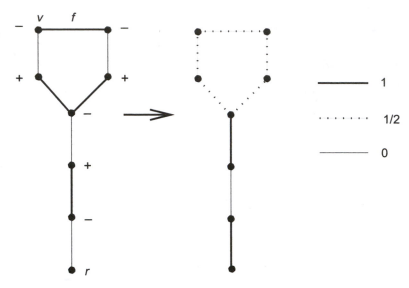

Figure 12.4 A 1/2–cycle augmentation from a "−" vertex.

T some more, until we can perform an augmentation. After an augmentation, all the "+" and "−" labels are removed, and a new tree is grown, as long as there remain vertices having $\bar{x}(\delta(v)) < 2$.

The algorithm can be implemented quite efficiently, as we report below, but it does not directly provide an optimal basis for the problem. The final solution is a good starting point for computing such a basis, however, as we now describe. This basis-finding routine was presented to us by W. H. Cunningham (private communication, 1995).

A basis for the problem corresponds to a set B of $|V|$ edges such that each connected component of the graph (B, V) contains exactly one odd circuit and no even circuit. A basis that corresponds to the solution $\bar{x}$ must include all edges that have been assigned the value 1/2, together with a subset of the remaining tight edges that have $\bar{z}_e = 0$. To construct such a basis, we will not need to alter $\bar{x}$, but we might need to adjust the dual solution $(\bar{y}, \bar{z})$.

Initially, set B equal to the set of edges that are assigned the value 1/2. If this is not a basis, then select a component D of the graph (B, V) that does not contain an odd circuit. Note that D is a tree, possibly containing only a single vertex. Pick a root of the tree and label the vertices "+" and "−" so that no edge in the tree has two "+" ends or two "−" ends. If there are any tight edges e having $\bar{z}_e = 0$ that join two vertices in D having the same label, then add e to B and stop processing this component (it now has an odd circuit). If there is a tight edge $e = \{u, v\}$ having $\bar{z}_e = 0$, such that u is in D and v is not in D, then add the edge to B and expand D to again be a connected component of (B, V), that is, add to D the edge e and the connected component that contains v. If D now has an odd circuit, then stop processing this component. Otherwise, expand the "+", "−" labeling to the new vertices in D and continue.

If there are no edges that can be added to D, then we carry out a dual change as in the main algorithm. If the dual change ϵ is not restricted by some edge, then we attempt to make a dual change with the roles of "+" and "−" reversed, that is, we add ϵ to the "−" vertices and subtract ϵ from the "+" vertices. If ϵ is still unrestricted, then the graph (E', V) does not contain a basis (a connected component of the graph is an even circuit). Otherwise, we can add to B a newly created tight edge having $\bar{z}_e = 0$ and continue.

In Table 12.3 we present the running times for this fractional 2-matching algorithm on several test instances. The two largest instances are random Euclidean TSPs; the remaining instances are from the TSPLIB. Notice that although the basis

Table 12.3 Fractional 2-matchings (Intel Xeon, 2.66 GHz).

Vertices	Edges	2-Match (seconds)	Basis	Total
4,461	7,211	0.01	0.01	0.02
11,849	16,797	0.03	0.02	0.05
13,509	21,777	0.01	0.03	0.05
33,810	53,395	0.08	0.03	0.12
100,000	168,314	0.46	0.26	0.83
1,000,000	1,695,220	3.47	1.61	5.67

finding routine does add some computation time to the algorithm, it is less than the time needed to find the original optimal $\bar{x}$ and $(\bar{y}, \bar{z})$.

12.1.4 Adding and deleting edges

The results in Table 12.2 indicate the growth of the cardinality of the core edge set as the combined cutting-plane and column-generation algorithm progresses. To help limit this growth, we are selective about the edges that are permitted to enter the core, and we also take steps to remove edges from the core if they do not appear to be contributing to the LP solution.

When an edge is found to have negative reduced cost in our pricing routine, it is not directly included in the core edge set, but rather it is added to a queue of edges that are waiting for possible inclusion in the core. The Concorde code will at irregular intervals (determined by the increase in the optimal value of the LP relaxation) remove the N edges from the queue having the most negative reduced cost (we use $N = 100$), and add these to the core LP. An LP solver is then used to obtain new optimal primal and dual solutions, and the reduced costs for the remaining edges in the queue are recomputed; any edge in the queue that has reduced cost greater than some small negative tolerance (we use negative 0.00001) is removed from the queue.

After an edge e is added to the LP, we monitor the corresponding component of x^* at each point when the LP solver produces a new solution. If for L_1 consecutive LP solves the value of x^*_e is less than some small tolerance ε_1, then we remove edge e from the core LP. (Typical values are $L_1 = 200$ and $\varepsilon_1 = 0.0001$.)

12.1.5 Adding and deleting cuts

As in the case for edges, when a cutting plane is found by a separation routine, it is attached to the end of a queue of cuts that are waiting to be added to the core LP. The Concorde code repeatedly takes the first cut from the queue for processing and checks that it is violated by the current x^* by at least some small tolerance (we use 0.002). If the cut does not satisfy this condition, then it is discarded; otherwise it is added to the core LP. (Note that the current x^* may perhaps not be the same as the vector that was used in the separation algorithm that produced the cut.)

After k cuts have been added to the LP (we use $k = 250$ in our code) or after the cut queue becomes empty, an LP solver is called to compute optimal primal and dual solutions for the new core LP. If the slack variable corresponding to a newly added cut is in the optimal LP basis, then the cut is immediately deleted from the LP. Otherwise, the cut remains in the LP and a copy of the cut is added to a pool of cuts that is maintained for possible later use by the separation routines. (Subtour inequalities are not added to the pool, since an efficient exact separation routine for them is available.)

Once a cut is added, after each further point where the LP solver produces new solutions, we check the dual variable corresponding to the cut. If for L_2 consecutive LP solves the dual variable is less than some fixed tolerance ε_2, then the cut is deleted from the LP. (Typical values are $L_2 = 10$ and $\varepsilon_2 = 0.001$.) This deletion condition is weaker than the standard practice of deleting cuts only if they are slack (that is, only if they are not satisfied as an equation by the current x^*); for our separation routines on large TSP instances, we found that the core LP would accumulate a great number of cuts that were satisfied as an equation by the current x^* if we only removed slack inequalities.

It may well happen in our computation that a cut is deleted from the LP, but then added back again after a number of further LP solves. Although this is obviously a waste of computing time, when working on large instances it seems necessary to be very aggressive in attempting to keep the core LP small, both for memory considerations and to help the LP solver.

Notice that the tolerances and constants we use for cuts and edges differ by large amounts. These values were obtained through computational experiments, and they are dependent on the method used to solve the core LP problems.

12.2 CUT STORAGE

The storage of cutting planes and their representation in an LP solver account for a great portion of the memory required to implement the cutting-plane algorithm for the TSP. In this section we discuss the methods used by Concorde to attempt to reduce this demand for memory when solving large problem instances.

There are three places in the computer code where we need to represent cuts: in the LP solver, in the external LP machinery, and in the pool of cuts. We discuss below representations for each of these components.

12.2.1 Cuts in the LP solver

The standard cutting planes we use in Concorde can all be written as

$$\sum (x(\delta(S) : S \in \mathcal{F}) \geq \mu(\mathcal{H}) \tag{12.3}$$

for a specified hypergraph $\mathcal{H} = (V, \mathcal{F})$, that is, $\mathcal{H} \circ x \geq \mu(\mathcal{H})$. This is how we like to think about the cuts, but we need not be restricted to this form when carrying out a computation. Indeed, the degree equations give us a means to dramatically alter the appearance of a cut. For example, we can write (12.3) as

$$\sum_{S \in \mathcal{F}} x(\{e : e \subseteq S\}) \leq I_{\mathcal{H}} \tag{12.4}$$

for some constant $I_{\mathcal{H}}$. The representation (12.4) can be further altered by replacing some sets S by their complements $V - S$. Moreover, starting with any form, we can add or subtract multiples of degree equations to further manipulate a cut's appearance. We make use of this freedom in selecting the representation of the cut in the LP solver.

The most important criterion for choosing the LP representation of a cut is the number of nonzero entries the representation adds to the constraint matrix. The amount of memory required by the solver is proportional to the total number of nonzeros, and, other things being equal, LP solvers are more efficient in solving LPs with fewer nonzeros.

We compare four different representations: (1) each cut in the form given in (13.1) (the "outside" form), (2) each cut in the form given in (12.4) (the "inside" form), (3) each cut in the inside form with individual sets S complemented if it decreases the number of nonzero coefficients among the columns in the core LP, and (4) individual cuts in the form (either outside, or inside with complemented sets) that gives the least number of nonzeros.

For each of these representations we consider three different algorithms for selecting multiples of the degree equations to add or subtract from the individual cuts to reduce the number of nonzeros among the columns in the core LP.

In the first method, for each cut we simply run through the cities in order, and subtract the corresponding degree equation from the cut if it is advantageous to do so (that is, if it will decrease the number of nonzeros).

The second method is a greedy algorithm that, for each cut, first makes a pass through all of the cities and counts the number of nonzeros that can be saved by subtracting the corresponding degree equation. It then orders the cities by nonincreasing values of this count and proceeds as in the first method.

The third method is similar to greedy, but adjusts the counts to reflect the current status of the cut. In this algorithm, we keep two priority queues, one containing cities whose corresponding degree equation can be advantageously subtracted from the cut and another containing cities whose degree equation can be advantageously added to the cut. Both priority queues are keyed by the number of nonzeros the operations save, that is, cities that save more nonzeros have priority over cities that save fewer nonzeros. At each step of the algorithm, we select the city having the maximum key and either subtract or add the corresponding equation. We then update the keys appropriately, possibly inserting new cities into the queues. The algorithm terminates when both queues are empty. Notice that this algorithm permits equations to be added or subtracted a multiple number of times.

We tested these algorithms and representations on core LPs taken from our computations on 21 small problems from the TSPLIB collection; the 21 instances range in size from 1,000 cities to 7,397 cities. The results are reported in Table 12.4, in multiples of the number of nonzeros in the outside representation.

Table 12.4 Nonzeros in LP representations.

Algorithm	Outside	Inside	Complemented	Best
None	1.00	2.81	2.07	0.90
Straight	0.54	1.40	1.11	0.49
Greedy	0.52	1.19	0.95	0.48
Queues	0.49	1.07	0.87	0.45

The immediate conclusions are that the outside form of cuts clearly dominates the inside form (even with complementation) and that the reduction routines appear to be worthwhile. Although the "best" form of cuts is slightly preferable to the outside form, we use the outside form in our implementation since this simplification leads to a more efficient routine for computing the reduced costs of columns that are not in the core LP.

For our reduction algorithm, we use the queue-based routine: it is efficient and is slightly better than the other two methods.

The representations chosen in earlier computational studies vary greatly from research team to research team. Dantzig et al. [151], Hong [270], Clochard and Naddef [130], Jünger et al. [300], and Naddef and Thienel [424] appear to have used the outside form of subtour inequalities, whereas Miliotis [394], Land [333], Padberg and Hong [451], Crowder and Padberg [147], Grötschel and Holland [232], and Padberg and Rinaldi [447] all appear to have used the inside form. Clochard and Naddef [130] and Naddef and Thienel [424] used the outside form of combs and more general inequalities, and Land [333] used a special outside form for blossoms, but each of the other studies used the inside form for all cuts other than subtours. The complementation of subtours was probably carried out in most of the studies that used the inside form of the inequalities, but Padberg and Rinaldi [447] is the only paper that mentions complementing other inequalities—they consider complementing the handles in combs.

12.2.2 External storage of cuts

It is not sufficient to use the LP solver's internal list of the current cutting planes as our only representation of the cuts. The trouble is that this internal representation does not support the computation of the reduced costs of the edges not present in the core LP. What is needed is a scheme for storing the cuts in their implicit form, that is, as hypergraphs $\mathcal{H} = (V, \mathcal{F})$, where $\mathcal{F}$ is a family of subsets of V. The most important criteria for evaluating such an external representation scheme are the total amount of storage space required and the ease with which the data structure supports a fast edge-pricing mechanism.

In our implementation, we choose a very compact representation of the cuts, one that fits in well with the pricing routine that we describe in Section 12.3. Before we discuss the representation, we present some alternative schemes that have appeared in earlier studies.

PREVIOUS WORK

The external representation of cuts is first treated in Land [333]. Land's technique for storing subtour inequalities is to pack a collection of pairwise disjoint subsets of V into a single vector of length $|V|$, where the subsets correspond to the vertex sets of the subtours. The entries of the vector provide a labeling of the vertices such that each of the subsets is assigned a distinct common label. This representation was particularly useful for Land since her separation routines (connectivity cuts and a version of subtour shrinking) naturally produced collections that were pairwise disjoint.

Land used this same technique for storing the handles of blossom inequalities. She got around the problem of storing the teeth of the blossoms by requiring that these edges be part of the core LP. This meant that the routines for pricing out edges outside the core were never needed to compute the reduced cost of a tooth edge.

Grötschel and Holland [232] also used column generation, but they did not report any details of the external representation scheme used in their study, writing only: "After some experiments we decided to trade space for time and to implement space-consuming data structures that allow us to execute pricing in reasonable time."

A similar philosophy of trading off space for time was adopted by Padberg and Rinaldi [447]. They used a full $|V|$-length vector to represent each of the cuts. The representation, however, allowed them to compute the reduced cost of an edge by making a very simple linear scan through the cuts, performing only a constant amount of work for each cut.

A more compact representation was used by Jünger et al. [300]. They store a cut for the hypergraph $\mathcal{H} = (V, \mathcal{F})$ by keeping the sets in $\mathcal{F}$ as an array that lists the vertices in each set, one after another. The lists are preceded by the number of sets, and each set in the list is preceded by the number of vertices it contains. In the typical case, the length of the array will be much less than $|V|$, but the extraction of the cut coefficient on an individual edge is more costly than in Padberg and Rinaldi's approach.

Our scheme is similar to Jünger et al., but uses a different representation for the individual sets in the cut. It suffers the same drawback in extracting individual edge coefficients, but the pricing mechanism we describe in Section 12.3 does not make use of single coefficient extraction, using instead a strategy that calls for the simultaneous pricing of a large group of edges. This simultaneous pricing allows us to spend a small amount of CPU time, up front, setting up an auxiliary data structure that will speed the coefficient generation for the edges in the pricing group.

This strategy allows us to take advantage of the compact cut representation without incurring a significant penalty in the performance of our pricing engine.

HYPERGRAPHS

Most of the space used in Jünger et al.'s cut representation is occupied by the lists of the vertices in individual sets. To improve on this, we must either find a way to write these lists more compactly or find a way to avoid writing them altogether. We postpone a discussion of list compression and first consider a method for reducing the number of lists.

The idea is simple: we reuse old representations. To carry this out, if we have a collection of cuts $\mathcal{H}_1 = (V, \mathcal{F}_1), \mathcal{H}_2 = (V, \mathcal{F}_2), \ldots, \mathcal{H}_r = (V, \mathcal{F}_r)$, we represent separately the sets $\mathcal{S} = \mathcal{F}_1 \cup \mathcal{F}_2 \cup \cdots \cup \mathcal{F}_r$ and the hypergraphs $\mathcal{H}_1, \mathcal{H}_2, \ldots, \mathcal{H}_r$. In this setup, the hypergraphs are lists of pointers to the appropriate sets, rather than lists of the sets themselves. The benefit of this split representation is that the number of sets in $\mathcal{S}$ is typically considerably less than the sum of the cardinalities of the individual $\mathcal{F}_i$'s. (This split representation also helps our pricing routine, as we describe in Section 12.3.)

When a cut $\mathcal{H} = (V, \mathcal{F})$ is added to the core LP, we add to $\mathcal{S}$ the sets from $\mathcal{F}$ that do not already appear in $\mathcal{S}$, and we build a list of pointers for $\mathcal{H}$ into $\mathcal{S}$. To delete a cut $\mathcal{H} = (V, \mathcal{F})$, we delete $\mathcal{H}$'s list of pointers as well as all sets in $\mathcal{S}$ that were only being referenced by $\mathcal{H}$. To do this efficiently, we keep track of the number of times each set in $\mathcal{S}$ is being referenced (increasing the numbers when we add a cut, and decreasing them we delete a cut), and remove any set whose reference number reaches 0. The elements of $\mathcal{S}$ are stored as entries in an array, using a hash table to check whether a prescribed set is currently in $\mathcal{S}$.

TOUR INTERVALS

In Chapter 6 we discussed that fact that as our core LP matures, the LP solution vector x^* approximates the incidence vector $\bar{x}$ of an optimal tour, and hence $x^*(S, V - S) \approx \bar{x}(S, V - S)$ for most subsets S of V. For this reason, the sets that appear in cutting planes can be expected to have a small $\bar{x}(S, V - S)$ value. An interpretation of this is that sets from our cutting planes can be expected to consist of a small number of intervals in an optimal tour.

We put this to the test, using a pool for the 7,397-city TSPLIB instance pla7397. The pool consists of 2,868,447 cuts and 1,824,418 distinct sets in $\mathcal{S}$. (The large pool of cuts was accumulated over a branch-and-cut run; working with this small TSP instance allowed us to build the pool in a reasonable amount of computing time.) The average number of vertices in the members of $\mathcal{S}$ is 205.4, but the sets can be represented using an average of only 5.4 intervals from a specific optimal tour—a savings of a factor of 38. This is typical of the compression ratios we obtained in other tests, so we adopt the interval list approach as our representation scheme, using the best available initial tour as a substitute for an optimal tour.

At the start of our TSP computation, we reorder the cities in the problem so that the best tour we have found up to that point is simply $0, 1, 2, \ldots, |V| - 1$. A set is then recorded as an array of pairs of integers giving the starting and ending indices of the intervals that make up the set, with a separate field giving the total number of intervals stored in the array.

12.2.3 Pool of cuts

Like the external representation for the LP, the pool needs to store cuts in the implicit (hypergraph) form. In this case, the important criteria for evaluating a representation are the amount of storage space required and the speed with which we can compute, for a prescribed x^*, the slacks of the individual cuts in the pool, that is, $\mu(\mathcal{H}) - \mathcal{H} \circ x^*$. Our approach is to use the methods we adopted in our external LP representation but take advantage of the distribution of the references into $\mathcal{S}$ to further reduce the storage requirements.

Consider again the pla7397 pool we mentioned above. The 2,868,447 cuts make a total of 33,814,752 references to sets, and the sets make a total of 696,150,108 references to cities. Representing each of the cuts as an $|V|$-length integer vector would require approximately 78 gigabytes of memory on machines that uses 4 bytes to represent an integer. The representation of Jünger et al. [300] lowers the storage requirement considerably, calling for a little over 1.3 gigabytes of memory. To achieve this, we use only 2 bytes for each reference to a city, taking advantage of the fact that the instance has less than 2^{16} cities in total. (Recall that a byte can represent 8 binary digits.) The split interval representation we described above would use 4 bytes for each interval, to specify its ends; 2 additional bytes for each set in $\mathcal{S}$, to specify the number of intervals used by the set; 3 bytes for each reference to a set in $\mathcal{S}$, using the fact that we have less than 2^{24} sets; and 2 additional bytes for each cut, to specify the number of sets. The collection $\mathcal{S}$ has a total of 1,824,418 sets that use a total of 9,877,792 intervals, so the pool can be stored in approximately 143 megabytes of memory with this representation.

REMOVING EDGE TEETH

The savings in storage for the split interval representation is roughly a factor of 9 over the set-list representation for the pla7397 pool. This magnitude of savings is a big step toward making the direct use of large pools possible, but it is also somewhat disappointing, given the factor of 38 compression we obtained in the representation of the sets in $\mathcal{S}$. The poor showing can be explained by examining the data in Table 12.5, giving the number of sets of each size used by the cuts in the pool. Nearly two-thirds of the sets have cardinality of just two, and over four-fifths of the sets have cardinality of five or less. This means that, for the majority of the set references, we are not winning very much over the set-list representation.

An aggressive way to turn this lopsided distribution of set sizes to our advantage is to simply remove all of the references to sets of cardinality two, using a teething-like algorithm to reconstruct the sets on the fly. We did not pursue this idea, however, due to the considerable time this would add to the algorithm for searching for violated cuts among the entries in the pool.

Table 12.5 Sets in the pla7397 pool of cuts.

Size	No. of Sets	Percentage
2	21,342,827	63.1%
3	3,255,535	9.6%
4	2,267,058	6.7%
5	736,881	2.2%
6	651,018	1.9%
7	261,641	0.8%
8	337,799	1.0%
9	140,427	0.4%
≥ 10	4,821,566	14.3%

SET REFERENCES

Another route for decreasing the space requirements of our pool representation is to be more efficient in specifying our references to sets in S. The straightforward method of using an integer (or 3-byte) value for each reference is clearly wasteful if one considers the distribution of the references among the cuts: there are 1,362,872 sets in S that are referenced only once, whereas the top 10 most used sets are each referenced more than 100,000 times. An efficient representation should use less space to reference the sets that appear most often in the cuts. One way to do this is to use only a single byte to reference the most frequently used sets, use two bytes for next most frequently used sets, and restrict the use of three or four bytes to the sets that have very few references. There are a number of techniques for implementing this strategy. For example, the first byte in a set reference could either specify a set on its own, or indicate that a second, third, or fourth byte is needed. This is the strategy we adopt. If the first byte of our set reference contains a number, K, between 0 and 127, then the set index is simply K. If K is between 128 and 191, then the difference $K - 128$ together with a second byte are used to represent the index. If K is between 192 and 223 then $K - 192$ plus two additional bytes are used, and if $K > 224$ then $K - 224$ plus three additional bytes are used. (In the pla7397 pool, no reference requires 4 bytes, but we want to have a representation that would also work for much larger pools.)

Using 3 bytes for each of the 33,814,752 sets contributes 97 megabytes to the storage requirements for the split interval representation of the pla7397 pool. The compressed form of the references brings this down to 62 megabytes.

This technique for expressing the set references using variable-length strings is a very rudimentary data compression technique. Better results can be obtained using an encoding developed by Huffman [275] (see Cormen et al. [139] or Knuth [320]). Working with a byte-level Huffman code, this would save about 1 megabyte from the 62-megabyte total. If we are willing to work at the bit level, then the set reference storage can be reduced to 54 megabytes for the pool. It should be noted, however, that working with binary encodings would result in some computational overhead to address the individual bits in the reference strings.

INTERVAL REFERENCES

The list of intervals accounts for approximately 38 megabytes of the storage requirement for the pla7397 pool. Since there are only 245,866 distinct intervals among the lists, adding another level of indirection, together with a method for compressing the interval references, could result in savings similar to what we achieved for the set references. Rather than doing this, however, we use a direct method to reduce the 4-byte per interval requirement that the straightforward technique of writing the ends of the intervals provides. The idea is to represent the interval from i to j as the number i together with the offset $j - i$, using 2 bytes to write i and either 0, 1, or 2 bytes to write j. This reduces the storage needed for the intervals down to 28 megabytes.

This offset technique can be pushed a little further, writing the entire list of intervals for a prescribed set as a sequence of 1- or 2-byte offsets (splitting any interval that contains city 0 into two smaller intervals, to avoid the issue of intervals wrapping around from city $|V| - 1$ to city 0). This is the method we use in our code, and it lowers the storage requirement for the interval lists down to approximately 23 megabytes, for the pla7397 pool.

SUMMARY

The memory requirements for the representations of the pla7397 pool are summarized in Table 12.6. The entries are progressive in the sense that "Compressed set

Table 12.6 Memory requirements for the pla7397 pool of cuts.

Representation	Size (megabytes)
$|V|$-length vector	80,940
List of sets	1,403
Split lists of cities	826
Split intervals	143
Compressed set references	109
Start plus offset intervals	99
List of interval offsets	96

references" works with the split-interval representation, "Start plus offset intervals" uses the compressed set references, and so on.

In our computer code, we use the split interval representation of the pool, with compressed set references and lists of interval offsets (the "List of interval offsets" entry from Table 12.6).

12.3 EDGE PRICING

Column generation for the TSP is a simple operation for small instances—we simply run through all of the edges not currently in the core LP, compute their reduced costs, and add any negative edges into the core. This method breaks down on larger instances, however, owing both to the time required to price the entire edge set and to the large number of negative reduced-cost edges that the pricing scans will detect. In this section we describe the techniques we adopt to handle these two difficulties.

12.3.1 Previous work

We begin with a discussion of some earlier TSP column-generation systems, starting with the paper of Land [333].

In Land's study, the problem instances are described by specifying the geometric coordinates of the cities. Thus, she did not explicitly store the complete set of edges, relying instead on the ability to compute the cost between any prescribed pair of cities in order to carry out a pricing scan. This implicit treatment of the edge set raised two problems. First, since her external LP representation did not list the teeth in blossom inequalities, the reduced costs for these edges could not be computed from the representation. Second, edges in the core LP that were not in the optimal basis but were set to the upper bound of 1 could appear to have negative reduced costs in the pricing scan. To deal with these "spurious infeasibilities," she maintained a data structure (consisting of an ordered tree and an extra list for edges that did not fit into the tree) to record the teeth edges and the nonbasic 1-edges, and skipped over these pairs of cities during the pricing scan. Every negative edge that was detected in a scan was added to the core until a limit of 500 core edges was reached. After that, any further negative edges were swapped with core edges that were both nonbasic and set to 0.

The pricing scans in Land's code were carried out at fixed points in the solution process, where the code moved from one phase to another. These transition points were determined by the completion of subtour generation, the completion of blossom generation, the detection of an integral LP solution, and the production of an LP optimal value that was above the cost of the best computed tour.

The issue of spurious edges was avoided by Grötschel and Holland [232], using an original approach to edge generation. Before starting the LP portion of their TSP code, they passed the entire edge set through a preprocessor that eliminated the majority of the edges from further consideration. Working with this reduced set, they maintained the core edges and the noncore edges as disjoint lists and restricted the pricing scans to the noncore set. Their scans were carried out after the cutting-plane phase of the computation ended without finding a new cut. If additional edges were brought into the core LP, their code reentered the cutting-plane phase and continued to alternate between cutting and pricing until all noncore edges had nonnegative reduced cost.

The preprocessor used by Grötschel and Holland takes the output of a run of the Held and Karp [254] lower-bound procedure, computes the implicit reduced costs, and uses the resulting values to indicate edges that cannot be contained in any tour that is cheaper than the best tour computed up to that point. These edges could thus be discarded from further consideration. For the instances they studied, this process resulted in sufficiently many deletions to permit them to implement their explicit edge list approach without running into memory difficulties.

Padberg and Rinaldi [447] did not use preprocessing but obtained a similar effect by implementing an approach that permitted them to quickly eliminate a large portion of the edge set based on the LP reduced costs. At the start of their computation, they stored a complete list of the edge costs, ordered in such a way that they could use a formula (involving a square root computation) to obtain the two endpoints of an edge with a prescribed index. Their pricing scans were carried out over this list, and possible spurious infeasibilities were handled by checking that candidate edges having negative reduced cost were not among the current nonbasic core edges that have been set to their upper bound of 1. Each of the remaining negative-reduced-cost edges were added to the core set after the scan. Taking advantage of this simple rule, Padberg and Rinaldi stopped the computation of the reduced cost of an edge once it had been established that the value would indeed be negative.

Edge elimination comes into the Padberg and Rinaldi approach once a good LP lower bound for the problem instance has been obtained. Their process works its way through the complete set of edges, adding to a "reservoir" any edge that could not be eliminated based on its reduced cost and the value of the best computed tour. If the number of edges in the reservoir reaches a prescribed maximum, then the elimination process is broken off and the index k of the last checked edge in the complete list is recorded. From this point on, pricing scans can be carried out by working first through the edges in the reservoir and then through the edges in the complete list starting at index k. Further elimination rounds attempt to remove edges from the reservoir and increase the number of preprocessed edges by working from k and adding any noneliminated edges to the free places in the reservoir.

Padberg and Rinaldi also introduced the idea of permitting the status of the LP to determine when a pricing scan should be carried out, rather than using fixed points in the solution process. Their stated purpose is to try to keep the growth in the size of the core edge set under control. To achieve this, they wanted to avoid the bad pricing scans that can arise after a long sequence of cutting-plane additions. Their strategy was simple: carry out a pricing scan after every five LP computations.

Jünger et al. [300] adopt the Padberg-Rinaldi reservoir idea, but they also keep a precomputed "reserve" set of edges and do not price over the entire edge set until each edge in the reserve set has nonnegative reduced cost. When their initial set of core edges is built from the k-nearest neighbors, the reserve set consists of the $(k+5)$-nearest graph. The default value for k is 5, but Jünger et al. also carried out tests using other values.

Following the Padberg and Rinaldi approach, Jünger et al. carry out a price scan after every five LP computations. They remark that, with this setup, the time spent on pricing for the instances in their test set (which included TSPs with up to 783 cities) was between 1% and 2% of the total running time.

12.3.2 Underestimating the reduced costs

A pricing mechanism for larger problem instances must deal with two issues that
were not treated in the earlier studies. First, in each of the above approaches, the
entire edge set is scanned, edge by edge, at least once, and possibly several times.
This would be extremely time-consuming for instances having a large number of
cities. Second, although the Padberg and Rinaldi approach of carrying out a pricing
scan after every five LP computations is aimed at keeping the size of the core set of
edges under control, early on in the computation of larger instances, far too many
of the noncore edges will have negative reduced costs (if we begin with a modestly
sized initial core edge set) to be able to simply add all of these edges into the core
LP. This latter problem is a subtle issue in column generation and we have no good
remedy, using only a simple heuristic for selecting the edges (see Section 12.1.4
and Section 12.3.4). The first problem, on the other hand, can be dealt with quite
effectively using an estimation procedure that allows us to skip over large numbers
of edges during a pricing scan, as we now describe.

Suppose we would like to carry out a scan with a core LP specified by the cuts
$\mathcal{H}_1 = (V, \mathcal{F}_1), \mathcal{H}_2 = (V, \mathcal{F}_2), \ldots, \mathcal{H}_r = (V, \mathcal{F}_r)$. The dual solution provided
by the LP solver consists of a value y_v for each city v and a nonnegative value
Y_j for each cut $\mathcal{H}_j$. Let $\mathcal{S} = \mathcal{F}_1 \cup \cdots \cup \mathcal{F}_r$ be the collection of member sets
of the hypergraphs and, for each S in $\mathcal{S}$, let $\pi_j(S)$ denote the number of times S
is included in $\mathcal{F}_j$, for $j = 1, \ldots, r$. (Recall that the members of $\mathcal{F}_j$ need not be
distinct.) For each S in $\mathcal{S}$, let

$$Y_S = \sum(\pi_j(S)Y_j : j = 1, \ldots, r).$$

The reduced cost of an edge $e = \{u, v\}$ having cost c_e is given by the formula

$$\alpha_e = c_e - y_u - y_v - \sum(Y_S : e \cap S \neq \emptyset, e - S \neq \emptyset).$$

The computational expense in evaluating this expression comes from both the number of terms in the summation and the calculations needed to determine the sets S
that make up this sum. A quick estimate can be obtained by noting that each of
these sets S must contain either u or v. To use this we can compute

$$\bar{y}_v = y_v + \sum(Y_S : v \in S \text{ and } S \in \mathcal{S})$$

for each city v and let

$$\bar{\alpha}_e = c_e - \bar{y}_u - \bar{y}_v.$$

For most edges, $\bar{\alpha}_e$ will be a good approximation to α_e. Moreover, since $\bar{\alpha}_e$ is
never greater than α_e, we only need to compute α_e if the estimate $\bar{\alpha}_e$ is negative.

The simplicity of the $\bar{\alpha}_e$ computation makes it possible to work through the
entire set of edges of a TSP instance in a reasonable amount of time, even in the
case of the 500 billion edges that make up an instance having 10^6 cities. It still
requires us to examine each edge individually, however, and this would not work
for instances much larger than 10^6 cities. Moreover, the time needed to explicitly
pass through the entire edge set would restrict our ability to carry out multiple price
scans. What we need is a scheme that is able to skip over edges without computing

the $\bar{\alpha}_e$'s. Although we cannot do this in general, for geometric problem instances we can accomplish this by taking advantage of the form of the edge cost function when carrying out a pricing scan. The approach we take is similar to that used by Applegate and Cook [20] to solve matching problems.

Suppose we have coordinates (v_x, v_y) for each city v and that the cost $c_{\{u,v\}}$ of an edge $\{u, v\}$ satisfies

$$c_{\{u,v\}} \geq t|u_x - v_x| \tag{12.5}$$

for some positive constant t, independent of u and v.

For these instances, we have

$$\alpha_e \geq \bar{\alpha}_e \geq t|u_x - v_x| - \bar{y}_u - \bar{y}_v.$$

So we only need to consider pricing those edges $\{u, v\}$ such that

$$t|u_x - v_x| - \bar{y}_u - \bar{y}_v < 0. \tag{12.6}$$

This second level of approximation allows us to take advantage of the geometry. (Condition (12.6) holds for each "EDGE_WEIGHT_TYPE" in the TSPLIB, other than the EXPLICIT and SPECIAL categories.)

At the start of a pricing scan, we compute $tv_x - \bar{y}_v$ for each city v and build a linked list of the cities in nondecreasing order of these values. We then build a permutation of the cities to order them by nondecreasing value of v_x. With these two lists, we begin processing the cities in the permuted order. While processing city v, we implicitly consider all edges $\{u, v\}$ for cities u that have not yet been processed. The first step is to delete v from the linked list. Then, since $u_x \geq v_x$ for each of the remaining cities u, we can write the inequality (12.6) as

$$tu_x - \bar{y}_u < tv_x - \bar{y}_v. \tag{12.7}$$

We therefore start at the head of the linked list and consider the cities u in the linked list order until (12.7) is no longer satisfied, skipping over all of the cities further down in the order. For each of the u's that we consider, we first compute $\bar{\alpha}_{\{u,v\}}$ and then compute $\alpha_{\{u,v\}}$ only if this value is negative.

This cut-off procedure is quite effective in limiting the number of edges that need to be explicitly considered. For the cost functions that are supported by kd-trees (see Bentley [56]), however, it would be possible to squeeze even more performance out of the pricing routine by treating the $\bar{y}$ values as an extra coordinate in a kd-tree (as David S. Johnson (private ommunication) proposed in the context of the Held-Karp lower bound procedure) and replacing the traversal of the linked list by a nearest-neighbor search.

12.3.3 Batch pricing

Coming out of the approximate pricing, we have edges for which we must explicitly compute α_e. As we discussed in Section 12.2.2, extracting these reduced costs from the external LP representation is considerably more time-consuming than with the memory-intensive representation used by Padberg and Rinaldi [447]. It is therefore important, in our case, to make a careful implementation of the pricing scheme.

Jünger et al. [300] were faced with a similar problem. Their method is to build a pricing structure before the start of a pricing scan, and use this to speed up their computations. The structure consists of lists, for each city v, of the hypergraphs that contain v in one of their member sets. Working with the inside form of cuts, they compute the reduced cost of an edge $\{v, w\}$ by finding the intersection of the two hypergraph lists and working through the sets to extract the appropriate coefficients.

We also build a pricing structure, but the one we use is oriented around edges, rather than cities. To make this work, we price large groups of edges simultaneously, rather than edge by edge. This fits in naturally with the pricing scan mechanism that we have set up, since we can just hold off on computing the necessary α_e's until we have accumulated some prescribed number of edges (say 20, 000).

In our setup, to compute the reduced costs of a prescribed set, U, of edges, we can begin with the $\bar{\alpha}_e$ values that have already computed. To convert $\bar{\alpha}_e$ to α_e, we need to add $2 * Y_S$ for each set S in $\mathcal{S}$ that contains both ends of e. The structure we build to carry this out consists of an incidence list for each city v, containing the indices of the edges of U that are incident with v. For each set S in $\mathcal{S}$ having $Y_S > 0$, we run though each city v in S and carry out the following operation. We consider, in turn, each edge in v's incidence list and check whether the other end of the edge is marked. If it is indeed marked, then we add $2 * Y_S$ to the edge's $\bar{\alpha}_e$ value. Otherwise, we simply move on to the next edge. After all of the edges have been considered, we mark the city v and move on to the next city in S. The marking mechanism can be handled by initially setting a mark field to 0 for all cities, and, after each set is processed, incrementing the value of the integer label that we consider to be a "mark" in the next round.

After all sets have been processed, the $\bar{\alpha}_e$ values have been converted to α_e for each $e \in U$. We then attach each of the edges having negative α_e to a queue that holds the edges that are candidates for entering the core LP (see Subsection 12.1.4). Finally, the incidence structure is freed, and we wait for the next set of edges that need to be priced.

12.3.4 Cycling through the edges

Our pricing strategy does not require us to implicitly consider the entire edge set during each pricing scan, but only that a good sampling of the potentially bad edges be scanned. We therefore use the pricing algorithm in two modes, as in Jünger et al. [300]. In the first mode, we only consider the k-nearest edges for some integer k (say $k = 50$), and in the second mode we treat the full edge set. In both cases, the routine works its way through the edges, city by city, accumulating the set U of edges having $\bar{\alpha}_e < 0$ that will be priced exactly. Each time it is called, the search picks up from the last city that was previously considered, wrapping around to city 0 whenever the end of the city list is reached.

The approximate pricing algorithm is reset whenever the y_v's and Y_j's are updated. A reset involves the computation of the new Y_S's and $\bar{y}_i$'s, as well as the creation of a new linked list order. If successive calls to the algorithm, without a reset, allow us to process all of the edges, then we report this to the calling routine and carry out no further approximate pricing until the next reset occurs; this provides information that can be used to terminate a pricing scan.

12.3.5 Permitting negative edges

During our column-generation procedure we take advantage of the fact that if z^* is the objective value of the LP solution and p is the sum of the reduced costs of all edges having negative reduced cost, then $z^* + p$ is a lower bound for the TSP instance. This observation follows from LP duality, using the dual variables corresponding to the $x_e \leq 1$ constraints for each edge e.

In our computations, we terminate the column-generation process when $|p|$ falls below some fixed value (0.1 in our code). This small penalty in the lower bound is accepted in order to avoid the bad behavior that can arise as we try to complete the column generation while maintaining a small core LP.

12.4 THE MECHANICS

The machinery built around the LP solver makes up a large part of our TSP code. In this section we describe how we partition the work that needs to be done and how the parts communicate.

12.4.1 Division of labor

We have attempted to build the code in such a way that the major components are logical units that can run independently. Besides making the code modular, this partition can allow the code to take advantage of multiple processors when they are available.

The components are the LP solver, the cutting-plane routines, the cut pool, the pricing routines, and the tour-finding heuristics. These components are tied together by a controller program that determines the order in which the various routines are called and provides the interface between them. An overview of the code is depicted in Figure 12.5.

The engine is the LP solver for the core LPs. The inner workings of the solver are invisible to the rest of the code, but a set of interface routines is available to add and delete rows and columns and to extract solution information.

The cutting-plane routines generate the new cuts to be added to the core LP. The routines consist of a number of *cutters* that implement the separation algorithms described in the earlier chapters, together with a small control program that determines which of the cutters should be called. The decisions of this cut-control program are based on the history of the calls up to that point, including the incremental amount that the LP optimal value has increased after earlier rounds of cuts were added to the core LP.

Access to the cut pool is provided by a search routine that returns violated cuts (or, if requested, cuts that hold as an equation for the given x-vector), as well as a routine for adding new cuts to the pool. There is also a routine that returns tight

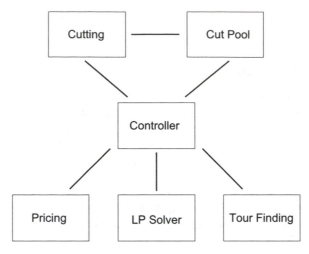

Figure 12.5 Components of the code.

(or nearly tight) sets, chosen from the edges of the hypergraphs that describe the inequalities. These sets are used by several of the cutters described in earlier chapters.

The pricing routines handle column generation by computing the reduced costs of edges not currently in the core LP and selecting some of these to possibly be added to the core, as we discussed in Section 12.3. This portion of the code also provides the edge-elimination mechanism to eliminate edges from the LP, that is, to conclude that the edges cannot be set to one in any solution to the LP that gives an objective value less than the cost of the best tour found thus far. It will also fix LP edges to one when possible. We discuss these elimination and fixing routines in Chapter 14.

The final component is the collection of tour-finding heuristics. These heuristics interact with the LP machinery in two ways. In one direction, the tours they find can be used to provide the upper bounds that permit edge elimination and edge and cut fixing. In the other direction, the current x-vector can be used to create an initial tour to start a local-search heuristic.

12.4.2 Communication

The division of the work space makes it necessary to set up a means to pass certain data back and fourth among the components. The crucial objects are the cutting planes and the LP primal and dual solutions; see Figure 12.6.

CUTS

The vehicle for passing the cutting planes found by the cutters over to the LP solver is a single queue of cuts that is maintained by the controller. When a cutter finds a new cut, it passes a pointer to the cut to a routine that will consider adding the cut to the existing queue; the cut is added if it is not already in the queue and if the

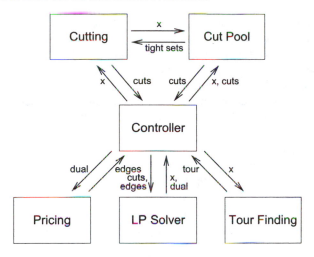

Figure 12.6 Communication.

queue has not exceeded some maximum size. The LP solver takes cuts from the queue when it is in a cut-processing mode.

EDGES

The edges found by the pricing routines are also communicated by means of a single queue maintained by the controller. The pricing routines add edges to the queue, and the LP solver takes edges from the queue when it is in an edge-processing mode.

x-VECTORS

All cutters take as their input a solution vector x that satisfies (12.1). In other words, they take a weighted graph where the sum of the edge weights over the edges meeting each vertex is two.

These x-vectors need to be fetched from the LP solver and passed over to the cutters. Since the cutters and LP solver are working independently, the controller periodically receives a new x-vector from the solver and keeps it on hand to pass to the cutters whenever cutting is called for. This means that if we are running in parallel, some cutters may be working with a vector that is out of date, in the sense that it is no longer a solution to the current core LP. This can be justified by the fact that the LP solver is usually the most time-intensive piece of the computation, which would make the solver a bottleneck if the cutters could not run on out-of-date information. Moreover, since the optimal solutions to the core LPs often do not change that much from step to step, the cuts that are generated from out-of-date vectors are nearly as useful as those generated from the current x-vector.

To help the cutters run efficiently, rather than passing the raw x-vector, the controller first eliminates edges having LP value 0 and performs shrinking steps to reduce the size of the graph. Since this process involves some computation, it is only carried out when the controller needs to call a cutter. Once it is carried out, the shrunk vector is stored and handed out directly to a cutter whenever further cutting is called for, until a new x-vector is received from the LP solver.

In this scheme, the cuts found by the cutters need to be converted from their shrunk form back into cuts for the original problem. To provide for this, the controller passes to the cutters not only the x-vector, but also the data needed to unshrink the vector. Namely, for each vertex v, the controller passes the list of original vertices that were shrunk to obtain v. The cutters use this information to unshrink the cuts before passing them over to the routine that adds them to the cut queue.

Similarly, the tour-finding heuristics can also work with the shrunk vector, using the extra information to convert the tours they find into tours for the original graph. Note, however, that we also permit the tour-finding routines to work with the raw x-vector, since the goals of the shrinking performed by these heuristics is different from that of the cutters.

Dual Solutions

The pricing component needs the LP dual solution in order to compute the reduced costs. Like the x-vector, the dual solution is fetched periodically from the LP solver and kept on hand to pass to the pricing routines when it is required. Since the vector may be out of date with the current LP, the dual solution must consist of representations of the cutting planes corresponding to the positive dual variables, together with the values on the variables. The representation of the cuts uses the external representation described in Section 12.2.2.

12.4.3 Controller

The control program for the LP machinery has the task of allocating the available computational resources between the components of the system. This engineering aspect of a TSP code has been approached in different ways in studies by Land [333], Grötschel and Holland [232], Padberg and Rinaldi [447], and Jünger et al. [300]. We provide below a high-level description of the Concorde controller; the design used in Concorde is based on experience with early versions of the code.

The goal of a controller is to provide a mechanism for maintaining a steady stream of useful cutting planes and edges to the LP solver. In this effort, it is important to interleave the cutting-plane and column-generation components, to avoid time-consuming work on LP relaxations that are not relevant to the overall computation. This is a problem that is observed, for example, when a great effort is spent on cutting-plane generation only to see the gains in the LP value disappear after an edge-pricing step.

The Concorde controller consists of an inner loop of cutters with edge pricing on the k-nearest edges (as described in Section 12.3.4), embedded in an outer loop with full edge pricing. The inner loop cutters are ordered by a rough estimate on the amount of time we expect them to take, starting with the cut-pool search and the subtour-inequality heuristics, and moving through the local-cut procedure.

The device used in Concorde to switch between cutting-plane generation and column generation is a simple evaluation function on the progress the solver is making toward moving the lower bound up to the value of the best tour we have found for the problem instance. The evaluation function sets a target for a cutting-plane phase as a fraction of the gap between a starting LP value and the cost of the best available tour; the target is set by default to 0.01 times the gap. If the target is met or exceeded, the cutting-plane phase in the inner loop is broken off, and we move to column generation. The idea behind the target value is that if the bound looks too good to be true, then it may well be due to an insufficient collection of edges in the core LP.

We also use a smaller fraction of the gap to determine if we should make an additional pass through the inner loop of cutters; this loop target is set to 0.001 times the gap. If the inner loop cutters fail to increase the LP value by at least this target value, then we move to the full-pricing phase in the outer loop.

The decision to terminate the outer loop itself is based on the change in the LP value that is created by column generation in the full-pricing phase. If the LP value decreases by more than the loop target, then we stay in the outer loop (making an additional pass through the inner loop cutters and the pricing routines); otherwise we break out of the loop and terminate the function. The idea here is that if the additional edges have significantly decreased the LP value, then there is a good chance that further cutting can recover some of the lost value.

Chapter Thirteen

The Linear Programming Solver

13.1 HISTORY

The computational histories of the TSP and linear programming (LP) are closely related. Central parts of the computational procedure of Dantzig et al. [151] depend in a fundamental way on our ability to solve LP problems, usually sequences of closely related LP problems. Conversely, developments of codes for LP have also been strongly influenced by the TSP. For example, the work described in this monograph has been the direct source of many of the important ideas behind the development of the callable library routines for the CPLEX commercial LP solver.

It is interesting at this point to note a statement from Dantzig et al. [151], referring to the explicit computational parts of their work: "... such a procedure has been relatively easy to carry through by hand." Since linear programming was a key part of this "procedure," these hand computations presumably included the solution of the associated LP problems, a conclusion confirmed by Dantzig (private communication, 1998). Of course, the associated LP problems were modest by today's standards; nevertheless, just imagine performing simplex iterations by hand on an LP problem with more than 50 constraints and 850 variables![1]

Though apparently manageable in the instance Dantzig, Fulkerson and Johnson considered, the absence of computers and computer codes for carrying out the LP computations necessary to apply their method obviously represented a major roadblock to the wider application of these ideas.

13.1.1 Computer codes for linear programming

The history of LP computation begins with Dantzig's introduction of the simplex method in 1947. The first paper on this method is an unpublished technical report Dantzig [149]. It is interesting to note that while there is some argument about the origins of the subject of LP as a whole, there seems to be no controversy concerning Dantzig's introduction of the simplex method.

The first computer implementation of the simplex method seems to have been developed at the National Bureau of Standards, the present-day National Institute of Standards and Technology, on the SEAC computer. Orden [439] and Hoffman

---

[1] In view of these numbers, one suspects that Dantzig, Fulkerson, and Johnson must have made early use of the idea of omitting some of the edges from the main part of the computation, checking the reduced costs of these omitted edges only later. These ideas were formalized much later; the first explicit mention we found was in Martin [378], who used the term "vertical truncation" to refer to leaving aside some edges and checking their reduced costs in a step external to the application of the LP solver.

et al. [266] report performing computational tests with this machine. One instance with 48 equations and 71 variables was solved in 18 hours and 73 simplex iterations.

William Orchard-Hays began his pioneering work on implementations of the simplex method in 1953–54 [438]. This work was the beginning of the development of comercially available LP codes. The computing machine used was an IBM "card programmable calculator," hardly a real computer by today's standards. In the words of Orchard-Hays, "The CPC was an ancient conglomeration of tabulating equipment, electro-mechanical storage devices, and an electronic calculator (with tubes and relays), long since forgotten. One did not program in a modern sense, but wired three patch-boards which became like masses of spaghetti." The first code implemented by Orchard-Hays used an explicit basis inverse, with the inverse freshly recomputed at each iteration. Again, in the words of Orchard-Hays, "One could have started an iteration, gone to lunch, and returned before it finished. ..." The initial results were not encouraging. However, in 1954, Dantzig recalled the idea of the product-form of the inverse, and this device led to a second, more efficient CPC implementation. Orchard-Hays [438] reports that the largest instance solved with this code had 26 constraints and 71 variables, and took "eight hours of hard work feeding decks to hoppers" to complete the solution.

As computers continued to get better, so did implementations of the simplex method. In 1956 a code named RLSL1 was implemented on an IBM 704, a machine with 4K of core storage, minuscule by today's standards. The maximum number of constraints was limited to 255, with an essentially "unlimited" number of variables. RSLP1 was distributed through the SHARE organization and was used by larger petroleum companies. In the period 1962–66 LP/90 was implemented for the IBM 7090, followed by LP/90/94 for the IBM 7094. The number of allowed constraints grew to 1,024. By that time, the use and application of LP had grown as well.

In 1966, IBM introduced a major advance in computing hardware, the family of IBM 360 computers. At about the same time, IBM commissioned the development of an LP system designed to run on these computers. That system became known as MPS/360. MPS/360 was followed by MPSX and later by MPSX/370. During that period, motivated by Kalan's [301] work on supersparsity, Ketron Corporation, with support from Exxon Corporation (Milt Guttermann, then at Exxon, was one of the principal proponents of this work), developed MPS III with Whizard. Tuned for the new, fast IBM platforms, these codes, most particularly MPS III, represented quantum leaps both in speed and problem size, accepting models with up to 32000 constraints. Other powerful systems were also developed during that period, including the UMPIRE system for the UNIVAC 1108, APEX III for CDC machines, and in the mid- to late 1970s LAMPS, written independently by John Forrest. With modest updates and machine improvements, these systems were the dominant LP computing environments into the late 1980s.

13.1.2 Linear programming and TSP

As we discussed in Chapter 4, from 1954 through 1972 there were scattered computational TSP studies, most involving rather small instances by today's standards.

The hybrid algorithm of Martin [378] used subtour inequalities in combination with Gomory cuts. Martin reported the solution of the 49-city instance of Dantzig, Fulkerson, and Johnson, using Gomory's algorithm three times, together with nine subtour inequalities. In his solution process he used a total of 353 simplex iterations. The total solution time on an IBM 7094 was around 5 minutes.[2] We note that since Martin did his work at C-E-I-R, and LP/90/94 was developed there during this period, this work may be the first application of a commercial LP package in the solution of TSPs; however, Martin makes no explicit comment in this regard.

The first work that actually used the Dantzig, Fulkerson, and Johnson approach with TSP-specific cutting planes was the thesis of Hong [270], an excellent though short dissertation that anticipated many of the ideas currently in use in TSP codes, including subtours, 2-matching constraints, and branching. For the solution of the LP problems, Hong did not use any existing code, choosing instead to implement his own "exact" simplex solver.

Miliotis [393], [394] published results using a method apparently similar to that of Martin, but with 10 years of computer hardware improvements behind him and a better implementation, the computational results were considerably better. As the LP solver he used the exact-arithmetic code of Land and Powell [335].[3] Later Land [333] added 2-matching constraints to the tool set (unaware of Hong's work) and managed to solve instances with up to 80 cities.

Padberg and Hong [451] used an IBM 370 and, once again, a home-grown, exact LP solver, implementing most of the ideas in Hong's thesis, but excluding branching. They did not use Gomory cuts. Without cuts and branching, there was no guarantee of an integral solution; their computations generally yielded only bounds for the optimal solution. Because of limitations in memory, for their largest instance, lin318, they were forced to reduce the number of edges to 30,938 from a total of 50,403 = 318(318-1)/2. Those edges not included were handled using reduced-cost arguments.

In 1980, papers started appearing that used the then state-of-the-art LP solvers. Crowder and Padberg [147] applied MPSX MIP, adding subtours to eliminate invalid integral solutions. They solved "10 large" TSPs, the largest of which was lin318. The reported solution time for lin318 was 6 minutes. Crowder (private communication, 1999) noted that writing a driver to add these subtours was not easy, requiring detailed knowledge of the workings of the MPSX system. In work published at about the same time, but taken from Grötschel's 1977 thesis, Grötschel [230], using MPSX as the LP solver, solved a 120-city instance. Cuts were found by hand and added interactively.

In 1991 two important papers appeared giving computational results for significantly larger instances. Both included important steps forward in developing the Dantzig, Fulkerson, and Johnson approach.

---

[2]By contrast, using Concorde, on a present-day machine, dantzig42 was solved with a total number of simplex iterations of 105, not hugely different from Martin's 353. The total time to solve the LP problems was too small to be accurately measurable.

[3]A private communication from Land gives some indication of why an exact LP solver was used: "It was very nice doing LP simplex calculations and knowing that the arithmetic was 100% accurate—no problem of setting tolerances."

Grötschel and Holland [232] used MPSX, solving a then very difficult "world tour" instance with 666 cities. Their paper is very interesting from an LP point of view. The authors note that "Some linear programs that arose were hard to solve, even for highly praised commercial codes like IBM's MPSX." Our later computational results will indicate that this phenomenon is certainly no longer present for instances of the size treated by Grötschel and Holland (for example, running Concorde with standard settings on gr666, the average time consumed per LP solve on the core LP problem was 0.1 seconds, and the total time spent on all LP-related computations was only 29% of total computation time); nevertheless, difficult LP problems still do arise. Indeed, difficult LP problems are the primary reason that TSPLIB instance d2103 remained unsolved much longer than many larger TSPLIB instances. Grötschel and Holland also note that if the LP package used had been "better suited for a row generation process than MPSX is, the total speed-up obtained by faster (cut) recognition procedures might be worth the higher programming effort." In our view, these experiences, and similar experiences apparent in other work on integer programming (see for example the paper of Crowder et al. [146]) were extremely important motivators in the LP developments of the 1990s.

The second important paper from 1991 was that of Padberg and Rinaldi [447]. It has become a standard reference in the subject of computation for the TSP. Several new instances were solved, including the then largest instance pr2392, and several smaller, though more difficult instances such as att532 and pr439. Their paper is also extremely important from an LP standpoint. In the initial part of their work, carried out in 1987, Padberg and Rinaldi made use of Marsten's [376] XMP code. XMP was built around the Harwell LA05 linear algebra routines and was apparently designed by Marsten for applications in integer programming. While this code did not represent a particularly strong implementation of the simplex method, it was adequate for many applications, and constituted a major step forward in terms of usability. Somewhat later, in 1989, Padberg and Rinaldi gained access to what was really the first LP code in the "new generation" of high-performance codes, succeeding MPS III and the other successful mainframe codes developed in the mid-1970s. The simplex routines for this new code, known as OSL (Optimization Subroutine Library), were written by John Forrest at IBM. The magnitude of improvement represented by OSL is illustrated by the following results taken from Padberg and Rinaldi [447]. Using XMP, gr666 was solved in 57,755 seconds on a VAX8700; on an IBM 3090/60 using OSL, with essentially the same TSP routines, gr666 was solved in 2,311 seconds. Similarly, gil249 solved in 1,292 seconds with XMP on a CYBER 205—a supercomputer in those days—while solving in 187 seconds with OSL.

One additional important event should be mentioned at this point. In 1984, the interior-point "revolution" was started with the publication of the paper of Karmarkar [306]. This work has had a profound impact on LP computation, and certainly served to help motivate advances in simplex implementations. At the same time, its effect on TSP computation seems to have been minor. This issue will be addressed more thoroughly later.

13.1.3 An example

We close this section with an example that illustrates some of the effect of improved LP solvers. It also serves as a brief introduction of the *dual simplex algorithm* with *steepest edge*, the algorithm of choice in almost all of our TSP computations. It will be one of our main goals to describe in detail various aspects of the implementation of this algorithm.

In the early 1990s there was considerable research as well as business interest in the LP community in solving some difficult LP instances arising from application in the airline industry. One particular such instance, with fairly wide distribution, was an early "fleet-assignment" model with the dimensions given in Table 13.1.

Table 13.1 A fleet-assignment model.

Constraints	4,420
Variables	6,711
Constraint nonzeros	101,377

An initial attempt to solve this model with CPLEX 1.0 on a Cray Y-MP failed miserably. Using a version of the primal simplex method, one incorporating no mechanism to deal with "stalling" (long sequences of pivots, with no change in the objective value—discussed later), after 7 full hours of computation, the code still had not found a primal-feasible solution. It had been stuck at the same objective value (measuring total infeasibility of the current basis) for a considerable time. The run was terminated at that point. Around the same time, Forrest and Tomlin [192] obtained the sequence of very impressive results given in Table 13.2 for that same model, with the last result produced in around 1992 using a "dual steepest

Table 13.2 A fleet-assignment model: Progress.

Code	CPU Seconds	Simplex Iterations	Machine
MPSX Version 1	31,020	302,357	IBM3090E
MPSX Version 2	4,980	48,858	IBM3090E
OSL Primal	1,500	36,050	IBM3090E
OSL Dual	660	11,410	IBM3090E
OSL Dual	450	4,827	RS6000/540

edge" algorithm.[4] By contrast, with a present-day version of CPLEX gives the results report in Table 13.3. We note that the above LP problem *is* relatively small by

---

[4]Based on J. J. H. Forrest and J. A. Tomlin (private communication, 1999), MPSX Version 1 refers to an older, unmodified version of MPSX/370, using presumably something like "partial pricing," Version 2 was a new, vectorized version using "devex." The first two OSL times refer to some version of devex pricing.

Table 13.3 A fleet-assignment model: Present-day code.

Code	Seconds	Iterations	Machine
CPLEX 7.1 Primal	349	48,301	800-MHz Pentium III
CPLEX 7.1 Primal	100	45,381	667-MHz Digital Alpha EV67
CPLEX 7.1 Dual	43	4,807	800-MHz Pentium III
CPLEX 7.1 Dual	13	4,900	667-MHz Digital Alpha EV67

today's LP standards. As such, it does not fully illustrate the kinds of improvements that have been recently achieved.

13.2 THE PRIMAL SIMPLEX ALGORITHM

Typical linear programming codes accept problems in the form

$$\text{Minimize } c^T x \tag{13.1}$$

$$\text{subject to}$$

$$Ax \asymp b$$

$$l \le x \le u$$

where $c \in \mathbf{R}^n$, $b \in \mathbf{R}^m$, $A \in \mathbf{R}^{m \times n}$, $l \in (\mathbf{R} \cup \{-\infty\})^m$, $u \in (\mathbf{R} \cup \{+\infty\})^m$, and $\asymp$ denotes that for each row a_i of A either $a_i x \le b_i$, $a_i x \ge b_i$, or $a_i x = b_i$. The term $c^T x$ is called the *objective function*, b the *right-hand side*, $Ax \asymp b$ the *constraints*, A the *constraint matrix*, l the vector of *lower bounds*, and u the vector of *upper bounds*. If for some i, $l_i = -\infty$ and $u_i = +\infty$, x_i is called a *free* variable. In the LP problems we encounter, typically $l_i = 0$ and $u_i = 1$ (for each i). Usually one of the first steps in any code is to internally append a nonnegative *slack variable* to each inequality in $Ax \asymp b$, yielding an equality system. It is sometimes also deemed necessary, in order to find a convenient, nonsingular matrix to serve as an initial basis, to introduce so-called *artificial* variables, variables x_j appearing in exactly one constraint and satisfying $l_j = u_j = 0$.

In the sections that follow we investigate certain aspects of the simplex algorithm in detail. We will, however, avoid certain issues not relevant to the TSP. For example, we will not investigate methods for constructing starting bases. In our typical computation, the given linear program has a readily available starting basis provided by the solution of the "previous" LP problem. For a description of how CPLEX constructs starting bases, see Bixby [62]. For more comprehensive references on the simplex method, see Chvátal [126] and Murtagh [415].

In most of our exposition we do not need the generality of (13.1); the following *standard form*

$$\text{Minimize } c^T x \tag{13.2}$$

$$\text{subject to}$$

$$Ax = b$$

$$x \geq 0$$

will suffice. Let B be an ordered set of m distinct indices $(B_1, \ldots, B_m)$ taken from $\{1, \ldots, n\}$. B is called a *basis header* and determines a *basis* for (13.2) if $\mathbf{B} = A_B$ is nonsingular. The variables x_B are the *basic variables*. N denotes the set of indices $\{1, \ldots, n\} \backslash B$ of the *nonbasic variables*. B is *primal feasible* if $\mathbf{B}^{-1}b \geq 0$. A generic iteration of the *standard primal simplex algorithm* for (13.2), sometimes called the *revised* simplex method, takes as input a primal feasible basis B, and vectors $X_B = \mathbf{B}^{-1}b$ and $D_N = c_N - A_N^T \mathbf{B}^{-T} c_B$, where $\mathbf{B}^{-T}$ is shorthand for the transpose of $\mathbf{B}^{-1}$. The steps of the iteration are the following:

Step 1. (*Pricing*) If $D_N \geq 0$, then B is optimal and we stop; otherwise, let $j = argmin\{D_k : k \in N\}$. The variable x_j is the *entering variable*.

Step 2. (*FTRAN*) Solve $\mathbf{B}y = A_j$.

Step 3. (*Ratio Test*) If $y \leq 0$, then (13.2) is unbounded and we stop; otherwise, let $i = argmin\{X_{B_k}/y_k : y_k > 0, k = 1, \ldots, m\}$. The variable x_{B_i} is the *leaving variable*.

Step 4. (*BTRAN*) Solve $\mathbf{B}^T z = e_i$, where $e_i \in \mathbf{R}^m$ is the ith unit vector.

Step 5. (*Update*) Compute $\alpha_N = -A_N^T z$. Set $B_i = j$. Update X_B (using y) and D_N (using α_N).

Step 1 describes what is sometimes called *full pricing*. This pricing step will be discussed in great detail later, but suffice it to say that to obtain a "valid" choice for entering variable, we could equally choose any x_j with $D_j < 0$. This idea leads to notions such as partial pricing, discussed later, and implies changes in Step 5, where it is then only necessary to compute the values for "some" components of α_N.

The first major topic we discuss is "degeneracy." Before embarking on that discussion, we begin with a few basic calculations related to the basis matrix $\mathbf{B}$. In the notation above, suppose that we have completed one full iteration of the primal simplex algorithm, with j denoting the index of the entering variable and i the index into B of the leaving variable. Let $\bar{B}$ denote the new basis header after B_i is replaced by j; thus, $\bar{B}_i = j$. Using the vector y computed in Step 2, it is easy to see that

$$\bar{\mathbf{B}} = \mathbf{B}(I + (y - e_i)e_i^T).$$

It follows that

$$\bar{\mathbf{B}}^{-1} = (I - \frac{1}{y_i}(y - e_i)e_i^T)\mathbf{B}^{-1}. \tag{13.3}$$

The element y_i is called the *pivot element*. Denoting the first matrix in the expression on the right-hand side by E^{-1} we have

$$\bar{\mathbf{B}}^{-1} = E^{-1}\mathbf{B}^{-1}.$$

Matrices of the form E are called *eta* matrices. If a sequence of k iterations is carried out, and we denote the corresponding eta matrices by $E_1, \ldots, E_k$, then the inverse of the basis matrix after these k iterations is given by

$$E_k^{-1} \cdots E_1^{-1} \mathbf{B}^{-1}.$$

This expression provides an *update* formula for the basis matrix, called the *product-form update*. In early implementations of the simplex method, product-form updates were the update of choice, and the indicated matrices were stored sequentially, first on punch cards and later on more advanced storage devices. A simple examination of Steps 2 and 4 reveals a convenient property of this product-form update, particularly so in the context of accessing slow, sequential storage devices; in solving the linear system in Step 2, we apply the terms of the update in the order $\mathbf{B}, E_1, \ldots, E_k$, and for the solve in Step 4, in exactly the reverse order. Hence, the term FTRAN, which stands for *forward transformation*, and the term BTRAN, which stands for *backward transformation*. Of course, modern codes no longer compute $\mathbf{B}^{-1}$, using instead an LU-factorization of $\mathbf{B}$ in its place. CPLEX computes this factorization using an implementation based on Suhl and Suhl [514]. The update used is a variant of the well-known Forrest-Tomlin [191] update (see Chapter 24 of Chvátal [126] for a description of this update). As the number of terms in the update grows, it eventually pays to compute a fresh LU-factorization. In CPLEX the interval between refactorizations is determined dynamically.

13.2.1 Primal degeneracy

A basic solution of (13.1) is called *(primal) degenerate* if $X_{B_i} = l_i$ or $X_{B_i} = u_i$ for one or more values of i. For a problem in standard form (13.2), this condition means $X_{B_i} = 0$ for some i.

Let us consider how a change in basis occurs. Suppose x_j has been selected as the entering variable, where $j \in N$, and let Θ be the new value of x_j, keeping $x_{N \setminus \{j\}} = 0$. The implied values of the basic variables are

$$x_B^\Theta = \mathbf{B}^{-1}(b - \Theta A_j) = X_B - \Theta y. \tag{13.4}$$

Letting w^Θ denote the value of the objective function as a function of Θ, we have

$$
\begin{aligned}
w^\Theta &= c^T x^\Theta \\
&= c_B^T x_B^\Theta + c_j \Theta \\
&= c_B^T X_B - \Theta c_B^T y + c_j \Theta.
\end{aligned}
$$

Note that

$$c_j - c_B^T y = c_j - c_B^T \mathbf{B}^{-1} A_j = D_j,$$

which implies

$$w^\Theta = z + \Theta D_j.$$

Thus, given $D_j < 0$, the larger we can make Θ, the better.

In the ratio test we determine the maximum possible value of Θ, subject to maintaining the nonnegativity of the basic variables. The formula for this maximum value, based on (13.4), is

$$\Theta_{\max} = min\{\frac{X_{B_k}}{y_k} : y_k > 0\}.$$

The problem with degeneracy is that it can cause $\Theta_{max} = 0$, and hence no progress in the value of the objective function at the given iteration.

As we shall see, the effects of degeneracy in some TSP LP problems can be significant. We have implemented several features to deal with degeneracy. To describe the first of these features, let ϵ_f be a small positive constant. The subscript "f" here stands for *feasibility*. In CPLEX, the default value of ϵ_f is 10^{-6}. Using ϵ_f, we replace the feasibility condition in the definition of a feasible basis by the numerical condition

$$X_B \geq -\epsilon_f e,$$

where e denotes the vector $e = (1, \ldots, 1)^T \in \mathbf{R}^m$. Why do we need such a condition?

In reality, linear programs typically include some degree of degeneracy; often they are highly degenerate. In the context of a real implementation using floating-point arithmetic, this fact means that at least some of the basic variables "want" to take on the value zero, and hence, computed numerically, will have an actual value that is near zero, in some cases slightly positive, in others slightly negative. Thus, if we insist on the condition $X_B \geq 0$, then we will force some bases that *are* feasible to appear infeasible. Beyond this fact, there are many other possible sources of error in real LP problems. For example, it is often true that the data specifying the LP problem have been rounded to some fixed number of digits, or, even without rounding, are only known to be accurate to a very small number of significant digits. In addition, as the simplex method runs, there are various linear systems to be solved, such as FTRAN and BTRAN, and there is no direct way to control the residuals from these solves.

At this point we make a general remark on "tolerances," of which ϵ_f is one example. It will come as no surprise that any real implementation of the simplex method involves many such tolerances. We do not propose to discuss each in detail. To do so would be a considerable diversion from the main theme. Thus, in some situations we will suppress the mention of a tolerance that is, strictly speaking, necessary to the discussion. One such instance occurs below, where we write $y_k > 0$ without specifying numerically what strictly positive means.

Returning to our discussion of the ratio test, we come to our first novel idea in the implementation of the primal simplex algorithm, due to Harris [250]: We exploit ϵ_f to introduce extra freedom in the selection of the leaving variable. In place of the computation in Step 3, we use the following two-step approach. Let

$$\bar{\Theta}_{max} = min\{\frac{X_{B_k} + \epsilon_f}{y_k} : y_k > 0\}$$

and define the index of the leaving variable by

$$i = argmax\{y_k : \frac{X_{B_k}}{y_k} \leq \bar{\Theta}_{max}, y_k > 0\}.$$

The idea is to replace the maximum step size Θ_{max} by a somewhat larger value $\bar{\Theta}_{max}$, thus increasing the set of possible choices for the leaving variable. Then, instead of simply picking any leaving variable that maximizes the step size, we choose a variable that respects this new, larger maximum step size and maximizes the size of the pivot element. Since the reciprocal of the pivot element appears as a multiplicative term in the computation of the basis inverse, keeping it as large as possible can be expected to yield generally smaller numbers in the inverse, and hence more stable numerical computations. Moreover, with this two-pass ratio test, the actual value of the step size will often deviate from the value Θ_{max}, thus leading to basic variables with values that differ from zero by a significant enough amount to make strictly positive step sizes more likely, inhibiting the effects of degeneracy.

To justify the above rule, note that when a variable leaves the basis, we do set it to zero. Thus, the true step size *will* be X_{B_i}/y_i. Hence, by the computation of $\bar{\Theta}_{max}$, for any $y_k > 0$, $k \neq i$, we have

$$\bar{X}_{B_k} = X_{B_k} - \frac{X_{B_i}}{y_i} y_k$$
$$\geq X_{B_k} - \frac{X_{B_k} + \epsilon_f}{y_k} y_k$$
$$= -\epsilon_f.$$

The observant reader will have noted at this point that $X_{B_i} < 0$ is possible; indeed, this new, modified ratio test seems to promote just that. What is the effect? First, the actual step size is then negative, $X_{B_i}/y_i < 0$; instead of decreasing the objective, as desired, it will increase it. Second, and perhaps worse, basic variables B_k with $y_k < 0$, previously ignored, then become relevant in the ratio test. The result is that after such a step, some basic variables may violate their bounds by an amount greater than ϵ_f. Of course, one can imagine explicitly trying a step in the opposite direction, but that also may not help. In earlier versions of CPLEX, this problem was dealt with by rejecting the current choice for entering variable, and trying, in the range from 5 to 35 other choices for entering variable, the number of attempted alternatives depending on m. If, after this additional work, no acceptable variable was found, a negative step was accepted, and measures were taken to remove any resulting infeasibility. It was not until the introduction of CPLEX 2.1 that a cleaner, more stable solution was found to this problem of negative steps. We return to that issue shortly.

Even with the measures just described, it may still happen that the simplex method *stalls*, that is, experiences long sequences of iterations in which there is "no progress" in the objective function. For an explicit instance taken from our TSP computations, see Section 13.4.

To take measures against stalling, we first introduce a notion of *progress*. In our approach we do not simply look at individual pivots and determine whether they are nondegenerate, that is, determine whether they yield "progress" in the objective function. One reason is that negative steps *are* possible, in spite of all attempts to the contrary.

Our criteria for progress are the following. If an artificial variable leaves the basis or a free variables enters, that is designated as progress. Otherwise, where N_p is a given positive integer, if no progress has occurred in N_p iterations, we compute the value of the objective function and compare it to a value for the objective recorded at or before the last iteration at which progress occurred. If the objective has not improved, then we declare that no progress has occurred, and introduce a *perturbation*, as described below; otherwise, progress is declared and the new value of the objective is recorded. The integer N_p is set dynamically by CPLEX depending on model size; it can also be set by the user.

To describe how a perturbation is introduced, for each $i = 1, \ldots, m$, let Z_i be randomly sampled from the uniform distribution on $[0, 1]$. Let ϵ_p be a small positive tolerance. In versions prior to CPLEX 6.0, the default value of ϵ_p was 10^{-4}. It was subsequently changed to 10^{-6}. The perturbation takes the following form. For each basic variable, replace its lower bound by

$$x_{B_i} \geq -\epsilon_p Z_i.$$

In subsequent iterations, if a variable enters the basis that has not been perturbed, we perturb its lower bound as indicated above and continue. Note that this perturbation preserves primal feasibility. In earlier versions of CPLEX (3.0 and earlier), all variables were perturbed simultaneously, typically resulting in the loss of primal feasibility.

Now let us assume that a perturbation has been introduced. At termination, since the bounds have changed, the LP problem that has actually been solved is different from the original LP problem. The original bounds must be restored. There are then several possibilities; we discuss the most common situation, when optimality has been proved. After restoring the original bounds, we recompute X_B. If $X_B \geq -\epsilon_f e$, clearly $D_N \geq 0$, and we are done; the basis B is optimal for the original LP problem. On the other hand, if feasibility is violated, we simply continue applying the simplex method, replacing ϵ_p by $\epsilon_p/10$. Typically the latter step is unnecessary, but instances have been observed where this step is repeated several times, resulting in reductions of ϵ_p to values as low as 10^{-13}. To our knowledge, this procedure has never failed to converge in practice.

We remark that after removing a perturbation, even though the basis may become primal infeasible, it will be "dual feasible," so a reasonable alternative to the approach described above would be to invoke the dual simplex algorithm—see the next section. This approach is used by the public domain code SoPlex (Wunderling [560]).

Now, returning to the ratio test, a solution to our nonmonotonicity problem is at hand. Since we are already considering changing bounds by perturbing, the instrumentation is in place to consider dealing with altered bounds in general. To deal with the nonmonotonicity issue, we introduce what can be viewed as a restricted kind of perturbation, called *bound shifting*. If a selected leaving variable B_i has $X_{B_i} < 0$, we replace the bound on X_{B_i} by $min\{-10\epsilon_f, X_{B_i}\}$ and recompute the leaving variable.

Unlike perturbation, we must be willing to shift a variable more than once, and it does seem possible that this could lead to nonconvergence. We have no proof that this cannot happen, though again, this phenomenon has never been observed in practice.

One measure that does seem to be dictated by shifting is the introduction of an *inner feasibility tolerance* ϵ'_f. Initially $\epsilon'_f = \epsilon_f$. If at termination a shift and/or perturbation is removed, and an infeasibility is introduced, we set $\epsilon'_f = \epsilon'_f/10$, in a manner similar to the manner in which the perturbation constant is reduced. Again, we have not seen an instance in which $\epsilon'_f < 10^{-13}$ was necessary to achieve convergence.

There is one more subtlety to be considered. When a perturbation is introduced, it is important that the feasibility tolerance that is applied is small relative to the size of the perturbation; otherwise, there are quantities of the same order of magnitude, both being used to adjust variable bounds, one of which is to be thought of as significant (the amount of the perturbation), and the other of which is not. This situation is dealt with by instrumenting the following condition:

$$\epsilon'_f \leq min\{\epsilon_f, \epsilon_p/100\}.$$

13.3 THE DUAL SIMPLEX ALGORITHM

The primal simplex algorithm, the topic of the previous section, is the standard starting point for discussions of the simplex method. It is also the natural algorithm for solving sequences of linear programs obtained by adding and deleting columns, keeping the set of constraints constant. Indeed, assuming that (1) no basic columns are deleted, and (2) no nonbasic columns set to nonzero values are deleted, the optimal basis for one linear program will naturally give a primal feasible starting basis for the subsequent linear program. However, for the TSP, and for cutting-plane methods in general, we must also consider the addition and deletion of constraints, keeping the set of columns constant. Assuming that the added and deleted constraints are all inequality constraints, and, for the deleted constraints, that their corresponding slacks are basic, we obtain a natural, "dual feasible" starting basis for the subsequent optimization, as follows: The slacks for the deleted constraints are deleted from the basis, and for the appended inequality constraints, the associated slacks are added to the basis. The appropriate algorithm for this situation is the dual simplex method, introduced by Lemke [344]

Consider a linear program of the form (13.1), and designate this problem the *primal*. The corresponding *dual* is

$$\text{Maximize } b^T \pi \qquad\qquad (13.5)$$
$$\text{subject to}$$
$$A^T \pi \leq c$$
$$\pi \quad \text{free.}$$

Assuming that x is feasible for the primal and π is feasible for the dual, we have, by a standard calculation,

$$c^T x \geq \pi^T A x = \pi^T b.$$

Hence, if $c^T x = \pi^T b$, x is optimal for the primal and π is optimal for the dual, a result commonly known as the *Weak Duality Theorem*. Note that for any x and π producing equality in the above expression, we must have

$$(c^T - \pi^T A)x = 0.$$

For feasible x and π, this can only happen if

$$x_j > 0 \quad \text{implies} \quad c_j - \pi^T A_j = 0.$$

Vectors x and π that satisfy the above conditions are called *complementary*.

Now adding explicit slacks to the dual problem, we obtain the following expanded dual problem:

$$\text{Maximize } b^T \pi \qquad\qquad (13.6)$$

$$\text{subject to}$$

$$A^T \pi + d = c$$

$$\pi \quad \text{free}, d \geq 0.$$

Note that (A^T, I), where I is an identity matrix of appropriate order, is the constraint matrix of this expanded dual. Note also that complementarity of x and π can now be expressed as $x_j > 0$ implies $d_j = 0$.

Let B be a basis of the primal. There is a naturally associated complementary basis for the dual. Since, for the given primal basis B, x_N are the only variables that are guaranteed to be 0, complementarity demands $d_B = 0$. Designating the variables d_B as dual nonbasic forces $d_B = 0$, and implies that all remaining variables π and d_N are basic. To see this, note that the expanded dual has n constraints, $m + n$ variables, and $|B| = m$. Denoting the corresponding (dual) basis matrix by $\mathbf{H}$, with basis header H, and after a possible rearrangement of the rows of $\mathbf{H}$, we have

$$\mathbf{H} = \begin{pmatrix} \mathbf{B}^T & 0 \\ A_N^T & I \end{pmatrix}.$$

Clearly $\mathbf{H}$ is nonsingular if and only if $\mathbf{B}$ is nonsingular. Hence, H is a valid dual basis, given that B is a valid primal basis. Note also that for any basis of the form H, applying the simplex method does not change its form. In particular, since the π variables are free, none can be limiting variables in the ratio test, and, hence, none can be selected as a leaving variable. Thus, any subsequent basis will continue to contain $n - m$ variables from d, and the rows of A^T not covered by these slacks must then necessarily specify an $m \times m$ nonsingular submatrix of A, yielding a basis for the primal.

For the basis H, it is easy to see that we can compute the values of the dual basic variables as follows. First, solving for π yields

$$\Pi = \mathbf{B}^{-T} c_B$$

and then substituting these values we obtain

$$D_N = c_N - A_N^T \Pi.$$

Note that these two quantities are also computed in the course of executing the primal simplex method; moreover, the termination criterion for optimality in the primal algorithm is $D_N \geq 0$. That is, if termination occurs in Step 1 of the primal algorithm, the basis B is primal feasible, and its corresponding dual basis H is dual feasible. Since B and H are complementary, it follows that both B and H are optimal. In other words, we have a proof of the correctness of one of the termination criteria in the primal simplex algorithm.

The above observations are the foundation of the dual simplex method. The idea is to assume a primal basis B is given, and that this basis is dual feasible, meaning $D_N \geq 0$. We then perform an iteration of the primal simplex method applied to H, yielding a new B. The computations are carried out on the primal representation. The details follow.

Take a nonbasic variable d_{B_i} in the dual, and consider increasing it from its current value of 0 to some new value $\Theta \geq 0$, leaving $d_{B \setminus \{B_i\}} = 0$. Letting π^Θ and d_N^Θ denote the values of the dual basic variables as a function of Θ, it is easy to see that

$$\pi^\Theta = \Pi - \Theta z,$$

where $z = \mathbf{B}^{-T} e_i$, and that

$$d_N^\Theta = D_N + \Theta A_N^T z.$$

The value of the dual objective for this new dual solution is given by

$$b^T \pi^\Theta = b^T \Pi - \Theta b^T z$$
$$= b^T \Pi - \Theta b^T \mathbf{B}^{-T} e_i$$
$$= b^T \Pi - \Theta X_{B_i}.$$

We're maximizing, so $X_{B_i} < 0$ is good, and if $X_B \geq 0$, as we have seen, the given basis is optimal (since it is primal and dual feasible). These observations lead to a rule for selecting the entering variable d_{B_i} in the dual:

$$i = argmin\{X_{B_k} : X_{B_k} < 0\}.$$

To determine the leaving variable we maximize Θ, subject to $d_N \geq 0$. Letting $\alpha_N = -A_N^T z$, we can write

$$d_N^\Theta = D_N - \Theta \alpha_N \geq 0,$$

from which we obtain the choice for the leaving variable

$$j = argmin\{\frac{D_k}{\alpha_k} : k \in N, \alpha_k > 0\}.$$

If $\alpha_N \leq 0$, the dual is unbounded and the primal infeasible, as is easy to show.

We may now state the *standard dual simplex algorithm*. A generic iteration for (13.6) takes as input a dual feasible basis B and vectors $D_N = c_N - A_N^T \mathbf{B}^{-T} c_B$ and $X_B = \mathbf{B}^{-1} b$, and runs through the following steps.

Step 1. *(Pricing)* If $X_B \geq 0$, then B is optimal and we stop; otherwise, let $i = argmin\{X_{B_k} : k = 1, \ldots, m\}$. The variable d_{B_i} is the *entering variable*.

Step 2. *(BTRAN)* Solve $\mathbf{B}^T z = e_i$. Compute $\alpha_N = -A_N^T z$.

Step 3. *(Ratio Test)* If $\alpha_N \leq 0$, then (13.5) is infeasible and we stop; otherwise, let $j = argmin\{D_k / \alpha_k : \alpha_k > 0, k \in N\}$. The variable d_j is the *leaving variable*.

Step 4. *(FTRAN)* Solve $\mathbf{B}y = A_j$.

Step 5. Set $B_i = j$. Update X_B (using y) and D_N (using z).

We describe two additional, important features of the CPLEX dual simplex implementation. The first is the treatment of "phase I": finding a dual feasible basis from a given, infeasible dual basis. This part of the algorithm will find application when dealing with column additions and deletions. Normally one would expect the primal simplex algorithm to be a more natural choice in this circumstance, but we will see later that for the LPs that arise in our TSP computations, the dual is a better choice.

A basic solution of (13.6) is called *(dual) feasible* if $D_N \geq 0$ ($D_B = 0$ by definition). If some of the components of D_N are negative, one approach to finding a feasible basis is to consider the problem

$$\text{Minimize } \sum_{d_j < 0} -d_j \tag{13.7}$$

$$\text{subject to}$$

$$A^T \pi + d = c$$

$$\pi \text{ and } d \quad \text{free.}$$

The objective function in (13.7) is piecewise linear. "Locally" it can be thought of as linear. For a given basis solution, and for each j, let

$$\kappa_j = \begin{cases} -1 & \text{if } D_j < 0, \text{ and} \\ 0 & \text{otherwise.} \end{cases}$$

The corresponding local problem is then

$$\text{Minimize } \kappa^T d \tag{13.8}$$

$$\text{subject to}$$

$$A^T \pi + d = c$$

$$\pi \quad \text{free}, d_I \leq 0, \text{ and } d_F \geq$$

where $I = \{j : D_j < 0\}$ and $F = \{j : D_j \geq 0\}$.

To apply a simplex algorithm to this problem, we first solve

$$x^T \mathbf{H} = \kappa^T,$$

which yields $X_N = \kappa_N$ and $X_B^T = -\kappa_N^T A_N^T \mathbf{B}^{-T}$. Clearly the reduced costs of the nonbasic variables, d_B, are simply $-X_B^T$. Suppose $X_B \leq 0$. Since $X_N = \kappa_N \leq 0$, we have $X \leq 0$. The dual constraints, leaving out the slacks, state that $A^T \pi \leq c$. Using $X \leq 0$, we obtain

$$X^T A^T \pi \geq X^T c.$$

But $X^T A^T = 0$, and

$$X^T c = X_B^T c + X_N^T c$$
$$= X_N^T (c_N - A_N^T \mathbf{B}^{-T} c_B)$$
$$= \kappa_N^T D_N$$
$$= -\sum_{D_j < 0} D_j > 0$$

proving that (13.5) has no feasible solution.

In the alternative case, suppose that $X_{B_i} > 0$ for some i and consider setting $d_{B_i} = \Theta \geq 0$, keeping $d_{B \setminus \{B_i\}} = 0$. Solving $\mathbf{B}^T pi^\Theta = c_B - \Theta e_i$ yields

$$pi^\Theta = B^{-T}(c_B - \Theta e_i) = \Pi - \Theta z,$$

where $\mathbf{B}^T z = e_i$, which implies

$$d^\Theta = c - A^T \Pi^\Theta$$
$$= c - A^T \Pi + \Theta A^T z$$
$$= D + \Theta A^T z.$$

Now computing the corresponding objective value objective, we have

$$\kappa^T d^\Theta = \kappa^T D + \Theta \kappa^T A^T z$$
$$= \kappa^T D - \Theta X_{B_i}.$$

Since $X_{B_i} > 0$, this quantity decreases as Θ increases. Thus, we have the ingredients for a simplex algorithm. In fact, since our real objective is to solve (13.7), rather than (13.8), we can do a little more.

Let us consider the usual ratio test for computing the step length Θ for the problem (13.8). Where $\alpha_N = -A_N^T z$, we can write $d_N^\Theta = d_N - \Theta \alpha_N$. Let

$$\Theta_1 = min\{D_j / \alpha_j : d_j \geq 0, \alpha_j > 0, j \in N\}$$
$$\Theta_2 = min\{D_j / \alpha_j : d_j < 0, \alpha_j < 0, j \in N\}$$
$$\Theta_{max} = min\{\Theta_1, \Theta_2\}.$$

Then Θ_{max} is the maximum value for d_{B_i} that does not violate the bounds of (13.8) on the variables d. However, as noted, (13.8) is not the real problem of interest. It is (13.7). We can certainly consider violating these bounds if the net effect is a further reduction in the objective of (13.7). To this end, define

$$N_1 = \{j : d_j / \alpha_j = \Theta_1, d_j \geq 0, \alpha_j > 0, j \in N\}$$
$$N_2 = \{j : d_j / \alpha_j = \Theta_2, d_j < 0, \alpha_j < 0, j \in N\}.$$

Then making Θ larger than Θ_{max} has the effect in (13.8) of changing κ_j to -1 from 0 for $j \in N_1$, and κ_j to 0 from -1 for $j \in N_2$. Given these changes, will it continue to reduce the total infeasibility if we further increase Θ? To determine that, we simply need to compute the updated values of x_B. Denoting these by $\bar{X}_B$, we have

$$\bar{X}_B = X_B + \sum_{j \in N_1} \mathbf{B}^{-1} A_j - \sum_{j \in N_2} \mathbf{B}^{-1} A_j,$$

which implies

$$\bar{X}_{B_i} = X_{B_i} + \sum_{j \in N_1} z^T A_j - \sum_{j \in N_2} z^T A_j$$
$$= X_{B_i} - \sum_{j \in N_1} \alpha_j + \sum_{j \in N_2} \alpha_j$$
$$= X_{B_i} + \sum_{j \in N_1 \cup N_2} |\alpha_j|.$$

In other words, we simply add to X_{B_i} the absolute values of the quantities α_j for those d_j that change sign if Θ exceeds Θ_{max}. Since the vector α_N has already necessarily been computed in order to update d_N, the computation is cheap. Note also, as a for instance, that if we had simply taken Θ_{max} as the step size, then some d_j ($j \in N_1 \cup N_2$) would have been selected as the leaving basic variable, and it is a simple calculation to see that α_j would have been the pivot element (see earlier discussion). Thus, if all $|alpha_j|$ ($j \in N_1 \cup N_2$) are small, then it would appear that all available choices of pivots are numerically undesirable. However, it is in precisely this case that the above update of X_{B_i} is more unlikely to change its sign.

Clearly the above procedure can be repeated until X_{B_i} is no longer positive. That is, after having completed the the update of X_{B_i}, assuming it stays positive, we can change the bounds on the relevant d_j, compute a new, enlarged Θ_{max}, compute the next update of X_{B_i}, and so on.

A similar procedure is applied in the primal.

We close this discussion by interpreting degeneracy, the Harris ratio test, perturbation, and bound shifting for the dual algorithm.

Degeneracy: A basis B is said to be *dual degenerate* if $D_j = 0$ for some $j \in N$. To deal with such degeneracies, let ϵ_o (o stands for "optimality") be a given positive tolerance. The default value for ϵ_o in CPLEX is 10^{-6}. We replace the dual feasibility condition $D_N \geq 0$ by $D_N \geq -\epsilon_o e$.

Harris Ratio Test: CPLEX uses a two-phase ratio test in the dual, exploiting ϵ_o in the same way that we exploit ϵ_f in the primal. Let

$$\bar{\Theta}_{max} = min\{\frac{D_k + \epsilon_o}{\alpha_k} : \alpha_k > 0, k \in N\}.$$

The leaving variable is selected as

$$j = argmax\{\alpha_k : \frac{D_k}{\alpha_k} \leq \bar{\Theta}_{max}, \alpha_k > 0\}.$$

Note that, where $By = A_j$, we have

$$y_i = e_i^T y = e_i^T \mathbf{B}^{-1} A_j = z^T A_j = -\alpha_j.$$

Thus, the choice of j above maximizes the size of the pivot element, as in the primal.

Perturbation: Where Z_i, $i \in N$, are randomly sampled from the uniform distribution on $[0, 1]$, and $\epsilon_p > 0$ is a given positive constant (the same one as used in the primal), replace the bound $d_k \geq 0$ by

$$d_k \geq -\epsilon_p Z_k = \delta_k.$$

In the CPLEX implementation, it is implicitly assumed that all reduced costs are nonnegative, so the above bound changes are effected by actually changing the corresponding objective function coefficient:

$$c_k \longleftarrow c_k + \delta_k.$$

Applying this transformation to c_N clearly preserves dual feasibility. In subsequent iterations, when a previously unperturbed variable leaves the basis B, its objective function coefficient is also perturbed.

Shifting: To avoid negative steps in the dual, we consider "shifting" the bounds of the dual slacks, where necessary. As in the case of perturbation, shifts are effected by appropriately increasing the corresponding objective coefficient.

13.4 COMPUTATIONAL RESULTS: THE LP TEST SETS

In our computational results, we make use of various instances of LP problems, both general LP problems and those arising from TSP computations.

The standard test set for work in LP is the *Netlib* set. Since these LP problems provide a convenient reference point, we do make use of them, even though they do not have a direct connection to the TSP. Here we will view the Netlib set as made of two distinct parts, one of which we actually designate as the Netlib set, made up of 94 models, and a second of which contains the 16 models often designated as the "Kennington" models. Table 13.4 gives detailed statistics for the Kennington models. See *http://netlib.lucent.com/netlib/lp/data/index.html.* for details on the full set of Netlib models.

Table 13.4 The Kennington Netlib models.

Name	No. of Constraints	No. of Variables	No. of Nonzeros
cre_a	3,516	4,067	14,987
cre_b	9,648	72,447	256,095
cre_c	3,068	3,678	13,244
cre_d	8,926	69,980	242,646
ken07	2,426	3,602	8,404
ken11	14,694	21,349	49,058
ken13	28,632	42,659	97,246
ken18a	105,127	154,699	358,171
osa007	1,118	23,949	143,694
osa014	2,337	52,460	314,760
osa030	4,350	100,024	600,144
osa060	10,280	232,966	1,397,796
pds-02	2,953	7,535	16,390
pds-06	9,881	28,655	62,524
pds-10	16,558	48,763	106,436
pds-20	33,874	105,728	230,200

For some of the testing we do, we find the above set of general (non-TSP) LP problems too restrictive, and perhaps not representative of the range of difficulties that occur in real-world LP problems. Thus, we make use of several proprietary test sets that have been collected over a period of more than 10 years. We designate these test sets as *P_10*, *P_100*, and *P_1000*. From among a total collection of over 500 test problems, these sets were obtained as follows. Running a fixed machine type, all problems with a solve time of one second or less were excluded. From the remaining models, those in P_10 are those for which both the default primal and dual algorithms had solve times of less than 10 seconds. From the remaining models, P_100 was defined by a 100-second time limit, and the remaining set by a 1,000-second time limit.

In Table 13.5 we give summary statistics for our non-TSP test sets. Listed are the number of models in each set and the average number of constraints, variables, and constraint nonzeros. As a calibration step for these test sets, we ran CPLEX 6.5 using the default primal and dual simplex algorithms. Runs were made on a 500-MHz Digital Alpha Ev5 computer. In Table 13.6 we list the total CPU times. Table 13.7 gives a more detailed listing of the solution times for the Kennington models.

Table 13.5 Summary statistics.

Test Set	No. of Models	Avg. No. of Constraints	Avg. No of Variables	Avg No. of Nonzeros
Netlib	94	919	2,237	13,243
Kennington	16	16,087	60,785	244,487
P_10	107	4,715	11,172	45,432
P_100	107	12,883	37,452	195,714
P_1000	109	20,281	53,800	269,384

Table 13.6 Total run times (CPU seconds).

Test Set	Dual	Primal
Netlib	392.0	548.7
Kennington	273.2	399.8
P_10	206.8	205.1
P_100	1,820.3	2,219.8
P_1000	10,865.9	16,446.1

We remark that the larger Netlib models—including several of the larger Kennington models, such as ken18 and pds-20—were considered quite difficult when introduced. That is not longer the case.

For our TSP LP problems we have collected various sets of LP problem instances. These may be classified roughly as

• Degree LP problems

Table 13.7 Run times—Kennington models (CPU seconds).

Model	Dual	Primal
cre_a	0.8	2.6
cre_b	18.3	13.9
cre_c	0.8	2.0
cre_d	8.4	12.0
ken07	0.1	0.2
ken11	1.7	2.9
ken13	5.8	22.4
ken18a	43.9	212.7
osa007	1.6	1.0
osa014	7.8	3.2
osa030	25.6	8.7
osa060	88.1	30.4
pds-02	0.5	0.5
pds-06	3.7	3.3
pds-10	8.4	7.9
pds-20	57.7	76.1

- Root LP problems

- Some hard LP problems

The degree LPs refer the set of initial LP problems generated in the runs that were used to generate the root LP problems. Root LP problems refer the final LP problems generated in standard runs of Concorde:

```
concorde -s 99 -C 15 -Z file_name
```

where file_name is the name of a ".tsp" file from TSPLIB. By hard LP problems we simply mean some difficult or interesting instances that have been generated and archived over the duration of our TSP computations. We note that with the current version of Concorde it is probably not possible to regenerate some of these instances. We should also remark that this "hard" set includes several "easier" LP instances from smaller than 2,000-city TSP instances. These easy LP instances were the hardest that were encountered in this size range. The idea of the "hard" set is that it represents what difficulties can arise in TSP computations, and, hence, what choices are appropriate in order to achieve robustness.

In Tables 13.8, 13.9, and 13.10 we list the hard TSP instances that have been collected. The naming convention used first gives the name of the corresponding TSPLIB instance, followed by either "r" or "c." The designator 'r' denotes that the given LP problem was obtained as part of a sequence of problems in which constraints were added in consecutive terms, while "c" corresponds to the case in which variables were added. For each instance, an accompanying basis was also archived, where that basis was optimal to the given problem before new constraints or variables were added. These bases will be used in several of our tests.

Table 13.8 Hard TSP LP problems (Part 1).

Problem	No. of Contraints	No. of Variables.	No. of Nonzeros
ts225.c1	487	803	12,807
ts225.r1	811	922	53,458
pr439.c1	747	1,269	13,867
pr439.r1	769	1,234	13,018
vm1084.c1	1,748	3,136	25,343
vm1084.r1	1,834	3,205	34,439
d1291.c1	2,421	5,773	234,244
d1291.c2	2,343	5,851	205,986
d1291.r1	2,318	5,777	206,127
d1291.r2	2,726	4,915	301,321
rl1304.c1	2,229	4,042	54,855
rl1304.r1	2,360	3,942	56,777
fl1577.c1	2,614	6,690	132,902
fl1577.c2	2,413	6,814	123,009
fl1577.r1	2,999	6,488	184,625
fl1577.r2	2,526	6,894	156,138

13.4.1 An asymmetry

Before examining the choices of feasibility and optimality tolerances, in the next section, there is an issue to discuss related to the management of tolerances. In CPLEX, there is an asymmetry in the way the feasibility and optimality tolerances are treated.

Let us consider the primal simplex algorithm. For this algorithm CPLEX interprets the feasibility tolerance as the more fundamental of the two tolerances. If a solution cannot be found that satisfies the given feasibility tolerance, then we want to declare the given problem as infeasible. On the other hand, if a feasible solution does exist, then we want to find the best possible feasible solution. The optimality tolerance is viewed primarily as a stopping criterion in that search.

Given the above point of view, if for some reason the primal simplex algorithm determines that the optimality tolerance cannot be achieved, we need a mechanism to relax this tolerance in order to achieve convergence. The implementation of this mechanism is quite complex. We describe only its two principal attributes.

Table 13.9 Hard TSP LP problems (Part 2).

Problem	No. of Contraints	No. of Variables.	No. of Nonzeros
d2103.c1	5,096	8,744	1,494,990
d2103.c2	5,055	8,763	1,463,930
d2103.c3	4,185	8,777	959,273
d2103.c4	4,312	8,813	1,041,862
d2103.r1	5,406	8,749	1,667,353
d2103.r2	5,481	8,775	1,732,026
d2103.r3	5,289	8,784	1,602,190
d2103.r4	4,180	8,834	961,098
d2103.r5	4,462	8,843	1,112,505
pr2392.c1	3,524	5,594	30,476
pr2392.r1	3,758	5,523	32,986
pcb3038.c1	4,362	6,656	73,626
pcb3038.c2	4,234	6,672	65,984
pcb3038.r1	4,522	6,651	89,710
pcb3038.r2	4,349	6,651	73,993
fl3795.c1	10,857	13,948	1,645,257
fl3795.c2	5,447	14,229	387,085
fl3795.c3	5,389	14,311	359,027
fl3795.r1	6,701	14,164	631,280
fl3795.r2	6,623	14,164	620,077
fl3795.r3	10,886	13,948	1,645,758

First, it may happen that even though a problem has been perturbed, no progress is being made (see the definition of *progress* in 13.2.1). In such cases, to achieve termination, the optimality tolerance is adjusted upward, where the amount of the adjustment is larger if the $||c_B - \mathbf{B}^T\Pi||$ residuals significantly exceed the optimality tolerance. A second possibility is that optimality has ostensibly been proved, at least twice, only to find that the computation of a fresh factorization and the recomputation of the reduced costs produces new reduced-cost violations. Again, the optimality tolerance is adjusted upward, taking into account appropriate residuals.

With the above mechanism, obviously it can happen that termination occurs relative to an expanded optimality tolerance. In that case, ϵ_o is reset to its original value, and the optimization is reinitiated. If, after two such restarts, the optimality tolerance is again expanded, convergence is declared and the optimization is terminated.

For the dual simplex method, a similar approach is used, where it is the optimality tolerance that is considered more fundamental, and the feasibility tolerance is

Table 13.10 Hard TSP LP problems (Part 3).

Problem	No. of Contraints	No. of Variables.	No. of Nonzeros
fnl4461.c1	6,445	9,386	83,301
fnl4461.c2	6,068	9,395	56,067
fnl4461.r1	6,572	9,386	90,727
fnl4461.r2	6,258	9,386	64,902
pla7397.c1	12,179	15,413	278,654
pla7397.c2	13,031	15,415	327,257
pla7397.c3	10,724	15,650	204,602
pla7397.c4	13,305	15,366	332,928
pla7397.r1	12,429	15,413	292,944
pla7397.r2	13,315	15,413	366,157
pla7397.r3	13,665	15,364	358,812
pla7397.r4	10,656	15,373	187,848
usa13509.c1	19,241	33,576	293,477
usa13509.c2	19,440	33,587	314,858
usa13509.r1	19,637	33,583	334,006
usa13509.r2	19,458	33,569	338,272
d18512.c1	24,966	40,932	294,520
d18512.c2	24,860	40,949	289,395
d18512.c3	24,841	40,958	299,373
d18512.c4	24,958	40,957	306,415
d18512.r1	25,052	40,926	303,694
d18512.r2	25,103	40,926	307,334
d18512.r3	24,958	40,947	297,937
d18512.r4	25,230	40,947	326,274

expanded to achieve convergence.

13.4.2 Feasibility and optimality defaults

The setting of the optimality and feasibility tolerances will turn out to play a significant role in our TSP computations. In this section we examine the values of these tolerances, by default both equal to 10^{-6} in CPLEX. This value is the same as is chosen in many commercial codes. Is it the correct choice?

To test the behavior as feasibility and optimality tolerances were varied, we made 90 distinct families of runs. For each of the five non-TSP LP problem test sets, and for each of the primal simplex and dual simplex algorithms, a run was made with the feasibility and optimality tolerances set to the same value t, where t varied from

10^{-2} to 10^{-10} by factors of 10. In this section we refer to this value as *the* tolerance value.

For the 433 models in the five test sets, there were 17 in total that, for some choice of tolerances, were determined to be infeasible by the primal, unbounded by the dual, or simply did not solve. The latter, failure to solve, occurred for just one model, with the tolerance set to 10^{-8} or smaller. In this one case, numerical singularities occurred unless some variables were excluded from the optimization.

A summary of the infeasibility/unboundedness results is given in Table 13.11 and Table 13.12. Neither infeasibility nor unboundedness occurred for the Netlib or Kennington sets.

Table 13.11 Number of models infeasible with the primal algorithm (with given tolerance).

Test	10^{-2}	10^{-3}	10^{-4}	10^{-5}	10^{-6}	10^{-7}	10^{-8}	10^{-9}	10^{-10}
P_10	2	2	2	3	5	5	5	5	5
P_100	1	1	1	1	1	4	5	5	5
P_1000	2	3	4	4	4	5	5	5	6

Table 13.12 Number of models unbounded with the dual algorithm (with given tolerance).

Test	10^{-2}	10^{-3}	10^{-4}	10^{-5}	10^{-6}	10^{-7}	10^{-8}	10^{-9}	10^{-10}
P_10	2	2	2	3	5	5	5	5	5
P_100	1	1	2	2	2	3	4	4	4
P_1000	2	2	4	4	4	6	6	7	7

Note that the numbers of infeasible and unbounded models in each category are not equal. That's no surprise given the asymmetry discussed in the previous section. Note also that with the primal, there is a significant increase in the number of infeasible models when moving form tolerance 10^{-6} to 10^{-7}.

We removed the 17 infeasible/unbounded models from further testing, leaving 416 models.

We next consider the issue of solution "accuracy." This issue is obviously important; however, without a clear definition of what is meant by accuracy, drawing hard conclusions for our data is not straightforward; moreover, giving such a definition in the context of practical linear programming appears to be almost impossible. If nothing else, the input data themselves typically have questionable accuracy.

We present two kinds of data. First, in Table 13.13 we count the numbers of models that failed to achieve the specified tolerance as it was varied from 10^{-2} to 10^{-10}. Thus, for values of the tolerance of 10^{-6} and higher, all 416 instances achieved the specified tolerance. On the other hand, using the primal simplex algorithm there were 11 models that failed to satisfy the optimality tolerance when it was set to 10^{-10}, along with the feasibility tolerance. For this point of view, 10^{-6} appears to be at least a reasonable setting.

Table 13.13 Number of models not achieving tolerance.

Algorithm	10^{-10}	10^{-9}	10^{-8}	10^{-7}
Primal	11	5	1	1
Dual	15	5	3	1

For our second set of accuracy data, we compared the objective function values at the various settings to those when the tolerance was set to 10^{-10}, in effect, considering the latter to be the most accurate setting.

The meaning of the numbers in Table 13.14 and Table 13.15 is the following. For example, in Table 13.14, the value "25" in the row labeled "7" and column labeled 10^{-3} means that with the tolerance set to 10^{-3}, there were 25 instances where only seven digits of accuracy were achieved relative to the 10^{-10} setting. Saying that seven digits of accuracy were achieved means that $10^{-7} \leq |(z_7 - z_{10})/z_{10}| < 10^{-6}$, where z_k denotes the optimal objective value for the tolerance setting of 10^{-k}.

Table 13.14 Primal simplex algorithm.

Digits of Accuracy	Tolerance							
	10^{-2}	10^{-3}	10^{-4}	10^{-5}	10^{-6}	10^{-7}	10^{-8}	10^{-9}
0	3	2						
1	9	3	5	2	2	1		
2	13	4	1	2				
3	18	12	6	2	1	1	3	1
4	26	11	3	2	2	1		
5	32	10	3	2		2		1
6	38	27	11	6	5	1	2	1
7	26	25	22	3	3	4	2	1
8	16	24	14	9	6	2	1	1
9	17	16	20	21	8	6	6	3
10	9	11	14	15	9	7	3	
11	7	10	11	12	13	6	3	6

Again, these results are very difficult to interpret, but if we were to take the point of view that eight digits of accuracy is reasonable, roughly single precision accuracy, then it seems clear that a tolerance of 10^{-5} is qualitatively worse than 10^{-6}. At least for the dual, in achieving eight digits of accuracy, there seems little advantage to a setting tighter than 10^{-8}.

We also examined solution times. These were generated by running all models with the default primal and dual simplex algorithms on a fixed machine type. We then computed the ratios for each of primal and dual and for each tolerance setting.

Table 13.15 Dual simplex algorithm.

Digits of Accuracy	Tolerance 10^{-2}	10^{-3}	10^{-4}	10^{-5}	10^{-6}	10^{-7}	10^{-8}	10^{-9}
0	2	1	1					
1	4	2	1	1	1			
2	16	10	5	3	2	1		
3	13	2	2	3	1		2	1
4	26	13	6	2	1	2		
5	32	14	7	3	1		2	
6	25	17	12	1	3	1	1	1
7	33	24	16	12	4	1		
8	22	29	15	10	5	1	1	1
9	14	14	11	11	13	8	3	1
10	4	14	13	14	8	8	6	6
11	3	4	6	5	8	8	6	4

The ratios were computed by dividing the time for the given tolerance setting into the time for the default tolerance, 10^{-6}. Thus, any ratio bigger than 1.0 means a solve time faster than defaults and any ratio smaller than 1.0 means a solve time slower than defaults. The entries in Table 13.16 were computed as geometric means of the ratios over all problem instances, with the exception that any instance having a solve time of less than 1.0 second for tolerance 10^{-6} was excluded. (The rows of the table correspond to "Primal," "Dual," and "Both.")

Table 13.16 Run time ratios (with given tolerance).

	10^{-2}	10^{-3}	10^{-4}	10^{-5}	10^{-6}	10^{-7}	10^{-8}	10^{-9}	10^{-10}
P	1.070	1.044	1.020	1.010	1.000	0.998	1.002	0.996	0.972
D	0.906	0.950	1.000	1.001	1.000	1.008	0.997	1.004	0.971
B	0.984	0.996	1.010	1.005	1.000	1.003	0.999	1.000	0.971

From the point of view of the solve times, any setting in the range 10^{-3} to 10^{-9} produces roughly equivalent overall results. If we restrict to the dual simplex algorithm, then any setting in the range 10^{-4} to 10^{-9} is acceptable. For the primal, the picture is quite different, for reasons that we do not fully understand. There, loosening the tolerance can significantly improve performance, but at the cost of what, for most users, would be an unacceptable degradation in accuracy—see the earlier results in this section.

Finally, we made a direct comparison of the primal and dual simplex algorithms. This comparison is potentially of interest in determining the proper setting for the optimality and feasibility tolerances, assuming we insist that the default values not vary with the choice of algorithm. The running time for the two algorithms do favor different tolerance settings. We find the comparison of general interest as well. We report below only results for the settings from 10^{-6} through 10^{-9}. These seem the only reasonable candidates for a default.

According to Table 13.17, for a default tolerance of 10^{-7}, dual is 82% faster on the P_1000 test set, and 25% faster overall. It's badly slower on the Kennington models, by about 14%. However, the Kennington set is small, and is even smaller when one considers that it involves only four distinct model types.

Table 13.17 Primal run time/dual run time geometric means (with given tolerance).

Test Set	10^{-6}	10^{-7}	10^{-8}	10^{-9}
Netlib	1.14	1.13	1.08	1.12
Kennington	0.93	0.86	0.89	0.92
P_10	0.98	0.99	1.00	0.98
P_100	1.17	1.18	1.18	1.19
P_1000	1.77	1.82	1.73	1.80
ALL	1.24	1.25	1.23	1.25

What do we conclude from the results of this section? The basic message is that accuracy improves gently as tolerances are reduced, and solution times are stable over a broad range of settings. One could conclude that 10^{-6} provides a reasonable middle ground. Larger settings produce what seem to us to be an unacceptable degradation in accuracy. At the other end of the spectrum, settings as tight as 10^{-9} could be justified, but do increase the number of infeasible/unbounded models. In summary, there is no clear enough advantage to any particular setting to justify a change in the current defaults, particularly given a large, existing customer base for a commercial code such as CPLEX. It is interesting to note the remarkable degree of overall robustness in behavior as the tolerance is varied over a considerable range. This means that for particular applications, where some particular criterion such as speed or accuracy is particularly important, the tolerances can be reset with reasonable confidence that overall performance will be robust. This fact will turn out to be useful in the TSP.

13.4.3 Shifting and the Harris ratio test

To test the effect of the Harris ratio test, we set the so-called inner tolerance (discussed in earlier sections) to zero, leaving the actual feasibility and optimality tolerances at their default values, 10^{-6}. This change has the effect of disabling the two-pass nature of the ratio test and any shifting. Results for these runs are summarized below in three separate tables.

In Table 13.18, we count the number of models that encountered numerical singularities when the Harris test was deactivated. No table is given for defaults, when the Harris test is active, since no singularities occurred. This table shows, for example, that there were 37 models in P_1000 that encountered singularities during the solution process with the primal simplex method.

Table 13.18 Singularities

Test Set	Primal	Dual
Netlib	2	8
Kennington	0	0
P_10	4	7
P_100	19	16
P_1000	37	14

Singularities are repaired as they occur.[5] If, however, it is necessary to repair a total of more than 10 factorizations (a settable parameter), the model itself is declared singular, and optimization is terminated. The number of models for which that occurred is recorded in Table 13.19 . Thus, when the Harris test was turned off, two models in P_1000 failed to solve with the dual simplex algorithm because an 11th singularity was encountered. Note that for both the number of models affected by singularities and the number that, as a result, actually failed to solve, the primal is much more strongly affected. At present we can offer no explanation for this characteristic.

Table 13.19 Singularity failures

Test Set	Primal	Dual
Netlib	1	0
Kennington	0	0
P_10	2	0
P_100	5	0
P_1000	15	2

Finally, we compared computation times for primal and dual algorithms and for all models that solved to completion both with and without the Harris ratio test. The numbers in Table 13.20 have the following meaning. For each of the test sets we divided the solve time without the Harris ratio test by the solve time for the default algorithm and took the geometric mean of these ratios. Thus, for the Netlib test set, the primal was about 12.6% slower with the Harris ratio test disabled. With the

---

[5]The factorization routine employs a singularity tolerance of 10^{-13}. If no pivot larger than this value can be found in the part of the matrix remaining to be factorized, the matrix is declared singular. All columns not yet factorized are removed from the basis and replaced by slack or artificial columns that cover the rows not yet covered by the factorization.

mean computed over all five test sets, the primal was about 17% slower. The dual was about 11% slower overall. Note that the Kennington models are numerically quite easy; any extra work to improve stability for these models seems to simply increase the solution time.

Table 13.20 Run time ratios with no Harris ratio test.

Test Set	Primal	Dual
Netlib	1.126	1.326
Kennington	0.983	0.897
P_10	1.065	1.119
P_100	1.243	1.177
P_1000	1.269	1.045
All	1.170	1.115

Overall, the value of the Harris test is clear, particularly for the primal algorithm. However, the relative insensitivity of the solution times prompted us to examine the default values of the inner tolerances, currently equal to the default value of the corresponding outer tolerance. Changing outer tolerances, an issue that was carefully examined in the previous section, affects user expectations and is thus a sensitive issue. However, the inner tolerance, at least ostensibly, affects only numerical stability issues and, potentially, the amount of "shifting" that is necessary to achieve convergence. With a greater inner tolerance, it is natural to expect more shifting and a greater likelihood of additional iterations being necessary to correct the objective for the effects of the shifting. The trade-off is presumably greater freedom during the optimization and some decrease in the number of iterations to proved optimality for the shifted problem.

In Table 13.21 and Table 13.22 we compare the solution times for the default algorithms (inner tolerance 10^{-6}) to the solution times when the inner tolerance is varied from 10^{-7} down to 10^{-13}. Again, these numbers are geometric means of ratios of solve times to default solve times. Numbers smaller than one indicate solution times better than with defaults.

Table 13.21 Dual run time ratios with Harris inner tolerance.

Test Set	10^{-7}	10^{-8}	10^{-9}	10^{-10}	10^{-11}	10^{-12}	10^{-13}
Netlib	1.024	1.055	1.002	1.011	1.033	1.061	1.125
Kennington	1.002	0.978	0.971	0.944	0.935	0.944	0.943
P_10	0.984	0.980	0.996	0.972	1.000	1.017	1.013
P_100	1.013	1.017	1.000	1.007	1.010	1.022	1.033
P_1000	0.964	0.964	1.003	0.976	1.022	0.993	1.040
All	0.990	0.991	0.998	0.985	1.009	1.011	1.031

Table 13.22 Primal run time ratios with Harris inner tolerance.

Test Set	10^{-7}	10^{-8}	10^{-9}	10^{-10}	10^{-11}	10^{-12}	10^{-13}
Netlib	1.002	0.988	1.001	0.992	0.993	0.986	1.000
Kennington	1.002	1.001	1.002	1.017	1.011	1.021	1.010
P_10	0.995	0.987	0.993	0.983	0.988	1.002	1.013
P_100	1.004	1.023	1.000	1.029	1.017	1.012	1.019
P_1000	0.975	0.964	1.000	0.985	0.954	0.975	1.003
All	0.992	0.991	0.998	0.999	0.987	0.996	1.011

As suspected, the solution times are very insensitive to the choice of the inner tolerance. Indeed, the given numbers indicate a slight overall advantage to choosing some value smaller than the current default. However, we did not view this advantage to be significant enough to merit a change. Beyond that, as the Harris inner tolerances are decreased, there is a noticeable trend toward more instability; for much larger models, for which a comprehensive test would be prohibitive, it is reasonable to expect this tendency to be more pronounced. At the same time, if for a particular class of models the Harris tolerance and resulting shifting do seem to cause difficulties, then one can expect to reset the parameters for that class, with relatively little concern for the overall stability of the code. This observation will be useful for the TSP.

Finally, we come to the issue of shifting. The frequency with which shifting occurs is indicated Table 13.23. For example, Table 13.23 shows that for the test set P_1000 and the primal algorithm, there were 62 models that required one removal of shifts before termination, there were 8 that resulted in shifts being removed twice, and so on. The total number of models affected by shifting was 75, close to two-thirds of the models in P_1000. Note that the Kennington test set is omitted in the table; no shifting was recorded for these models.

Finally, we examined the effect of disabling shifting. Table 13.24 compares running times of the default primal and dual algorithms to the running times for these algorithms with shifting disabled. To obtain each number in the table we computed the ratio of the time without shifting divided by the time with shifting for each problem in the indicated test set and took geometric means of these ratios. A mean ratio bigger than one indicates that disabling shifting degraded performance. For the "P" test sets, the degradation increases with model difficulty.

13.4.4 Stalling

Consider pcb3038.c1. In the corresponding Concorde run, this LP problem was generated by starting with a smaller LP problem, solving it to optimality, and then adding 11 new variables with reduced-costs violating optimality. With these variables at value 0, the primal feasibility of the previous optimal basis is preserved. Using this starting point, we ran CPLEX 6.5 default primal using a 400-MHz (Ev5) Digital Alphaserver 4100. The solution time was 453 seconds with an iteration

Table 13.23 Number of shift removals.

Algorithm	Number				
Primal	1	2	3	4	5
Netlib	15				
P_10	19	5	3		
P_100	43	7	4		2
P_1000	62	8	3	1	1
Dual					
Netlib	15	2			
P_10	33	2			
P_100	35	1			
P_1000	39	3	1		

Table 13.24 Run time ratios with no shifting.

Test Set	Primal	Dual
P_10	1.006	1.069
P_100	1.080	1.101
P_1000	1.152	1.185
Netlib	1.194	1.246
All	1.085	1.126

count of 18,356. The solution time was 27 seconds, with 1,518 iterations, using the version of the simplex algorithm Concorde normally uses (dual simplex with "steepest edge").

Based on the above solution time, pcb3038.c1 would not be considered a difficult instance. However, by turning off perturbation, the following behavior was observed. The initial value of the objective function was 137704.4810. At iteration 438, the objective value had reached 137703.470790. It stayed at that value until iteration 1,193,919, at which point it dropped to 137703.459069, staying at that value until iteration 4,781,469, then dropping to 137703.457333, followed by steady decreases in the objective to the optimum of 137700.75465 at iteration 4,794,956. The total run time was 68,493 seconds, more than 150 times slower than with default primal, with the objective function having stayed constant over one span of 3,587,550 iterations! Thus, stalling is indeed a real phenomenon. Some measure for dealing with it, such as perturbation, was absolutely essential to our work.

13.5 PRICING

In the simplex method, primal or dual, a basis is given, and an iteration is effected by choosing two variables, one in the basis and one out, and exchanging these variables, making the basic variable nonbasic, and the nonbasic variable basic. In the primal the first variable selected, the one that really determines the iteration, is the entering variable. In the dual, the roles are reversed, and the variable leaving the primal representation of the basis is selected first. In each case we refer to the process of selecting this first variable as *pricing*. In our standard versions of each algorithm, pricing is carried out as follows:

Primal: $\quad j = argmin\{D_k : k \in N, D_k < 0\}$

Dual: $\quad i = argmin\{X_{B_k} : k = 1, \ldots, n, X_{B_k} < 0\}.$

As we shall see, neither of these two variable selection rules works very well. This standard version for the dual is particularly bad. In this section we begin our description of alternative approaches. We use as our vehicle the primal simplex algorithm, probably the more familiar of the two. In addition, the dual can be handled as a special case of the primal. The reverse is not true.

The primal pricing paradigm given above, where we compute the most negative reduced cost among the nonbasic variables, is sometimes called *full pricing*. To use this approach, at each iteration we must compute

$$e_i^T \mathbf{B}^{-1} A_N$$

and then update all of D_N. Note that this update approach is much more efficient than directly recomputing

$$D_N^T = c_N^T - (c_B^T \mathbf{B}^{-1}) A_N$$

since $e_i^T \mathbf{B}^{-1}$ is typically much sparser than $c_B^T \mathbf{B}^{-1}$ (e_i is a unit vector and c_B is often close to dense—in the TSP the only zero entries correspond to slack variables) so that both the solve and the subsequent vector-matrix products are faster. Moreover, we do not compute $e_i^T \mathbf{B}^{-1} A_N$ in what might seem the natural way, by column, but rather store A_N in an auxiliary, row-oriented data structure so that the computation is carried out as a sum of scalar multiples of rows of A_N, where this multiplication is performed only for the nonzero entries of $e_i^T \mathbf{B}^{-1}$. Thus we avoid examining the nonzero entries of A_N that occur in other rows.

Even given an intelligent update approach to maintaining the vector of reduced costs, it is sometimes too expensive to update all of D_N at each iteration, when in fact only one negative D_j is needed to carry out an iteration. Given this fact, one can easily imagine several alternatives that avoid computing all of D_N. Two standard ones are the choices *multiple pricing* and *partial pricing*. We describe simple versions of each, followed by a description of the approach used by CPLEX.

13.5.1 Multiple pricing

Given a basic, feasible solution determined by a basis B, compute all of D_N and select a "small subset" $N' \subseteq N$ such that $D'_N < 0$. Let $S = B \cup N'$ and solve

$$\text{Minimize } c_S^T x_S$$

subject to

$$A_S x_S = b$$

$$x_S \geq 0. \tag{13.9}$$

Since S is small, full pricing, or even more expensive approaches that explicitly examine the step lengths of several alternative entering variables, become feasible. After the above problem is solved, a new N' is computed, and the process is repeated until $D_N \geq 0$ following a subproblem optimization.

The primary advantage of the multiple-pricing approach is that it can be used in cases where memory is limited, as it was in early implementations of the simplex method. Thus, if the bulk of the matrix A was stored on some slow auxiliary storage device, then accessing A_N needed to be infrequent, and when it occurred it seemed wise to get as much good information as possible. The result was multiple pricing. The obvious drawback of the multiple-pricing approach is that variables outside of S that become good candidates for the basis while optimizing (13.9) can be ignored for long periods of time.

13.5.2 Partial pricing

Let $P_1, \ldots, P_K$ be a partition of $\{1, \ldots, n\}$. Suppose at the start of an iteration we are given $1 \leq k \leq K$. Compute D_{P_l} for $l = k, \ldots, K, 1, \ldots, k-1$ until the first $D_{P_l} \not\geq 0$ is found. The entering variable is given by

$$j = argmin\{D_p : p \in P_l\}.$$

Set $k = l \bmod K + 1$ and repeat at the next iteration. Note that for $K = 1$, partial pricing reduces to full pricing, and for $K = n$, it reduces to the simple (and *very* bad) rule of selecting the first variable encountered with negative reduced cost.

Partial pricing is obviously simple to implement, and for appropriate choices of K can be quite effective. However, it also seems clear that it can be improved by, in some way, making use of the negative reduced-cost variables that were found on previous pricing rounds but were not selected as the entering variable. The version of partial pricing used by CPLEX is an attempt to do just that.

13.5.3 Partial multiple pricing

Among the pricing paradigms employed in CPLEX is the following approach, which can be viewed as a hybrid of partial and multiple pricing; it is called here *partial multiple pricing*. Let $M > 0$ be a positive integer and let $P_1, \ldots, P_K$ be a partition of $\{1, \ldots, n\}$ such that $|P_k| \leq M$ for $k = 1, \ldots, K$. Let $E \subseteq \{1, \ldots, n\}$ and assume $|E| \leq 2M$. Finally, let $1 \leq p \leq K$. Initially $E = \emptyset$ and $p = 1$. Define *pricing a group* as follows: Given a block, or *group*, P_l from the indicated partition, compute D_{P_l} and add to E all $k \in P_l$ such that $D_k < 0$. Now at a particular iteration of the primal simplex algorithm proceed as follows:

1. If $|E| > M$, remove from E all but the M variables with the most negative reduced costs from the previous iteration.

2. Compute D_E and remove from E any k such that $D_k \geq 0$.

3. Price groups in the order $P_p, P_{p+1}, \ldots, P_K, P_1, \ldots, P_{p-1}$ until $|E| \geq M$ or all groups have been priced. Reset p to the index of the next group that would have been priced.

4. Select as entering variable

$$j = argmin\{D_k : k \in E\},$$

and remove j from E.

Specifically, CPLEX takes

$$K = \lceil \frac{n}{M} \rceil$$

and defines

$$P_k = \{i, i+1, \ldots, i + K \lceil \frac{n-k}{K} \rceil\}$$

for $k = 1, \ldots, K$. If not set by the user, M is determined based on problem size, with, for example, $M = 5$ for $n < 1000$, and $M = 10$ for $1000 \leq n < 10000$. The selected value of M (in the current implementation) never exceeds 60.

13.5.4 Steepest edge

Partial multiple pricing works well for a broad range of linear programs, especially early in the solution process, when choices for "good" entering variables abound. However, as the solution approaches optimality, something approaching full pricing is unavoidable, and partial multiple pricing is a very inefficient way to do that full pricing. Even more important, for more "difficult" linear programs, simply using the reduced cost as a criterion for selecting the entering variable seems to be inadequate. A much better approach, better in the sense of total iterations required to solution, is provided by the very natural idea of "steepest-edge pricing." The key references for this idea are Harris [250], Goldfarb and Reid [213], and Forrest and Goldfarb [190].

Consider an LP problem in the form (13.2). Let B be a feasible basis and let $j \notin B$, where the reduced cost of j relative to B is given by D_j. If we consider increasing the value of x_j to $\Theta \geq 0$, keeping the values of all other nonbasic variables at 0, the change in the objective function is given by ΘD_j. In other words, D_j is the rate of change of the objective per unit change along the x_j-axis. However, as this change in x_j occurs, and the implied changes in the values of the basic variables occur, the trajectory of the full solution vector x is not along the x_j-axis, but rather along some "edge" of the polyhedron of feasible solutions for (13.2). If this edge has a large angle with respect to the x_j-axis, then a one-unit change in x_j can mean a very long step along the actual edge, and hence can be very misleading: all other things being equal, there is no reason to expect that the "length" of this edge, the distance to the next vertex (basic) solution, is particularly greater than any other edge emanating from the current vertex solution. Thus, it seems reasonable that a better criterion for the choice of entering variable can be obtained by taking the edge direction into account.

To determine the *edge* generated by the nonbasic variable x_j relative to B, let $X_j^\Theta = \Theta$ and $X_{N\backslash\{j\}}^\Theta = 0$. Then clearly

$$X_B^\Theta = \mathbf{B}^{-1}(b - \Theta A_j) = X_B - \Theta y,$$

where $\mathbf{B}y = A_j$, as in the statement of the standard primal simplex algorithm. Let $r^\Theta = X^\Theta - X$. Then $r_B^\Theta = -\Theta y$, $r_j^\Theta = \Theta$, and $r_{N\backslash\{j\}}^\Theta = 0$. Defining η^j by

$$r^\Theta = \Theta\eta^j,$$

it follows that η^j is the desired edge direction for x_j relative to B. Note that η_j has a 1 in coordinate j, and that the other nonzero entries are given by the coordinates of the vector $-y$ spread over the positions occupied by the basic indices $\{B_1, \ldots, B_m\}$.

Now, in terms of the above notation, if $\Theta = 1$, then the solution vector x moves a distance given by $||\eta^j|| = \sqrt{(\eta^j)^T\eta^j}$, the L_2 norm of η^j. Since the objective changes by D_j in this case, the rate of change per unit movement along the edge is given by $D_j/||\eta^j||$. The corresponding *steepest-edge* pricing rule is thus given by

$$j = argmin\left\{\frac{D_j}{||\eta_j||} : j \in N\right\}.$$

The idea of using such an approach is quite natural and dates back at least to Kuhn and Quandt [329]. However, a straightforward implementation is completely impractical—the computation of each $||\eta^j||$ involves the solution of the linear system $\mathbf{B}y = A_j$, and would be prohibitively expensive if it were necessary at each iteration. However, Goldfarb and Reid [213] showed that there is an effective update formula for η^j, and it is this update that we now present.

Suppose for basis B, x_j is selected as the entering variable, and x_{B_i} is selected as the leaving variable. Denote the new basis by $\bar{B}$ and the new set of nonbasic indices by $\bar{N}$. Let $\bar{\eta}_k$ ($k \in \bar{N}$) denote the new edge vectors. Now consider a particular $k \in \bar{N}\backslash\{B_i\}$. By (13.3) we have

$$\bar{\mathbf{B}}^{-1}A_k = (I - \frac{1}{y_i}(y - e_i)e_i^T)\mathbf{B}^{-1}A_k$$

$$= w - \delta_k y + \delta_k e_i$$

where $\mathbf{B}w = A_k$ and $\delta_k = w_i/y_i$ (w depends on k). Note that the ith component of this vector is δ_k; it is otherwise equal to $w - \delta_k y$. Hence $\bar{\eta}^k$ is given by the formula

$$\bar{\eta}_l^k = \begin{cases} -\delta_k & \text{if } l = j, \\ -w_p + \delta_k y_p & \text{if } l = B_p, p \neq i, \\ 1 & \text{if } l = k, \text{ and} \\ 0 & \text{otherwise.} \end{cases}$$

Now consider the vector $v = \eta^k - \delta_k \eta^j j$. We have

$$v_j = 0 - \delta_k \cdot 1 = -\delta_k,$$

$$v_k = 1 - 0 = 1, \text{ and}$$

$$v_{B_i} = -w_i - (w_i/y_i)(-y_i) = 0.$$

Clearly the only other nonzero components of v have indices B_p for $p \neq i$; clearly $v_{B_p} = -w_p + \delta_k y_p$ for such indices. We have proved the following result.

LEMMA 13.1 $\bar{\eta}^k = \eta^k - \delta_k \eta^j$. *for $k \in N \backslash \{j\}$*   □

We can also show the following identity.

LEMMA 13.2 $\bar{\eta}^{B_i} = -\frac{1}{y_i} \eta^j$.

Proof. By (13.3) we have

$$\bar{\mathbf{B}}^{-1} A_{B_i} = (I - \frac{1}{y_i}(y - e_i)e_i^T)e_i$$

$$= (1 + \frac{1}{y_i})e_i - \frac{1}{y_i}y.$$

The ith component of this vector is $1/y_i$, and otherwise it is equal to $-y/y_i$.   □

We now readily obtain the Goldfarb-Reid update formula. Where n^k denotes $||\eta^k||^2$, $\bar{n}^k$ denotes $||\bar{\eta}^k||^2$, and $k \in N \backslash \{j\}$, we have

$$\bar{n}^k = (\bar{\eta}^k)^T \bar{\eta}^k \tag{13.10}$$
$$= (\eta^k)^T \eta^k - 2\delta_k(\eta^k)^T \eta^j + \delta_k^2 (\eta^j)^T \eta^j$$
$$= n^k - \delta_k^2 n^j - 2\delta_k y^T (\mathbf{B}^{-1} A_k)$$
$$= n^k + \delta_k^2 n^j - 2\delta_k (y^T \mathbf{B}^{-1}) A_k$$

In this formula n^k and n^j are known from the previous iteration and y was computed to carry out the just completed simplex iteration. For the other terms in this computation, note that

$$\delta_k = \frac{e_i^T \mathbf{B}^{-1} A_k}{y_i} = \frac{-\alpha_k}{y_i}$$

where α_N is as computed in Step 5 of the primal simplex algorithm. Hence, δ_k is available. For the additional computations needed to implement (13.10), we are left with the following: solve $\mathbf{B}w = y$ for w, and use w to compute $w^T A_{N \backslash \{j\}}$. In other words, to implement steepest edge, the additional work is (1) the solution of one additional linear system for a vector w, and (2) the computation of an additional set of inner products on the order of $w^T A_N$. We note that y is typically quite sparse, but often not nearly as sparse as a column of A; hence, w can be quite dense, making the solution of $\mathbf{B}w = y$ more expensive, and the computation of the inner products *very* expense. Some computation time can be saved by combining the computations of y and w, avoiding some of the work involved in traversing twice the updated factorization of $\mathbf{B}$; in addition, we need compute $w^T A_k$ only for k such that $\delta_k \neq 0$. Nevertheless, even these reduced computations can be quite expensive; in addition, there is the problem of computing the initial n^k values, also potentially a nontrivial expense. The net result is, in our experience, that the primal steepest-edge algorithm is not used much in practice. However, an approximate version of the steepest-edge algorithm, introduced by Harris [250] several years before the Goldfarb-Reid [213] paper, is used extensively. We note that the connection described below between steepest edge and the method of Harris is taken from [213]. For Harris' motivation, see [250].

We have described steepest edge in terms of the full polyhedron of feasible solutions. However, there is nothing to prevent us from using a different polyhedron, so long as we can derive an appropriate update formula for that polyhedron, and, of course, so long as we keep enough information about the nature of the problem that the edge directions we obtain are useful. The primary motivation for considering such an alternative approach is to simplify some of the computations necessary to implement the norm updates.

Let $R \subseteq \{1, \ldots, n\}$ be a given set of indices. The set R is called a *reference system*. The new polyhedron we consider is the projection onto the components in R. Let the edge directions in the new polyhedron be denoted by

$$\eta_R^k$$

for $k \in N$. It is then not hard to see that the same update formula holds as before:

$$\bar{\eta}_R^k = \eta_R^k - \delta_k \eta_R^j \quad (k \in N \backslash \{j\}). \tag{13.11}$$

Harris' idea, which she called *devex* (the Latin word for steep) pricing was the following. At some iteration, take $R = N$. Then η_R^k is a unit vector for each $k \in N$, and so we trivially obtain $n_R^k = 1$. Hence, there is no initialization cost. Harris then went one step further and used an update formula that may be viewed as taking the maximum of the norms of the two terms on the right-hand side of (13.11):

$$\bar{n}_R^k \longleftarrow max\{n_R^k, \delta_k^2 n_R^j\}$$

for $k \in N \backslash \{j\}$. Note that this update requires no significant auxiliary computations other than the application of the formula itself. However, the update is only an approximation, and the values it yields for each nonbasic variable are evidently monotone nondecreasing. Thus, the difference between these approximate values and the correct values can grow significantly. We control that growth as follows. At each iteration, we readily obtain the exact value for $\bar{\eta}_R^{B_i}$. If this value differs by more than a factor of two from the value obtained through the update, we reset the reference system R to the current N (and reset all the norms to one).

13.5.5 A hybrid

The default primal pricing in CPLEX is a hybrid of partial multiple pricing and devex. A typical primal simplex solve will begin with partial multiple pricing and, depending on the solution path, switch to devex dynamically at some later iteration. Once devex is activated, it stays active until termination. The criteria for the switch are based on the number of iterations relative to problem size (after a large enough number of iterations, all LP problems will eventually switch to devex) and an estimate of the cost per iteration of devex relative to partial multiple pricing. If the estimate is that devex is no more than twice as expensive, the switch is made.

Chapter Fourteen

Branching

The branch-and-cut algorithm embeds the cutting-plane method within a branch-and-bound search. An outline of the general framework is given in Section 5.7. In this chapter we describe details of its specialization to the TSP.

14.1 PREVIOUS WORK

The first computer implementation of the branch-and-cut algorithm is due to Hong [270]. Branching is carried out in his work by selecting a fractional variable x_e^* from the LP solution x^*, and creating child subproblems with the additional constraints $x_e = 0$ and $x_e = 1$. The choice of the variable x_e is guided by Hong's implementation of the dual simplex algorithm. Concerning this rule, Hong [270] writes the following.

> Now, we will branch in order to get a maximum possible change in the dual objective value in the next iteration.

After selecting e, Hong processes the subproblem that leads to this maximum change, and returns to the second subproblem only after the first branch has been fully explored. Thus, the tree of subproblems is searched in a depth-first order.

PADBERG-RINALDI AND JÜNGER-REINELT-THIENEL

Hong's branching mechanism is adopted in the work of Padberg and Rinaldi [447], but with a different approach for selecting a branching variable x_e. Given a non-integral LP solution x^*, let

$$\alpha = \text{maximum } (x_e^* : e \in E, x_e^* \leq 0.5)$$

and

$$\omega = \text{minimum } (x_e^* : e \in E, x_e^* \geq 0.5)$$

and define a set of edges

$$I = \{e \in E : 0.5\alpha \leq x_e^* \leq \omega + 0.5(1.0 - \omega)\}.$$

Padberg and Rinaldi examine I to see if there is a significant split between a subset $I_2 \subseteq I$ having large costs and the remaining edges $I_1 = I - I_2$. A split (I_1, I_2) such that $c_e \leq c_f$ for all $e \in I_1, f \in I_2$ is considered significant if

$$\mu_1 + 3.5\sigma_1 < \mu_2 - 3.5\sigma_2$$

where μ_j and σ_j are the mean and standard deviation of the edge costs $\{c_e : e \in I_j\}$, for $j = 1, 2$. If such a split is found, then the candidate set of branching edges J is set to I_2; otherwise J is set to I. Letting

$$t = \text{minimum } (|x_e^* - 0.5| : e \in J),$$

Padberg and Rinaldi choose to branch on the variable corresponding to the edge in

$$\{e \in J : t - 0.05 \leq x_e^* \leq t + 0.05\}$$

having the greatest cost c_e.

A simplified version of this branching rule is adopted in the computational study of Jünger et al. [299]. The Jünger et al. rule is to choose the variable corresponding to the edge in

$$\{e \in E : 0.75\alpha \leq x_e^* \leq \omega + 0.25(1.0 - \omega)\}$$

having the greatest cost c_e, where α and ω are defined as above.

Both Padberg and Rinaldi [447] and Jünger et al. [299] choose to explore the subproblem having the minimum LP bound among all leafs of the current search tree. In contrast to Hong's depth-first-search approach, this *best-bound search* aims to increase the overall lower bound at each step of the branch-and-cut algorithm.

Interestingly, Padberg and Rinaldi do not record the cutting planes used to obtain the LP bound associated with a subproblem; they rely on their separation routines, and the use of a cut pool, to reconstruct an LP relaxation when the subproblem is selected for processing. This design decision is reversed in Jünger et al., who append to the storage of a subproblem a list of the cutting planes included in its LP relaxation. On large instances, a great amount of time can be saved in this way, avoiding the multiple rounds of the cutting-plane algorithm that are required to reconstruct a comparable LP relaxation.

Clochard and Naddef's Branching on Subtours

An important extension of the standard mechanism of branching on a variable is introduced in Clochard and Naddef [130]. To create two subproblems, Clochard and Naddef propose to select a proper subset $S \subseteq V$ such that $x^*(\delta(S))$ is close to some odd value, say $2s + 1$. Since any tour x makes $x(\delta(S))$ even, every tour must satisfy either $x(\delta(S)) \leq 2s$ or $x(\delta(S)) \geq 2s + 2$. This disjunction is used to create two child subproblems by adding first one inequality, then the other, to the LP relaxation of the parent subproblem. The process is called *subtour branching*. In their implementation of this idea, Clochard and Naddef search for branching sets S by considering the handles of path inequalities recorded in their pool of cutting planes.

Subtour branching generalizes branching on variables, since choosing the set $S = \{u, v\}$, and letting $s = 1$, is equivalent to branching on x_e with $e = \{u, v\}$. The extra freedom given by subtour branching can be important in obtaining subproblems that yield lower bounds exceeding the bound of the parent problem.

14.2 IMPLEMENTING BRANCH AND CUT

The Concorde code makes use of a variety of ideas from the studies cited in the previous section, including best-bound search and branching on subtours. We describe below the details of the implementation of these ideas.

ACCURACY OF LP BOUNDS

The search process in Concorde is centered on the procedure discussed in Section 5.4 for obtaining valid lower bounds from LP relaxations, independent of the accuracy of the linear programming solver. This bounding procedure is used whenever a decision is made to discard a subproblem based on its LP bound or based on the infeasibility of the LP relaxation. This accurate *pruning* of subproblems ensures that the code will return an optimal tour if the search process is run to completion.

VARIABLE FIXING

In a preliminary step, the initial *root LP* relaxation is used to permanently fix variables to 0 or 1 when this is possible. To be successful, a good upper bound on the optimal TSP value is needed; this is usually provided by a tour T^*, found with a heuristic algorithm. Let z^* denote the objective value of the root LP relaxation and define the *integrality gap* as

$$\Delta = c(T^*) - z^*.$$

If Δ is sufficiently small, then a large portion of the variables can typically be fixed to 0 or 1; fixing a variable x_e to 0 allows edge e to be removed from further consideration, thus reducing the computational effort and storage required to carry out the branch-and-cut search. This reduction in the variable set is an important tool in modern TSP codes, as we discussed in Section 12.3.

The variable-fixing process follows an outline given in Dantzig et al. [151]. For each edge $e \in E$, let γ_e denote the reduced cost of the variable x_e in the solution to the root LP. From the form of the dual LP problem, it is easy to see that if $\gamma_e > 0$ for some edge e, then adding the constraint $x_e = 1$ will increase the dual objective to at least $z^* + \gamma_e$. Similarly, if $\gamma_e < 0$ for some edge e, then adding the constraint $x_e = 0$ will increase the dual objective to at least $z^* - \gamma_e$.

We assume the edge costs $(c_e : e \in E)$ are integers, so any tour T of cost less than T^* must satisfy $c(T) \leq c(T^*) - 1$. It follows that

if $\gamma_e > \Delta - 1$, then we may fix $x_e = 0$,

and

if $\gamma_e < -(\Delta - 1)$, then we may fix $x_e = 1$.

In applying this rule, we again use the procedures in Section 5.4, to avoid problems of numerical inaccuracy in the computations of z^* and γ_e.

BRANCH SELECTION

Concorde uses a combination of branching on variables and subtour branching. The explicit use of branching on variables yields a small improvement in efficiency, since setting x_e to 0 or 1 can be accomplished by adjusting the upper or lower bounds on the variable, rather than adding a new constraint to the LP relaxation. More important, considering edges and subtours separately allows us to apply separate selection procedures that are tuned for the two structures.

To select candidate branching edges, we adopt a version of Hong's [270] dual-simplex-based measure. For each fractional variable x_e, we make an estimate of the change in the objective value that would result from setting x_e to either 0 or 1 and making a single dual simplex pivot. Let z_0 denote the estimate obtained for $x_e = 0$, and let z_1 denote the estimate obtained for $x_e = 1$. The pair of estimates is evaluated as

$$p(z_0, z_1, \gamma) = \frac{\gamma \cdot \text{minimum}\, (z_0, z_1) + \text{maximum}\, (z_0, z_1)}{\gamma + 1} \qquad (14.1)$$

where we set $\gamma = 10$. We compute this quantity for each fractional variable x_e, ranking the branching candidates in nonincreasing order. The motivation behind (14.1) is to promote candidates that lead to an increase in the smaller of the two bounds for the child relaxations. This point is emphasized by giving the smaller bound γ times the weight given to the larger bound.

For subtour branching we have two procedures for selecting candidates. The first is similar to the selection of branching variables, using information from the simplex method to obtain estimates on the impact of a single pivot after the addition of the two inequalities. To choose candidates to evaluate in this way, we search for sets S in the family

$$\mathcal{S} = \mathcal{F}_1 \cup \mathcal{F}_2 \cup \cdots \cup \mathcal{F}_r$$

where $\mathcal{H}_1 = (V, \mathcal{F}_1), \mathcal{H}_2 = (V, \mathcal{F}_2), \ldots, \mathcal{H}_r = (V, \mathcal{F}_r)$ are the hypergraph inequalities in our cut pool. Within $\mathcal{S}$, we evaluate the $1,000 + 100K_{sub}$ sets having $x^*(\delta(S))$ as close as possible to 3.0. A typical value for K_{sub} is 25. The sets are ranked in nonincreasing order of the evaluation function $p(z_0, z_1, 10)$.

A second method for selecting candidates for subtour branching is based on combinatorial conditions, rather than LP information. Consider a set $S \subseteq V$. Let β_0 be the maximum value of

$$\text{minimum}\, (c_e : x_e^* > 0, e \in \delta(v), e \subseteq S)$$

over all $v \in S$, and let β_1 be the value

$$\text{minimum}\, (c_e : x_e^* > 0, e \in \delta(S)).$$

We use the quantities $\delta_0 = \beta_0(x^*(\delta(S)) - 2)/3$ and $\delta_1 = \beta_1(4 - x^*(\delta(S))/10$ as rough measures of the potential improvement that can be obtained by subtour branching on the set S. (The chosen values of the denominators are based on computational experiments.) A shrinking algorithm is used to locate a collection of K_{sub} sets having a wide distribution of values of (δ_0, δ_1).

The above quick selection procedures are used only to produce a list of candidate branching structures, rather than narrowing the choice to a single selection as in earlier studies on the TSP. The candidate list is made up of the top K_{var} choices for variable branching and the top K_{sub} choices for subtour branching; a typical value for K_{var} is 5. The final selection for branching is determined by a more careful evaluation procedure, described in Section 14.3.

Our code is designed to run on either a single computer or in a distributed parallel environment, using a master-worker framework. The module that controls the search tree can be run as a separate process, to serve as a master when the code is run in parallel. The controller hands out tasks upon receiving requests for work. The tasks are of three types:

- branch—select a set of branching structures,

- split—create the subproblems corresponding to a branching structure,

- cut—apply the cutting-plane method to a subproblem.

The best-bound search method is used to select subproblems for processing. In a parallel environment, the task corresponding to an unassigned subproblem with the smallest lower bound is handed out in response to a work request; if all subproblems are already assigned to workers, then a message is returned indicating that the requester should wait t seconds before requesting work again. ($t = 60$ is a typical value.)

A full description of each subproblem is stored (on disk) as a file, including the list of cutting planes, the list of variables set by the branching steps, and information to warm-start the LP solver. This storage allows the controller to pass a subproblem to a worker by simply specifying the name of the corresponding file.

14.3 STRONG BRANCHING

Our evaluation of branching candidates is guided by simple principles. First, in a best-bound search, the dominant criterion for preferring one structure over another is the increase obtained in the lower bounds of the child subproblems. Other criteria, such as quickly locating a portion of the search tree that contains a high-quality tour, are of much less importance, given the goal of exactly solving the TSP. Second, a poor choice of branching structure can essentially double the amount of work needed to solve a particular subproblem. Indeed, a poor branch will create children that are nearly equivalent to the parent, so we now have two copies of the subproblem to tackle rather than one. We are thus willing to spend a significant amount of computation time in the evaluation procedure, going well beyond the fast selection rules that have been used in previous studies.

The process we propose relies on the use of an LP solver to produce good estimates of the lower bounds that can be achieved in the child subproblems. For each of the candidates, create a pair of LP relaxations by adding to the parent LP the $x_e = 0$ and $x_e = 1$ bounds or the $x(\delta(S)) \leq 2s$ and $x(\delta(S)) \geq 2s + 2$ constraints. We thus obtain a list of relaxations

$$(P_0^1, P_1^1), (P_0^2, P_1^2), \ldots, (P_0^k, P_1^k),$$

where k is the number of branching candidates we are considering. Starting with a basic solution obtained from the parent LP, we carry out a limited number of dual steepest-edge simplex pivots on the child LP relaxations. For each $j = 1, \ldots, k$, let (z_0^j, z_1^j) denote the pair of dual objective values that are obtained by these computations; if an LP relaxation is shown to be infeasible, set the corresponding z_d^j equal to $c(T^*)$, where T^* is the best tour we have found for the given TSP. In evaluating the pairs

$$(z_0^1, z_1^1), (z_0^2, z_1^2), \ldots, (z_0^k, z_1^k),$$

we search for estimates that indicate a substantial improvement in the bounds of both children. We therefore choose to evaluate the pairs using function (14.1). In this case we set $\gamma = 100$, to give even greater emphasis to increasing the smaller of the two bounds. As our branching structure, we select the candidate that maximizes $p(z_0^j, z_1^j, 100)$.

We call the above process *strong branching*. A variant is to simply solve the LPs for each pair of child subproblems and evaluate the results with (14.1). This gives more accurate information, but obtaining the optimal LP values on large TSP instances can be very costly in terms of computation time. Such long computations would force a reduction in the number of branching candidates that can be evaluated and may cause the process to overlook a desirable branching structure. In applying strong branching, we must handle the trade-off between obtaining accurate LP information, using a large number of pivots, versus a wider search with a smaller number of pivots.

This trade-off is managed in Concorde with a staged approach. Initially, $K_{var} = 5$ branching variables and $K_{sub} = 25$ branching sets are evaluated with 100 dual simplex pivots. From these candidates, the top $K'_{var} = 2$ variables and $K'_{sub} = 5$ sets are selected, and strong branching is applied again, this time using 500 dual simplex pivots. The final branching selection is the best of these candidates, evaluated with $p(z_0^j, z_1^j, 100)$.

GENERAL APPLICATIONS

The theme of strong branching is not limited to the TSP. Indeed, since its introduction in Concorde, strong branching has been widely adopted in the solution of general problems in discrete optimization. The many areas of application include the following.

- **Mixed-integer programming.** Strong-branching rules are available in leading commercial codes for the solution of mixed-integer programming problems, including ILOG's CPLEX solver [276] and Dash Optimization's solver Xpress MP [154]. Computational testing of strong branching for MIP is reported in Achterberg et al. [2], Bixby et al. [63], Ferris et al. [173], Linderoth and Savelsbergh [358], Margot [373], and Martin [377].

- **Quadratic assignment problem.** A version of strong branching for the quadratic assignment problem is described in Anstreicher et al. [14] and Anstreicher [13]. It is used in their solution of several very difficult instances of the problem.

- **Airline crew scheduling.** Klabjan et al. [317] adopt strong branching in software for assigning airline crews to flight segments.

- **Steiner trees and max-cut.** Barahona and Ladányi [38] use strong branching in an implementation of a branch-and-cut framework based on the LP approximation algorithm of Barahona and Anbil [37]. They specialize their framework for the Steiner tree problem in graphs and the max-cut problem.

- **Vehicle routing.** Ralphs et al. [469] include strong branching in their code for the capacitated vehicle routing problem. Strong branching is also used in the branch-and-cut-and-price implementation of Fukasawa et al. [198]; their computer code is the most successful to date, solving for the first time a large number of challenge instances from the literature.

- **Traveling tournament problem.** Easton et al. [160] use strong branching in their software for the solution of a tournament scheduling problem.

- **Resource-constrained scheduling.** Strong branching is adopted in the work of Jain and Grosssman [283], solving a series of scheduling problems arising in the development of new products.

- **Quadratic programming.** Vandenbussche and Nemhauser [530] test a version of strong branching for the solution of quadratic programming problems with box constraints.

A common finding in these computational studies is that strong branching reliably increases the range of instances solved within the various algorithmic frameworks.

14.4 TENTATIVE BRANCHING

For large and difficult instances of the TSP, we employ an extension of the strong-branching rule that involves considerably more computational effort but can result in more effective choices of branching structures. In this process we use strong branching to select a list of candidates, and for each of these structures we tentatively carry out the branching step and apply a restricted version of our cutting-plane algorithm to the pair of children. This *tentative-branching rule* combines a wider search for candidates with much more accurate estimates of the bounds that can eventually be achieved in the subproblems.

To run k-way tentative branching, where we test k candidates, we multiply the strong-branching constants K_{var} and K_{sub} by k and apply strong branching as above. Rather than selecting a single branching structure, however, we rank the candidates in nonincreasing value of $p(z_0^j, z_1^j, 100)$ and return the top k choices. For each $j = 1, \ldots k$, we solve the LP relaxations corresponding to the jth choice, and apply our cutting-plane and column-generation routines to obtain a pair of lower bounds $(\bar{z}_0^j, \bar{z}_1^j)$. Our final selection of a branching structure corresponds to the pair of bounds that maximizes $p(\bar{z}_0^j, \bar{z}_1^j, 10)$; here our computational tests indicate that it is preferable to return to the $\gamma = 10$ weight, rather than $\gamma = 100$. The cutting-plane routines applied to the tentative subproblems consist of our default separation methods, but with some of the more time-consuming algorithms, such as local cuts, turned off. After we make our final branching choice, the cutting-plane method is continued from the relaxation created in the tentative-branching step.

TEST ON PLA85900

The great amount of CPU time required to carry out tentative branching makes it unsuitable for instances of modest size, where strong branching typically behaves very well. For large instances, where our full cutting-plane algorithm can require many days of computation, the additional precision of tentative branching is well worth the effort. Indeed, for a number of the examples solved by Concorde, tentative branching appears to be a crucial component of the solution process.

To illustrate the power of tentative branching, we consider the instance pla85900, the largest problem in the TSPLIB. In Chapter 16 we describe in detail the computations used to obtain a very strong root LP relaxation for pla85900, yielding an integrality gap that is under 0.0009% of the cost of an optimal tour. The actual lower bound is 142,381,405, while the optimal tour has cost 142,382,641.

In Table 14.1 we report the times required for three branch-and-cut runs, starting from the same 0.0009% root LP. The tests were carried out on a network of Sun Microsystems V20z compute nodes, each with two 2.4-GHz AMD Opteron 250 processors; the CPU times are the sum of the times used on the individual processors. Using 4-way and 32-way tentative branching, the search completed successfully, using 1,147 search nodes and 243 search nodes, respectively. When our standard strong-branching rule is used, that is, 1-way branching, the search tree grew to 3,000 nodes, with 1,365 active nodes waiting to be processed, after a total of four CPU years. It is very unlikely that this strong-branching run could be completed with our existing parallel computing platforms; given the slow progress in the lower bounds of the subproblems, the total number of search nodes may well grow to 1 million or more.

Table 14.1 Branch-and-cut search for pla85900.

Tentative Branching	No. of Search Nodes	CPU Time
1-way	$\gg 3,000$	$\gg 4$ years
4-way	1,149	1.8 years
32-way	243	0.8 years

Drawings of the search trees for the three runs are given in Figures 14.1, 14.2 and 14.3. In these figures, the vertical locations of the nodes correspond to the lower bounds of the subproblems they represent; a scale giving the values of the bounds is located on the left side of each figure. For the strong-branching run depicted in Figure 14.3, the solid nodes indicate active subproblems waiting to be processed. It is apparent in the figure that there is large frontier of active nodes around the

bound 142,382,250; it is this dense frontier that suggests the run will be difficult
to complete. Note that in each figure, if a child problem is pruned, then we do not
draw the corresponding child node.

The decrease in the size of the search tree as we increase the level of tentative
branching suggests that perhaps the tree can be further improved by an even more
intensive search for good branching structures. To test this, we made a partial
run using 256-way tentative branching. In Table 14.2 we report the percentage of
the integrality gap closed by branching on the structure selected in this manner,
compared with the percentage of the gap closed by the branching structure selected
at the root LP in the earlier three runs. The results show a striking improvement in

Table 14.2 Integrality gap closed in pla85900 subproblems after first branch.

Tentative Branching	Child 0	Child 1
1-way	36%	2%
4-way	36%	2%
32-way	Pruned	5%
256-way	35%	54%

the bounds for the new run.

The branching structure selected in the 256-way run consists of a set S of five
cities in a vertical line, located in the lower right-hand corner of the point set. The
precise position of S is indicated by the arrow in Figure 14.4. It is quite interesting
that the seemingly modest restrictions that $x(\delta(S)) \leq 2$ and $x(\delta(S)) \geq 4$ have
such a major impact on the LP lower bounds.

A drawing of the first two levels of the search tree obtained with 256-way branching is given in Figure 14.5. The lower bound achieved by this 7-node tree already exceeds the bound given by the 3,000-node tree using the standard strong-
branching rule.

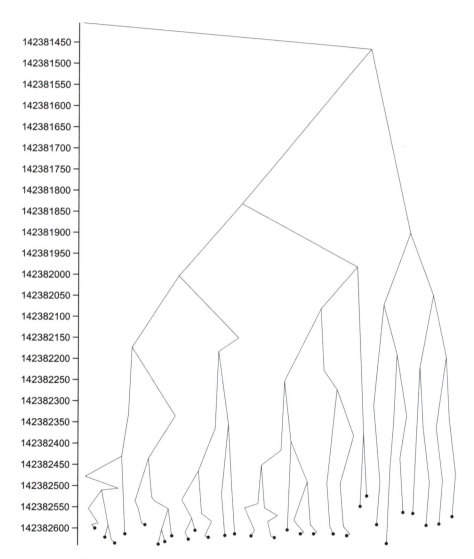

142381450 –
142381500 –
142381550 –
142381600 –
142381650 –
142381700 –
142381750 –
142381800 –
142381850 –
142381900 –
142381950 –
142382000 –
142382050 –
142382100 –
142382150 –
142382200 –
142382250 –
142382300 –
142382350 –
142382400 –
142382450 –
142382500 –
142382550 –
142382600 –

Figure 14.1 pla85900 search tree with 32-way tentative branching.

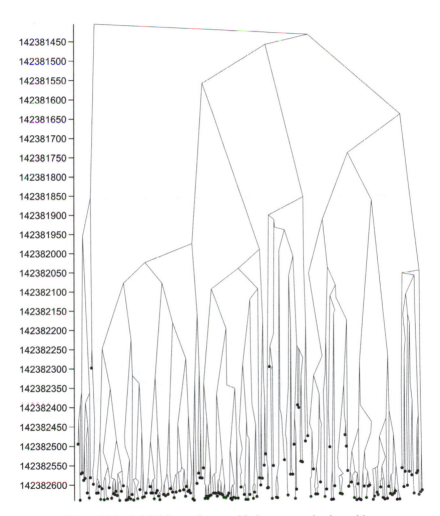

Figure 14.2 pla85900 search tree with 4-way tentative branching.

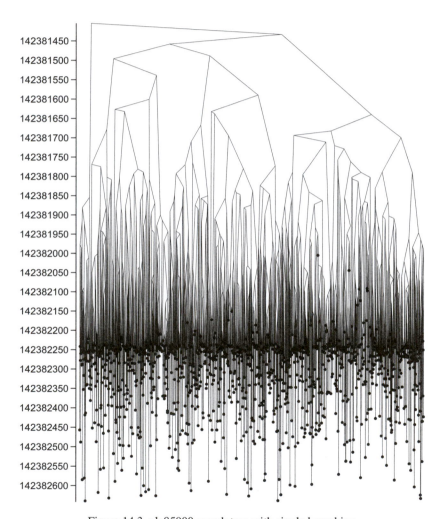

Figure 14.3 pla85900 search tree with single branching.

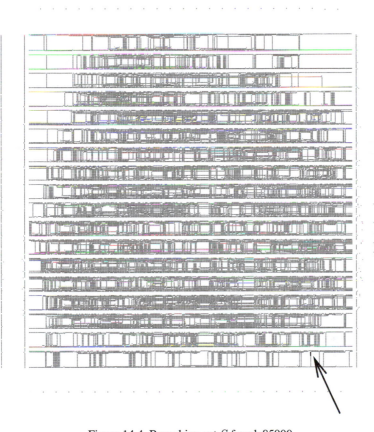

Figure 14.4 Branching set S for pla85900.

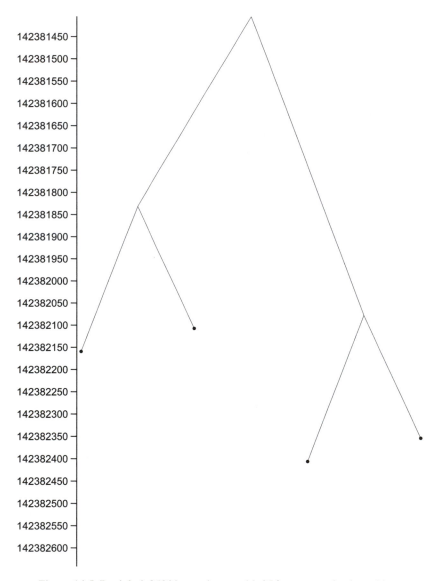

Figure 14.5 Partial pla85900 search tree with 256-way tentative branching.

Chapter Fifteen

Tour Finding

The study of heuristic algorithms for the TSP is a popular topic, having a large and growing literature devoted to its various aspects. Our treatment is restricted to tour finding that is applicable to solution methods for the TSP, namely, finding near-optimal tours within a reasonable amount of computing time. Other topics, such as finding tours very quickly, can be found in Bentley [56], Johnson and McGeoch [291], and Reinelt [475]; an important theoretical result concerning the approximate solution of Euclidean instances of the TSP can be found in Arora [24].

15.1 LIN-KERNIGHAN

At the heart of the most successful tour-finding approaches to date lies the simple and elegant algorithm of Lin and Kernighan [357]. This is remarkable, given the wide range of attacks that have been made on the TSP in the past three decades, and even more so when one considers that Lin and Kernighan's study was limited to problem instances having at most 110 cities (very small examples by today's standards). We begin by describing briefly some of the work leading up to their approach.

Shortly after the publication of Dantzig, Fulkerson, and Johnson's [151] classic paper, Flood [182] studied their 49-city example from a tour-finding perspective. He began by solving the assignment problem relaxation to the TSP, obtaining the dual solution $(u_0, \ldots, u_{48})$. He used these values to compute a reduced-cost matrix $[c_{ij}]$ by subtracting $u_i + u_j$ from the original cost of travel for each pair of cities (i, j). Note that this does not alter the set of optimal solutions to the TSP, but it may help in finding a good tour. Working with these costs, Flood found a *nearest-neighbor tour* by choosing a starting city (in his case, Washington, D.C.) and then always proceeding to the closest city that was not already visited. He followed this with a local improvement phase, making use of the observation that in any optimal tour, $(i_0, \ldots, i_{n-1})$, for an n-city TSP, for each $0 \le p < q < n$ we have

$$c_{i_{p-1}i_p} + c_{i_q i_{q+1}} \le c_{i_{p-1}i_q} + c_{i_p i_{q+1}} \tag{15.1}$$

where subscripts are taken modulo n. A pair (p, q) that violated (15.1) was called an "intersecting pair," and his method was to fix any such pair, as illustrated in Figure 15.1, until the tour became intersectionless.

Croes [145] used Flood's intersectionless tours as a starting point for an exhaustive search algorithm for the TSP. He also described a procedure for finding an intersectionless tour by a sequence of "inversions." The observation is that if (p, q)

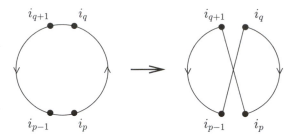

Figure 15.1 Fixing an intersecting pair.

intersect in the tour

$$(i_0, \ldots, i_{p-1}, i_p, \ldots, i_q, i_{q+1}, \ldots, i_{n-1}),$$

then the pair can be fixed by inverting the subsequence $(i_p, \ldots, i_q)$, that is, moving to the tour

$$(i_0, \ldots, i_{p-1}, i_q, i_{q-1}, \ldots, i_{p+1}, i_p, i_{q+1}, \ldots, i_{n-1}).$$

We call this operation $flip(p, q)$; we assume tours are oriented, so $flip(p, q)$ is well-defined.

A strengthening of Croes' inversion method was proposed and tested by Lin [356], building on earlier studies by Morton and Land [403] and Bock [69]. Rather than simply flipping a subsequence $(i_p, \ldots, i_q)$, he also considered reinserting it (either as is, or flipped) between two other cities that are adjacent in the tour, if such a move would result in a tour of lesser cost. This increases the complexity of the algorithm, but Lin showed that it produces much better tours. (Reiter and Sherman [478] studied a similar method, but included a specific recipe for which subsequences to consider and allowed arbitrary reorderings of the subsequence, rather than just flips.)

Lin [356] also provided a common framework for describing intersectionless tours and tours that are optimal with respect to flips and insertion. He called a tour *k-optimal* if it is not possible to obtain a tour of lesser cost by replacing any k of its edges (considering the tour as a cycle in a graph) by any other set of k edges. Thus, a tour is intersectionless if and only if it is 2-optimal. Moreover, it is not difficult to see that a tour is optimal with respect to flips and insertions if and only if it is 3-optimal. Croes' algorithm and Lin's algorithm are commonly referred to as *2-opt* and *3-opt*, respectively.

A natural next step would be to try k-opt for some greater values of k, but Lin found that despite the greatly increased computing time for 4-opt, the tours produced were not noticeably better than those produced by 3-opt. As an alternative, Lin and Kernighan [357] developed an algorithm that is sometimes referred to as "variable k-opt." The core of the algorithm is an effective search method for tentatively performing a possibly quite long sequence of flips, such that each initial subsequence appears to have a chance of leading to a tour of lesser cost. (It may help in understanding their algorithm to note that while any k-opt move can be realized as a sequence of flips, some of the intermediate tours may have cost greater

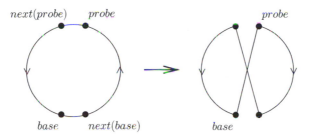

Figure 15.2 $flip(next(base), probe)$.

than that of the initial tour, even in an improving k-opt move.) If the search is successful in finding an improved tour, then the sequence of flips is made and a new search is begun. Otherwise, the tentative flips are discarded before we begin a new search, and we take care not to repeat the same unsuccessful sequence. The procedure terminates when every starting point for the search has proven to be unsuccessful.

We now describe the search method. The algorithm we present differs somewhat from the one given in Lin and Kernighan [357], but the essential ideas are the same. We describe the method in sufficient detail to have a basis for discussing our computational study in later sections.

Suppose we are given a TSP with $c(i, j)$ representing the cost of travel between vertex i and vertex j. Let T be a tour and let $base$ be a selected vertex. We will build a sequence of flip operations, and denote by $current\_tour$ the tour obtained by applying the flip sequence to T. For any vertex v, let $next(v)$ denote the vertex that comes immediately after v in $current\_tour$ and let $prev(v)$ denote the vertex that comes immediately before v. (Since a tour is oriented, $next$ and $prev$ are well-defined.) The only flips that will be considered are those of the form $flip(next(base), probe)$, for vertices $probe$ that are distinct from $base$, $next(base)$, and $prev(base)$. Such a flip will replace the edges $(base, next(base))$ and $(probe, next(probe))$ by $(next(base), next(probe))$ and $(base, probe)$, as illustrated in Figure 15.2. $current\_tour$ would be improved by such a flip if and only if

$$c(base, next(base)) + c(probe, next(probe)) > \qquad (15.2)$$
$$c(next(base), next(probe)) + c(base, probe),$$

as in the 2-opt algorithm. Rather than demanding such an improving flip, Lin-Kernighan requires that

$$c(base, next(base)) - c(next(base), next(probe)) > 0.$$

This is a greedy approach that tries to improve a single edge in the tour. The idea can be extended as follows: Let $delta$ be a variable that is set to 0 at the start of the search and is incremented by

$$c(base, next(base)) - c(next(base), next(probe)) +$$
$$c(prob, next(probe)) - c(probe, base)$$

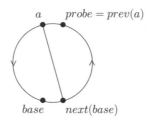

Figure 15.3 Finding a promising vertex.

after each $flip(next(base), probe)$. Thus, $delta$ represents the amount of local improvement we have obtained thus far with the sequence of flips; the cost of $current\_tour$ can be obtained by subtracting $delta$ from the cost of the starting tour T. In a general step, we require that

$$delta + c(base, next(base)) - c(next(base), next(probe)) > 0. \qquad (15.3)$$

Thus, we permit $delta$ to be negative, as long as it appears that we might be able to later realize an improvement. We call $probe$ a *promising vertex* if (15.3) holds.

A rough outline of the search method is the following:

$delta = 0$
while there exist promising vertices
do   choose a promising vertex $probe$;
        $delta = delta + c(base, next(base)) - c(next(base), next(probe)) +$
                    $c(probe, next(probe)) - c(probe, base)$;
        add $flip(next(base), probe))$ to the flip sequence;
end

If we reach a cheaper tour, we record the sequence of flips, but continue on with the search to see if we might find an even better tour.

Notice that $probe$ is promising if and only if the cost of the edge

$$(next(base), next(probe))$$

is small enough, since the other two terms in (15.3) do not depend on $probe$. So an efficient way to check for a promising vertex is to consider the edges incident with vertex $next(base)$, ordered by increasing costs. When we consider edge $(next(base), a)$, we let $probe = prev(a)$. See Figure 15.3.

Just selecting the first edge that produces a promising $probe$ is too shortsighted and often leads to long sequences that do not result in better tours. Instead, Lin-Kernighan also considers a third term from inequality (15.2), choosing the edge $(next(base), a)$ that maximizes

$$c(prev(a), a) - c(next(base), a). \qquad (15.4)$$

To avoid computing this quantity for all edges incident with $next(base)$, only a prescribed subset of vertices a are considered. We refer to this prescribed subset as the set of *neighbors* of a vertex; a typical example of a neighbor set is the set of k closest vertices, for some fixed integer k. The price we pay for using only the neighbors of $next(base)$ is that we may overlook a promising flip operation. This is outweighed, however, by the greatly reduced time of the steps in the search. We call a neighbor, a, of $next(base)$ promising if $probe = prev(a)$ is a promising vertex. The outline of the search becomes:

$delta = 0;$
while there exist promising neighbors of $next(base)$
do   let a be the promising neighbor of $next(base)$ that
        maximizes (15.4);
        $delta = delta + c(base, next(base)) - c(next(base), a) +$
                    $c(prev(a), a) - c(prev(a), base);$
        add $flip(next(base), prev(a)))$ to the flip sequence;
end

To increase the chances of finding an improving sequence, a limited amount of backtracking is used. For each integer $k \geq 1$, let $breadth(k)$ be the maximum number of promising neighbors we are willing to consider at level k of the search. Rather than just adding the flip involving the most promising neighbor, we will consider, separately, adding the flips involving the $breadth(k)$ promising neighbors having the greatest values of (15.4). Lin and Kernighan set $breadth(1) = 5$, $breadth(2) = 5$, and $breadth(k) = 1$ for all $k > 2$. Setting $breadth(k) = 0$ for some k provides an upper bound on the length of any flip sequence that will be considered.

Mak and Morton [369] proposed another method for increasing the breadth of a search. Their idea is to try flips of the form $flip(probe, base)$, as well as those that we normally consider. This can be accomplished by considering the neighbors a of $base$, other than $next(base)$ and $prev(base)$, that satisfy

$$delta + c(base, next(base)) - c(base, a) > 0.$$

In this case, the vertices are ordered by decreasing values of

$$c(a, next(a)) - c(base, a) \qquad\qquad (15.5)$$

and after a $flip(next(a), base)$, the value of $delta$ is incremented by

$$c(base, next(base)) - c(base, a) + c(a, next(a)) - c(next(a), next(base)).$$

These details are analogous to those for the usual flips. The final piece of a Mak-Morton move is to change $base$ to be the vertex that was $next(a)$ before the flip. This means that $next(base)$ is the same vertex before and after the flip, analogous to the fact that $base$ remains the same in the usual case. See Figure 15.4.

There is no need to consider the Mak-Morton moves separately from the usual flips, so we can create a single ordering consisting of the permitted neighbors of $next(base)$ and $base$, sorted by nonincreasing values of (15.4) and (15.5), respectively. (Some vertices may appear twice in the ordering.) Call this the lk-$ordering$ for $base$. At each step of the search, the vertices will be processed in this order.

To give an outline of the full search routine, incorporating backtracking and Mak-Morton moves, it is convenient to use the recursive function $\texttt{step}$ defined in Algorithm 15.1. This function takes as arguments the current $level$ and the current

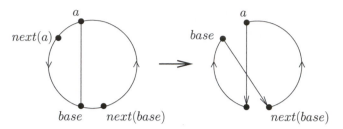

Figure 15.4 A Mak-Morton move.

Algorithm 15.1 step(*level, delta*).

---

create the *lk-ordering* for *base*;
$i = 1$;
while there exist unprocessed vertices in the *lk-ordering*
    and $i \leq$ breadth(*level*)
do   let a be the next vertex in the *lk-ordering*;
    if a is specified as part of a Mak-Morton move
    then $g = c(base, next(base)) - c(base, a) +$
        $c(a, next(a)) - c(next(a), next(base))$;
      $newbase = next(a)$;
      $oldbase = base$;
      add $flip(newbase, base)$ to the flip sequence;
      $base = newbase$;
      call step(*level* $+ 1, delta + g$);
      $base = oldbase$;
    else $g = c(base, next(base)) - c(next(base), a) +$
        $c(prev(a), a) - c(prev(a), base)$;
      add $flip(next(base), prev(a))$ to the flip sequence;
      call step(*level* $+ 1, delta + g$);
    end
    if an improved tour has been found
    then return;
    else delete the added flip from the end of the
      flip sequence;
      increment i;
    end
end

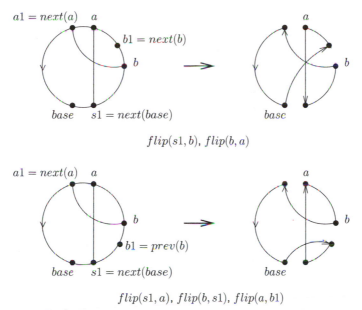

$$flip(s1, b), \; flip(b, a)$$

$$flip(s1, a), \; flip(b, s1), \; flip(a, b1)$$

Figure 15.5 Alternative first step, case 1.

delta. A search from *base* is then just a call to step(1,0). Note again that at any point, the cost of *current_tour* can be computed using the cost of the initial tour and *delta*. It is easy, therefore, to detect when an improved tour has been found.

Lin-Kernighan increases the breadth of a search in a third way, by considering an alternative first step. The usual $flip(next(base), prev(a))$ removes the edge $(prev(a), a)$ from the tour. The alternative (actually a sequence of flips) will remove $(a, next(a))$. To accomplish this, we select a neighbor b of $next(a)$. There are two cases, depending on whether or not b lies on the segment of the tour from $next(base)$ to a. If b lies on this segment, then two sequences of flips are considered, the first removes $(b, next(b))$ from the tour and the second removes $(prev(b), b)$ from the tour. These moves are illustrated in Figure 15.5, together with the flips needed to realize them. If b does not lie on the segment from $next(base)$ to a, then we select a neighbor d of $prev(b)$, such that d lies on the segment from $next(base)$ to a. We again consider two sequences of flips, the first removing $(d, next(d))$ and the second removing $(prev(d), d)$. These moves are illustrated in Figure 15.6.

To permit backtracking in this alternate first step, let $breadth_A$ be the maximum number of vertices a that we are willing to try, let $breadth_B$ be the maximum number of pairs $(b, b1)$ (where $b1$ is either $next(b)$ or $prev(b)$), and let $breadth_D$ be the maximum number of pairs $(d, d1)$ (where $d1$ is either $next(d)$ or $prev(d)$). In selecting a, we consider only the promising neighbors of $next(base)$. These are ordered by decreasing value of $c(next(a), a) - c(next(base), a)$. Call this the *A-ordering*. In selecting $(b, b1)$, we consider the neighbors of $next(a)$, distinct

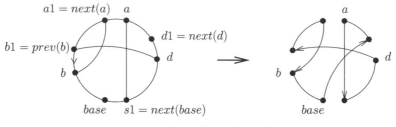

$$flip(s1, b1),\ flip(b1, d1),\ flip(a1, s1)$$

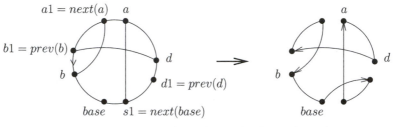

$$flip(s1, d1),\ flip(d, a),\ flip(a1, b1)$$

Figure 15.6 Alternative first step, case 2.

from $base$, $next(base)$, and a, that satisfy

$$c(next(a), b) < c(a, next(a)) + c(base, next(base)) - c(next(base), a).$$

(This is the analog of inequality (15.3).) The pairs are ordered by decreasing values of $c(b1, b) - c(next(a), b)$. Call this the B-*ordering*. Finally, in selecting $(d, d1)$, we consider the neighbors of $b1$, distinct from $base$, $next(base)$, a, $next(a)$, and b, that satisfy

$$c(b1, d) < c(b, b1) + c(base, next(base) - c(next(base), a)$$
$$+ c(a, next(a)) - c(next(a), b).$$

In this case, the pairs are ordered by decreasing values of $c(d1, d) - c(b1, d)$. Call this the D-*ordering*.

   With this terminology, we can write the function `alternate_step` as in Algorithm 15.2.

   Putting the pieces together, we can write the function `lk_search` that takes as arguments a vertex v and a tour T, as in Algorithm 15.3.

   To apply this, we mark all vertices, then call `lk_search(v)` for some vertex v. If the search is unsuccessful, we unmark v and continue with some other marked vertex. If the search is successful, then it is possible that some unmarked vertices may now permit successful searches. For this reason, Lin and Kernighan again mark all vertices before continuing with the next search. This was an appropriate strategy for the problem instances they studied, but it is too time-consuming for larger instances. The trouble is that unmarked vertices that are not close to the improving sequence of flips have little chance of permitting successful searches.

Algorithm 15.2 `alternate_step`.

$s1 = next(base)$;
create the *A-ordering* of the neighbors of $next(base)$;
i = 1;
while there exist unprocessed vertices in the *A-ordering* and
    $i \leq breadth_A$
do let a be next vertex in the *A-ordering*;
    $a1 = next(a)$;
    create the *B-ordering* from the neighbors of $next(a)$;
    $j = 1$;
    while there exist unprocessed pairs in the *B-ordering* and
        $j \leq breadth_B$
    do let $(b, b1)$ be the next pair in the *B-ordering*;
        if b lies of the tour segment from $next(base)$ to a
        then add the flips listed in Figure 15.5 to the flip sequence
            and set *delta* to the difference of the weight of the
            deleted edges and the weight of the added edges;
            call `step`(3, *delta*);
            if an improved tour has been found
            then return;
            else delete the added flips from the flip sequence;
        else create the *D-ordering* from the neighbors of $b1$;
            k = 1;
            while there exist unprocessed pairs in the
                D-ordering and $k \leq breadth_D$
            do let $(d, d1)$ be the next pair in the *D-ordering*;
                add the flips listed in Figure 15.6 to the flip;
                sequence and set *delta* as above;
                call `step`(4, *delta*);
                if an improved tour has been found
                then RETURN;
                else delete the added flips from the
                    flip sequence;
                    increment k;
            end
        end
    end
    increment j;
  end
  increment i;
end

Algorithm 15.3 lk_search(v, T).

> initialize *current_tour* as T;
> initialize an empty flip sequence;
> $base = v$;
> call step(1, 0);
> **if** an improved *current_tour* has been found
> **then** return the improving flip sequence;
> **else** call alternate_step();
>> **if** an improved *current_tour* has been found
>> **then** return the improving flip sequence;
>> **else** return with an unsuccessful flag;
> **end**

To deal with this issue, Bentley [55] (in the context of 2-opt) proposed to mark only those vertices that are the end vertices of one of the flips in the improving sequence. This leads to somewhat worse tours, but greatly improves the running time of the algorithm. We summarize the method in the function lin_kernighan described in Algorithm 15.4; lin_kernighan takes as an argument a tour T.

Algorithm 15.4 lin_kernighan(T).

> initialize *lk_tour* as T;
> mark all vertices;
> **while** there exist marked vertices
> **do**   select a marked vertex v;
>> call lk_search(v, *lk_tour*);
>> **if** an improving sequence of flips is found
>> **then while** the flip sequence is nonempty
>>> **do**   let $flip(x, y)$ be the next flip in the sequence;
>>> apply $flip(x, y)$ to *lk_tour* to obtain a new *lk_tour*;
>>> mark vertices x and y;
>>> delete $flip(x, y)$ from the flip sequence;
>> **end**
>> **else** unmark v;
> **end**
> return tour *lk_tour*;

The Lin-Kernighan routine consistently produces good-quality tours on a wide variety of problem instances. Computational results on variations of the algorithm can be found in Bachem and Wottawa [28], Bland and Shallcross [66], Codenotti et al. [131], Grötschel and Holland [232], Johnson [289], Johnson and McGeoch

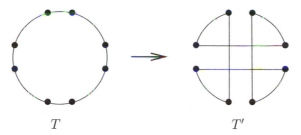

$$T \qquad\qquad\qquad\qquad T'$$

Figure 15.7 A double-bridge.

[291], Jünger et al. [296], Mak and Morton [369], Padberg and Rinaldi [447], Perttunen [460], Reinelt [475], Rohe [484], Schäfer [490], Verhoeven et al. [537], Verhoeven et al. [536], as well as the original paper of Lin and Kernighan [357].

An important part of Lin and Kernighan's overall tour-finding scheme is the repeated use of lin_kernighan. The idea is simple: for as long as computing time is available, we have a chance of finding a tour that is better than the best we have found thus far by generating a new initial tour and applying lin_kernighan. This worked extremely well in the examples they studied and it remained the standard method for producing very good tours for over 15 years. The situation changed, however, with the publication of the work of Martin, Otto, and Felten [379] (see also Martin et al. [380] and Martin and Otto [381]). Their new idea was to work harder on the tours produced by lin_kernighan, rather than starting from scratch on a new tour. They proposed to *kick* the tour found by lin_kernighan (that is, to perturb it slightly), and apply lin_kernighan to the new tour. Their kick was a sequence of flips that produces the special type of 4-opt move, called a *double-bridge*, that is illustrated in Figure 15.7. (There are many other natural candidates for kicking, but this particular one appears to work quite well.) The resulting algorithm, known as *Chained Lin-Kernighan*, starts with a tour S and proceeds as described in Algorithm 15.5.

Chained Lin-Kernighan is a substantial improvement over the "Repeated Lin-Kernighan" approach. Computational results comparing the two schemes can be found in Codenotti et al. [131], Hong et al. [269], Johnson [289], Johnson and McGeoch [291], Jünger et al. [296], Martin et al. [380], Neto [430], and Reinelt [475].

In the remainder of the chapter, we will discuss the computational issues involved in implementing and using Chained Lin-Kernighan. It should be remarked that Martin, Otto, and Felten described a more general scheme than the one we have outlined. They proposed to use a simulated annealing-like approach and replace T by T' (even if T' is not a better tour) with a certain probability that depends on the difference in the costs of the two tours.

We call Martin, Otto, and Felten's algorithm "Chained Lin-Kernighan" to match the "Chained Local Optimization" concept introduced in Martin and Otto [381], and to avoid a conflict with Johnson and McGeoch's [291] use of the term "Iterated Lin-Kernighan" to mean the version of the algorithm when random double-bridge

Algorithm 15.5 Chained Lin-Kernighan.

call lin_kernighan(S) to produce the tour T;
while computing time remains
do   find a kicking sequence of flips;
        apply the kicking sequence to T;
        call lin_kernighan(T) to produce the tour T';
        if T' is cheaper than T
        then replace T by T';
        else use the kicking sequence (in "reverse") to convert
                T back to the old T (the one we had before
                applying the kick);
end
return T

moves are used as kicks and no simulated annealing approach is used. (See also Johnson [289].)

15.2 FLIPPER ROUTINES

The main task in converting the outline of Chained Lin-Kernighan into a computer code is to build data structures to maintain the three tours: the *lk_tour* in lin_kernighan, the *current_tour* in lk_search, and the overall tour (T in Algorithm 15.5). If these are not implemented carefully, then operations involving them will be the dominant factor in the running time of the code.

Note that the three data structures need not be distinct, since additional flip operations can be used to undo the flips made during an unsuccessful lk_search, as well as to undo all of the flips made during an unsuccessful call to the function lin_kernighan.

An abstract data type sufficient to represent all three tours should provide the functions

- flip(a, b)—inverts the segment of the tour from a to b

- next(a)—returns the vertex immediately after a in the tour

- prev(a)—returns the vertex immediately before a in the tour

- sequence(a, b, c)—returns 1 if b lies in the segment of the tour from a to c, and returns 0 otherwise

as well as an initialization routine and a routine that returns the tour represented by the data structure (we are following Applegate et al. [19]). In the outline of Chained Lin-Kernighan, whenever flip(a, b) needs to be called, *prev*(a) and

$next(b)$ are readily available (without making calls to prev and next). This additional information may be useful in implementing flip, so we allow our tour data structures to use the alternative

- flip(x, a, b, y)—inverts the segment of the tour from a to b (where $x = prev(a)$ and $y = next(b)$)

if this is needed.

Asymptotic analysis of two tour data structures can be found in Chrobak et al. [124] and Margot [372]. In both cases, the authors show that the functions can be implemented to run in $O(\log n)$ time per function call for an n-city TSP. We present a detailed computational study of practical versions of these two structures as well as several alternatives.

An excellent reference for tour data structures is the paper by Fredman et al. [195]. Their study works with a slightly different version of flip: they allow the function to invert either the segment from a to b or the segment from $next(b)$ to $prev(a)$. Along with computational results, Fredman et al. [195] established a lower bound of $\Omega(\log n/\log \log n)$ per function call on the amortized computation time for any tour data structure in the cell probe model of computation.

Test Data

To provide a test for the tour data structures, we need to specify both a problem instance and a particular implementation of Chained Lin-Kernighan. Problem instances are readily available through Reinelt's TSPLIB, with sizes ranging from 14 to 85,900 cities. From this library, we have selected two instances arising in applications and one instance obtained from the locations of cities on a map. These examples are listed in Table 15.1. The version of Chained Lin-Kernighan we use

Table 15.1 Problem instances.

Name	Size	Cost Function	Target Tour
pcb3038	3,038	Rounded Euclidean	139,070
usa13509	13,509	Rounded Euclidean	20,172,983
pla85900	85,900	Ceiling Euclidean	143,564,780

is one that is typical of those studied in Section 15.3 of this chapter. The various parameters that must be set in Chained Lin-Kernighan have an impact on the relative performance of the tour data structures, but the trends will be apparent with this test version. For each test problem, we run Chained Lin-Kernighan until a tour is found that is at least as good as the "Target Tour" listed in Table 15.2. These values are at most 1% greater than known lower bounds for the problem instances. Since the operation of Chained Lin-Kernighan is independent of the particular tour data structure that is used, each of our test runs will be faced with exactly the same set of flip, next, prev, and sequence operations. Some important statistics for these operations are given in Table 15.2. The lin_kernighan, lk_search,

Table 15.2 Statistics for test data.

Function	pcb3038	usa13509	pla85900
lin_kernighan	141	468	1,842
lin_kernighan winners	91	261	1,169
flips in a lin_kernighan winner	61.0	99.1	108.3
flips in a lin_kernighan loser	42.5	88.2	86.4
lk_search	19,855	95,315	376,897
lk_search winners	1,657	9,206	29,126
flips in an lk_search winner	4.7	4.8	6.3
flip	180,073	1,380,545	5,110,340
undo flips	172,396	1,336,428	4,925,574
size of a flip	75.6	195.3	607.9
flips of size at most 5	67,645	647,293	1,463,090
next	662,436	6,019,892	14,177,723
prev	715,192	4,817,483	13,758,748
sequence	89,755	773,750	263,7757

flip, next, prev, and sequence rows give the number of calls to the listed function; "lin_kernighan winners" is the number of calls to lin_kernighan that resulted in a tour that was at least as good as the current best tour; "flips in a lin_kernighan winner" is the average number of flips performed on lk_tour in lin_kernighan winners; "flips in a lin_kernighan loser" is the average number of flips performed on lk_tour in calls to lin_kernighan that resulted in a tour that was worse than the current best tour; "lk_search winners" is the number of calls to lk_search that returned an improving sequence; "flips in an lk_search winner" is the average length of an improving sequence of flips; "undo flip" is the number of flip operations that are deleted in lk_search; "size of a flip" is the average, over all calls $flip(a, b)$, of the smaller of the number of cities we visit when we travel in the tour from a to b (including a and b) and the number of cities we visit when we travel from $next(b)$ to $prev(a)$ (including $next(b)$ and $prev(a)$); "flips of size at most 5" is the total number of flips of size 5 or less.

The codes tested in this section are written in the C Programming Language (Kernighan and Ritchie [314]), and run on a workstation equipped with a 500-MHz Compaq Alpha EV56 processor. The times reported are in seconds of computing time, including the time spent in computing the starting tour and the neighbor sets; the important point here is the relative speed of the various implementations of the algorithms, rather than the absolute times on the Alpha processor.

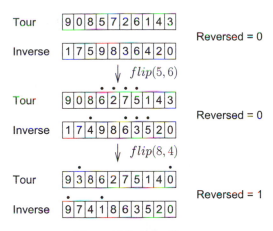

Figure 15.8 Array flipper.

ARRAYS

A natural candidate for a tour data structure is to keep an array, called *tour*, of the cities in the order they appear in the tour. To locate a given city in *tour*, we use another array, called *inverse*, where the *i*th item in *inverse* contains the location of city *i* in *tour*. The top pair of arrays in Figure 15.8 illustrate the data structure for the tour 9-0-8-5-7-2-6-1-4-3.

This data structure is particularly easy to implement. The functions `next` and `prev` are provided by the formulas

$$next(v) = tour(inverse(v) + 1)$$
$$prev(v) = tour(inverse(v) - 1)$$

where the indices are taken modulo n. To provide `sequence`, we can determine whether or not city b lies on the tour segment from a to c by examining the values of $inverse(a)$, $inverse(b)$, and $inverse(c)$. The time-consuming operation is `flip`. To carry out $flip(a, b)$, we need to work our way through the a to b segment of the tour array, swapping the positions of a and b, $next(a)$ and $prev(b)$, and so on, until the entire segment has been inverted. At the same time, we need to swap $inverse(a)$ and $inverse(b)$, $inverse(next(a))$ and $inverse(prev(b))$, and so on. This operation is illustrated in the middle pair of arrays in Figure 15.8. The entries with the dots are the ones that were swapped.

It is too time-consuming to create copies of *tour* and *inverse* with each call to `lk_search`, so we will use the same pair of arrays to represent both *lk_tour* and *current_tour*. To do this, we use an additional call to `flip` whenever we delete an item from the current flip sequence. (To delete $flip(a, b)$, we call $flip(b, a)$.)

The performance of this array data structure is quite poor. For the three test problems, the Alpha workstation times (in seconds) are

pcb3038	usa13509	pla85900
7.2	246.6	10,422.5

respectively.

Not surprisingly, the above running times are dominated by the times for the flip operations: 91.7% for pcb3038, 97.4% for usa13509, and 99.4% for pla85900. A simple way to improve this is to maintain a *reversed* bit that indicates the orientation of the tour. If *reversed* is set to 0, then *tour* gives the proper order of the cities, and if it is set to 1, then *tour* gives the cities in the reverse order. The big advantage is that we can carry out flip(a, b) by inverting the tour segment from $next(b)$ to $prev(a)$ and complementing the *reversed* bit, if the segment from $next(b)$ to $prev(a)$ is shorter than the segment from a to b. The *reversed* bit must be examined when we answer next, prev, and sequence, but this extra computation is more than compensated by the much lower time for flip. The resulting code improves the running times to

pcb3038	usa13509	pla85900
1.6	21.6	265.9

on our test set. In this version, flip is fast enough that we can lower the running times a bit more by also representing the overall best tour with the same pair of arrays that is used for *lk_tour* and *current_tour*. This means that we must keep a list of all of the successful flip sequences made during a given call to lin_kernighan, and "undo" all of these flips: working in reverse order, we call flip(b, a) for each $flip(a, b)$. The slightly better running times are

pcb3038	usa13509	pla85900
1.6	21.5	242.2

for our test problems.

A breakdown of the time spent in the tour operations for this last version of the array-based code is given in Table 15.3. The rapidly growing time for flip means

Table 15.3 Profile for arrays.

Function	pcb3038	usa13509	pla85900
flip	51%	74%	88%
next	1%	1%	0%
prev	1%	1%	0%
sequence	0%	0%	0%

that the data structure is probably not acceptable for instances that are much larger than pla85900. On the other hand, given the ease with which the computer code can be written, the good performance of arrays on the two smaller problems indicates that arrays may be the method of choice in many situations.

LINKED LISTS

A second natural tour data structure is a doubly linked list, where each city has pointers to its two neighbors in the tour. (See Figure 15.9.) With this structure, it is

Figure 15.9 Linked list tour.

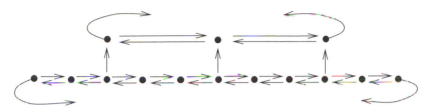

Figure 15.10 Linked list with a second level.

convenient to use the alternative $flip(x, a, b, y)$ form of the flip function, since this allows us to implement flip by manipulating four pointers:

replace $x \rightarrow a$ by $x \rightarrow b$,

replace $a \rightarrow x$ by $a \rightarrow y$,

replace $y \rightarrow b$ by $y \rightarrow a$,

replace $b \rightarrow y$ by $b \rightarrow x$.

The orientation of the tour can be maintained by choosing one or more cities and marking which of their two neighbors is *next* and which is *prev*. Call such a marked city a *guide*. Initially we can select any subset of the cities as guides, since the orientation of the starting tour is known, and after $flip(x, a, b, y)$ we can directly store the information to use any one or more of a, b, x, and y as a guide (the old guides are invalidated by the flip operation). To implement $next(v)$, we start at city v and trace the tour until we arrive at a guide, which will indicate whether or not the first city we visited after v was $next(v)$. If it was indeed $next(v)$, then we return it. Otherwise we return the other neighbor of v. The same procedure can be used to implement $prev(v)$. To implement $sequence(a, b, c)$, we first trace the tour from a until we reach a guide, in order to learn the orientation of the neighbors of a, then trace the tour in the forward direction from a. If we reach c before we reach b, then we return 0. Otherwise, we return 1.

A difficulty with this straightforward implementation is that we will often traverse large portions of the tour in calls to next, prev, and sequence. Margot's [372] answer is to include additional information in the linked list to allow the traversals to skip over large blocks of cities. (A similar method was proposed by Shallcross (private communication, 1990).) The simplest version is to include a second doubly linked list consisting of a subset of $\sqrt{n}$ of the cities, and require that each of these cities be guides. (See Figure 15.10.) As long as the selected cities are spread out roughly evenly, this requires only $O(\sqrt{n})$ additional work in $flip$: we trace the tour from a in the direction of y until we reach one of the selected cities and fix its orientation, we then use the extra doubly linked list to fix the orientation of each of the other selected cities. Furthermore, with the large supply of guides, next and prev will run in $O(\sqrt{n})$ time, and with some additional work, sequence can also be implemented in $O(\sqrt{n})$ time (using the extra doubly linked list to trace from a to c).

Margot [372] takes this idea to its natural conclusion, adding not one extra doubly linked list but $\log n$ additional lists, each one a subset of the previous list. He also gives a construction for explicitly maintaining a balance condition that keeps the cities spread out roughly evenly, and thus obtains an $O(\log n)$ running time bound for each of the tour functions. We will study how this idea of having additional linked lists works in practice.

To begin, let us consider the straightforward implementation of a single doubly linked list. Given the low cost (in terms of CPU time) for the flip operations, we will use the same list to represent the overall tour, *lk_tour*, and *current_tour*. As guides, we use the two ends of the most recent flip.

The running times are

pcb3038	usa13509	pla85900
50.8	5,929.4	> 50,000

for our test problems. (The run on pla85900 was not completed after 50,000 seconds on the Alpha workstation.) This performance is even worse than the original (no *reversed* bit) array implementation. The profile of the runs on the two smaller problems, given in Table 15.4, indicates that the search for guides is taking nearly

Table 15.4 Profile for linked lists.

Function	pcb3038	usa13509
flip	0%	0%
next	41%	37%
prev	51%	56%
sequence	6%	6%

all of the CPU time. The main reason for this is simply that the tour segments that need to be traced can be quite long. To attack this issue, we modified the code to carry out the search for a guide simultaneously in both directions around the tour. If we are working from random cities, then this change should not have much of an affect on the running times, since, on average, we would be still be visiting the same number of cities per search. The important point, however, is that we are not working from random locations: the next, prev, and sequence queries tend to be from cities that are near to the previous flip (due to the use of neighbor sets). Indeed, the change improves the running times to

pcb3038	usa13509	pla85900
15.7	426.3	24,047.9

for our test problems.

The profile for the improved code is given in Table 15.5. Although the running times are better, it is clear that guide searching is still taking too long. A brute force

Table 15.5 Profile for linked lists with simultaneous searches.

Function	pcb3038	usa13509	pla85900
flip	2%	1%	0%
next	31%	35%	28%
prev	32%	35%	23%
sequence	19%	24%	49%

way to deal with this problem is to give up the constant-time flip operations, and explicitly store the two neighbors of each city as *prev* and *next*. To maintain this information, we will need to swap *prev* and *next* for each city along the segment that is flipped, so flips will become much more expensive. As in the array implementation, it is important to use a *reversed* bit to permit us to flip the shorter of the two segments in a flip operation. This give the more respectable

pcb3038	usa13509	pla85900
2.8	113.6	2,109.5

performance.

A difficulty with this code (as opposed to the array implementation) is that we need to do extra traversals of the tour in order to determine whether it is better to invert the segment from a to b or the segment from $next(b)$ to $prev(a)$, in a call to $flip(a,b)$. We can get around this by maintaining an *index* for each city that gives its relative position in the tour. The indices are consecutive integers, starting from some city. With such indices, we can immediately determine which of the two possible segments in a flip operation is the shorter. As a byproduct, we can also use the indices to implement sequence without traversing the tour. The downside is that the index for each city in the tour segment that is inverted in a flip operation needs to be updated. The resulting code is indeed faster, however, as is shown by the running times

pcb3038	usa13509	pla85900
1.8	65.6	697.3

for our test instances.

The profile for these runs is given in Table 15.6. It indicates that the only way to make substantial improvements in the code is to somehow speed up the flip operations, without completely sacrificing next and prev as we did earlier. Fortunately, this is exactly the type of improvement provided by Margot's idea of keeping additional linked lists as "express lanes." In our implementation, we follow the lessons learned from the single list implementations: we use simple linked lists for all lower levels of the data structure and a linked list with explicit *next* and *prev*, reversed bit, and indices for the top level. Rather than choosing $n^{(D-k)/D}$ cities for the kth level in a D-level data structure (where k runs from 0 up to $D-1$), we allow a bit more flexibility, choosing first some constant γ and then selecting $\gamma^k n^{(D-k)/D}$ cities at the kth level. In Table 15.7, we report the running times

Table 15.6 Profile for linked lists with explicit *next* and *prev*.

Function	pcb3038	usa13509	pla85900
flip	54%	79%	94%
next	1%	1%	0%
prev	1%	1%	0%
sequence	0%	0%	0%

Table 15.7 Running times for multilevel linked lists.

Structure	pcb3038	usa13509	pla85900
2 levels	1.7	16.0	104.1
3 levels	2.3	18.5	81.4
4 levels	2.5	20.3	82.1

using two, three, and four levels. The values of γ were 8, 5, and 3, respectively. (These were the (positive integer) values giving the best results for the particular data structure.) The profile for the three-level data structure is given in Table 15.8. The running times are a big improvement over the best times for our single list im-

Table 15.8 Profile for three-level linked lists.

Function	pcb3038	usa13509	pla85900
flip	42%	37%	38%
next	5%	8%	10%
prev	6%	10%	9%
sequence	2%	3%	4%

plementations, and are also significantly better than the array implementation (with a *reversed* bit). On the other hand, the times point out that the theoretically attractive method of using $\log n$ levels is probably too aggressive to use in practice; for example, at 85,900 cities, three levels is preferable to four or more levels.

TWO-LAYERED LISTS

Multilevel linked lists traded off very fast next and prev times for improved performance in flip operations. A different approach, proposed by Chrobak et al. [124], aims at getting improvements in flip while maintaining the constant time operation of next and prev. The idea is to divide the tour into blocks of size roughly $\sqrt{n}$. Each block has an associated *parent* node, and the parents are

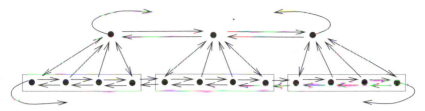

Figure 15.11 Two-layered list.

organized in a cycle that gives the ordering of the blocks in the tour. An important concept is that each parent p has a bit, $p.reversed$, that indicates whether or not the tour segment in the associated block is reversed in the tour represented by the data structure. These bits allow us to invert a block of cities in constant time, and this leads to a fast implementation of flip. Chrobak, Szymacha, and Krawczyk call their data structure a *two-layered list*. It is studied in detail in Fredman et al. [195] (under the name "two-level trees"), and we follow their implementation.

The tour segments making up the blocks in the data structure are represented as doubly linked lists with explicit *next* and *prev* pointers and with indices giving the relative location of the cities in the segment. The cities that are ends of a block also have a pointer to the neighboring city in the tour that is not a member of their block. Each city has a pointer to the parent of its block and each parent has pointers to the two cities that are the ends of its associated tour segment. The cycle of parents is represented by a doubly linked list with explicit *next* and *prev*, location indices, and a *reversed* bit. A sketch of the data structure is given in Figure 15.11.

The structure does indeed allow constant time next and prev operations, since, for example, $next(a)$, for a given city a, is the city in a's *next* pointer if *reversed* and the reversed bit of a's parent are equal, and otherwise $next(a)$ is the city in a's *prev* pointer. (The "if" test can be avoided if we store the *next* and *prev* pointers as a two-element array, and index the array by the exclusive or of *reversed* and a's parent's reversed bit.) The sequence operations can also be provided in constant time, making use of the indices on the cities to determine the relative order within a segment and the indices for the parents to determine the relative order of the blocks. If the size of the blocks is kept to roughly $\sqrt{n}$, then flip can be implemented to run in $O(\sqrt{n})$. Chrobak, Szymacha, and Krawczyk accomplish this in the following way. At the start, each block is of size between $\sqrt{n}$ and $2\sqrt{n}$. To perform flip(a, b), we examine city a. If a is not the first city in its block (or the last city if the block is reversed), then we remove the portion of the block that precedes a and merge it with the preceding block. If the merged block now has size larger than $2\sqrt{n}$, then it is split into two blocks of (nearly) equal size. Next, we merge a's block with the block following it, and again split the merged block in two if it is too large. In a similar way, we make city b the last city in its block (or first city, if the block is reversed). Now the tour segment from a to b can be inverted by inverting the segment in the parent cycle from the parent of a to the parent of b, and complementing the reversed bit of each parent node that is involved in the flip.

Fredman, Johnson, McGeoch, and Ostheimer speed up the practical performance of this `flip` procedure in three ways. First, they give up the idea of explicitly keeping the size of the segments balanced; this could have a detrimental effect on the data structure in the long run, but it seems to be the appropriate choice in practice. Second, they make a the first city in its block by either merging the portion of the segment preceding a with the preceding block or merging the portion of the segment starting at a with the following block (depending on which of the two segments is shorter), rather than performing two merges. Third, if a and b are in the same block and are not too far apart in the tour segment (this can be determined using the indices), then the segment from a to b is inverted directly in the linked list structure for the block, rather than performing the merges that make a and b the ends of the block.

In the Fredman et al. implementation, the initial blocks are chosen to contain approximately K cities each, where K is some constant. If cities a and b are in the same block and no more than $.75K$ cities apart, then $\text{flip}(a, b)$ is performed directly in the linked list for the block. Although the running times are affected by the choice of K, the dependence is fortunately not that strong. For this reason, Fredman et al. choose to use $K = 100$ in their code for all problem instances having at most 100,000 cities. In our implementation, we set $K = \alpha\sqrt{n}$ to allow the code to be somewhat more robust over a wider range of problem sizes. The default value in our code is $\alpha = 0.5$.

Fredman et al. point out that the choice of $.75K$ as the cutoff for performing `flip` operations directly in a block helps keep the size of the blocks roughly in balance, since only the larger blocks are likely to be split by `flip` operations involving two cities in the same block. In our computer code, the operations necessary to perform a merge are considerably faster than those for inverting a segment of the linked lists, so it is worthwhile to consider cutoffs βK for smaller values of β. Based on a series of tests, we have chosen $\beta = 0.3$ as our default value.

Using the same data structure for lk_tour, $current\_tour$, and the overall tour, the code gives the running times

pcb3038	usa13509	pla85900
1.2	10.1	43.9

for our test instances. A profile of the test runs is given in Table 15.9.

Table 15.9 Profile for two-layered lists.

Function	pcb3038	usa13509	pla85900
flip	19%	19%	26%
next	4%	4%	3%
prev	3%	4%	3%
sequence	1%	1%	1%

The performance of two-layered lists is very good—the test results are nearly a factor of 2 better than those for three-level linked lists. Furthermore, the profile reports that on the three test instances, the time spent on the tour operations is less than one third of the total running time of the code. Nonetheless, it certainly should be possible to improve the performance of the data structure with some additional tweaks or new ideas. One possibility would be to replace the linked list used to represent the parent cycle by something more effective. We made one attempt in this direction, using an array with *reversed* bit for the parents, but the running times

pcb3038	usa13509	pla85900
1.2	10.3	46.0

are not as good as those for the linked list structure (due to the extra dereferencing that was needed).

BINARY TREES

Along with two-layered lists, Chrobak et al. [124] proposed an elegant method for obtaining an $O(\log n)$ bound on tour operations using binary trees. The key idea is to attach a *reversed* bit to each node of the tree indicating how the subtree rooted at that node should be interpreted.

Let B be a rooted binary tree on n vertices, with each vertex associated with a unique city. If all of the *reversed* bits are set to 0, then the tour represented by B is that given by an in-order traversal of the cities. Setting the *reversed* bit of a vertex v to 1 inverts the tour segment associated with the subtree rooted at v.

In general, suppose some subset of the *reversed* bits are set to 1. For each vertex v, let $parity(v)$ denote the parity of the number of *reversed* bits that are equal to 1 on the path from v up to the root of B (including both v and the root). The tour represented by B is given by traversing the tree, starting at the root, using a rule that depends on $parity(v)$ for each vertex v. If $parity(v)$ is 0, then we traverse the subtree rooted at v by first traversing the subtree rooted at the left child of v, then v itself, followed by a traversal of the subtree rooted at the right child of v. If $parity(v)$ is 1, then we traverse the subtree in the reverse order, first traversing the subtree rooted at the right child of v, then v itself, followed by the subtree rooted at the left child of v. As an example, the tree given in Figure 15.12 represents the tour 9-0-8-5-7-2-6-1-4-3.

It is clear from this definition that the tour represented by B is unchanged if we pick some vertex v, complement its *reversed* bit and the *reversed* bits of its children, and swap v's left child with v's right child. We call this procedure *pushing* v's *reversed* bit.

The role of the *reversed* bit is to allow us to implement flip efficiently. To carry out flip(a, b), we partition B into into several components to isolate the $[a, b]$ segment, we flip this segment by complementing a *reversed* bit, and we glue the tree back together. The partitioning and gluing can be handled by the standard *split* and *join* operations for binary trees, as described, for example, in Knuth [318] or Tarjan [520]. A simple way to handle the complications of the *reversed* bits is to push any bit that is set to 1 on the path from x to the root of the tree, starting at the root and working backward. (It is more efficient not to do this, but the difference is quite small.)

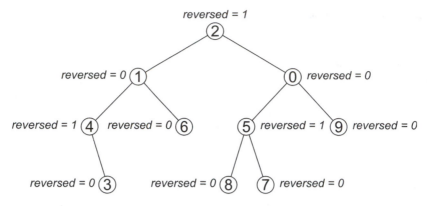

Figure 15.12 Binary tree for 9-0-8-5-7-2-6-1-4-3.

Similarly, we can implement $sequence(a, b, c)$ by splitting B into components to isolate $[a, c]$, and checking which component contains b.

Finally, given a tree B, we can find $next(a)$ and $prev(a)$ by searching B starting at the vertex associated with a. The amount of work needed for the search is bounded by the maximum depth of the tree.

To obtain the O($\log n$) result, Chrobak et al. [124] used balanced binary trees (AVL trees in their implementation). We did not implement this feature, but our code does not appear to be hurt by the fact that the trees are not being explicitly balanced. Indeed, for each of the four tour operations we computed the depth of the vertices corresponding to the cities involved in the operation (at the time the function was called). The average value over all operations was

pcb3038	usa13509	pla85900
6.0	6.3	6.7

for our set of test problems.

The running times for the implementation were

pcb3038	usa13509	pla85900
1.4	12.6	52.9

respectively. In this test, we used the same data structure to maintain lk_tour, $current\_tour$, and the overall tour. Although the times are slightly worse than the times for the two-layered list implementation, the binary-tree data structure is much more natural to code than two-layered lists, as well as being more efficient on larger instances. A profile for the runs is given in Table 15.10.

An alternative approach for implementing flip was proposed by Applegate et al. [19] and Fredman et al. [195]. Rather than splitting the tree into components, we can perform *splay* operations (see Sleator and Tarjan [502] and Tarjan [520]) on B, to make the vertices in the flipping segment appear together as a single subtree of the tree. We then complement the *reversed* bit associated with the root of the subtree. A detailed treatment of this approach can be found in Fredman et al. [195].

Table 15.10 Profile for binary trees.

	pcb3038	usa13509	pla85900
flip	17%	16%	17%
next	7%	10%	13%
prev	9%	12%	11%
sequence	5%	6%	6%

The running times for our implementation were

pcb3038	usa13509	pla85900
1.5	12.7	50.5

for the test instances. These times are similar to those for the split and join version of the data structure.

SUMMARY

From the tests, it is clear that the array data structure is adequate for many applications, particularly when the instances are relatively small (say at most 10,000 cities). The multilevel linked-list implementations perform better than arrays on larger instances, but they are dominated in performance by the two-layered list data structure. Binary trees (including the splay tree variants) perform slightly worse than two-layered lists on TSPLIB problems, but they are very easy to code and should be the data structure of choice in applications involving large instances. Indeed, on an example having 5 million cities, the binary-tree codes were nearly a factor of 2 faster than two-layered lists.

We did not include, in our tests, the *segment-tree* data structure proposed by Applegate et al. [19]. Details of this implementation can be found in Fredman et al. [195] and Verhoeven et al. [536]. Segment trees suffer from several drawbacks: they are not easy to code efficiently and they do not scale up well to large problem instances. Moreover, in our implementations, segment trees performed somewhat worse than two-layered lists on TSPLIB instances.

Further work on tour data structures is presented in Gamboa et al. [199] and in the doctoral thesis of Nguyen [432].

15.3 ENGINEERING LIN-KERNIGHAN

With a tour data structure in hand, it is not difficult to get a working version of Chained Lin-Kernighan. There are, however, a wide variety of implementations that are consistent with the basic algorithm that we have outlined. The decisions that must be made in an implementation can, moreover, make or break the performance of the algorithm. We discuss these design issues in this section.

To keep our presentation down to a manageable size, we will limit our computational reports to the single TSPLIB instance usa13509. In each of our tables, we give results on running specific implementations of Chained Lin-Kernighan until we reach a tour of value at most 20,082,519, which is within 0.5% of the optimal value for usa13509. Reports on other TSPLIB instances and other tour qualities will be presented in the next section.

BREADTH OF lk_search

In our description of lk_search, we included the parameters

$$breadth(k), breadth_A, breadth_B, breadth_D,$$

specifying the maximum number of promising neighbors to be considered. These parameters are the principal means for controlling the breadth of the search, and it is obvious that some care must be taken in selecting their values.

In Table 15.11, we report results for several choices of $breadth$. Each row of the table summarizes 10 runs using distinct random seeds. The column labeled "Main" contains the values $(breadth(k), k = 1, \ldots, t)$, where $breadth(k) \leq 1$ for all $k > t$. The column labeled "Alternate" contains the triple $(breadth_A, breadth_B, breadth_D)$. The running times are reported in seconds on the 500-MHz Alpha EV56 workstation described in the previous section. The "Mean Steps" column gives the average number of calls to lin_kernighan in the 10 runs.

Table 15.11 Varying the breadth of the search (10 trials).

Main	Alternate	Mean Time	Max Time	Mean Steps
(1)	(1, 1, 1)	99.9	138.0	14,659
(3)	(2, 1, 1)	40.1	51.1	5,442
(10)	(10, 1, 1)	39.4	62.8	3,854
(5, 2)	(5, 2, 1)	31.9	40.1	2,867
(4, 3, 2)	(5, 2, 1)	34.9	42.5	2,389
(10, 5)	(10, 5, 2)	41.1	73.2	2,519
(10, 5, 3)	(10, 5, 2)	41.2	55.5	1,734
(12, 8, 4, 2)	(10, 5, 2)	56.3	72.2	1,468
(4, 3, 2, 2, 2, 2, 2)	(5, 2, 2)	53.0	71.4	1,372
(8, 6, 4, 2, 2, 2)	(10, 5, 2)	75.8	97.8	1,108

The rows of Table 15.11 are ordered according to the total breadth of the search. Not surprisingly, the number of steps required is almost uniformly decreasing as we move down the table. The running times, however, favor a modest amount of backtracking, spread out over the first two or three levels of the search.

The algorithm used to obtain these results includes Mak-Morton moves, but only for levels k such that $breadth(k) = 1$. We can further manipulate the breadth of the search by either including Mak-Morton moves at all levels or by excluding them entirely. Moreover, we have the option of performing another type of move developed by Reinelt [475]. His moves are called *vertex insertions* since they correspond to taking a vertex from the tour and inserting it at another point in the tour. In Table 15.12, we report on a number of combinations of these moves, with *breadth* set at (5, 2) and the alternate *breadth* set at (5, 2, 1).

Two things are apparent in Table 15.12. First, vertex-insertion moves decrease the number of steps required, but in our implementation this is more than offset by the extra time needed to handle these moves. Second, Mak-Morton moves appear to be a good idea later in a search, but not at the first several levels. We must remark that Mak and Morton [369] originally developed their moves as an alternative to using the `alternate_step` function (motivated by the complication this function adds to the implementation of `lk_search`), and thus it is not surprising that using both `alternate_step` and early Mak-Morton moves is not advantageous.

Table 15.12 Mak-Morton moves and vertex insertions (10 trials).

Mak-Morton	Vertex-Insertion	Mean Time	Max Time	Mean Steps
No	No	43.7	60.2	5,329
Partial	No	31.9	40.1	2,867
Yes	No	42.9	62.2	2,986
No	Yes	48.1	60.6	4,607
Partial	Yes	38.1	45.0	2,756
Yes	Yes	50.2	64.1	2,925

For the remaining tests in this section, we use partial Mak-Morton moves, we use no vertex-insertion moves, and we set *breadth* and the alternate *breadth* at (5, 2) and (5, 2, 1), respectively.

THE NEIGHBOR SETS

Our description of `lk_search` makes use of a prescribed set of neighbors for each vertex in the TSP. The choice of these neighbor sets directly affects the quality of the moves in `lk_search`, since we consider only flips that involve a vertex and one of its neighbors.

Rather than treating the neighbors as subsets of vertices, we can consider the *neighbor graph* consisting of the vertex set of the TSP, with edges joining vertices to their neighbors. Indeed, we define our neighbor sets in terms of this graph: if two vertices are adjacent, then we make each a neighbor of the other.

There are many choices for the neighbor graph. An obvious candidate is the *k-nearest* graph, consisting of the k least costly edges meeting each vertex. This works well on many examples, but it can cause problems on geometric instances where the points are clustered, since it does not tend to choose intercluster edges. To help in these cases, Miller and Pekny [397] proposed the *k-quad-nearest* graph, defined as the k least costly edges in each of the four geometric quadrants (for two-dimensional geometric instances) around each vertex. Miller and Pekny studied this graph in the context of 2-matching algorithms, and Johnson and McGeoch [291] have shown that it is an effective neighbor graph for Chained Lin-Kernighan.

For geometric instances, another good choice is the *Delaunay triangulation* of the point set, as described, for example, in Aurenhammer [25], Edelsbrunner [162], and Mehlhorn [388]. This triangulation has been proposed as a neighbor graph in Jünger et al. [296] and Reinelt [474]. It has the nice property of being very sparse, while still capturing well the structure of the point set.

The results reported in Table 15.11 and in Table 15.12 were obtained with the 3-quad-nearest graph. In Table 15.13, we report results for a number of other choices.

The point set for usa13509 is reasonably well distributed, and thus the k-nearest graph works well for modest choices of k. Superior results were obtained, however, using the Delaunay graph. The triangulation was computed using the computer code "sweep2" by Fortune, based on the sweepline algorithm described in Fortune [193].

Table 15.13 Choosing the neighbor graph (10 trials).

Neighbor Graph	Mean Time	Max Time	Mean Steps
5-nearest	53.6	114.0	7,148
10-nearest	31.7	41.6	3,142
20-nearest	35.5	43.7	3,152
1-quad-nearest	36.0	50.1	5,392
2-quad-nearest	33.6	42.1	3,257
3-quad-nearest	31.9	40.1	2,867
4-quad-nearest	35.9	43.9	3,131
5-quad-nearest	40.4	52.7	3,266
Delaunay	26.8	35.2	3,546

In the remainder of this section, we will use the Delaunay graph to determine our neighbor sets.

DEPTH OF lk_search

In lk_search we attempt to construct a sequence of flips that results in an improved tour. Lin and Kernighan [357] proposed a straightforward method for ensuring that these sequences are bounded in length: they forbid flips that add edges to *current_tour* that have previously been deleted in the search, as well as forbidding flips that delete edges that have previously been added. We have incorporated this idea into our implementation, but we have found it useful to take further measures to limit the depth of the search, as we describe below.

To begin, we can use the *breadth* parameters to set a hard limit on the depth by fixing $breadth(k) = 0$ for some k. In the previous tables, our implementations had $breadth(50) = 0$; in Table 15.14, we compare this choice with several others. For our test instance, it appears that a smaller bound performs better and we will set $breadth(25) = 0$ for the remaining tables in this section.

Table 15.14 Varying the maximum depth (10 trials).

Max Depth	Mean Time	Max Time	Mean Steps
5	37.6	67.5	9,110
10	28.7	37.0	5,111
25	22.1	33.9	3,093
50	26.8	35.2	3,546
100	26.5	34.8	3,248
∞	25.8	31.9	3,205

The advantage of a bounded depth search is that it prevents us from considering long sequences of flips that eventually involve vertices that are quite distant from the original *base* vertex. A particularly disturbing case of this is when we have already found an improved tour on the given search, but we continue even though

$$delta + c(base, next(base)) - c(next(base), next(probe)) \qquad (15.6)$$

is less than the amount of the improvement. To handle this situation, we tighten the definition of a promising vertex by insisting that (15.6) be at least as large as any improvement we have found thus far in `lk_search`, rather than requiring only that it be nonnegative. As indicated in Table 15.15, this gives slightly better

Table 15.15 Restrictions on promising neighbors (10 trials).

Restriction	Mean Time	Max Time	Mean Steps
Nonnegative	25.4	34.2	3,457
Max improvement	22.1	33.9	3,093

performance for the algorithm.

VERTEX MARKING

The number of times we call `lk_search` in a single run of `lin_kernighan` is controlled by the strategy we adopt in our Bentley-marking scheme. Recall that we have marks on our vertices; we begin searches only from marked vertices; we unmark a vertex after an unsuccessful search; and we mark the vertices that are the ends of the flips in the sequences found by successful searches. By marking additional (or fewer) vertices after a successful search, we can increase (or decrease) the number of `lk_search` calls.

In Table 15.16, we consider several possibilities: we mark either just the flip ends, the flip ends plus their adjacent vertices in *current_tour*, the flip ends plus the vertices that are at most two away from them in *current_tour*, or the flip ends plus their neighbor sets. The running times indicate that the best choice is to simply mark the flip ends as Bentley [56] proposed. We carried out a final experiment, where we marked each of the flip ends with probability 0.5, but this performed very poorly.

Table 15.16 Marking vertices (10 trials).

Marks	Mean Time	Max Time	Mean Steps
Flip ends	22.1	33.9	3,093
Tour 1-neighbors	24.7	31.6	3,303
Tour 2-neighbors	26.1	32.1	3,458
Graph neighbors	28.3	42.3	3,296
Flip ends (probability 0.5)	47.4	72.7	7,323

In contrast to these results, it does appear to be worthwhile to mark more than just the flip ends after applying a kicking sequence in Chained Lin-Kernighan. Our default strategy is to mark, after a kick, the flip ends as well as their neighbor sets and the vertices that are at most 10 away in T, the overall tour. A comparison of this approach with several other strategies is given in Table 15.17.

Table 15.17 Marking nodes after kicks (10 trials).

Marks	Mean Time	Max Time	Mean Steps
Flip ends	26.4	37.6	3,777
Graph neighbors	25.0	35.0	3,280
Graph & tour 5-neighbors	24.6	28.7	3,090
Graph & tour 10-neighbors	22.1	33.9	3,093
Graph & tour 25-neighbors	23.7	28.8	2,289
Graph & tour 50-neighbors	31.3	46.9	2,548

Complementing the marking strategy, we need to determine the order in which we process the marked vertices. Two simple choices are to use a stack ("last-in-first-out") or a queue ("first-in-first-out") to store the marked vertices and thus control the processing order. Another possibility, used successfully by Rohe [484] in a Lin-Kernighan heuristic for matching problems, is to order the marked vertices by some measure of the likelihood that a search from the vertex will be successful. The measure proposed by Rohe is to compute the nearest neighbor, $near(v)$, to each vertex v, and order the marked vertices by decreasing values of

$$c(v, next(v)) - c(v, near(v)). \tag{15.7}$$

The motivation is that vertices with high values of (15.7) appear to be out of place in the tour (they travel along an edge of cost much greater than the cost to visit their nearest neighbor). To implement this ordering, we store the marked vertices in a priority queue.

Table 15.18 Processing marked vertices (10 trials).

Marks	Mean Time	Max Time	Mean Steps
Stack	29.9	40.2	3,635
Queue	22.1	33.9	3,093
Priority queue	33.3	47.5	3,489

In Table 15.18, we compare the three different approaches for processing the vertices. Both the running times and the number of steps clearly favor the queue implementation. One factor contributing to this is that a queue will tend to distribute the searches around the tour, rather than concentrating the effort on a small tour segment where recent successes have occurred (as in the stack approach) or on a set of vertices with consistently high values of (15.7) (as in the priority queue approach).

INITIAL TOUR

Lin and Kernighan [357] use (pseudo) random tours to initialize their search procedure. This remains a common choice in implementations of Lin-Kernighan and Chained Lin-Kernighan. Random starting tours have the nice feature of permitting repeated calls to lin_kernighan without explicitly building randomization into the algorithm. It is possible, however, to initialize lin_kernighan with tours produced by any tour construction heuristic, and for very large examples (over a million cities) the choice can have a great impact on the performance of the algorithm. For smaller instances, however, random tours perform nearly as well as any other choice we have tested.

In Table 15.19, we report results for a number of different initial tours. "Farthest Addition" and "Spacefilling Curve" are tour construction heuristics proposed by Bentley [56] and Bartholdi and Platzman [41], respectively. "Greedy" is a heuristic developed by Bentley [56] (he called it "multiple fragment" and used it in implementations of 2-opt and 3-opt; it was used in Chained Lin-Kernighan by Codenotti et al. [131]; a description of the algorithm can be found in Johnson and McGeoch [291]). "Greedy + 2-opt" and "Greedy + 3-opt" are Greedy followed by 2-opt and 3-opt, respectively, implemented as described in Bentley [56].

The final line in Table 15.19 reports results for our default initial tour. This heuristic is called *Quick-Borůvka*, since it is motivated by the minimum-weight spanning tree algorithm of Borůvka [81]. In Quick-Borůvka, we build a tour edge by edge. The construction begins (for geometric instances) by sorting the vertices of the TSP according to their first coordinate. We then process the vertices in order, skipping those vertices that already meet two edges in the partial tour we are building. To process vertex x, we add to the partial tour the least costly edge

Table 15.19 Initial tours (10 trials).

Tour	Mean Time	Max Time	Mean Steps
Random	23.4	28.2	3,175
Nearest neighbor	23.4	31.4	3,369
Farthest addition	24.0	51.9	3,151
Spacefilling curve	34.2	71.5	4,987
Greedy	23.6	40.7	3,331
Greedy + 2-opt	21.6	28.1	3,038
Greedy + 3-opt	22.2	32.0	3,081
Quick-Borůvka	22.1	33.9	3,093

meeting x that is permissible (so we do not consider edges that meet vertices having degree 2 in the partial tour, nor edges that create subtours). This procedure can be implemented efficiently using kd-trees. (For a discussion of kd-trees, see Bentley [56].)

Quick-Borůvka gives tours of slightly worse quality than Greedy, but it requires slightly less time to compute and it appears to work well together with Chained Lin-Kernighan.

For further results on the initial tour selection, see Bland and Shallcross [66], Codenotti et al. [131], Johnson [289], Perttunen [460], and Rohe [484].

KICKING STRATEGY

A standard choice for a kicking sequence is the double-bridge kick we described in Section 15.1. This sequence was proposed in the original Chained Lin-Kernighan papers of Martin et al. [379]. In their computations, Martin et al. generated double-bridges at random, but they used only those that involved pairs of edges of relatively small total cost. Johnson [289] and Johnson and McGeoch [291] dropped this restriction on the cost of the double-bridge and simply generated them at random. An argument in favor of this later strategy is that using purely random kicks permits Chained Lin-Kernighan to alter the global shape of the tour on any iteration, whereas most of the cost-restricted kicks will tend to be local in nature and might cause the algorithm to get stuck in some undesirable global configuration. On the other hand, we can expect that lin_kernighan will be much faster after a restricted move than after a random move, and thus in the same amount of computing time we can perform many more iterations of the algorithm.

It is important to notice that for instances with as many vertices as usa13509, finding a cheap double-bridge by taking random samples is inefficient—most kicks will be rejected at any reasonable cutoff point. To get around this, we consider below two alternative methods for obtaining cheap double-bridges. In both of our procedures, we employ a method proposed by Rohe [484] for selecting the first edge of a kick. Like the lk_search method we described earlier, Rohe's idea is to start the double-bridge at a vertex v that appears to be out of place in the tour. This is accomplished by considering a small fraction of the vertices as candidates for v and choosing the one that maximizes (15.7). The first edge of the double-bridge will be $(v, next(v))$. To complete the construction, we choose the remaining three edges to be close to v, as we describe below.

Our first selection procedure examines, for some constant β, a random sample of βn vertices. We then attempt to build a double-bridge using three edges of the form $(w, next(w))$, for vertices w that are among the six nearest neighbors of v, distinct from $next(v)$, in the random sample. We call the double-bridges found by this procedure *close kicks*. Note that as we increase β, the kicks we obtain with this method are increasingly local in nature.

A second, perhaps more natural procedure, is to complete the double-bridge from edges of the form $(w, next(w))$, where w is chosen at random among the k vertices nearest to v. By varying k, we can get very local kicks or kicks similar to those generated purely at random. Notice, however, that these kicks are time-consuming to compute in general instances, since we would be required to examine every vertex in order to obtain the k nearest vertices. In geometric instances, however, we can use kd-trees to efficiently examine the nearest sets. We call the double-bridges found in this way *geometric kicks*.

In Table 15.20, we compare random, close, and geometric kicks, as well as random kicks where we use Rohe's rule for choosing the initial edge. The results indicate a clear preference for the restricted-cost kicks. We shall see in the next section, however, that for some small instances the situation is reversed.

Table 15.20 Kicks (10 trials).

Tour	Mean Time	Max Time	Mean Steps
Random	99.6	116.3	4,056
Random, long first edge	103.6	146.5	4,167
Close ($\beta = 0.01$)	24.3	36.0	3,366
Close ($\beta = 0.03$)	23.5	28.7	4,469
Close ($\beta = 0.10$)	52.0	75.0	10,916
Geometric ($k = 50$)	32.6	44.6	9,712
Geometric ($k = 250$)	22.1	33.9	3,093
Geometric ($k = 1,000$)	23.5	27.7	3,078

There is no strong argument for favoring double-bridges over other kicking sequences, but in our tests they appear to work at least as well as any alternatives that we have tried. For some interesting studies of general kicks, see Codenotti et al. [131] and Hong et al. [269]. A study of kicking strategies aimed at very large-scale instances is given in Applegate et al. [21].

15.4 CHAINED LIN-KERNIGHAN ON TSPLIB INSTANCES

In this section, we report on tests involving Reinelt's TSPLIB, and make observations on the choices of neighbor sets and kicking strategies for instances of varying size. The computational study was carried out on a Sun Microsystems V20z compute node, with a 2.4-GHz AMD Opteron 250 processor.

QUICK SOLUTIONS

We begin our discussion with the modest goal of obtaining tours that are at most 1% more costly than an optimal tour. In Table 15.21, we compare three versions of our code, reporting results for each of the TSPLIB instances having at least 1,000 cities. In the first two versions, we use the 3-quad-nearest graph to define our neighbor sets, but vary the kicking strategy, using random kicks in one case and geometric kicks (or close kicks for nongeometric instances) in the other. The table entries give the average CPU time over 10 distinct runs of Chained Lin-Kernighan. For these tests we set a CPU limit of 100 seconds; an entry followed by "[k]" means that k of the trials failed to produce an acceptable tour in the allotted time.

As we mentioned in the previous section, a number of the smaller instances do not behave well with the local kicks provided by the geometric or close procedures. Indeed, on the instances having less than 10,000 cities, we reached the time limit in 15 cases when using geometric kicks, while random kicks always returned the target 1% tour. On the other hand, the average running time for each of the larger instances was significantly better when using geometric kicks. We therefore propose a hybrid method, where we use random kicks on instances having less than 10,000 cities, but we switch to geometric kicks on instances having more than 10,000 cities. We report on this approach in the third set of results in Table 15.21, using the Delaunay triangulation as our nearest-neighbor graph. (There is no entry for si1032, since this is not a Euclidean example.) We recorded no failures in this third set of tests, and, for the larger instances, the average running times were slightly better than the best of the 3-quad-nearest results.

In Figure 15.13, we plot the running times for the entire set of 110 TSPLIB instances. Each of the marks in the figure represents the mean CPU time over 10 independent runs of Chained Lin-Kernighan. In these tests, we adopted the hybrid random/geometric approach, and used the Delaunay graph for Euclidean instances, the 3-quad-nearest graph for non-Euclidean geometric instances, and the 12-nearest graph for nongeometric instances. The target tour was obtained in less than 0.01 seconds for all instances having at most 200 cities, and the target tour was obtained in under 0.1 seconds for all instances having at most 1,000 cities.

HIGH-QUALITY SOLUTIONS

The above discussion points out that Chained Lin-Kernighan is a suitable method for obtaining reasonable-quality tours in small amounts of CPU time. A strength of the algorithm, however, is that it can effectively use larger amounts of CPU time to obtain higher-quality results.

In Table 15.22, we report on 10 longer runs for the three TSPLIB instances

Table 15.21 1% Optimality gap (10 trials).

Instance	Random	Geometric/Close	Split-Delaunay
dsj1000	0.19	0.11	0.15
pr1002	0.09	0.08	0.11
si1032	0.12	0.12	nongeometric
u1060	0.10	0.08	0.09
vm1084	0.07	0.11	0.04
pcb1173	0.05	0.05	0.06
d1291	0.24	15.52	0.08
rl1304	0.50	9.30	0.13
rl1323	0.15	0.55	0.06
nrw1379	0.05	0.03	0.04
fl1400	0.88	2.29	0.33
u1432	0.11	0.12	0.13
fl1577	25.40	NO TOUR	0.50
d1655	0.80	2.50	0.27
vm1748	0.07	0.06	0.05
u1817	0.30	0.26	0.22
rl1889	0.61	0.49	0.18
d2103	1.02	23.11 [1]	1.49
u2152	0.39	0.36	0.46
u2319	0.08	0.07	0.07
pr2392	0.25	0.27	0.32
pcb3038	0.23	0.18	0.26
fl3795	3.63	42.35 [4]	1.52
fnl4461	0.27	0.15	0.25
rl5915	1.49	2.02	0.78
rl5934	1.22	0.67	0.95
pla7397	0.89	0.97	1.06
rl11849	2.71	1.50	1.15
usa13509	2.68	1.46	0.98
brd14051	1.74	0.91	0.88
d15112	1.96	1.09	0.95
d18512	2.24	1.08	0.92
pla33810	5.04	3.69	2.87
pla85900	5.84	4.11	3.35

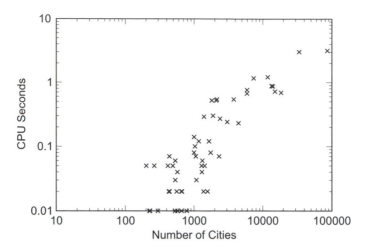

Figure 15.13 1% of optimality for TSPLIB.

pcb3038, usa13509, and pla85900. These examples are representative of the larger instances in the library. We set time limits for the tests at 100 seconds for pcb3038, 250 seconds for usa13509, and 1,000 seconds for pla85900. The average costs of the tours obtained were within 0.13% of the optimal value for pcb3038, within 0.18% of the optimal value for usa13509, and within 0.18% of a known lower bound for pla85900. If we take the best of the 10 tours for each of the instances, then the costs are within 0.05% for pcb3038, 0.14% for usa13509, and 0.13% for pla85900.

Table 15.22 Longer runs (10 trials).

Instance	CPU Seconds	Mean Cost	Max Cost	Min Cost
pcb3038	100	137,874	138,097	137,756
usa13509	250	20,017,878	20,024,418	20,010,799
pla85900	1,000	142,631,460	142,688,231	142,561,793

To exhibit how the quality of the tour improves over time, in Figure 15.14 we plot the cost of the tour against the number of CPU seconds used by Chained Lin-Kernighan over a single run of pla85900. The curve shown in the figure is typical for runs on TSPLIB instances: we get sudden jumps in the tour quality when a kicking sequence permits a global change in the tour structure, and these are followed by a number of smaller improvements as the kicks provide additional local optimization.

The speed with which Chained Lin-Kernighan can identify good-quality solutions for pla85900 suggests that the algorithm could be applied to even much larger examples. This was investigated by Applegate et al. [21], using Concorde's implementation of Chained Lin-Kernighan and with kicking rules selected for good

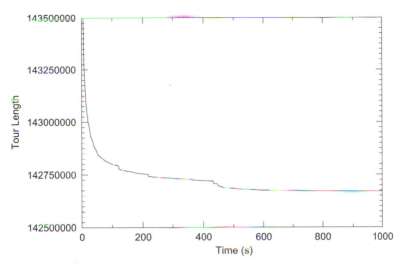

Figure 15.14 Chained Lin-Kernighan on pla85900.

performance on very large instances. A plot of their results for a 25,000,000-city problem is given in Figure 15.15. The data set for this example consists of integer points drawn uniformly from the $n \times n$ square, where $n = 25,000,000$. We do not have good lower bounds for the problem, so as a very rough guess on the tour quality we consider the empirical estimate $\beta = 0.7124 \pm 0.0002$ for the Beardwood et al. [45] constant, obtained by Johnson et al. [293]. (A similar estimate was also obtained by Percus and Martin [459].) Using this guess of the optimal value, the 25,000,000-city run produced, in 8 CPU days, a tour that is approximately 0.3% greater than that of an optimal tour. Applegate et al. carried out this test on an IBM RS6000, Model 43-P 260 workstation with 4 Gbyte of memory.

15.4.1 Optimal solutions

One of the points stressed by Lin and Kernighan [357] is that, in many cases, their algorithm can be adopted to find optimal TSP solutions. Indeed, in the abstract to their classic paper, the authors write: "The procedure produces optimum solutions for all problems tested, 'classical' problems appearing in the literature, as well as randomly generated test problems, up to 110 cities." Their results were obtained using the "Repeated Lin-Kernighan" approach of making independent runs on a given problem instance and selecting the best of the tours. We have of course argued that this process is inferior to Chained Lin-Kernighan's approach of iterating the calls to `lin_kernighan`. But if we examine the cost curve given in Figure 15.14, it is clear that with our implementation of Chained Lin-Kernighan, there comes a point where additional running time is unlikely to lead to significant further improvements in the quality of the tour produced. Johnson [289] and Johnson and McGeoch [291] reached this same conclusion with their implementations; they proposed to go back to making independent runs, but this time using Chained Lin-

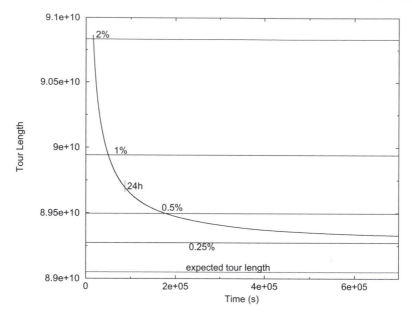

Figure 15.15 Chained Lin-Kernighan on a 25,000,000-city Euclidean problem.

Kernighan as the core algorithm. In Table 15.22, we have seen examples of how this technique can lead to improved tours, and we will now discuss this in detail.

Let us begin with the very small examples from TSPLIB. For each of the 49 instances having at most 200 cities, we made 10 independent runs of Chained Lin-Kernighan, with the optimal values as the target tours and with a CPU limit of 1 second (on a 2.4-GHz AMD Opteron 250 compute node). We performed the test three times, using different selections for the neighbor graph. In the first test, we used the Delaunay graph for Euclidean instances, the 3-quad-nearest graph for non-Euclidean geometric instances, and the 12-nearest graph for nongeometric instances. In the second test, we used the 3-quad-nearest graph for all geometric instances and the 12-nearest graph for nongeometric instances. In the final test, we used the 5-quad-nearest graph for geometric instances and the 20-nearest graph for nongeometric instances. The number of times the codes failed to find optimal tours were

Delaunay	3-Quad-Nearest	5-Quad-Nearest
6	3	12

for the sets of 490 test runs. The instance d198 accounted for 12 of the 21 failures, including all three in the 3-quad-nearest test, so we choose it to discuss in greater detail.

In Table 15.23, we report on a series of additional tests on d198. In these runs, we varied the CPU limit from 0.1 seconds up to 5 seconds. For each value, we made 1,000 independent trials of Chained Lin-Kernighan, recording the mean CPU times between occurrences of optimal solutions. These times are reported in seconds in Table 15.23. On this instance, the Delaunay graph performed better than the (denser) 3-quad-nearest graph. Using the Delaunay graph, the number of times we actually found optimal solutions in the 1,000 trials ranged from 138, with the 0.1-second limit, up to the full 1,000, when we allotted 5 seconds for the runs. With the 3-quad-nearest, we were able to obtain an optimal solution 43 times when we limited the runs to 0.1 seconds, and 979 times with the 5-second limit.

Table 15.23 Optimal solutions for d198 (1,000 trials).

Time Limit	Time per Tour (Delaunay)	Time per Tour (3-Quad)
0.1	0.74	2.51
0.2	0.54	1.38
0.5	0.55	1.09
1.0	0.61	1.10
2.0	0.62	1.14
5.0	0.62	1.17

We consider next the remaining TSPLIB instances having less than 1,000 cities. These examples range in size from 202 cities up to 783 cities, including 27 instances in total. For examples in this range, it becomes difficult to produce optimal solutions using the straight version of Lin-Kernighan. For example, Johnson [289] reported that a good implementation of Lin-Kernighan was unable to find an optimal solution to att532 in 20,000 independent trials. Using the extra power of Chained Lin-Kernighan, however, Johnson [289] and Johnson and McGeoch [291] report that optimal solutions for a number of the examples could be found in reasonable amounts of CPU time, including the instances lin318, pcb442, and att532.

In our tests on these midsized instances, we set time limits of 1, 5, and 10 seconds on the individual calls to Chained Lin-Kernighan. We carried out 1,000 independent runs on each instance and for each time limit, using the 3-quad-nearest graphs. In Table 15.24, we again report the mean time between occurrences of optimal solutions. Although we could indeed find optimal solutions for each of the test instances, we need to point out that in a number of cases the required CPU times are far above those required for solving the given instances with Concorde.

In several of the cases requiring large amounts of computation, one of the problems lies with the choice of the neighbor graph. For these instances, to keep the algorithm from frequently getting trapped in nonoptimal solutions, we need a broader collection of neighbors than those provided by our relatively sparse graphs. For example, with the nongeometric instance si535, if we switch from using the 12-nearest graph to using the 40-nearest graph, then the average time to find an optimal tour drops from 75.21 seconds down to 28.69 seconds, when employing runs with a 10-second limit.

In other cases, such as gr666, the difficulty appears to be more fundamental. Increasing the size of the neighbor sets in gr666 does not improve the time needed for obtaining optimal solutions. We will come back to this example later in the chap-

Table 15.24 Optimal solutions, time per tour (1,000 trials).

Instance	1 Second	5 Seconds	10 Seconds
gr202	0.58	0.58	0.58
ts225	0.03	0.03	0.03
tsp225	0.05	0.05	0.05
pr226	0.31	0.30	0.31
gr229	2.14	6.15	9.11
gil262	0.08	0.08	0.08
pr264	0.17	0.17	0.17
a280	0.02	0.02	0.02
pr299	0.37	0.38	0.37
lin318	0.90	1.89	2.32
rd400	5.73	14.36	25.16
fl417	3.41	3.36	3.60
gr431	20.17	8.87	8.42
pr439	0.63	1.22	1.86
pcb442	0.68	0.70	0.70
d493	35.71	39.26	51.96
att532	12.38	12.52	16.69
ali535	3.79	6.25	7.06
si535	NO TOUR	238.11	75.21
pa561	4.02	8.00	12.96
u574	2.80	9.41	15.29
rat575	111.85	35.25	29.06
p654	6.74	7.12	10.14
d657	11.69	17.67	24.45
gr666	NO TOUR	217.42	252.19
u724	31.17	44.61	65.66
rat783	1.75	1.43	1.48

ter, when we discuss a more robust procedure for using multiple runs of Chained Lin-Kernighan.

We now move on to the TSPLIB instances in the range of 1,000 cities up to 2,000 cities. In Table 15.25, we report results for this set of 17 instances, running 100 trials of Chained Lin-Kernighan with a time limit of 25 seconds and using the 3-quad-nearest graph. As before, we report the number of seconds between

Table 15.25 Optimal solutions with 25-second limit (100 trials).

Instance	Time per Tour
dsj1000	1,241.10
pk1002	11.19
si1032	11.85
u1060	195.90
vm1084	129.21
pcb1173	103.94
d1291	211.73
rl1304	47.64
rl1323	401.34
nrw1379	2,488.17
fl1400	13.05
u1432	105.48
fl1577	2,495.71
d1655	107.90
vm1748	2,483.79
u1817	NO TOUR
rl1889	2,486.36

occurrences of optimal tours. Unlike the tests on small instances, however, in one case we failed to find an optimal tour and in three other cases we were successful only once in the 100 trials. Nonetheless, it is interesting that such a straightforward use of Chained Lin-Kernighan can easily produce optimal solutions on more than three-quarters of the instances of this size.

When we proceed to even larger instances, we encounter even more failures. In Table 15.26, we report on the remaining instances having less than 4,000 cities. Using 100 trials with the 3-quad-nearest graph and with a 100-second time limit, we found optimal solutions for only three of the six instances. In two of the three successful cases, only a single optimal tour was found in the 100 trials.

We remark that, using ad hoc methods, it is possible to coax Chained Lin-Kernighan into producing optimal tours for those instances in Tables 15.25 and 15.26 where our multiple short runs failed. (For example, increasing the range of

Table 15.26 Optimal solutions with 100-second limit (100 trials).

Instance	Time per Tour (3-Quad)
d2103	NO TOUR
u2152	9981.0
u2319	NO TOUR
pr2392	250.6
pcb3038	9984.8
fl3795	NO TOUR

the *breadth* parameters, using longer runs, using Martin, Otto, and Felten's simulated annealing-like approach for accepting tours, etc.) We were not able to do this,
however, on any of the examples having greater than 4,000 cities.

15.5 HELSGAUN'S LKH ALGORITHM

Chained Lin-Kernighan is an effective method for obtaining very good TSP solutions, even for problems of exceptionally large scale. As our computational
tests demonstrate, however, Chained Lin-Kernighan cannot usually produce optimal tours for modest-sized instances (with several thousand cities), even with a
large number of multiple applications of the algorithm. Further closing the gap to
optimal tours can play an important role in the exact solution of large-scale TSP instances, where a sharp estimate on the optimal value helps guide the branch-and-cut
search.

   In the domain of heuristics for producing near-optimal tours, an important set
of modifications to the basic Lin-Kernighan algorithm was introduced by Helsgaun [260] with his LKH code. The work of Helsgaun is a milestone in the evolution of the Lin-Kernighan heuristic, following the major improvement by Martin
et al. [379] with their introduction of Chained Lin-Kernighan. The LKH code is
usually more time-consuming than Chained Lin-Kernighan, but it offers a dramatic
increase in the quality of tours. We outline below the major ideas adopted in LKH.

EXTENDING THE BASIC SEARCH STEP

The main step in Lin-Kernighan is a search for a sequence of flips that, when applied one after the other, leads to a reduction in the cost of the current tour. This
can be viewed as a process for searching for a sequence of 2-opt moves, since a
flip operation corresponds to the exchange of two edges in the tour. A main feature
of Helsgaun's LKH code is the replacement of 2-opt moves by 5-opt moves as the
basic step in the algorithm.

   It is easy to see that any k-opt move can be realized as a sequence of flips,
and thus it can be decomposed into a sequence of 2-opt moves. This does not

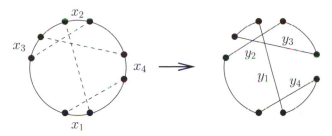

Figure 15.16 Sequential 4-opt move.

imply, however, that Lin-Kernighan's search procedure will consider all possible k-opt moves for some $k \geq 4$, even if the *breadth* parameters are set to allow full backtracking. The reason for this is twofold. First, in building the sequence of flips, the Lin-Kernighan search can be viewed as creating an alternating path of edges $x_1, y_1, x_2, y_2, \ldots, x_k, y_k$ such that $x_1, \ldots, x_k$ are removed from the current tour and $y_1, \ldots, y_k$ are added to the current tour. An alternating path is illustrated in Figure 15.16; at each step the algorithm selects an edge x_i to be deleted and an edge y_i to be added. Lin and Kernighan [357] call a k-opt move *sequential* if it can be created by such a path of edges. A classic example of a nonsequential 4-opt move is the double bridge, illustrated in Figure 15.7; the basic Lin-Kernighan algorithm cannot produce such nonsequential moves. This is not, however, the point where LKH gains power over Lin-Kernighan; owing to concerns for speed, Helsgaun also limits the 5-opt moves he considers to those that can be realized sequentially.

A second reason Lin-Kernighan can miss spotting an improving k-opt move is through the use of the *gain* criterion (15.3). This criterion is critical for limiting the complexity of the search, but it is a greedy method that does not permit the selection of very poor 2-opt moves that may in the end lead to an improving k-opt move. In directly using 5-opt moves in its search, LKH avoids this greediness (at least for 5-opt) and thus broadens the overall search for an improved tour.

To help offset the fact that the main step of LKH is limited to sequential moves, Helsgaun also considers an extended set of 5-opt moves consisting of particular "nonfeasible" 2-opt and 3-opt moves, where nonfeasible refers to an exchange of edges that results in the creation of two subtours. This is a more powerful version of a postprocessing step proposed by Lin and Kernighan [357], where a search is made for double-bridge moves that can improve the final tour. Note, however, that LKH uses the nonsequential moves as a part of its main search process, rather than as a separate postprocessing phase.

CHOICE OF NEIGHBORS

Given the complexity of the basic moves considered in LKH, it is important to have a very sparse neighbor graph to limit the search steps. To accomplish this, as a preliminary step, Helsgaun [260] carries out the iterative 1-tree scheme of Held and Karp [253]. The purpose of the calculation is not to obtain a lower bound for the TSP but rather to approximate the reduced costs given in the subtour relaxation.

The neighbor graph is then chosen as the five smallest reduced-cost edges meeting each vertex, adopting a strategy similar to the one proposed by Flood [182] in his work with the reduced costs for the assignment problem relaxation.

The execution of the Held-Karp method can be time-consuming, but Helsgaun has shown that the reduced cost of an edge provides a good indicator of its likelihood of appearing in an optimal tour. Moreover, the reduced costs give a natural order for considering the candidate edges meeting each vertex when searching for a basic move in the LKH algorithm.

REPEATED TRIALS

In Lin and Kernighan [357], pseudorandom starting tours are used to permit repeated application of their local-search procedure. Besides just taking the best of the tours that are produced, Lin and Kernighan propose to use the intersection of the edge sets of the tours as a means to guide further runs of their algorithm. Their idea is to modify the basic search procedure so that it will not delete any edge that has appeared in each of the tours that has been found up to that point (they start this restricted search after a small number of tours have been found [between two and five in their tests]).

Helsgaun [260] adopts a modified version of this rule, where the best tour found thus far in a run serves as a guide. A run of his algorithm consists a series of k trials, where k is set by default to the number of cities in the problem being considered. Each trial is initialized with a tour found by a greedy-like algorithm, starting at a random city. The search is pruned by forbidding any basic move that begins by removing an edge in the current tour that also belongs to the best tour. In this way, the best tour is used to pull the algorithm toward its neighborhood, in contrast to the Chained Lin-Kernighan idea of using a kick to obtain a starting tour in the neighborhood of the best tour.

For large instances, Helsgaun [260] adopted a second pruning rule to guide long searches on examples that were not solved with the default algorithm. In these cases, he first ran LKH with the default setting of n trials. The runs were then continued with an additional rule forbidding any nonimproving 5-opt move such that the last deleted edge in the move is in the best tour.

COMPUTATIONAL RESULTS

The LKH algorithm substantially raises the quality level of tours that can be produced by local-search heuristics. Indeed, Helsgaun [260] reports on finding optimal solutions for all TSPLIB instances solved up to the year 2000, including the 13,509-city instance usa13509.

To give a sample of the quality of results that can be achieved with LKH, we ran several tests with Version 1.3 of Helsgaun's implementation, available at

www.akira.ruc.dk/~keld/research/LKH/.

The experiments were carried out on the 2.4-GHz AMD Opteron 250 compute node that was used in the study reported in the previous section.

In 10 trials on usa13509, the following tour values were obtained

Mean Cost	Max Cost	Min Cost
19,984,516	19,985,530	19,983,681

with an average CPU time of 7,391 seconds per run, after the initial preprocessing time of 673 seconds. The best of the 10 tours found by LKH is within 0.0042% of the optimal value for usa13509; this is well beyond the results that can be achieved with standard runs of Chained Lin-Kernighan.

We saw in the previous section that Chained Lin-Kernighan can easily produce solutions that are within 1% of an optimal tour. With LKH this can be lowered to 0.1%, as shown by the results given in Table 15.27 for all TSPLIB instances having at least 1,000 cities; the times are given in seconds, split between the preprocessing time and the time for the main search routine of the algorithm. The only failure in these runs occurred on the 85,900-city instance, where the default rule of n repeated trials produced a solution within 0.112% of the best-known lower bound; the running time in this case was 1.3 million seconds.

The growth in the running time of LKH as the problem size increases makes it difficult to directly apply the code to very large instances. Helsgaun (private communication, 2001), however, has had success in using decomposition methods to extend the reach of the algorithm, including the computation of the currently best-known solution to the 1.9 million-city World TSP.

15.6 TOUR MERGING

In the quest for ever better approximations to optimal TSP solutions, the standard practice to improve the results produced by Chained Lin-Kernighan and LKH is to run the algorithms multiple times, varying the input tour or varying parameters in the implementations of the algorithms. Although this simple technique does produce better solutions than single long runs of the algorithms, by choosing only the best of a large collection of tours, we may be discarding a great deal of valuable information. A challenge is to utilize the combined content of the tours to produce a final tour that is superior to any single one in the collection.

This task also confronts designers of genetic algorithms for the TSP, as described, for example, in Oliver et al. [436], Mühlenbein et al. [405], and Ulder et al. [529]. In our context, however, the quality of the population of tours is such that the simple randomized techniques used in successful genetic algorithms are very unlikely to produce tour improvements. Instead, we propose a *tour-merging* procedure that attempts to produce the best possible combination of tours by solving a TSP restricted to the edges that are present in at least one tour in the collection.

To illustrate the procedure, consider the TSPLIB instance fl1577. As reported in Table 15.25, with standard runs of Chained Lin-Kernighan it is difficult to produce an optimal solution for this problem. Attacking fl1577 with tour merging, we carried out 25 runs of Chained Lin-Kernighan, each with a 4-second time limit, on a Compaq ES40 server, equipped with a 500-MHz EV6 Alpha processor.[1] The

---

[1]The computational results in this section were carried out on several different models of computers

Table 15.27 0.1% Optimality gap with LKH.

Instance	Preprocessing Time	Search Time
dsj1000	2.16	6.88
pr1002	2.61	0.08
si1032	2.32	0.04
u1060	2.16	2.14
vm1084	2.40	2.26
pcb1173	2.74	0.17
d1291	3.17	25.46
rl1304	3.56	1.45
rl1323	3.72	0.90
nrw1379	4.04	0.15
fl1400	4.17	151.54
u1432	4.27	0.12
fl1577	5.29	24.53
d1655	5.79	0.52
vm1748	6.91	0.27
u1817	6.87	0.40
rl1889	8.29	0.65
d2103	9.11	1.46
u2152	9.90	6.50
u2319	12.94	0.14
pr2392	12.64	0.36
pcb3038	20.88	0.70
fl3795	34.10	297.90
fnl4461	52.63	0.59
rl5915	98.90	4.91
rl5934	102.00	42.51
pla7397	262.19	4.59
rl11849	602.75	245.28
usa13509	821.98	20.22
brd14051	882.39	7.79
d15112	1,047.13	3.19
d18512	1,576.23	20.74
pla33810	4,914.54	1,902.51
pla85900	36,387.19	NO TOUR

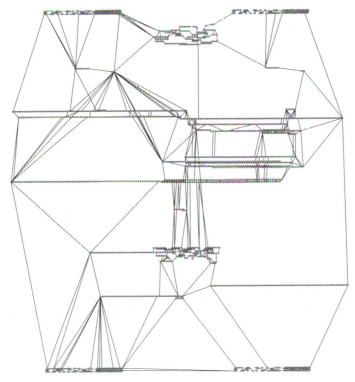

Figure 15.17 Union of 25 Chained Lin-Kernighan tours for fl1577.

cost of the best of the 25 tours is 0.018% above the cost of an optimal tour. Taking the union of the edge sets of the tours, we assemble the graph represented in Figure 15.17. This graph has 2,391 edges in total, so on average each vertex has approximately three neighbors. The crucial point is that this sparse graph contains an optimal tour for the full TSP, that is, by piecing together segments from the 25 tours we can produce an optimal solution to fl1577.

As a second example, the graph in Figure 15.18 consists of the union of 10 tours for the 5,934-city TSPLIB instance rl5934; the 10 tours were obtained by running the LKH code of Helsgaun [260]. While the cost of the best of the 10 LKH tours is only 0.006% above the optimal value for rl5934, the best tour in the union is again an optimal solution for this instance.

These two examples illustrate the potential power of the tour-merging idea, but it is far from obvious how to solve the TSP over the restricted graph G. A first attempt, considered below, is to simply apply Concorde with the problem initialized to the sparse edge set. Using this approach, the optimal solution to the restricted problem for the fl1577 example was found in 188 seconds.

using Compaq Alpha processors. To permit direct comparisons, we have scaled the running times to the performance of a Compaq ES40 with a 500-MHz EV6 Alpha processor. The EV6 Alpha is roughly four times slower than the 2.4-GHz AMD Opteron that was used in the study reported in the previous section.

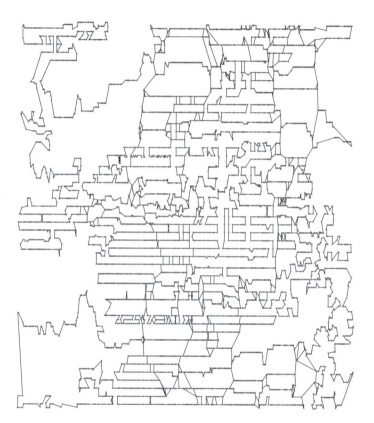

Figure 15.18 Union of 10 LKH tours for rl5934.

The Concorde-based method can be used to improve Chained Lin-Kernighan on small instances, but it is often very time-consuming on larger examples. As an alternative, Cook and Seymour [137] propose a "branch-width" algorithm as a practical means for merging collections of high-quality tours. Branch-width is a graph invariant introduced by Robertson and Seymour [482]; it provides a natural framework for implementing dynamic programming algorithms to solve optimization problems on sparse graphs. The work of Cook and Seymour [137] is one of the first practical applications of this approach.

Returning to the rl5934 example, the average cost of the 10 tours is 0.089% above optimum, and LKH required an average of 1,973 seconds to produce each tour. The union of the 10 tours has 6,296 edges, and the branch-width algorithm required 2.10 seconds to determine the best solution in the restricted graph (and thus produce an optimal solution for the instance). This example is typical of the performance of the merging algorithm on LKH-generated tours, where the consistent high quality of the LKH heuristic produces a very sparse union of edges. In general, the branch-width algorithm is sensitive to the density of the graph G, although the practical performance exceeds that which would be predicted by a worst-case analysis of the underlying dynamic programming method.

15.6.1 Merging with Concorde

We begin our discussion of Concorde-based merging by considering the collection of TSPLIB instances having less than 1,000 cities. From this set, all instances having less than 200 cities are solved handily with Chained Lin-Kernighan, so we will not consider further these problems. In Table 15.24, we reported results for the remaining examples using multiple 10-second runs of Chained Lin-Kernighan. Of the 27 instances considered in Table 15.24, the five requiring the greatest times to obtain optimal solutions were d493, si535, rat575, gr666, and u724. The average times for these five instances varied from 29 seconds for rat575 up to 217 seconds for gr666 (on the 4x faster AMD Opteron used in the previous section). We will use these five examples as our initial test bed for tour merging.

For each of the five instances, we made 10 trials of merging 25 tours found with Chained Lin-Kernighan, giving each Chained Lin-Kernighan run a 1-second time limit. In these tests, we used the 3-quad-nearest graph to determine the neighbor sets for Chained Lin-Kernighan, and we used the nearest-neighbor algorithm to produce the starting tours. The results are reported in Table 15.28, again with times on a 500-MHz EV6 Alpha processor. The column labeled "LK Quality" gives the average ratio of the cost of the best of the 25 Chained Lin-Kernighan tours in each trial, to the cost of an optimal tour; the column labeled "Merge Quality" gives the average ratio of the cost of the final merged tour for each trial, to the cost of an optimal tour; "LK-opt" reports the number of trials for which the best Chained Lin-Kernighan tour was optimal, and "Merge-opt" reports the number of times the final merged tour was optimal.

In 37 of the 50 trials the merged tour was indeed optimal, and the average tour quality over all trials was 1.00004, that is, on average the tours produced were within 0.004% of the optimal values. In these tests, the sparse TSP instances were

Table 15.28 Tour merging with 25 tours and 1-second limit (10 trials).

Instance	LK Quality	Merge Quality	LK-opt	Merge-opt	Time (seconds)
d493	1.00039	1.00000	0	10	67.1
si535	1.00120	1.00007	0	5	34.0
rat575	1.00025	1.00009	1	4	50.2
gr666	1.00061	1.00006	0	8	39.2
u724	1.00044	1.00000	0	10	54.1

solved in an average time of under 25 seconds with Concorde.

The combination of $N = 25$ tours and a time limit of $T = 1$ seconds on Chained Lin-Kernighan gives a reasonable compromise between tour quality and total running time for the under 1,000-city examples. In Table 15.29 we compare these results (over the 50 total trials), with the results for several other combinations of N and T. Note that the average running time for $N = 25$ and $T = 0.4$ is higher than

Table 15.29 Tour merging with five instances having less than 1,000 cities.

No. of Tours	Time Limit (seconds)	Tour Quality	Time (seconds)
5	2	1.00042	15.7
10	1	1.00045	22.0
10	2	1.00014	29.7
25	0.4	1.00017	54.8
25	1	1.00004	48.9
25	2	1.00003	68.1

the time for $N = 25$ and $T = 1$, despite the much lower time for computing the collection of tours. The explanation is that the lower-quality Chained Lin-Kernighan tours lead to a lesser number of repeated edges in the tour population, and therefore a denser (more difficult) TSP instance to be solved.

Let us now consider the TSPLIB instances in the range of 1,000 cities up to 2,000 cities. The results for multiple 25-second Chained Lin-Kernighan runs for this class of 17 examples were presented in Table 15.25. In our tour-merging tests, we will treat 13 of the more difficult instances, including the four examples that were solved only once in the 100 trials, and also u1817, for which no optimal tours were found. The results for this problem set, using $N = 25$ and $T = 10$, are given in Table 15.30. With these settings, optimal solutions were found in 95 of the 130 trials, and every instance was solved at least two times. The running times are modest, with the exception of u1817, which required an average of over 4,000 seconds to solve the sparse TSPs. A direct precaution against such behavior is to set an upper bound on the time allowed in the TSP solver, reporting a failure if

Table 15.30 Tour merging with 25 tours and 4-second limit (10 trials).

Instance	LK Quality	Merge Quality	LK-opt	Merge-opt	Time (seconds)
dsj1000	1.00094	1.00032	0	4	142.5
u1060	1.00070	1.00000	0	10	207.6
pcb1173	1.00013	1.00000	4	8	154.4
d1291	1.00091	1.00000	0	10	166.2
rl1304	1.00143	1.00038	0	7	105.5
rl1323	1.00114	1.00011	0	6	129.8
nrw1379	1.00046	1.00000	0	8	183.3
u1432	1.00099	1.00000	0	10	534.9
fl1577	1.00123	1.00021	0	2	243.3
d1655	1.00102	1.00002	0	8	180.0
vm1748	1.00057	1.00000	0	10	139.1
u1817	1.00253	1.00017	0	7	4,492.4
rl1889	1.00214	1.00002	0	5	180.6

the bound is reached. For example, setting a bound of 2,000 seconds lowers the average running time to 1780.7 seconds for u1817, while still reaching the optimal solution on five of the 10 trials.

In Table 15.31 we compare several different choices of N and T for the set of 13 instances considered in Table 15.30. The values reported are the averages over all 130 trials. The quality of the final tour drops quickly as we decrease the size of the

Table 15.31 Tour merging with 13 instances having between 1,000 and 2,000 cities.

No. of Tours	Time Limit (seconds)	Tour Quality	Time (seconds)
5	20	1.00098	135.8
10	4	1.00045	168.9
10	10	1.00045	190.5
25	4	1.00009	528.4

tour population, but there is a corresponding decrease in the time required to solve the resulting sparse TSP instances.

Proceeding to larger examples, let consider the five of the six TSPLIB instances under 4,000 cities that we treated in Table 15.26. These are all TSPLIB instances in the range of 2,000 to 4,000 cities, with the exception of d2103 (at the time the test was made, the optimal value for d2103 had not yet been determined). In the tests reported in Table 15.26, optimal solutions were found only for instances u2152, pr2392, and pcb3038, requiring an average of 9,981 seconds, 251 seconds, and

9,984 seconds, respectively. In Table 15.32, we report tour-merging results for 10 trials on each of the five test instances, using $N = 25$ and $T = 4$. Optimal solutions

Table 15.32 Tour merging with 25 tours and 4-second limit (10 trials).

Instance	LK Quality	Merge Quality	LK-opt	Merge-opt	Time (seconds)
u2152	1.00199	1.00000	0	10	2,410.6
u2319	1.00172	1.00000	0	10	11,239.2
pr2392	1.00124	1.00000	0	10	121.5
pcb3038	1.00190	1.00001	0	8	6,478.4
fl3795	1.00592	1.00031	0	3	1,923.2

were obtained in 41 of the 50 trials, and each example was solved at least three times. This is certainly an improvement over our multiple Chained Lin-Kernighan runs, but the running times are very long. The u2319 example can be solved easily using other methods, so we select pcb3038 for further discussion.

In Table 15.33 we report results of 10 trials on pcb3038 using a variety of settings for N and T. These results illustrate several common properties of tour-merging

Table 15.33 Tour merging on pcb3038 (10 trials).

Settings	LK Quality	Merge Quality	Merge-opt	Time (seconds)
$N = 25, \ T = 4$	1.00190	1.00001	8	6,478.4
$N = 25, \ T = 20$	1.00103	1.00001	7	4,017.0
$N = 25, \ T = 40$	1.00096	1.00002	5	2,802.6
$N = 25, \ T = 100$	1.00071	1.00000	10	4,527.2
$N = 25, \ T = 200$	1.00063	1.00002	7	6,458.2
$N = 25, \ T = 400$	1.00059	1.00000	9	11,512.5
$N = 10, \ T = 40$	1.00145	1.00024	0	1,452.7
$N = 10, \ T = 100$	1.00090	1.00009	1	1,611.8
$N = 10, \ T = 200$	1.00095	1.00008	3	2,610.6
$N = 10, \ T = 400$	1.00059	1.00008	3	4,385.2

runs. First, the quality of the final tour depends heavily on the size of the population of tours, but only to a lesser degree on the time limit given to Chained Lin-Kernighan (as long as the limit is reasonably large). Second, the solution time for the sparse TSP decreases dramatically as the quality of the tours in the population is improved, that is, for fixed N, the sparse TSP time decreases as T is increased. To select the best settings of N and T, we must trade off the time used to produce the tour population versus the time needed to solve the sparse TSPs. This is a difficult task on large instances, since the performance of the Concorde code is difficult to predict without running experiments.

To apply Concorde-based tour merging to even larger instances, we need to take great care in the selection of the tour population. For example, 10 trials on the instance fnl4461 produced the following results

LK Quality	Merge Quality	LK-opt	Merge-opt	Time (seconds)
1.00078	1.00000	0	10	34,368.4

using $N = 25$ and $T = 100$. The fact that the optimal tour was obtained in each of the trials is a very positive sign, but the average running time for the sparse TSPs is very high (the full instance can be solved in similar time with Concorde). It is possible, however, to use Chained Lin-Kernighan to obtain tour populations that are much more likely to be amenable to sparse TSP solution routines. What is needed is more homogeneous collection of tours. To achieve this, rather than using the nearest-neighbor algorithm to produce different starting tours for each run, we can start the Chained Lin-Kernighan runs from a common starting tour that is itself the result of a run of Chained Lin-Kernighan. This adds a third setting for our runs, namely, the number of seconds S used in the initial run to build the common starting tour.

In Table 15.34 we report results for fnl4461, where we fix N and T, and permit S to vary between 2, 4, and 10. As one would expect, as we increase the quality of

Table 15.34 Tour merging on fnl4461 with fixed start (10 trials).

Settings	LK Quality	Merge Quality	Time (seconds)
$N = 25$, $T = 40$, $S = 2$	1.00099	1.00065	1,564.6
$N = 25$, $T = 40$, $S = 4$	1.00106	1.00074	1,131.2
$N = 25$, $T = 40$, $S = 10$	1.00111	1.00087	1,076.2
$N = 25$, $T = 20$, $S = 2$	1.00128	1.00076	1,550.0
$N = 25$, $T = 20$, $S = 4$	1.00121	1.00082	672.5
$N = 25$, $T = 20$, $S = 10$	1.00123	1.00092	541.1

the common starting tour, the time needed to solve the sparse TSP drops, but at the expense of lowering the average quality of the final merged tour.

The added flexibility of using a high-quality common starting tour allows us to apply Concorde-based tour merging to a much wider range of instances. To illustrate this, we consider pla85900, the largest instance in the TSPLIB. On an instance of this size, there is a large variance in the quality of tours produced by individual runs of Chained Lin-Kernighan, so before we invest the time to obtain an entire collection of tours, it is important to make an effort to have a common starting tour that is likely to lead to a high-quality population. To achieve this, we can use a short multiple Chained Lin-Kernighan run to obtain the common tour, rather than simply using a single run. In our case, making five runs of Chained Lin-Kernighan, with a time limit of 400 seconds, we obtain tours of the following costs

142880000, 142744923, 142732749, 142754147, 142866379.

Now, making five further Chained Lin-Kernighan runs using the best of these as a starting tour (142732749), and using a 2,000-second time limit, we obtain the population

$$142628651, 142603285, 142613880, 142625393, 142624904.$$

With this collection, the merged tour can be computed in 61,214 additional seconds, resulting in a tour of cost 142575637. The TSP solve time is extremely high, but this example does indicate that tour merging is a viable approach, even for instances of this size.

Tour merging can be used in a wide variety of ways, and it is likely that ideas drawn from the field of genetic algorithms can be combined with tour merging to produce a powerful class of heuristics. On large examples, our use of the Concorde-based procedure has been confined primarily to ad hoc methods, but, as we report later in this section, combining the branch-width algorithm with the more powerful LKH heuristic can produce very good results.

15.6.2 Branch decompositions

For a graph $G = (V, E)$, any partition of E into two sets (A, B) is called a *separation* of G. Let us assume that each vertex in G has degree at least one (the degree of a vertex is the number of edges it meets). Then any separation (A, B) partitions the vertices of G into three classes: those that meet only edges in A, those that meet only edges in B, and those that meet both edges in A and edges in B. We call these three classes the $left(A, B)$ vertices, the $right(A, B)$ vertices, and the $middle(A, B)$ vertices, respectively. Note that no vertex in $left(A, B)$ has a neighbor in $right(A, B)$. Consequently, $middle(A, B)$ does indeed "separate" $left(A, B)$ from $right(A, B)$. The cardinality of $middle(A, B)$ is called the *order* of the separation.

Let T be a tree having $|E|$ leaves and in which every nonleaf vertex has degree at least three. Associate with each leaf v of T one of the edges of G, say $\sigma(v)$, in such a way that every edge of G is associated with a distinct leaf. The tree T, together with the function σ, is called a *partial branch decomposition* of G. If each nonleaf vertex of T has degree exactly three, then (T, σ) is a *branch decomposition*.

Let (T, σ) be a partial branch decomposition of G. If e is an edge of T, then the graph obtained from T by deleting e has exactly two connected components, and consequently the set of leaves of T is partitioned into two subsets. Since each leaf of T is associated with an edge of G, there corresponds a separation (A_e, B_e) of G. The *width* of (T, σ) is the maximum of the order of (A_e, B_e) over all edges e of T. The smallest k such that G has a branch decomposition of width k is called the *branch-width* of G.

Branch-width was introduced in Robertson and Seymour [482] as part of their graph minors project and it played a fundamental role in their proof of the Wagner conjecture. In our study, we make use of the algorithmic possibilities branch-width creates via dynamic programming. To motivate the definition of branch-width in this context, we describe a possible algorithm for the TSP.

If G has a separation (A, B) of small order, then there are only a small number of different ways in which a tour in G can hit $middle(A, B)$, as illustrated in

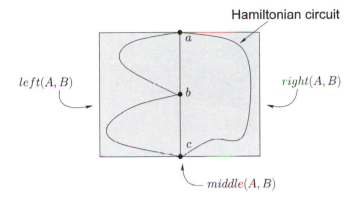

Figure 15.19 A separation of order three.

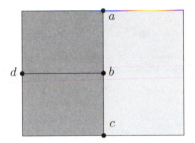

Figure 15.20 A second pair of separations.

Figure 15.19. For a given pattern of hitting $middle(A, B)$, we can independently solve the problems of finding collections of paths in A and B that realize the pattern, since the only thing one side needs to know about the other side is the way it behaves on $middle(A, B)$. This implies that we can reduce the problem of finding a minimum-cost tour to that of solving a list of subproblems in A and B. To handle the subproblems, we can repeat the separation process on each of the two pieces. Suppose we find two separations of G that together split A into two smaller pieces, as indicated in Figure 15.20. Using these separations, we could solve problems in the smaller pieces in order to create the solution for the larger piece. Now to solve these smaller problems, we could split the pieces again, as indicated in Figure 15.21. These steps can be summarized by means of a "separation tree," as indicated in Figure 15.22. In this figure, we have labeled the edges with the middle of the corresponding separation. If we continued the splitting process until each piece consists of a single edge, then the separation tree would represent the tree T in a branch decomposition of the graph, where σ is given by the edges in the final pieces of the decomposition.

This description makes it clear how we use branch decompositions in an optimization procedure: starting at the leafs of the tree, we work our way through the vertices, gluing together solutions to the pieces as we go along. This is a dynamic programming procedure. To carry it out, we need to be able to encode the hitting

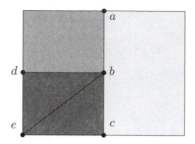

Figure 15.21 A third pair of separations.

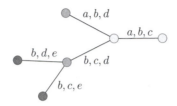

Figure 15.22 The separation tree.

patterns and to be able to combine two hitting patterns into a pattern for the middle of the next separation.

Branch-width is closely related to another graph decomposition scheme known as "tree-width." Suppose we have a graph G and a tree $T = (V(T), E(T))$. Associate with each vertex v of T a subset $\omega(v)$ of the vertices of G. The pair (T, ω) is called a *tree decomposition* of $G = (V, E)$ if

(*i*) $\bigcup \{\omega(v) : v \in V(T)\} = V$;

(*ii*) for each edge $(v, w) \in E$, there exists a vertex $t \in V(T)$ such that $\{v, w\} \subseteq \omega(t)$;

(*iii*) for all $u, v, w \in V(T)$, if v is on the path from u to w in T, then $\omega(u) \cap \omega(w) \subseteq \omega(v)$.

The *width* of a tree decomposition is the maximum of $|\omega(v)| - 1$, taken over all $v \in V(T)$. The smallest k such that there exists some tree decomposition of G with width k is called the *tree-width* of G. A number of equivalent characterizations of tree-width can be found in Bodlaender [73].

Robertson and Seymour [482] have shown that every graph of branch-width at most k has tree-width at most $3k/2$, and every graph of tree-width at most k has branch-width at most $k + 1$. So having small branch-width is equivalent to having small tree-width.

Tree-width is studied in a great many papers; surveys of the literature can be found in Bodlaender [71] and computational studies can be found in Koster et al. [323] and in Koster et al. [324]. For many discrete optimization problems that deal with edge sets of graphs, however, branch-width is a more natural framework for carrying out dynamic programming.

The computational complexity of solving a problem with an algorithm based on branch decompositions grows rapidly with the width of the given decomposition. In this context, therefore, we can accept only a small margin of error in our calculation of the branch-width of a graph.

In the special case of planar graphs, Seymour and Thomas [497] have shown that branch-width can be computed in polynomial time, and Hicks [263] has developed an implementation of the Seymour-Thomas algorithm that is practical for instances having up to several thousand vertices. In general, however, Seymour and Thomas [497] proved that determining if a graph has branch-width $\leq k$ is $\mathcal{NP}$-complete, when k is part of the input to the problem.

Robertson and Seymour [482] proposed a polynomial-time algorithm to estimate branch-width within a factor of three, but this rough approximation yields decompositions that are of little value in a practical implementation. The Robertson-Seymour approximation algorithm was improved in Matoušek and Thomas [384], Lagergren [331], Reed [471], Bodlaender [72], and Bodlaender and Kloks [74], but each of these methods is either impractical or yields results that are too far from optimal decompositions. In Bodlaender and Thilikos [75] there is a linear-time algorithm, for any constant k, that decides if a given graph has branch-width at most k and, if so, it produces the corresponding branch decomposition, but the growth of the running time makes this method impractical for even modest values of k. Bodlaender and Thilikos [76] give a practical method for the case $k = 3$, but extending the results to larger values is an open research question.

To work with branch decompositions in practice, Cook and Seymour [137] propose a heuristic algorithm that attempts to construct a decomposition that is close to optimal, although no good performance guarantees can be made. The algorithm is based on the eigenvector methods of Alon [9] and others for finding small separators in graphs. The heuristic appears to work well in practice and it is adopted in the computational tests reported in this section.

15.6.3 Dynamic programming

Let G be a 2-connected graph with edge costs $(c_e : e \in E)$ and let (T, σ) be a branch decomposition of G. The idea of the dynamic programming algorithm is to start at the leaves of the tree T and work inward, processing the corresponding separations of G. For fixed branch-width k, this approach provides linear-time algorithms for the TSP and many other problems. It must be noted, however, that the constant of proportionality usually grows very quickly with k.

To make the notion of "inward" precise, we *root* the branch decomposition by selecting an arbitrary edge (a, b) of the tree and adding new tree vertices r and s, and new edges (a, s), (s, b), and (s, r), and removing the edge (a, b); let T' denote the tree we obtain in this way and call r the root vertex of T'. No edge of G is assigned to vertex r, so the separations corresponding to (a, s) and (s, b) are the same as the separation for the old edge (a, b); the separation corresponding to (s, r) is $(E, \emptyset)$.

Each vertex v of T', other than r, meets a unique edge that lies on the path from v to r in the tree, and if v is not a leaf it also meets two edges that do not lie on this path; we refer to the unique edge as the *root edge* of v and we refer to the other two edges as the *left* and *right* edges of v (the choice of which is left and which is right is arbitrary). Since each tree edge e is the root edge of precisely one vertex v, when we have processed the separation corresponding to e we say that we have processed the vertex v. We say that a tree vertex is *ready* to be processed if either it is a leaf of the tree (and it is not the root vertex), or both its left edge and its right edge have already been processed. The overall procedure is then to select any ready vertex and process it, stopping when r is the only remaining vertex. To specialize this general algorithm, we need to describe how to "process" a separation when solving the TSP.

Theoretical studies of general dynamic programming algorithms with branch decompositions and tree decompositions can be found in Bern et al. [59], Courcelle [141], Arnborg et al. [22], Borie et al. [80], and elsewhere. Despite a large body of theoretical work, little in the way of practical computation has been reported in the literature.

ENCODING PARTIAL TOURS

Let $(A, \bar{A})$ be a separation of G, where $\bar{A} = E - A$. A *partial tour* in A is a subset $R \subseteq A$ consisting of the union of a collection of edge sets of paths having both ends in the middle of $(A, \bar{A})$, such that every vertex in $left(A, \bar{A})$ is included in exactly one of the paths and every vertex in $middle(A, \bar{A})$ is included in at most one of the paths.

Let R be a partial tour in A, and let $n_0, n_1, \ldots, n_k$ be the vertices in the middle set of $(A, \bar{A})$. A middle vertex n_i meets either zero, one, or two edges in R. We say that n_i is *free* if it meets zero edges, *paired* if it meets one edge, and *used* if it meets two edges. If we follow the path in R meeting a paired middle vertex n_i, we will reach another paired middle vertex n_k. We can specify how R "hits" the middle of $(A, \bar{A})$ by listing the pairs (n_i, n_k) of paired vertices and listing the free vertices and the used vertices. We refer to such an encoding as a *pairing*. To process the separation $(A, \bar{A})$, we find the list of pairings corresponding to the partial tours in A, and for each pairing we record the corresponding partial tour having the least cost, that is, the partial tour S that realizes the pairing and minimizes $\sum(c_e : e \in S)$.

PROCESSING SEPARATIONS

Let us first consider a separation corresponding to a leaf v of T' (other than the root vertex). Since G is simple and 2-connected, the middle set of the separation consists of two vertices, n_0 and n_1 (the ends of the edge $\sigma(v)$). The only pairings in this case are either to have both n_0 and n_1 free (of total cost 0), or to have n_0 paired with n_1 (of total cost $c_{\sigma(v)}$).

Now consider an internal vertex v of T'. The graph obtained by deleting v from T' has three connected components, giving a partition of the leaves of T' into three sets; let L, R, and N be the corresponding subsets of E, where L corresponds to the component of $T' - \{v\}$ that meets v's left edge, R corresponds to the component that meets v's right edge, and N corresponds to the component that meets v's root edge. So $(L, \bar{L})$ is the separation associated with the left edge of v, $(R, \bar{R})$ is the separation associated with the right edge of v, and $(L \cup R, \overline{L \cup R})$ is the separation associated with the root edge of v. Note that $N = \overline{L \cup R}$.

Each vertex in the set $middle(L \cup R, \overline{L \cup R})$ is either in $middle(L, \bar{L})$ or in $middle(R, \bar{R})$. This makes it possible to compute the pairings for $L \cup R$ by studying only the pairings for L and the pairings for R, we do not need to know anything else about the graph G. Any partial tour in $L \cup R$ must arise as the disjoint union of a partial tour in L and a partial tour in R, but not all such unions give partial tours in $L \cup R$. It easy to check, however, if a pair of pairings is *compatible* (that is, the union of their corresponding partial tours gives a partial tour in $L \cup R$) by examining only the vertices in the middle sets of $(L, \bar{L})$ and $(R, \bar{R})$. So we can proceed by running through all pairs of a pairing for L and a pairing for R and computing if the pair gives a valid pairing for $L \cup R$. This procedure involves some analysis (following paths to find the new paired vertices), but as long as the middle sets are small it can be carried out quickly.

MERGING LISTS OF PAIRINGS

The main computational difficulty in the dynamic programming algorithm arises from the fact that the lists of pairings for the left and right edges of a vertex may be quite long, and we need to run through each possible left-right pair in order to obtain the list for the root edge of the vertex. The worst-case complexity (counting the number of possible pairings on a middle set of cardinality k) indicates that one cannot hope to to solve every instance of width, say, 20 or higher. We attempt to solve, however, at least a reasonable portion of the examples that arise in practice by taking some simple steps to speed up the list merging, as we indicate below. The computational tests of Cook and Seymour [137] reported in later in the section indicate mixed results; although the code was usually successful, in some cases a time or memory limit was reached, even when the width was under 20.

Let k denote the width of the branch decomposition (T, σ) and suppose we have lists of pairings for L and R (using the notation as above) and we need to compute the implied list for $L \cup R$. The set Q of *local vertices* that we use in the computation consists of the union of the middle sets for the three separations $(L, \bar{L})$, $(R, \bar{R})$, and $(L \cup R, \overline{L \cup R})$. Since each vertex in Q appears in at least two of the three middle sets, we have $|Q| \leq 3k/2$. Let us order the vertices in Q as $q_0, q_1, \ldots, q_{t-1}$, where $t = |Q|$.

For a pairing γ, let $d(\gamma)$ denote the *degree sequence*
$$(d_0(\gamma), d_1(\gamma), \ldots, d_{t-1}(\gamma))$$
where for each $i = 0, \ldots, t - 1$, we have
$$d_i(\gamma) = \begin{cases} 0 & \text{if } q_i \text{ is free in } \gamma \text{ or if } q_i \text{ is not in the middle set of the separation.} \\ 1 & \text{if } q_i \text{ is paired in } \gamma. \\ 2 & \text{if } q_i \text{ is used in } \gamma. \end{cases}$$
If a pairing α for L and a pairing β for R are compatible, then for each vertex q_i that is in both $middle(L, \bar{L})$ and $middle(R, \bar{R})$ but not in $middle(L \cup R, \overline{L \cup R})$

we must have

$$d_i(\alpha) + d_i(\beta) = 2$$

since q_i will be in the left set of $(L \cup R, \overline{L \cup R})$. Also, for each vertex q_i that is in all three middle sets we must have

$$d_i(\alpha) + d_i(\beta) \le 2.$$

(This condition on the remaining local vertices will hold trivially.) Therefore, we can avoid explicitly checking many incompatible left-right pairs by keeping the pairing lists grouped according to their degree sequences; to merge the lists we run through the lists of degree sequences and check the sublists of pairings only if the degree sequences are compatible.

As we build the list of pairings for $L \cup R$, we use a hash table to store the collection of degree sequences, making it simple to determine if we have already encountered a sequence in the list-merging process. We encode each degree sequence as a single number (by considering the sequence as ternary digits) to make it easy to compare if two sequences are identical (when they hash to the same value).

If we determine that the degree sequence for the pairing we are considering is already in our collection, then we need to determine if the pairing itself is also one we have previously encountered. To handle this, it is possible to use a hash table to store the sublists of pairings for each degree sequence, but we found it satisfactory just to keep the sublists sorted by a numerical encoding of the pairings. If the current pairing is not in the sublist, we add it. Otherwise, we compare the cost of the current pairing to the cost of the identical pairing already in the sublist. The cost of the matching is just the sum of the costs of the matchings in the left-right pair we are merging. If the current pairing has lower cost, we record it in place of the existing copy. The record for a pairing is its cost together with links to the left pairing and to the right pairing we are merging; the left-right links are used to work backward to gather the edges in the optimal tour after we process the entire tree, permitting us to avoid explicitly storing the partial tours associated with the pairings in our list.

15.6.4 Tour-merging results

The sensitivity of dynamic programming to the width of the branch decompositions makes it clear that care needs to be taken in generating the tours to be merged. A simple rule is that as the problem size increases, the quality of the generated tours must also increase. This rule is followed in the computational study, using Chained Lin-Kernighan tours for small to medium instances, and using Helsgaun's LKH heuristic for larger instances.

MERGING CHAINED LIN-KERNIGHAN TOURS

We consider the combination of Chained Lin-Kernighan and tour merging on all TSPLIB instances having at least 1,000 cities and at most 10,000 cities. In each case, we report the average results over four trials, where we merge 10 tours in each trial. The tours are generated using the default parameters of Concorde's `linkern`, running n iterations of kicks, where n is the number of cities in the test instance.

The test results are presented in Table 15.35. The column "Best CLK" gives the percentage gap (to the optimal tour value) for the best of the 10 Chained Lin-Kernighan tours, and the "Merged" column gives the percentage gap for the merged tour. In the "Failures" column we report the number of times that the merge-step failed in the four trials. The running times are again given in seconds on a 500-MHz EV6 Compaq Alpha processor; the "CLK time" is the total time to generate the 10 tours; the "Merge Time" includes both the time to find the branch decomposition and the time to run the dynamic programming algorithm. The average width of the branch decompositions is reported in the "Width" column.

In two of the instances, u2319 and pla7397, the tour-merging code failed in all four trials. In both cases, the failures were due to decompositions of width greater than 20 (the maximum allowed in our dynamic programming algorithm). For fnl4461, the code failed in three of the four trials; in the three failures, decompositions of width 19 were found, but the run of the dynamic programming algorithm exceeded the 2-Gbyte memory limit of our computer. In Table 15.35, we report the "Width" and "Merge Time" only for the successful fn4461 trial.

Averaging over all trials in our tests, tour merging reduced the percentage gap (to the optimal tour value) from 0.21% down to 0.07%.

The tests give just one sample point of the many choices for the number of tours and the many possible settings for Chained Lin-Kernighan, but the results are typical of the type of improvements that are possible with tour-merging.

MERGING LKH TOURS

We use Helsgaun's LKH code in the tests on large-scale instances. Other recent high-end TSP heuristics that could also be considered include Balas and Simonetti [33], Ergun et al. [167], Schilham [493], and Walshaw [548], but LKH appears to exceed the performance of all TSP heuristic algorithms proposed to date.

In the merge tests, we consider only those TSPLIB instances having at least 5,000 cities. For the smaller instances, Helsgaun [260] reports that the default version of LKH finds optimal solutions in at least one run out of ten in each case. We also exclude the instance pla7397, since LKH routinely finds optimal solutions in this case. Finally, we exclude pla85900, since multiple runs of LKH on this instance would exceed the available computing time.

Our first test on the remaining eight TSPLIB instances is summarized in Table 15.36. Each run in this test consists of merging 10 LKH tours generated with the default settings (n repeated trials for each n-city TSP). For each instance we made four of these runs, and we report the average results in Table 15.36. The information in the table is organized in the same manner as in Table 15.35. The column "Best LKH" gives the percentage gap (to the best available lower bound) for the best of the 10 LKH tours, and the "Merged" column gives the percentage gap for the merged tour (in the case of rl11849, the optimal solution was found in each of the four trials). In the "Failures" column we report that the merge step failed in two of the four d18512 trials as well as in all four pla33810 trials; the

Table 15.35 Merging 10 Chained Lin-Kernighan tours: Average over 4 trials.

Name	Best CLK	Merged	Fail	CLK Time	Merge Time	Width
dsj1000	0.08%	0.02%	0	56.67	1.04	6.8
pr1002	0.22%	0.05%	0	22.05	1.20	8.0
si1032	0.09%	0.04%	0	18.78	0.14	6.8
u1060	0.14%	0.01%	0	33.65	1.70	7.5
vm1084	0.02%	0.00%	0	29.18	0.37	7.2
pcb1173	0.21%	0.01%	0	18.19	3.64	11.0
d1291	0.22%	0.11%	0	33.90	1.33	8.5
rl1304	0.34%	0.30%	0	40.80	0.42	8.5
rl1323	0.18%	0.07%	0	32.52	0.42	8.0
nrw1379	0.13%	0.03%	0	21.97	3.28	10.8
fl1400	0.00%	0.00%	0	199.62	1,582.47	9.5
u1432	0.20%	0.03%	0	34.49	282.46	8.8
fl1577	0.03%	0.01%	0	116.42	2.07	8.0
d1655	0.37%	0.12%	0	45.28	4.39	9.8
vm1748	0.10%	0.00%	0	55.22	1.47	8.5
u1817	0.38%	0.12%	0	32.33	10.72	11.5
rl1889	0.25%	0.05%	0	74.02	1.94	10.2
d2103	0.47%	0.41%	0	81.06	11.88	10.2
u2152	0.24%	0.14%	0	38.57	67.33	11.8
u2319	0.11%	–	4	168.15	–	> 20
pr2392	0.25%	0.04%	0	44.94	12.65	11.5
pcb3038	0.16%	0.03%	0	55.50	346.28	14.0
fl3795	0.41%	0.11%	0	271.61	65.19	9.5
fnl4461	0.17%	0.06%	3	96.46	(38.94)	(14)
rl5915	0.30%	0.03%	0	210.47	475.99	13.5
rl5934	0.27%	0.02%	0	228.74	47.11	15.2
pla7397	0.25%	–	4	404.04	–	> 20

Table 15.36 Merging 10 LKH tours: Average over 10 trials.

Name	LKH	Merged	Fail	LKH Time	Merge Time	Width
rl5915	0.0166%	0.0054%	0	16,298	1.55	7.50
rl5934	0.0133%	0.0081%	0	23,134	2.07	7.75
rl11849	0.0051%	Optimal	0	161,621	15.49	7.25
usa13509	0.0047%	0.0010%	0	242,118	29.22	9.25
brd14051	0.0086%	0.0036%	0	419,079	156.93	14.25
d15112	0.0047%	0.0007%	0	494,044	134.20	15.50
d18512	0.0161%	0.0075%	2	926,215	(311.71)	(17.50)

Table 15.37 Merging 40 LKH tours.

Name	Best LKH	Merged	Total LKH Time	Merge Time	Width
rl5915	0.0087%	Optimal	63,954	2.21	8
rl5934	0.0065%	Optimal	91,264	3.15	8
rl11849	0.0023%	Optimal	646,483	36.26	7
usa13509	0.0031%	0.0001%	968,473	62.06	13
brd14051	0.0060%	0.0030%	1,676,314	444.34	17
d15112	0.0029%	Optimal	1,976,174	3,944.32	19
d18512	0.0139%	Failed	3,704,858	-	> 20
pla33810	0.0335%	Failed	43,632,379	-	> 20

failures were due to decompositions of width greater than 20. The running times are given in seconds on a 500-MHz EV6 Compaq Alpha processor.

In Table 15.37 we report the results of merging 40 LKH tours for each of the test instances (only a single trial). The two largest instances failed, but in four of the remaining six instances an optimal solution was found. Note that although the total LKH time is quite high, merging the tours requires only a modest amount of additional computing time. The time for merging can be reduced, moreover, if we merge only the 10 best from each set of 40 tours, as we indicate in Table 15.38. This 10-out-of-40 test produced the same tours as the all-40 tests in the six smaller instances, and it also produced a solution to d18512 (but still failed on pla33810). The average branch-width was reduced by 26% in the 10-out-of-40 test, resulting in a large decrease in the merging time.

The results demonstrate the additional power tour merging brings to the task of finding near-optimal solutions to the TSP. The strong computational results have created interest in the development of heuristic algorithms aimed at achieving the performance of the exact tour-merging methods, but at a lower computational cost. Such heuristics have been proposed by Helsgaun (private communication, 2006) and Tamaki [518].

Table 15.38 Merging best 10 out of 40 LKH tours.

Name	Merged	Merge Time	Width
rl5915	Optimal	1.68	7
rl5934	Optimal	0.17	5
rl11849	Optimal	4.64	6
usa13509	0.0001%	16.54	8
brd14051	0.0006%	103.48	12
d15112	Optimal	107.21	15
d18512	0.0009%	278.90	13
pla33810	Failed	–	> 20

Chapter Sixteen

Computation

The algorithmic components described thus far in the book are brought together in the Concorde computer code for the TSP. The design of Concorde is focused on the solution of larger, more difficult, instances of the problem, rather than, say, concentrating on the fastest solution of small examples. This decision is based on our belief that by attempting to extend the reach of TSP solvers we can best continue the success of the problem as an engine of discovery, as well as making a direct contribution to the long-term research of the TSP itself.

In this chapter we describe Concorde and report on computational studies leading to the solution of the full set of TSPLIB instances, as well as extensive tests on the classical challenge of random Euclidean TSPs.

16.1 THE CONCORDE CODE

The development of Concorde is the center of our TSP research, providing the platform for testing each algorithmic idea. The software contains approximately 130,000 lines of code in the C Programming Language. The full source code is available at

www.tsp.gatech.edu

along with executable versions of Concorde for a variety of computing platforms. The code supports distributed parallel processing, via NFS sockets, in evaluating nodes of the branch-and-cut tree, and shared-memory parallel processing in routines for searching the pool of cuts.

Concorde includes implementations of each of the main separation routines discussed in the book, with the exception of the parametric connectivity algorithm from Section 6.1 and the gluing method from Section 10.4. These two routines were dropped from the code after tests of preliminary implementations, indicating better overall performance with the tour-interval approach for generating subtours and the tighten approach for modifying cutting planes. Also, the domino-parity separation algorithm is not part of the core Concorde code but rather an add-on module developed by Cook et al. [135]. This module is adopted in work reported here for the TSPLIB instances d18512, pla33810, and pla85900, but is otherwise not used in the computational tests.

The separation routines included in Concorde are organized into a cutting-plane loop, with the routines ordered according to rough estimates of their computational requirements, starting with the connectivity test and ending with the local-cuts procedure. The loop implements the strategy described in Section 12.4.3 for interleaving the cutting-plane and column-generation components of the computation. After the cuts from each routine are added to the LP relaxation, if the total improvement in the LP bound for the entire round of cutting planes is above a threshold value β_{next}, we break off the round, apply column generation, and go back to the start of the loop. If the total improvement is less than β_{next}, we continue the round of cutting planes by moving to the next separation routine in our list. Letting u_0 denote the value of the best tour known for the given TSP instance and letting z_0 denote the value of the LP relaxation when the cutting-plane loop is called, we set $\beta_{next} = 0.01(u_0 - z_0)$. The loop is terminated if after an entire round of cuts the total improvement in the bound is less than $\beta_{round} = 0.001(u_0 - z_0)$

SELECTION OF SEPARATION ROUTINES

We cannot claim to have made a thorough study of the impact of the great number of interrelated parameters that guide the control programs and the individual separation routines of Concorde. At the highest level, however, we can make a quick evaluation of the performance of the components in the cutting-plane loop, at least for instances of modest size. To carry out such tests, we consider 15 examples

dsj1000	pr1002	si1032	u1060	vm1084
pcb1173	rl1304	rl1323	nrw1379	fl1400
u1432	fl1577	d1655	vm1748	pr2392

from the TSPLIB, chosen from the full set of 21 instances having between 1,000 and 2,500 cities. The test set contains all but one of the instances in the 1,000-city to 2,500-city range that are solved within 1,000 seconds with our default code. The single exception is the instance u2319; the challenge in u2319 is to find the correct tour, rather than finding a good set of cutting planes, since its optimal value is equal to the bound given by its subtour relaxation.

In comparing the impact of including an additional separation routine or excluding one of the default routines, we summarize the set of 15 test runs by two statistics. The first is simply the sum of the CPU times in the tests. This is a standard measure, but it can be misleading, since the results can be skewed by a poor performance on a single example. To give a more balanced assessment, we also compute for each instance the ratio of the CPU time for the test code and the CPU time for the default code, and report the geometric mean of these ratios. The geometric mean of numbers $\alpha_1, \ldots, \alpha_n$ is defined as $\sqrt[n]{\alpha_1 \alpha_2 \cdots \alpha_n}$; it dampens the impact of outliers in the test results, in comparison with the usual arithmetic mean. The geometric mean of CPU ratios is also used in Chapter 13, in comparing implementations of the simplex algorithm.

Our tests are carried out on a single processor of a Sun Microsystems V20z compute node, equipped with two 2.4-GHz AMD Opteron 250 processors and 4 Gbyte of memory. On this platform the LP solver is ILOG CPLEX 9.1.

To begin, consider the three remaining subtour-finding routines described in Chapter 6. Subtour inequalities are the backbone of the Dantzig, Fulkerson, and Johnson approach to the TSP, and our code cannot operate without some form of subtour separation. We can, however, remove any one of the three routines and measure the change in the performance of the code. The results for these three experiments are reported in Table 16.1. The geometric mean of the running time

Table 16.1 Impact of subtour-generation routines.

Modification	Total CPU Seconds	Ratio
Default	2,379.2	1.00
Remove connectivity cuts	2,787.3	0.92
Remove tour-interval cuts	3,731.6	0.92
Remove exact subtour cuts	5,279.5	0.98

ratios actually decreases by dropping each of the routines, although the total CPU time increases. We have chosen to leave all three subtour-finding methods in the default settings, since it is useful on very large instances to be nearly in the subtour polytope when applying the separation algorithms for more complex classes of cutting planes.

The next class of separation routines are the blossom and block comb algorithms described in Chapter 7. By default, Concorde uses the fast blossom heuristics and a collection of block comb heuristics based on shrinking operations, but it does not use the exact separation algorithm for blossom inequalities. Test results for these three collections of routines are given in Table 16.2. For the range of instances

Table 16.2 Impact of blossom and block routines.

Modification	Total CPU Seconds	Ratio
Default	2,379.2	1.00
Remove fast blossoms	2,421.5	0.86
Add exact blossoms	6,816.0	1.41
Remove block combs	1,880.6	0.82

in the test set, the code performs somewhat better without the blossom and comb heuristics, and considerably worse if the exact blossom algorithm is adopted. We nonetheless choose to include the blossom and comb heuristics—the fast running times and the ability to produce large families of cuts allow Concorde to delay the use of more time-consuming heuristics on large instances. The exact blossom algorithm is not often adopted in our computations.

Two additional comb heuristics are presented in Chapters 8 and 9, derived from a study of the consecutive ones property and from the structure of tight dominoes. The impact of dropping these heuristics is reported in Table 16.3. Particularly striking is the factor of two increase in the ratio when both routines are removed from the mix of cutting planes. For very large instances the domino comb algorithm

Table 16.3 Impact of consecutive ones and domino combs.

Modification	Total CPU Seconds	Ratio
Default	2,379.2	1.00
Remove consecutive ones	3,483.9	1.30
Remove domino combs	2,721.3	1.09
Remove both	7,908.1	2.06

can become time-consuming as the cut pool grows in size, but this is not reflected in the test set.

A variety of routines for modifying existing cutting planes are presented in Chapter 10. The most important of these for large instances is the tighten algorithm. Removing tighten from the default code shows a small decrease in performance on the test set, as reported in Table 16.4. This default version of tighten applies the

Table 16.4 Impact of cut metamorphoses.

Modification	Total CPU Seconds	Ratio
Default	2,379.2	1.00
No tighten	3,343.2	1.03
Add tighten cut pool	2,295.7	0.91
No teething	2,829.7	0.97
No path inequalities	18,165.8	1.05
Add clique-tree inequalities	7,145.3	1.17
Add star inequalities	3,173.5	1.00

improvement algorithm to each nearly-tight cutting plane in the current LP relaxation. A further improvement in the test results is obtained by also applying the algorithm to inequalities in the cut pool, but this can be time-consuming for large instances and it is not turned on by default.

Results for the teething method described in Section 10.2 and for the path-inequality heuristic from Section 10.3 are also presented in Table 16.4. A small improvement is made by not teething combs, while a large increase in running time is incurred if path inequalities are dropped. In two further tests, we considered adding routines for clique trees and general star inequalities, based on outlines from Naddef and Thienel [424]; neither routine improved the performance of the code on these modest-sized instances.

Finally, consider the local-cuts method described in Chapter 11. A crucial choice in applying local cuts is the setting of the parameter t, used in the process for selecting the linear mapping ϕ described in Section 11.2. The linear mapping is defined by a partition $V_0, V_1, \ldots, V_k$ of the vertex set V, where each set in the partition is shrunk to a single point. The parameter t controls the number of sets in the partition, in the sense that $t - 3 \leq k \leq t$. A larger value of t allows the local-cuts code to detect more complex flaws in the LP solution vector x^*, but at greater computational expense than with smaller values of t. We therefore adopt the strategy of starting with $t = 8$ and gradually increasing its value up to a specified t_{max}. This process is implemented as a loop; after applying the cuts found with $t = k$, if the total improvement in the LP bound remains less than β_{next}, we set $t = k + 1$ and repeat the search for local cuts as long as $k < t_{max}$.

In Table 16.5 we report results for values of t_{max} from 0 up to 28; the default in Concorde is $t_{max} = 16$. On these instances, a small improvement is obtained

Table 16.5 Impact of local cuts.

Modification	Total CPU Seconds	Ratio
$t_{max} = 0$	2,509.0	1.01
$t_{max} = 8$	2,384.1	1.05
$t_{max} = 12$	2,187.2	0.97
$t_{max} = 16$	2,379.2	1.00
$t_{max} = 20$	2,757.7	1.19
$t_{max} = 24$	5,395.8	1.90
$t_{max} = 28$	14,096.9	3.44

with local cuts of size 12 and 16, but larger values are clearly too time-consuming to be effective. Tests on more challenging TSPLIB examples are reported in Section 16.3, where local cuts of size 32 or higher are an important component of the solution process.

To conclude this study, it is important to point out that the simple on/off tests we have reported do not capture the interplay between the various separation routines. Indeed, adopting a greedy strategy, the above tests suggest that, by default, the Concorde code should remove the fast-blossom, block combs, and teething routines, add the routine to tighten inequalities from the cut pool, and set $t_{max} = 12$. The combined impact of these modifications, however, gives the results

Total CPU Seconds	Ratio
2,553.2	1.22

showing a small decrease in the overall performance, rather than the expected improvement.

16.2 RANDOM EUCLIDEAN INSTANCES

Numerous early studies of the TSP rely on randomly generated Euclidean instances to test the proposed computational techniques. Such problems pose a significant challenge to TSP codes, unlike instances generated by assigning random travel costs to each pair of cities, as we discussed in Section 4.7. Some of the larger random Euclidean problems considered in the literature include

- a 64-city instance, solved by Held and Karp [254],

- a 67-city instance, solved by Camerini et al. [102],

- three 80-city instances, solved by Miliotis [394],

- ten 15-city instances, nine 30-city instances, five 45-city instances, four 60-city instances, and three 75-city instances, solved by Padberg and Hong [451],

- ten 100-city instances, solved by Land [333],

- five 100-city instances, two 150-city instances, and two 200-city instances, created by Krolak et al. [325] and solved by Crowder and Padberg [147], Grötschel and Holland [232], and Padberg and Rinaldi [447].

In this section we discuss the solution of a large collection of instances, ranging in size from 100 cities up to 2,500 cities.

Our test set is created with the RngStream generator of L'Ecuyer [340], using a single stream of numbers to produce 463,000 TSP instances, one after another. Each instance consists of n cities with integer (x, y)-coordinates drawn uniformly between 0 and $10^7 - 1$. The travel costs are defined by the TSPLIB EUC_2D-norm, that is, for each pair of cities (x_i, y_i) and (x_j, y_j), the cost c_{ij} is

$$\sqrt{(x_i - x_j)^2 + (y_i - y_j)^2}$$

rounded to the nearest integer. The test set consists of

- 10,000 examples for each $n = 100, 110, \ldots, 500$,

- 10,000 examples for each $n = 600, 700, \ldots, 1000$,

- 1,000 examples for each $n = 1500, 2000, 2500$.

Concorde successfully solved each of these problems, using the default settings. The runs were carried out on a cluster of compute nodes, each equipped with two 2.66-GHz Intel Xeon processors and 2 Gbyte of memory; the LP solver on this platform is ILOG CPLEX 6.5. In Table 16.6 we report the mean running times for a selection of the test sets, using a single Intel Xeon processor. A plot of the full set of mean running times is given Figure 16.1, using a log scale for the number of CPU seconds.

The mean values are only a very rough estimate of the performance of Concorde on these instances, since for a given n the individual times can vary greatly. For example, in the 10,000 tests with $n = 1,000$, seventeen of the runs completed in under 10 seconds, while eighteen instances required over 10,000 seconds. A histogram of the running times for $n = 1,000$ is displayed in Figure 16.2; the figure shows a considerable tale in the distribution at the high end.

Table 16.6 Running times for random Euclidean tests.

n	Samples	Mean CPU Seconds
100	10,000	0.7
200	10,000	3.6
300	10,000	10.6
400	10,000	25.9
500	10,000	50.2
600	10,000	92.0
700	10,000	154.5
800	10,000	250.4
900	10,000	384.7
1,000	10,000	601.6
1,500	1,000	3290.9
2,000	1,000	14065.6
2,500	1,000	53737.9

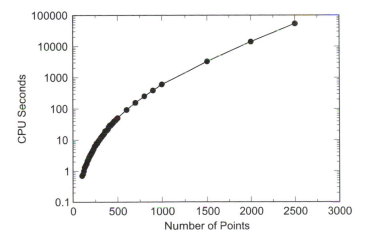

Figure 16.1 Average running times in CPU seconds.

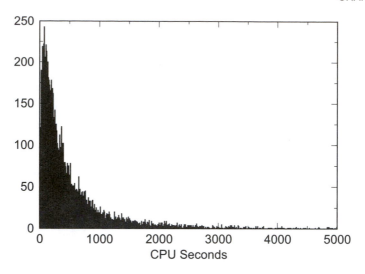

Figure 16.2 Histogram of running times for 1,000-city instances.

The plot of the mean values in Figure 16.1 indicates that the running times are increasing as an exponential function of n, but the default parameters in Concorde are clearly adequate for the solution of the wide range of instances in our test set. If we go out to much larger examples, however, it would be a great challenge to find optimal tours. Nonetheless, Concorde can be used to obtain very strong lower bounds on such problems. To demonstrate this, we consider an additional test set of 100 instances with $n = 100,000$. For these instances, we compute upper and lower bounds on the optimal tour values. The upper bounds were found by running a version of LKH, kindly provided by K. Helsgaun for our test; the lower bounds were found using Concorde's cutting-plane routines, without resorting to a branch-and-cut search.

In the Concorde runs, we adopt the following iterated strategy. To fully exploit local cuts with small t before moving to consider larger values, repeated calls to the full cutting-plane loop are made with increasing values of t_{max}. The process begins with $t_{max} = 16$ and increases its value by four with each call to the cutting-plane loop, terminating the entire run when t_{max} reaches a specified maximum value t^*; in these tests we set $t^* = 36$. With this procedure, the tolerances β_{next} and β_{round} are reset to smaller values with each successive call to the cutting-plane loop, as the LP bound improves with the additional cuts. These smaller tolerances permit the code to work harder in obtaining cuts before breaking off the search.

Each run of the Concorde code completed in under 14 CPU days, with the average being 11.2 days. For each of the 100 instances, the lower bound produced by Concorde was within 0.037% of the cost of the LKH tour; the average gap was 0.030%. The small integrality gaps show both the effectiveness of Concorde on large instances, and the excellent quality of the LKH heuristic.

THE TSP CONSTANT β

The number and size of Euclidean instances solved in our tests suggest the work could possibly give insights into the value of the Beardwood et al. [45] constant we discussed in Section 1.3. The TSP instances considered in this setting have n points chosen uniformly from the unit square, with travel costs given by the actual Euclidean distances; our test set approximates this by using a much larger square and rounding distances to the nearest integer. Recall that Beardwood et al. prove that with probability 1, as n approaches infinity, the optimal tour length divided by $\sqrt{n}$ approaches a constant value β. They show analytically that

$$0.44194\sqrt{2} \le \beta \le 0.6508\sqrt{2},$$

but the precise value of β is not known.

A nice discussion of the Beardwood et al. theorem can be found in Steele [510], who writes:

> There is a natural curiosity to the constants of the BHH theorem, and considerable computation has been invested in the estimation of $\beta_{TSP}(d)$.

The d in Steele's remark refers to the dimension of the Euclidean space; most work has gone into the case with $d = 2$, which is the setting we consider in our tests.

The computational investigation of β began in the original paper of Beardwood et al. [45]. The analytical result mentioned above yields $0.6249 \le \beta \le 0.9204$. To give an indication of where in this range β actually lies, Beardwood et al. computed heuristic tours for two examples, having 202 cities and 400 cities, by hand. The tours had lengths $0.744\sqrt{202}$ and $0.0758\sqrt{400}$, leading to an estimate of $\beta \approx 0.75$. Beardwood et al. write:

> Of course, the Monte Carlo experiment is very slight and gives very little information. With only two experiments, the 95% confidence limits for β_2 are 0.467 and 0.595. We hope someone will be encouraged to make a more precise statement.

Their β_2 is our β divided by $\sqrt{2}$, so the estimated range is $0.66 \le \beta \le 0.84$.

This theme was continued in work by Stein [511], Bonomai and Lutton [77], Krauth and Mézard [313], Ong and Huang [437], and Lee and Choi [342], relying mainly on heuristic algorithms to provide upper bounds on the optimal tour lengths. The most recent estimates are

- $\beta \approx 0.7120 \pm 0.0002$ by Percus and Martin [459] and

- $\beta \approx 0.7124 \pm 0.0002$ by Johnson et al. [293],

both obtained with indirect methods. In the Percus and Martin paper, a version of the problem on the torus, rather than the square, is considered and a large number of experiments with instances up to 100 cities are carried out using the Chained Lin-Kernighan heuristic. The use of the torus eliminates the effect of points near to the boundary of the square, helping to compensate for the restricted size of the test sets that are considered. Johnson et al. use a combination of Concorde and a version of Chained Lin-Kernighan to obtain an estimate of the gap between the optimal tour value and the subtour relaxation, allowing them to estimate β via an experimental estimate of the asymptotic value of the subtour bound.

Turning to our experimental data, in Figures 16.3 and 16.4 we display histograms of the 10,000 tour lengths obtained in the tests for $n = 100, 200, \ldots, 1000$. For each n the data are a good fit to the normal distribution, and it is clear that the individual values are increasingly concentrated around their mean. The histograms displayed in Figure 16.5 for the 1,000 tour lengths obtained in the $n = 1500, 2000, 2500$ tests also show increasingly concentrated distributions, drawn on a different scale than the earlier figures.

In Table 16.7 we give the mean values and standard deviations of the tour lengths for a portion of our test set. The mean values for all test cases are plotted in Fig-

Table 16.7 Results of random Euclidean tests.

n	Samples	Mean	Standard Deviation
100	10,000	0.7764689	0.0230018
200	10,000	0.7563542	0.0152578
300	10,000	0.7477629	0.0121529
400	10,000	0.7428444	0.0103775
500	10,000	0.7394544	0.0091352
600	10,000	0.7369409	0.0083160
700	10,000	0.7349902	0.0076233
800	10,000	0.7335751	0.0071642
900	10,000	0.7321114	0.0066941
1,000	10,000	0.7312235	0.0063176
1,500	1,000	0.7279798	0.0048828
2,000	1,000	0.7256264	0.0042759
2,500	1,000	0.7241373	0.0038671

ure 16.6. The data points give a nice downward curve and provide compelling computational evidence that the mean tour length is decreasing as n is increasing.

The data clearly show that estimates of β above 0.725 are too high, given the downward trend in the mean values. Any further claims, however, must be based on speculation about the form of the function describing the mean tour lengths. Least squares fits of the data under various assumptions put β in the range of 0.712 or 0.713, but we can neither confirm nor refute the more precise estimates of 0.7120 and 0.7124 proposed by Percus and Martin [459] and Johnson et al. [293].

One additional interesting data point is provided by the study of 100 instances with $n = 100,000$. In this test, the average lengths of the LKH tours is 0.7142 and the average lower bound from the Concorde runs with $t^* = 36$ is 0.7140. The

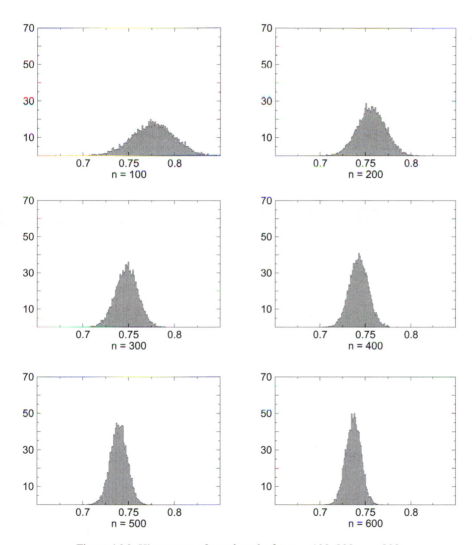

Figure 16.3 Histograms of tour lengths for $n = 100, 200, \ldots, 800$.

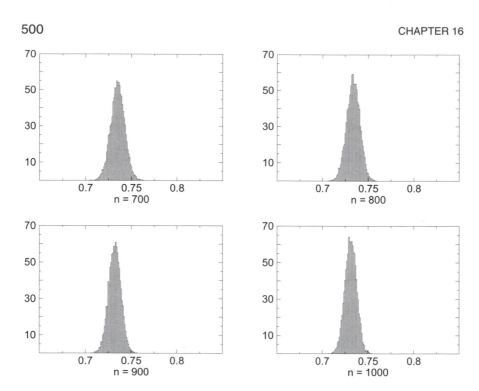

Figure 16.4 Histograms of tour lengths for $n = 900, 1000$.

standard deviations of 0.000593 for the tour lengths and 0.000594 for the lower bounds again points out the increasing concentration of values around their means as n increases. Moreover, assuming the mean tour lengths are decreasing with increasing n, the results for $n = 100,000$ indicate that the ultimate value of β is very likely to be under 0.714.

16.3 THE TSPLIB

Since its creation in 1990 by Reinelt [472], the TSPLIB collection of test instances has been far and away the most widely studied problem set in computational work on the TSP. The instances range in size from 14 cities up to 85,900 cities, covering a variety of industrial applications, geographic point sets, and academic challenge problems. In this section we consider all TSPLIB instances having less than 10,000 cities; a discussion of the larger instances is taken up in Section 16.4.

To begin, the TSPLIB instances having under 1,000 cities are readily solved with Concorde's default settings, with all solution times under one minute on a 2.4-GHz AMD Opteron processor. A scatter plot of the running times versus the number of cities, for a single trial on each of the 76 instances, is given in Figure 16.7; a log scale is used for the running times. Note that all instances under 400 cities are solved in less than 10 seconds of CPU time.

The next block of test instances consists of all TSPLIB examples having between

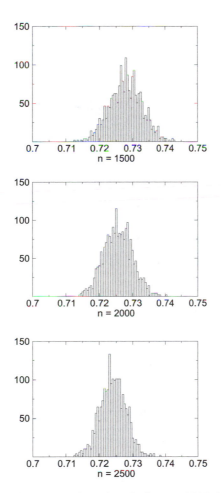

Figure 16.5 Histograms of tour lengths for $n = 1500, 2000, 2500$.

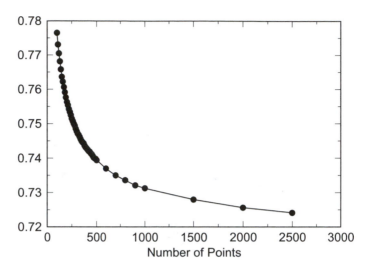

Figure 16.6 Average tour length scaled by $\sqrt{n}$.

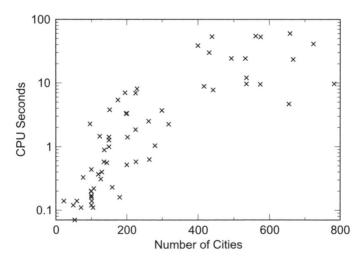

Figure 16.7 Solution times for small TSPLIB instances.

1,000 cities and 2,500 cities. The results for the individual problems are reported in Table 16.8, again with Concorde's default settings. Note the exceptionally long

Table 16.8 Midsized TSPLIB instances with default code.

Problem	Optimal Value	Search Nodes	CPU Seconds
dsj1000	18,660,188	5	57.4
pr1002	259,045	1	5.7
si1032	92,650	1	97.1
u1060	224,094	11	73.1
vm1084	239,297	7	132.8
pcb1173	56,892	11	62.7
d1291	50,801	11	2,858.0
rl1304	252,948	1	34.3
rl1323	270,199	31	734.2
nrw1379	56,638	13	78.3
fl1400	20,127	3	186.7
u1432	152,970	1	19.0
fl1577	22,249	5	568.1
d1655	62,128	3	43.9
vm1748	336,556	7	264.2
u1817	57,201	125	13,999.9
rl1889	316,536	73	1,854.8
d2103	80,450	2,079	18,226,404.4
u2152	64,253	45	3,345.3
u2319	234,256	1	21.5
pr2392	378,032	1	21.8

running time for d2103. Using the code with different settings can reduce the time for this example, but it appears to be genuinely more difficult than the other problems, at least for the Concorde code. The point set for d2103 displayed in Figure 16.8 shows considerable structure, with a large number of colinear points. At first glance, this would appear to be an easier problem than a distribution of points such as that of pr2392 displayed in Figure 16.9, but the running time is nearly 1,000,000 times longer.

For the next set of examples, including all remaining TSPLIB instances having under 10,000 cities, we employ the iterated version of the local-cuts loop, described in the previous section, increasing t_{max} by four after each call to the loop, starting with $t_{max} = 16$ and continuing to a maximum value of $t^* = 32$. This time-consuming process is used only for the root LP relaxation; once the branch-and-cut search is begun, the value of t_{max} is set to 16.

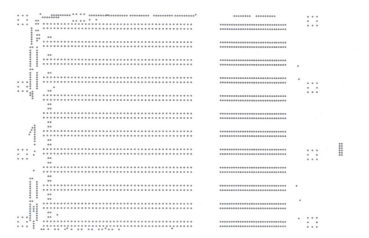

Figure 16.8 Point set for d2103.

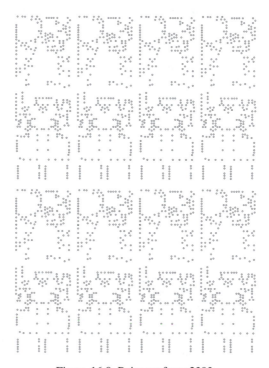

Figure 16.9 Point set for pr2392.

The results for the six test problems are reported in Table 16.9. Each of the

Table 16.9 Large TSPLIB instances with local cuts of size 32.

Problem	Optimal Value	Root Gap	Search Nodes	CPU Seconds
pcb3038	137,694	0.020%	57	7,022.4
fl3795	28,772	Optimal	1	18,377.6
fnl4461	182,566	0.008%	119	7,633.7
rl5915	565,530	0.020%	73	68,742.3
rl5934	556,045	0.022%	185	176,959.9
pla7397	23,260,728	0.028%	85	164,535.8

instances solved in under 50 hours of CPU time. Note that running times can vary considerably with small changes to the code, but these problems are clearly within the range of instances that can be readily solved with the straightforward application of Concorde.

In Tables 16.10 and 16.11 we compare the performance of Concorde on pcb3038 and fnl4461 when t^* varies between 0 and 48. In both cases, the lower bound provided by the root LP steadily increases as we raise the value of t^*, while the number of nodes in the search tree decreases. For pcb3038, the running time for the $t^* = 32$ trial is roughly half the time needed when local cuts are not employed. This is only a modest improvement, however, when compared with the results for fnl4461. In this later case, the run with $t^* = 32$ is a factor of 115 times faster than the test with local cuts turned off.

Table 16.10 Local cuts on pcb3038.

t^*	Root Gap	Search Nodes	Root CPU Seconds	Total CPU Seconds
0	0.048%	351	165.0	16,422.8
16	0.035%	139	373.3	8,953.8
20	0.029%	177	828.4	15,745.0
24	0.025%	99	1,115.6	8,638.3
28	0.022%	67	1,576.3	6,240.6
32	0.020%	57	2,760.3	7,022.4
36	0.017%	49	5,489.8	8,768.0
40	0.017%	53	9,710.7	13,448.6
44	0.017%	27	10,952.7	13,267.0
48	0.013%	31	34,321.7	36,905.8

Table 16.11 Local cuts on fnl4461.

t^*	Root Gap	Search Nodes	Root CPU Seconds	Total CPU Seconds
0	0.039%	22,083	119.3	881,239.2
16	0.022%	1,177	259.1	51,165.9
20	0.013%	409	649.5	18,881.9
24	0.010%	183	917.5	10,338.0
28	0.008%	223	1,309.2	9,682.6
32	0.008%	119	1,883.1	7,633.7
36	0.007%	97	2,462.9	6,808.9
40	0.007%	135	3,059.9	7,711.4
44	0.006%	101	6,534.4	10,301.7
48	0.006%	71	8,448.8	10,978.3

16.4 VERY LARGE INSTANCES

Much of our effort has been devoted to increasing the scale of problems that can be attacked with Concorde, often using networks of distributed computers to carry out branch-and-cut searches on very large instances. Of course, the notion of a "very large" instance changes as algorithms and computing hardware improve over time. Indeed, our initial solutions of pcb3038, fnl4461, and pla7390 in the early 1990s were carried out on large networks of state-of-the-art workstations, using several years of total CPU time per problem. These instances can now be readily solved with Concorde on a commodity workstation, due in large part to improvements driven by the demands of a collection of more difficult instances.

At the current time, our notion of "very large" refers to problems having over 10,000 cities. Concorde was successful in solving all seven of the TSPLIB instances in this range. The individual problems are as follows.

- rl11849—A point set arising in a circuit board application with 11,849 holes; first solved in October 1999 on a network of Compaq Alpha workstations at Rice University.

- usa13509—The locations of all 13,509 cities in the continental United States having a population of 500 or more; solved in May 1998 on a network of Digital Alphaservers, Intel Pentium II workstations, and Sun Microsystems Sparc workstations at Rice University.

- brd14051—A collection of 14,051 cities from the former West Germany; solved in March 2004 on a cluster of Intel Xeon compute nodes at Georgia Tech.

- d15112—The locations of 15,112 cities in Germany; solved in April 2001 on a network of Compaq Alphaservers, Intel Pentium III workstations, and

AMD Athalon compute nodes at Princeton University and Rice University.

- d18512—A second German problem, with 18,512 cities, including the point set of brd14051; solved in March 2005 at Georgia Tech on a cluster of Intel Xeon compute nodes, using Concorde with the domino-parity module.

- pla33810—The point set for a VLSI application with 33,810 cities arising at Bell Laboratories in the late 1980s; one of three instances contributed to the TSPLIB by D. S. Johnson in 1991; solved in October 2004 at Georgia Tech on a cluster of Intel Xeon compute nodes, using Concorde with domino-parity cuts.

- pla85900—A denser VLSI instance from Bell Laboratories, having 85,900 cities; solved in April 2006 at Georgia Tech on clusters of AMD Opteron compute nodes and Intel Xeon compute nodes, using Concorde with domino-parity cuts.

Two additional large instances were also solved, taken from a collection of problems at

www.tsp.gatech.edu/countries.html

giving locations of cities for a number of countries throughout the world.

- it16862—The point set of all 16,682 populated places in Italy; solved in December 2002 on a network of Compaq Alphaservers, Intel Pentium III workstations, and AMD Athalon compute nodes at Princeton University and Rice University.

- sw24978—All 24,879 populated places in Sweden; solved in May 2004 on a cluster of Intel Xeon compute nodes at Georgia Tech.

Details of the solution of these nine instances are given below. The large amounts of CPU time used in this work make these computations far from routine, but the variety and number of solved instances demonstrates the suitability of the cutting-plane method even for examples of this size.

FIVE INSTANCES HAVING OVER 10,000 CITIES

We begin our discussion with the five smaller instances, ranging in size from 11,849 cities up to 16,862 cities. A number of these problems were first solved in computations as the Concorde code was under construction. In these initial computations, ad hoc settings of parameters were adopted to push the work forward as a general strategy for very large instances was developed.

One of the lessons learned in the early computations is that a powerful additional source of cutting planes can be created through a preliminary branch-and-cut computation. In this effort, the inequalities found while processing nodes of the search tree are collected into a large cut pool that can be reapplied to the root LP relaxation. In many cases, this supply of cuts can greatly improve the quality of the root lower bound, permitting a second (or third, or fourth) branch-and-cut run to establish a proof of optimality for the TSP instance. This observation was also made by Warme et al. [549] in their study of the Euclidean Steiner-tree problem (D. M. Warme, private communication, 2003). It indicates some weakness in our separation routines, since the cutting planes are available, but not detected, in the root LP computation. Nonetheless, the observation can be turned into a planned strategy, where initial rounds of branch-and-cut searches are geared specifically toward producing a good collection of cuts, rather than seeking only to push the overall lower bound.

We adopt this repeated branch-and-cut strategy in a specified manner to solve each of the five instances under discussion. The computations consist of an initial cutting-plane phase using $t^* = 48$, that is, local cuts starting with $t_{max} = 16$ are applied, followed by additional rounds of cuts with t_{max} incremented by four each round, all the way up to $t_{max} = 48$. (The use of even larger values of t_{max} is not productive, due to the frequent occurrence of overflows in the internal calculations of the local-cuts procedure.) The root LP produced in this way is used to start a branch-and-cut run, where each search node is processed using $t_{max} = 48$ and where the entire run is terminated if it reaches 1,000 active subproblems. The cut pool collected in this branch-and-cut run is used in a further cutting-plane computation, starting with the existing root LP and again using $t^* = 48$. This is followed by second and third branch-and-cut runs, again stopping when the trees reach 1,000 active subproblems and applying the collected cuts to the existing root LP. The result of this long computation is a hopefully strong LP relaxation for the TSP instance. A fourth branch-and-cut run is then aimed at solving the problem, using eight-way tentative branching to improve the quality of the search tree.

A summary of the initial cutting-plane computation and three branch-and-cut runs is presented in Tables 16.12, 16.13, and 16.14. The first of the tables reports the improving value of the root LP bound as each additional branch-and-cut run adds to the overall cut pool; the "+1,000 BCx" notation refers to the x-th branch-and-cut run, terminated at 1,000 active subproblems. In Table 16.13 these root LP values are given as percentage gaps to the optimal tour values. The bounds obtained by the limited branch-and-cut runs are presented in Table 16.14, again as percentage gaps to the optimal values. Note that the runs for rl11849 and usa13509 were terminated early, since the first, respectively second, branch-and-cut runs solved the instances without reaching the 1,000-subproblem limit.

The number of search nodes used in the final branch-and-cut runs for each problem are reported in Table 16.15. Each of these runs was carried out on a cluster of 2.66-GHz Intel Xeon processors. The running times are reported as the cumulative time for a single processor; the "CPU Days" row gives the number of days used in the final branch-and-cut run, while the "Total CPU Days" gives the number of days in the entire computation, including the root LP work and the repeated branch-and-cut runs. The most time-consuming run in this set is the solution of d15112, using slightly more than 34 years in total.

The five tests were initialized with upper bounds of one greater than the optimal values, so the final branch-and-cut run for each instance produced an optimal tour. Drawings of these tours can be found in Figures 16.10 through 16.14.

Table 16.12 Root LP bounds.

	rl11849	usa13509	brd14051	d15112	it16862
Optimal	923,288	19,982,859	469,385	1,573,084	557,315
$t^* = 48$	923,053	19,979,209	469,321	1,572,863	557,233
+1,000 BC1		19,981,249	469,355	1,572,959	557,284
+1,000 BC2			469,357	1,572,967	557,287
+1,000 BC3			469,359	1,572,974	557,288

Table 16.13 Root LP integrality gaps.

	rl11849	usa13509	brd14051	d15112	it16862
$t^* = 48$	0.025%	0.018%	0.014%	0.014%	0.015%
+1,000 BC1		0.008%	0.006%	0.008%	0.006%
+1,000 BC2			0.006%	0.007%	0.005%
+1,000 BC3			0.006%	0.007%	0.005%

Table 16.14 Integrality gaps from branch-and-cut runs.

	rl11849	usa13509	brd14051	d15112	it16862
BC1	Optimal	0.003%	0.004%	0.006%	0.004%
BC2		Optimal	0.003%	0.005%	0.002%
BC3			0.002%	0.004%	0.002%
BC4 (8-way)			Optimal	Optimal	Optimal

Table 16.15 Final branch-and-cut runs.

	rl11849	usa13509	brd14051	d15112	it16862
Search Nodes	681	4,063	26,553	392,255	2,681
CPU Days	189.8	208.1	1,273.9	12,066.9	2,712.0
Total CPU Days	193.5	554.8	1,974.7	12,442.9	3,971.2

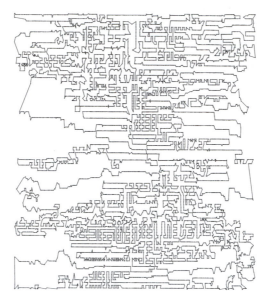

Figure 16.10 Optimal tour for a 11,849-city printed circuit board problem.

Figure 16.11 Optimal tour of 13,509 cities in the United States.

Figure 16.12 Optimal tour of 14,051 cities in West Germany.

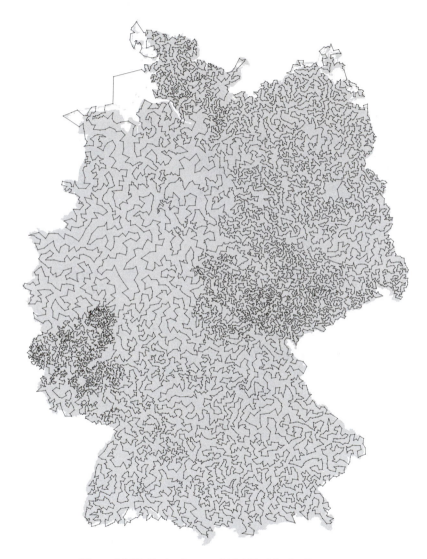

Figure 16.13 Optimal tour of 15,112 cities in Germany.

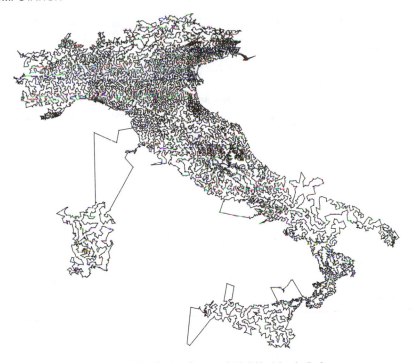

Figure 16.14 Optimal tour of 16,862 cities in Italy.

TOUR OF GERMANY—D18512

In February 2004 we attempted to extend the method used in the above computations to the larger, and at the time unsolved, TSPLIB instance d18512. The best reported heuristic result for this problem was a tour of length 645,238 found by Tamaki [518]. Our computations were carried out on the cluster of Intel Xeon processors, with ILOG CPLEX 6.5 as the LP solver.

The initial cutting-plane run and three branch-and-cut runs are reported in Table 16.16. The "Root LP" column gives the value of the LP lower bound obtained

Table 16.16 Initial d18512 runs.

	Root LP	Root Gap	B&C Gap	Total CPU Days
$t^* = 48$	645,166	0.0112%		3.1
+1,000 BC1	645,199	0.0060%	0.0045%	532.8
+1,000 BC2	645,201	0.0057%	0.0037%	765.1
+1,000 BC3	645,202	0.0056%	0.0036%	812.8

after the cut pool from each branch-and-cut run is applied in an additional round of cutting planes with $t^* = 48$; the "Root Gap" column gives the percentage gap between the LP and the optimal tour value; "B&C Gap" gives the percentage gap obtained by the limited branch-and-cut search; "Total CPU Days" reports the cumulative CPU time for the sequence of runs.

Each of the first three runs was made with the default one-way branching. In a fourth run we adopted eight-way branching in an attempt to solve the problem, starting with the 645,202 root LP relaxation. Unlike the runs in the previous section, however, the progress in the overall lower bound provided by the search tree did not indicate that the branch-and-cut run could be completed successfully in a feasible amount of CPU time. We therefore adopted a staged approach, permitting the tree to increase rapidly in size to see how high the bound could be pushed, not knowing at this point the distance to the actual optimal value. In this effort, the value of t_{max} was decreased as the number of active subproblems increased, and the run was switched from eight-way branching to one-way branching when it reached 1,000 active subproblems. The details of the run are summarized in Table 16.17; the "CPU Days" column reports the time used in the run, it does not include the time needed to construct the root LP. After a total of 8.3 CPU years, the search was terminated with 11,000 active subproblems and a bound of 645,230.

Table 16.17 Fourth d18512 run.

Active Nodes	Local Cuts	Branching	B&C Bound	CPU Days
1,000	$t_{max} = 48$	8-way	645,225	2,031.3
5,000	$t_{max} = 32$	1-way	645,228	2,505.5
11,000	$t_{max} = 16$	1-way	645,230	3,045.8

Applying the large cut pool obtained in the fourth branch-and-cut run allowed Concorde to raise the root LP bound to 645,206. This was further improved to 645,209 by applying the domino-parity module developed by Cook et al. [135] in an additional cutting-plane run with $t^* = 48$. This LP was the starting point for a long, but successful, branch-and-cut run, using 424,241 search nodes. The staged approach used in the final run is summarized in Table 16.18; the notation "DP" stands for domino-parity; the "CPU Days" column reports the time in the final run, it does not include the CPU time used in constructing the root LP.

Table 16.18 Final d18512 run.

Active Nodes	Local Cuts	Branching	B&C Bound	CPU Days
5,000	$t_{max} = 48$, DP	8-way	645,233	8,524.9
10,000	$t_{max} = 32$, DP	4-way	645,235	11,377.8
Complete	$t_{max} = 16$	1-way	Optimal	17,113.2

The entire solution process took approximately 57.5 CPU years on the cluster of 2.66-GHz Intel Xeon processors. The optimal value is 645,238, matching the

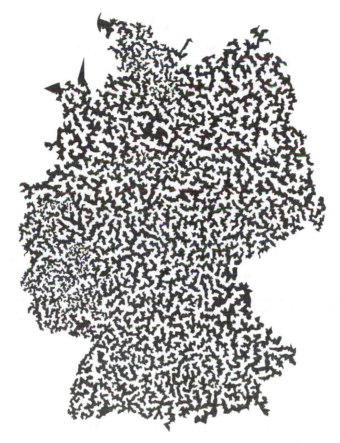

Figure 16.15 Optimal tour of 18,512 cities in Germany.

length of the tour found by Tamaki. The runs were given a starting upper bound of one greater than Tamaki's heuristic tour, so an optimal tour was also produced in the branch-and-cut search. The tour is displayed in Figure 16.15, drawn as a solid curve.

TOUR OF SWEDEN—SW24978

The computation for the 24,978-city tour of Sweden took place throughout 2003, with the final run completed in January 2004. This work was used to develop the repeated branch-and-cut strategy adopted in the tests reported above. The starting upper bound for the computation was provided by K. Helsgaun; his heuristic tour of length 855,597 was shown to be optimal in this work.

A summary of the initial cutting-plane run and four branch-and-cut runs is presented in Table 16.19. In these computations, the parameters for local cuts in the branch-and-cut trees was increased in each successive run and tentative branching was adopted in the third and fourth runs; a summary of the settings is given in

Table 16.19 Initial sw24978 runs.

	Root LP	Root Gap	B&C Gap	Total CPU Days
$t^* = 48$	855,392	0.0240%		2.8
+1,000 BC1	855,457	0.0164%	0.0122%	399.6
+1,000 BC2	855,488	0.0127%	0.0094%	1,218.1
+1,000 BC3	855,521	0.0089%	0.0064%	4,202.3
+1,000 BC4	855,535	0.0072%	0.0039%	8,151.1

Table 16.20. In the third run, the eight-way branching was switched to one-way branching when the tree reached 353 active subproblems. This same switch occurred at 357 active subproblems in the fourth run.

Table 16.20 Parameters used in sw24978 runs.

Run	Local Cuts	Branching
BC1	$t_{max} = 32$	1-way
BC2	$t_{max} = 36$	1-way
BC3	$t_{max} = 40$	(8,1)-way
BC4	$t_{max} = 44$	(8,1)-way

These initial runs produced a root LP relaxation yielding a bound of 855,535. This was the starting point of the final branch-and-cut search, proceeding in the six stages summarized in Table 16.21. In each of these stages, besides changing

Table 16.21 Final sw24978 run.

Active Nodes	Local Cuts	Branching	B&C Bound	CPU Days
353	$t_{max} = 48$	8-way	885,566	2,954.7
1,000	$t_{max} = 48$	4-way	885,576	4,871.9
1,872	$t_{max} = 48$	2-way	885,583	6,881.5
3,548	$t_{max} = 24$	2-way	885,587	7,482.4
18,608	$t_{max} = 16$	2-way	885,594	12,098.3
Complete	$t_{max} = 0$	1-way	Optimal	16,306.8

the size of local cuts and the number of tentative branches, we also redoubled the value of the tolerances β_{next} and β_{round} to limit the long cutting-plane computations. The run eventually produced the optimal tour, after a total of 167,263 search nodes; the search tree is displayed in Figure 16.16. The total time used in all computations was approximately 84.8 CPU years, on the cluster of 2.66-GHz Intel Xeon processors.

It must be pointed out that the long Sweden computation benefited greatly from the accurate upper bound provided by Helsgaun's tour. Although the branch-and-cut run can produce the optimal tour, the settings in the final attack were guided by the fact that we knew the bound only needed to be increased a small amount to complete the search. Indeed, the final stage that improved the lower bound from 855,594 up to 855,597 required 11.5 years of CPU time; without the knowledge of the 855,597 tour we would have likely not made the decision to carry out this final computation. A drawing of the optimal tour is given in Figure 16.17.

A summary of the distribution of the 14,827,429 non-subtour cutting planes that were generated and used in the branch-and-cut runs is given in Table 16.22. Our

Table 16.22 Cutting planes in sw24978 computation.

Type	Number
Combs	4,479,995
Paths and stars	7,196,548
Bipartition	28,010
Other (local cuts)	3,122,876

checking process first classified the cuts into the known classes; the *bipartition* inequalities are a generalization of clique trees introduced by Boyd and Cunningham [88]. This accounted for 11.7 million of the cutting planes. The final 3.1 million cuts were verified with direct TSP computations on the skeleton of the hypergraph inequalities, computing the right-hand-side $\mu(\mathcal{H})$ for each hypergraph $\mathcal{H}$; these tiny TSP instances were solved with an implementation of the Held and Karp [253] method when possible, and with a simplified version of branch-and-cut when Held-Karp failed to terminate in a set time limit.

VLSI APPLICATION—PLA33810

Our computations on pla33810 were begun in January 2004. At the start of the computation, the best-known tour was due to K. Helsgaun, having cost 66,050,499. Unlike the previous two examples, in this case the heuristic tour did not turn out to be optimal; the new tour produced in the final branch-and-cut run has cost 66,048,945. The improvement in the tours was thus a little over 0.002% of the optimal value.

A summary of the initial Concorde runs is given in Table 16.23. After three branch-and-cut runs, the LP gap remained a rather large 0.0459%. Applying the domino-parity module to the LP made a nice improvement, however, bringing the root LP to 66,030,718, for a gap of 0.0276%

Starting with the new root LP, we carried out three additional branch-and-cut runs, using the domino-parity module and $t_{max} = 32$. The results for these runs are reported in Table 16.24; the "Total CPU Days" column reports the cumulative

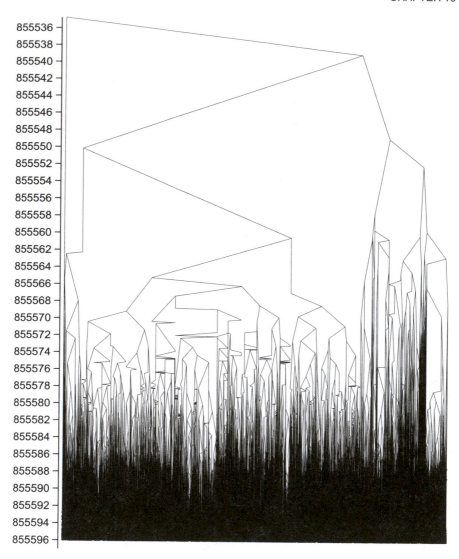

Figure 16.16 Branch-and-cut search tree for Sweden TSP.

Table 16.23 Initial pla33810 runs.

	Root LP	Root Gap	B&C Gap	Total CPU Days
$t^* = 48$	65,972,887	0.1154%		0.8
+1,000 BC1	66,012,387	0.0554%	0.0461%	196.8
+1,000 BC2	66,016,484	0.0492%	0.0311%	394.2
+1,000 BC3	66,018,619	0.0459%	0.0270%	608.3

Figure 16.17 Optimal tour of 24,978 cities in Sweden.

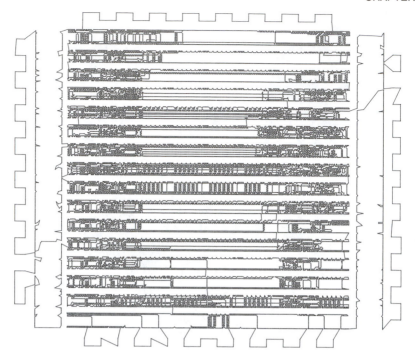

Figure 16.18 Optimal tour of 33,810 cities from a VLSI application.

CPU time, including the 608.3 days from the initial three runs; the "Search Nodes" column reports the total number of nodes in each of the trees.

Table 16.24 pla33810 runs with domino-parity cuts.

	Search Nodes	Branching	Root Gap	B&C Gap	Total CPU Days
BC4	6,097	1-way	0.0243%	0.0107%	2,391.4
BC5	7,959	(8,1)-way	0.0171%	0.0056%	5,506.1
BC6	577	8-way	0.0168%	Optimal	5,739.0

The sixth branch-and-cut run solved the problem, using eight-way branching and 577 search nodes. The total CPU time was approximately 15.7 years on the Intel Xeon cluster. A drawing of the optimal tour is given in Figure 16.18.

We carried out one additional run, starting with the 0.0168% LP obtained by applying the final cut pool to the root LP relaxation. This run used eight-way branching and 135 search nodes to verify the optimal solution. The search tree for this computation is displayed in Figure 16.19.

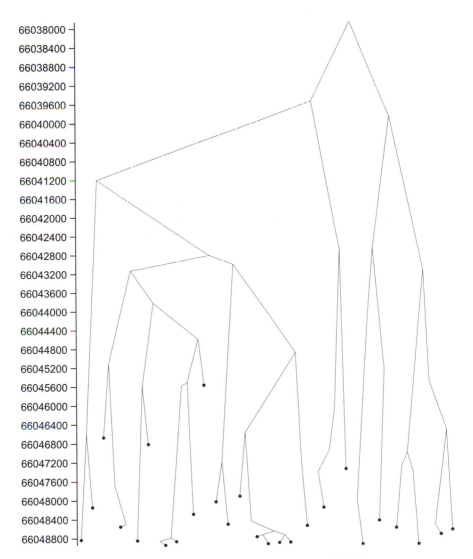

Figure 16.19 Branch-and-cut search tree for pla33810.

VLSI APPLICATION—PLA85900

The pla85900 instance from Bell Labs is the largest example in the TSPLIB. Its solution in April 2006 involved a very difficult computation with Concorde, consuming over 136 CPU years on clusters of 2.4-GHz AMD Opteron 250 compute nodes and 2.66-GHz Intel Xeon compute nodes, having a total of 64 Opteron processors and 192 Xeon processors. Remarkably, the optimal tour matched in value a heuristic tour found by K. Helsgaun, which itself was the result of a long sequence of studies by a number of research groups, as summarized in Table 16.25.

Table 16.25 History of pla85900 tours.

Date	Length	Heuristic Algorithm
07.06.1991	142,514,146	D. S. Johnson, Iterated LK
23.09.1997	142,482,068	Tour merging
14.10.1998	142,416,327	Helsgaun [260]
22.10.1999	142,409,553	Tour merging
27.06.2001	142,405,532	K. Helsgaun, LKH variant
31.08.2001	142,395,130	Tour merging with LKH
14.12.2001	142,393,738	K. Helsgaun, LKH variant
15.09.2002	142,385,237	Tamaki [518]
12.12.2002	142,383,704	K. Helsgaun, LKH variant
19.03.2003	142,383,467	Nguyen et al. [433]
28.04.2003	142,383,189	K. Helsgaun, LKH variant
02.05.2004	142,382,641	K. Helsgaun, LKH variant

The initial cutting-plane run on pla85800, using $t^* = 48$, produced an LP relaxation having a gap of just over 0.08%. The combination of a large gap and the very large point set creates a storage problem, since the variable-fixing phase is unable to eliminate a sufficient number of edges from further consideration. This was handled by providing the initial branch-and-cut run with an artificial upper bound of 142,320,000; the run achieved this bound after 45 search nodes. Similarly, the second branch-and-cut run achieved an artificial upper bound of 142,340,000 after 363 search nodes. After gathering cutting planes with these two computations, the root LP bound was strong enough to permit us to move ahead to valid branch-and-cut runs. As usual, the runs were halted when the search trees reached 1,000 active nodes.

A summary of the first four branch-and-cut computations is given in Table 16.26. The additional cutting planes lowered the LP integrality gap to 0.0304%. The modest improvements in the lower bounds provided by the search trees, however, strongly suggest the gap is still far too large to be closed by a branch-and-cut search of any reasonable size.

As a next step, the domino-parity module was applied to the 0.0304% root LP.

Table 16.26 Initial pla85900 runs.

	Root LP	Root Gap	B&Cut Gap	Total CPU Days
$t^* = 48$	142,268,559	0.0802%		22.6
+ BC1	142,317,776	0.0456%	0.0441%	144.0
+ BC2	142,326,532	0.0395%	0.0300%	227.3
+1,000 BC3	142,336,550	0.0324%	0.0251%	492.8
+1,000 BC4	142,339,384	0.0304%	0.0239%	740.0

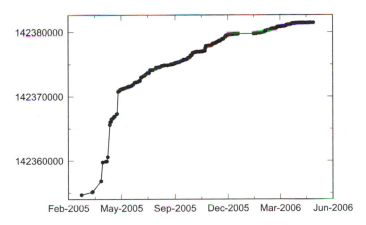

Figure 16.20 Progress in root LP bound for pla85900.

A summary of this run is given in Table 16.27. The 0.0281% LP produced in this

Table 16.27 Adding domino-parity inequalities to pla85900.

Root LP	Root Gap	CPU Days	Total CPU Days
142,342,669	0.0281%	11.9	751.9

way was the starting point for an intense sequence of computations, where over the course of more than one calendar year all separation algorithms in Concorde were repeatedly applied to drive up the value of the root bound; a plot of the progress in the bound is given in Figure 16.20. This work produced an LP relaxation yielding a bound of 142,381,405, for a root gap of 0.00087%; this is easily the most accurate relaxation among all very large instances we have considered. The distribution of cutting planes contained in the relaxation is given in Table 16.28.

Following our standard practice, the main tool for pushing the LP lower bound was the repeated use of branch-and-cut runs. The sequence of such runs carried out from December 2004 through April 2006 is summarized in Table 16.29. The

Table 16.28 Cutting planes in pla85900 root LP.

Type	Number
Subtours	1,030
Combs	2,787
Paths and stars	4,048
Bipartition	41
Domino-parity	164
Other (local cuts)	809

"Date" column gives the day when the run terminated, in day.month.year format; the "Branching" column describes the tentative branching used, with staged runs given multiple values; "Search Nodes" reports the total number of nodes in the search trees; "Gap" is the percentage gap provided by the lower bound from the search tree; "CPU Days" gives the number of days for the individual branch-and-cut runs, on a mix of Opteron and Xeon processors.

The pla85900 instance was first solved with 32-way branching and a search tree having 1,239 nodes. Two further branch-and-cut runs with 32-way branching verified the optimal value, with the final run taking 243 search nodes. A detailed discussion of the impact of tentative branching on this example can be found in Section 14.4.

Our branch-and-cut runs were given an upper bound of one larger than the cost of Helsgaun's tour, so they again produced optimal tours in the search trees. A drawing of one such optimal tour is given in Figure 16.21.

16.5 THE WORLD TSP

In this section we study the World TSP instance mentioned in Chapter 1. This challenge problem has 1,904,711 cities and is available at

www.tsp.gatech.edu/world

The World TSP was created in 2001, using data from the *National Imagery and Mapping Agency* [427] and from the *Geographic Names Information System* [203] to locate all populated points throughout the world. The cities are specified by their latitude and longitude, and the cost of travel between cities is given by an approximation of the great circle distance on Earth, treating Earth as a ball; this cost function is a variation of the TSPLIB GEO-norm, scaled to provide the distance in meters rather than in kilometers.

The size of the World TSP puts it well out of range for exact solution by Concorde, but it poses an interesting target for methods aimed at obtaining sharp upper and lower bounds on extremely large instances.

On the upper bound side, the World TSP has been considered by researchers

Table 16.29 Branch-and-cut runs on pla85900.

Date	Branching	Search Nodes	Gap	CPU Days
07.12.2004	8-way	155	0.0209%	269.5
14.03.2005	8-way	317	0.0178%	187.8
02.05.2005	8-way	7,359	0.0150%	3,539.0
01.07.2005	8-way	6,415	0.0060%	5,160.6
06.07.2005	128-way	27	0.0073%	750.9
14.07.2005	8-way	1,913	0.0054%	1,499.1
16.09.2005	(32,16,1)-way	14,797	0.0036%	7,702.7
20.08.2005	8-way	659	0.0049%	347.1
06.09.2005	32-way	257	0.0044%	527.2
02.10.2005	(32,1)-way	5,939	0.0030%	1,996.0
12.11.2005	32-way	1,807	0.0027%	4,305.7
09.11.2005	32-way	361	0.0022%	1,181.6
24.11.2005	32-way	569	0.0016%	1,961.1
06.12.2005	32-way	317	0.0019%	1,396.1
14.12.2005	32-way	181	0.0016%	1,317.0
06.03.2006	(32,4,1)-way	16,267	0.0007%	7,748.9
27.03.2006	(32,8,4,1)-way	9,253	0.00004%	3,698.0
13.04.2006	32-way	1,239	Optimal	2,719.5
16.04.2006	32-way	275	Optimal	512.9
20.04.2006	32-way	243	Optimal	286.2

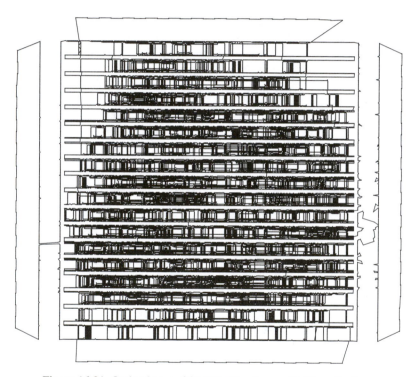

Figure 16.21 Optimal tour of 85,900 cities from a VLSI application.

developing heuristic algorithms over the past several years. A summary of the history of the best reported tours is given in Table 16.30; the progress has been dominated by K. Helsgaun and his continually improving LKH code. The record tour length currently stands at 7,516,043,366 meters.

Table 16.30 History of world TSP tours.

Date	Length	Heuristic Algorithm
31.10.2001	7,539,742,312	Chained Lin-Kernighan
18.11.2001	7,532,949,688	K. Helsgaun, LKH variant
16.04.2002	7,524,170,430	K. Helsgaun, LKH variant
18.03.2003	7,518,528,361	K. Helsgaun, LKH variant
02.06.2003	7,518,425,644	Nguyen et al. [433]
16.09.2003	7,517,285,610	K. Helsgaun, LKH variant
26.07.2004	7,516,122,185	K. Helsgaun, LKH variant
29.01.2006	7,516,043,366	K. Helsgaun, LKH variant

To study the cutting-plane method on this large instance, we use a version of Concorde with separation algorithms restricted to those that can easily scale up to problems having several million cities. The selected separation algorithms are

- the three subtour separation routines (Chapter 6),

- the fast blossom heuristics (Section 7.1),

- tightening cuts in the LP and in the cut pool (Section 10.1),

- teething combs in the LP (Section 10.2),

- the heuristic algorithm for path inequalities (Section 10.3),

- local cuts with a fixed valued of t (Chapter 11).

These routines are organized into a loop, with column generation carried out after each full pass through the separation algorithms.

In our tests, we make repeated runs of the restricted version of Concorde, slowly increasing the size of the local cuts that are considered. Each successive run in this study is initialized with the LP and cut pool produced in the previous run.

The results are reported in Table 16.31, where t ranges from 0 up to 18. The "Bound" column gives the final lower bound that was achieved by each run and the "Gap" column reports the percentage gap to the cost of the tour found by Helsgaun. The cumulative CPU time is reported in days, using a Compaq ES40 server equipped with a 500-MHz EV6 Alpha processor; the LP solver is ILOG CPLEX 7.1.

As t increases, we see steady progress in the value of the lower bound, with the final value of 7,511,705,615 demonstrating the superb quality of the Helsgaun tour,

Table 16.31 Concorde bounds on the world TSP.

Local Cuts	Bound	Gap	Total CPU Days
$t = 0$	7,500,743,582	0.204%	12.0
$t = 8$	7,504,218,236	0.158%	22.0
$t = 12$	7,508,333,052	0.103%	77.9
$t = 14$	7,510,154,557	0.079%	163.6
$t = 16$	7,510,752,016	0.071%	256.1
$t = 18$	7,511,705,615	0.058%	401.7

yielding a gap of only 0.058%. The running time of over one year is problematic, however, since this is a sequential computation. Indeed, approximately 98% of the CPU time in this test is used by the LP solver to produce optimal primal and dual solutions after the addition of cutting planes and after the addition of edges to the core LP; currently there are no effective methods for parallelizing the dual simplex algorithm on LP problems of the form that arise in these computations.

The rapidly increasing running time with larger values of t has prohibited further progress in the lower bound with this direct approach. What is needed is a strategy to greatly reduce the number of large LP problems that need to be solved. In Section 17.3 we briefly describe a possible line of research in this direction, using decomposition techniques to preselect a strong family of cutting planes.

Million-City Euclidean TSP

A second large test problem from the research literature is the 1,000,000-city random Euclidean instance created by D. S. Johnson in 1994. This problem is studied in Johnson and McGeoch [291], and it is included in the DIMACS [157] challenge test set under the name E1M.0.

The test used for the World TSP was repeated on E1M.0, but in this case the smaller running times permit us to continue the runs up to $t = 28$. The results are reported in Table 16.32, with the cumulative CPU time reported in days on the 500-MHz EV6 Alpha processor. The "Gap" column gives the percentage gaps to the best reported tour for E1M.0, also found by K. Helsgaun, having length 713,208,677.

As in the World TSP tests, the steady improvement in the lower bound is very encouraging, but the total CPU time again suggests that other approaches will be needed in order to make significant further progress. The CPU usage is once more dominated by the time spent in the LP solver after the addition of cutting planes and after the addition of edges. In this case, the portion of time spent solving LP problems is approximately 97% of the total time.

Table 16.32 Concorde bounds on 1,000,000-city Euclidean TSP.

Local Cuts	Bound	Gap	Total CPU Days
$t = 0$	711,088,074	0.299%	2.1
$t = 8$	712,120,984	0.153%	11.0
$t = 10$	712,651,618	0.079%	23.6
$t = 12$	712,746,082	0.065%	29.2
$t = 14$	712,813,323	0.055%	38.9
$t = 20$	712,903,086	0.043%	72.1
$t = 24$	712,954,779	0.036%	154.2
$t = 28$	713,003,014	0.029%	308.1

Chapter Seventeen

The Road Goes On

The solution of the TSPLIB test problems is by no means the end of the journey for the TSP. The general model will no doubt continue to inspire researchers and productively serve as an engine of discovery in applied mathematics. In this chapter we give a short survey of recent work by various research groups, aimed at pushing the limits of TSP computation.

17.1 CUTTING PLANES

The quality of the root LP bounds obtained by Concorde is very high, but the solution process for the largest instances involves the repeated application of branch-and-cut runs to gather cuts at great computational expense. This points out that TSP separation remains a clear target for development work. We discuss below a number of new frameworks that could be used in this effort.

CARR'S LIFTED CUTS

The study of the facial structure of the convex hull of tours has provided many of the templates that are adopted in modern codes for solving the TSP. One of the general tools in this work is the lifting of inequalities from small instances of the TSP to large instances. This line of research has been led by Naddef and Rinaldi [421], working with facet-inducing inequalities for both the TSP and the graphical TSP; further results are described in Carr [110] and Oswald et al. [441].

  In an interesting paper, Carr [111] has shown that the theoretical tool of lifting can be turned into an algorithmic tool with the help of linear programming. The focus of his study is on *path liftings*, defined as follows. An inequality $\alpha^T x \geq \beta$ for the graphical TSP on vertex set V is a path lifting of an inequality $\bar{\alpha}^T \bar{x} \geq \beta$ for the graphical TSP on $B \subset V$ if, for each $u, v \in B$, the shortest path in $G = (V, E)$ with edge weights $(\alpha_e : e \in E)$ has length at least $\bar{\alpha}_{uv}$. Carr proved that if $\bar{\alpha}^T \bar{x} \geq \beta$ is satisfied by all graphical TSP tours through B, then the path-lifted inequality $\alpha^T x \geq \beta$ is satisfied by all graphical TSP tours through V. Moreover, Carr showed that it is possible to separate over all possible path liftings of $\bar{\alpha}^T \bar{x} \geq \beta$ by solving a single LP.

  Carr uses this LP result to give a polynomial-time separation algorithm for path liftings of classes of inequalities defined on a *backbone* set B of fixed cardinality. This algorithm is unlikely to have a direct impact on practical computation, but the LP framework could possibly be adopted in computer codes. For example, hypergraph inequalities having a small number of nonempty atoms could be candidates for the lifting LP, providing a type of "super" tightening procedure, as an alternative to the algorithms described in Chapter 10.

SMALL INSTANCE RELAXATIONS

Another branch of polyhedral research forms the basis of a separation framework proposed by Christof and Reinelt [119] and extended in a detailed study in the Ph.D thesis of Wenger [551]. Beginning with work by Kuhn [328] and Norman [435] in the 1950s, researchers have sought to obtain complete linear descriptions of the convex hull of tours through small numbers of cities. Kuhn considered the asymmetric TSP, while Norman gave initial results for the symmetric TSP. Boyd and Cunningham [88] prove that Norman's list of inequalities completely describes the TSP polytope on six cities, and they introduce a class of facet-inducing inequalities that completes the description of the seven-city polytope. This work was followed by complete descriptions for the eight-city and nine-city TSP polytopes by Christof et al. [118] and Christof and Reinelt [120], respectively. The limit of these results, for now, was reached by Christof and Reinelt [120], who present a list of facet-inducing inequalities for the 10-city TSP that they conjecture to be complete. The inequalities for all of these polyhedra can be found on the web page SMAPO [503].

Let (V_k, E_k) denote the complete graph on k vertices and let x^k denote a vector indexed by E_k. Consider a hypergraph inequality $\mathcal{H}' \circ x^k \geq \beta$ that is valid for all tours though V_k. For a TSP defined on (V, E), we obtain, via Theorem 5.11, an entire class of valid inequalities $\mathcal{H} \circ x \geq \beta$ by assigning nonempty, pairwise disjoint, subsets of V to each nonempty atom of $\mathcal{H}'$. Thus, the facet-inducing inequalities discovered in the above line of research provide a rich source of templates for the TSP. Christof and Reinelt [119] and Christof [117] harness this source by adopting a general template-matching algorithm, combining a shrinking step to reduce (V, E) to k vertices and a method for assigning each shrunk vertex to a point in V_k. In Christof and Reinelt [121] this approach is called a *small-instance relaxation* of the TSP, and the framework is discussed in a context that also covers a number of other models in combinatorial optimization. When restricted to the TSP, Wenger [551] improves many aspects of the initial implementation by Christof and Reinelt, and we follow his approach in our discussion.

Let x^* be a solution vector for an LP relaxation of the TSP on (V, E). In the first step of the separation algorithm, a k-partition of V into sets $S_1, \ldots, S_k$ is identified, such that the value

$$\sum_{j=1}^{k} x^*(\delta(S_j))$$

is small. To select the k-partition, Wenger considers versions of the neighborhood methods used by Christof and Reinelt [119] and adopted in the local-cuts procedure described in Chapter 11, as well as considering an algorithm based on randomized shrinking of edges, following the ideas of Karger and Stein [305]. An additional source of candidate k-partitions is obtained via an algorithm of Wenger [551] that lists all partitions such that each set S_j is a minimum-weight cut in the support graph G^*.

In the second step of the separation algorithm, individual facet-inducing inequalities for the k-city polytope are considered as candidate templates for graphs obtained by shrinking, in turn, each of the selected k-partitions. Note that the lists of inequalities for small TSP polytopes are exceptionally long; Christof and Reinelt [120] give over 51 billion inequalities for the 10-city TSP. These lists can be greatly reduced, however, by considering the equivalence classes under permutations of the vertex set V_k. In the case of the 10-city TSP, the reduction results in 15,379 representatives. Using these as candidates, Wenger observed that the most productive inequalities are those such that the induced facet contains a large number of tour vectors. Therefore, to obtain a manageable set, Wenger proposes to consider only those facets that contain at least 100 tours; this drops the number of candidates to exactly 500 inequalities. Wenger works with these inequalities in tight-triangular form

$$\sum_{e \in E_k} \alpha_e x_e \geq \beta, \tag{17.1}$$

as defined by Naddef and Rinaldi [421]; see Section 11.9.4.

Suppose we are given a k-partition and a candidate facet-inducing inequality (17.1). Let x' denote the vector obtained from x^* by shrinking each S_j to a single vertex. In the second step, each of the sets S_j is assigned to a vertex of V_k, that is, each vertex j in the shrunk graph G' is assigned a vertex $\pi(j)$ in V_k. The goal in selecting the permutation π is to minimize the left-hand side

$$\sum_{e \in E_k} \alpha_e x'_{\pi(e)}$$

of (17.1), where for each edge $e = \{i, j\}$ we write $\pi(e)$ as shorthand for the edge having ends $\{\pi(i), \pi(j)\}$. This task is an instance of the *quadratic assignment problem*. It is handled by Wenger using an exact branch-and-bound algorithm of Burkard and Derigs [100] and a heuristic method of Pardalos et al. [454].

In computational tests with TSPLIB instances ranging in size from 225 cities up to 724 cities, Wenger [551] demonstrates that small-instance-relaxation cuts improve LP bounds obtained with the Naddef and Thienel [423] separation routines.

The small-instance-relaxation framework can be adopted directly as an additional source of cuts for the TSP, but it can also serve as a means to focus attention on the most productive templates from the small TSP polytopes. In this way, the framework could drive further work on template-specific separation algorithms.

SEARCHING FOR LOCAL CUTS

In Chapter 11 we presented powerful machinery for the separation of TSP inequalities with the local cuts framework. A possible weakness of the study is the rather simple approach taken to constructing the linear mappings σ. In our tests we only consider variations of the Christof-Reinelt theme of using neighborhoods of vertices to choose the shrinking sets that define the mapping.

A direct approach for further study is to consider the k-partition methods of Wenger [551] as techniques for creating possible maps σ. These ideas need to be adjusted to allow for the fact that local-cut mappings can effectively handle partitions having a single S_j with large value $x^*(\delta(S_j))$, whereas the small-instance-relaxation methods aim for partitions having each $x^*(\delta(S_j))$ small.

K-DOMINO INEQUALITIES

The planar-separation algorithm for domino-parity inequalities, proposed by Letchford [347] and described in Section 9.3, was shown to be an effective tool for improving LP lower bounds for the TSP in computational studies by Boyd et al. [85] and Cook et al. [136]. This suggests the study of further abstractions of classes of known inequalities, following the process Letchford used to move from combs to domino-parity constraints.

This line of research was considered by Cook et al. [135], who propose a generalization of dominoes to include families of sets with more than two members. They define a k-*domino* as a structure $(T_1, \ldots, T_k; T)$ such that T is a proper subset of V and for each $j = 1, \ldots, k$ we have $\emptyset \neq T_j \subseteq T$. Using a k-domino as an abstraction of a tooth, the components T_j can relate the k-domino to multiple handles. Cook et al. [135] use this to define a class of inequalities that generalize domino-parity constraints, and they provide planar-separation algorithms for a generalization of k-handled clique-trees under certain conditions. Computational tests of Cook et al. show a modest improvement in LP bounds with the addition of heuristic versions of the separation algorithms.

In comparison with standard clique-tree inequalities, simple heuristic algorithms for multiple-handle star inequalities have proved to be more successful in computational tests by Naddef and Thienel [423] and with the Concorde code. An interesting possibility, therefore, is to study extensions of the k-domino separation algorithms to include generalizations of stars. Such algorithms may provide a useful tool in attacking even larger instances of the TSP.

17.2 TOUR HEURISTICS

Tour finding continues to attract the attention of a wide range of researchers. A good summary of ongoing activities can be found on the DIMACS TSP Challenge web site [157], where contributors compare the performance of their computer codes on a common set of test instances. We describe below several projects aimed at providing ever better starting tours that can be used to guide algorithms for the exact solution of the TSP.

EXTENDING LKH

Since its introduction in 2000 by Helsgaun [260], the LKH variant of the local-search algorithm of Lin and Kernighan [357] has set the standard for high-end TSP heuristics. The algorithm has continued to evolve since the publication of [260]; Helsgaun's new versions of the LKH code dominate the lists of best-known tours for the National and VLSI collections available at

www.tsp.gatech.edu/data

These test problems were created to complement the TSPLIB collection, particularly in the realm of heuristic algorithms.

Recall from the description in Section 15.5 that one of the main ingredients of LKH is the use of a 5-opt move as a basic search step, rather than the 2-opt moves used in standard versions of Lin-Kernighan. An obvious direction for improvement is to extend the algorithm to adopt k-opt moves for larger values of k. A problem, however, is the resulting complexity of the computer code needed to realize the larger moves as a basic step in the search process. In Table 17.1 we list the number of cases that must be handled to consider all sequential k-opt moves for k ranging from 2 up to 10. Handling all 148 cases of 5-opt in the original version of LKH was

Table 17.1 Cases for k-opt moves.

k	No. of Cases
2	1
3	4
4	20
5	148
6	1,358
7	15,104
8	198,144
9	2,998,656
10	51,290,496

a complex task, involving a large amount of computer code; moving up to larger values of k with the same approach would be extremely difficult.

An alternative strategy for handling the explosion of cases was considered by D. Applegate and W. Cook. The regularity of the instructions needed to handle individual cases suggests that the code itself could possibly be generated by a computer, rather than the usual approach of writing code by hand. Adopting this strategy, Applegate and Cook produced computer-generated code for 6-opt, 7-opt, and 8-opt moves; the number of lines of code to handle these moves is given in Table 17.2. Although the overall process was successful, the rapid growth in the size of the resulting computer code makes it impractical to handle in this way k-opt moves for larger values of k.

To proceed further, Applegate and Cook modified the code-generation method to produce the steps for each case on the fly, rather than precomputing the instructions in a source code listing. By calling this procedure as a subroutine, the same search can be carried out without an explicit listing of the codes to handle individual cases. The method is still limited by the computing time required to execute the resulting search steps, but it does permit the use of larger values of k-opt moves.

Table 17.2 Computer-generated source code for k-opt moves.

k	No. of Lines
6	120,228
7	1,259,863
8	17,919,296

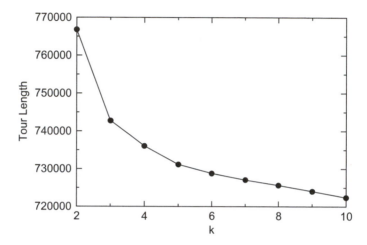

Figure 17.1 k-opt on a 10,000-city Euclidean TSP.

As a test of the utility of large basic moves, the on-the-fly approach was used to build a version of a standard k-opt code. In Figure 17.1 we plot the lengths of tours obtained by running the code on a 10,000-city random Euclidean TSP, with k varying from 2 up to 10. Although the rate of improvement decreases as we move past 5-opt, the plot shows that larger values of k can still add power to the search procedure.

A version of the k-opt technique was adopted in LKH , as reported by K. Hels-gaun (private communication, 2003). In this implementation, the user can specify the value of k to be used in the basic LKH moves of the search algorithm; Helsgaun writes that in a test on the Sweden TSP, the optimal tour was produced in a run that included the use of 10-opt moves. This use of high values of k-opt moves is one of a number of important improvements in LKH made by Helsgaun.

HYBRID GENETIC ALGORITHM

A combination of techniques is used by Nguyen [432] and Nguyen et al. [433] to obtain a very effective parallel heuristic for the TSP. The method is a hybrid of a genetic algorithm and a version of the Lin-Kernighan local-search algorithm. A population of tours is created in parallel with a randomized version of the Quick-Borůvka construction heuristic from Section 15.3; each of the tours is improved with a Lin-Kernighan variant that includes the 5-opt move of Helsgaun's LKH code [260]. Pairs (A, B) from the population are then used to create new tours by choosing a subsequence from A and extending it to a tour using, when possible, edges of B or edges of A, with preference given to the edges of B; this *crossover* technique of Nguyen et al. is based on the MPX2 crossover proposed by Gorges-Schleuter [222]. A version of Chained-Lin-Kernighan is then applied to each of the new tours to create a new population.

This algorithm of Nguyen et al. has produced the best-known tours for a number of instances from the National and VLSI collections. Moreover, in 2003 the algorithm was used to obtain a new best tour for the World TSP, which was later improved by Helsgaun. The excellent performance of the Nguyen et al. code on these challenge problems suggests such hybrid methods as a possible focus area for further work on high-end heuristics.

MORPHING TOURS

The tour-merging process described in Section 15.6.2 is a powerful tool for combining the output of multiple runs of a heuristic algorithm, but its computation time grows rapidly with the density of the union of tours that must be merged. A potentially useful strategy is to preprocess the tour collection in an attempt to modify any pair of tours (T_1, T_2) that differ in the choice of two subpaths of equal cost. Such differences can be detected as a circuit C in the symmetric difference $T_1 \triangle T_2$, such that the edges in $C \cap T_1$ have the same total cost as the edges in $C \cap T_2$. The Lin-Kernighan heuristic can be modified to search for such alternating circuits; we refer to the resulting method as *tour morphing*, since we gradually change T_1 to more closely resemble T_2.

In some cases, the morphing step can dramatically reduce the number of edges in the tour union. For example, running a version of LKH on the instance pm8069 from the National test set, consisting of 8,069 cities in Panama, we obtain 40 tours having a total of 21,352 edges and a branch decomposition of width 13. After pairwise morphing the collection of tours, the union graph was reduced to 9,162 edges, having a branch decomposition of width 9; in Figures 17.2 and 17.3 we show a portion of the graph before and after morphing. The computing time needed for the tour-merging step in this example was reduced from 620,598 seconds down to 24 seconds.

APPROXIMATE TOUR MERGING

Alternating circuits feature prominently in an interesting approach to tour merging proposed by Tamaki [518]. Rather than directly optimizing over the union of a collection of tours $\mathcal{T}$, Tamaki considers a heuristic method that attempts to improve a target tour $T_0 \in \mathcal{T}$ by applying a set of alternating circuits from $T_0 \triangle T_j$, where T_j ranges over the remaining tours in $\mathcal{T}$. To locate subsets of alternating circuits that can lower the cost of T_0, Tamaki uses a heuristic greedy algorithm and an exact procedure based on dynamic programming.

Tamaki used these ideas to create a high-performing heuristic for the TSP, adopting LKH as an engine to produce the initial collection of tours. Tamaki's code was

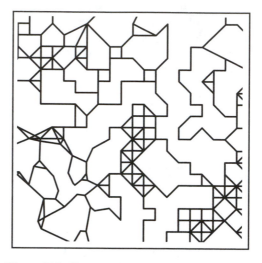

Figure 17.2 Close-up view of 40 tours for pm8069.

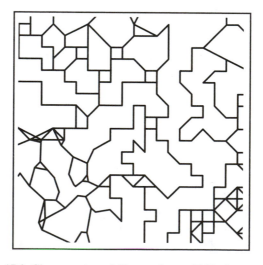

Figure 17.3 Close-up view of 40 tours for pm8069 after morphing.

the first to discover the tours for brd14051 and d18512 that were proved to be optimal by Concorde; the code also produced, in 2002, a new best-known tour for pla85900, which was later improved by K. Helsgaun.

17.3 DECOMPOSITION METHODS

Instances such as the World TSP can serve as an excellent platform for developing decomposition techniques to attack general large-scale problems in discrete optimization. This area offers many opportunities for new research efforts.

ROHE'S REOPTIMIZATION HEURISTIC

Tools for partitioning geometric points sets have been commonly applied in heuristic algorithms for Euclidean instances of the TSP, where subtours for individual regions are patched together to form a global tour. The well-known probabilistic analysis of Karp [309] adopts this approach, and, for large instances, good-quality tours can be obtained in practice when the partition contains a relatively small number of regions. An advantage of partition-based algorithms is the ease with which parallel computing platforms can be utilized; this is explored in work by Schäfer [490].

An interesting practical improvement for partition-based algorithms was introduced by Rohe [484]. The global tour produced by the patching phase of such an algorithm often behaves poorly along the borders of the geometric regions that make up the partition. Rohe's idea is to reoptimize the tour in geometric windows of the point set, concentrating at first on the border areas. A *window W* is a rectangular subregion of the Euclidean space; Rohe takes the intersection of the global tour with W and joins pairs of vertices on the border of W to create a tour T_W through all cities in the subregion. Chained Lin-Kernighan is applied to T_W, and the improvements are transferred to the global tour. Rohe's parallel implementation of this strategy produces high-quality tours on TSP instances having over 10 million cities. (Note that care needs to be taken when working in parallel, since combined improvements in two or more regions can create subtours in the global tour.)

CUTTING PLANES FROM SUBDIVISIONS

Partitioning strategies used in tour heuristics can also be adopted in the cutting-plane algorithm, to gather a global collection of inequalities by first searching in each region of the partition. A straightforward approach is to apply the cutting-plane algorithm to individual regions, and add to a global pool the inequalities used in the final regional LPs; Rohe's window technique can also be applied to discover useful inequalities that cross the borders of the partition. This partition-based approach can greatly reduce the number of global LPs that need to be solved, since the final LPs from the regions should provide a strong set of initial cuts for the global LP.

Figure 17.4 A sample of 100,000 points from the Star TSP.

The process can be extended to extremely large TSP instances, where it is infeasible to directly solve the global LP. The strategy here is to combine the dual solutions from the regional LPs into a feasible dual solution for the gloabl LP. This can be done by adding an additional vertex to each regional problem to represent the set of vertices not in the region; by choosing the costs appropriately for the edges meeting the new vertex, we can ensure that the regional dual solutions maintain global feasibility.

THE STAR TSP

To provide a challenge for the development of approximation algorithms for extremely large problems in discrete optimization, we created a TSP instance consisting of the 526,280,881 celestial objects in the United States Naval Observatory (USNO)–A2.0 Catalog; a description of the catalog can be found in Monet [401]. The objects are given as locations in the sky, so the TSP models the movement of a telescope to view the objects, rather than modeling a traveler attempting to visit the actual stars and galaxies. A drawing of a subset of 100,000 of the points is displayed in Figure 17.4.

An initial attack on this instance by D. Applegate, W. Cook, K. Helsgaun, and A. Rohe adopts the decomposition methods outlined above. A simple implementation of the partition-based heuristic and cutting-plane algorithms produced a tour T and a lower bound that proves the cost of T to be within 0.796% of the cost of an optimal tour.

Bibliography

[1] Aarts, E., J. K. Lenstra. 2003. *Local Search in Combinatorial Optimization*. Princeton University Press, Princeton, New Jersey, USA.

[2] Achterberg, T., T. Koch, A. Martin. 2005. Branching rules revisited. Operations Research Letters **33**, 42–54.

[3] Adleman, L. M. 1994. Molecular computation of solutions to combinatorial problems. Science **266**, 1021–1024.

[4] Agarwala, R., D. L. Applegate, D. Maglott, G. D. Schuler, A. A. Schäffer. 2000. A fast and scalable radiation hybrid map construction and integration strategy. Genome Research **10**, 350–364.

[5] Aho, A. V., J. E. Hopcroft, J. D. Ullman. 1974. *The Design and Analysis of Computer Algorithms*. Addison-Wesley, Reading, Massachusetts, USA.

[6] Aho, A. V., J. E. Hopcroft, J. D. Ullman. 1976. On finding lowest common ancestors in trees. SIAM Journal on Computing **5**, 115–132.

[7] Ahuja, R. K., T. L. Magnanti, J. B. Orlin. 1993. *Network Flows: Theory Algorithms, and Applications*. Prentice Hall, Englewood Cliffs, New Jersey, USA.

[8] Aigner, M., G. M. Ziegler. 1999. *Proofs from The Book*. Springer, Berlin, Germany.

[9] Alon, N. 1986. Eigenvalues and expanders. Combinatorica **6**, 83–96.

[10] Alpert, C. J., A. B. Kahng. 1997. Splitting an ordering into a partition to minimize diameter. Journal of Classification **14**, 51–74.

[11] Amos, M., G. Paun, G. Rozenberg, A. Salomaa. 2003. Topics in the theory of DNA computing. Theoretical Computer Science **287**, 3–38.

[12] Anderson, R. J., J. C. Setubal. 1993. Goldberg's algorithm for maximum flow in perspective: A computational study. D. S. Johnson, C. C. McGeoch, eds. *Network Flows and Matching: First DIMACS Implementation Challenge*. DIMACS Series in Discrete Mathematics and Theoretical Computer Science **12**. American Mathematical Society, Providence, Rhode Island, USA. 1–18.

[13] Anstreicher, K. M. 2003. Recent advances in the solution of quadratic assignment problems. Mathematical Programming **97**, 27–42.

[14] Anstreicher, K. M., N. Brixius, J.-P. Goux, J. Linderoth. 2002. Solving large quadratic assignment problems on computational grids. Mathematical Programming **91**, 563–588.

[15] Applegate, D., R. Bixby, V. Chvátal, W. Cook. 1995. Finding cuts in the TSP (A preliminary report). DIMACS Technical Report 95-05. DIMACS, Rutgers University, New Brunswick, New Jersey, USA.

[16] Applegate, D., R. Bixby, V. Chvátal, W. Cook. 1998. On the solution of traveling salesman problems. Documenta Mathematica Journal der Deutschen Mathematiker-Vereinigung, International Congress of Mathematicians. 645–656.

[17] Applegate, D., R. Bixby, V. Chvátal, W. Cook. 2001. TSP cuts which do not conform to the template paradigm. M. Jünger, D. Naddef, eds. *Computational Combinatorial Optimization*. Springer, Heidelberg, Germany. 261–304.

[18] Applegate, D., R. Bixby, V. Chvátal, W. Cook. 2004. Concorde. Available at www.tsp.gatech.edu .

[19] Applegate, D., V. Chvátal, W. Cook. 1990. Lower bounds for the travelling salesman problem. TSP '90, CRPC Technical Report CRPC-TR90547. Center for Research in Parallel Computing, Rice University, Houston, Texas, USA.

[20] Applegate, D., W. Cook. 1993. Solving large-scale matching problems. D. S Johnson, C. C. McGeoch, eds. *Algorithms for Network Flows and Matching*. American Mathematical Society, Providence, Rhode Island, USA. 557–576.

[21] Applegate, D., W. Cook, A. Rohe. 2003. Chained Lin-Kernighan for large traveling salesman problems. INFORMS Journal on Computing **15**, 82–92.

[22] Arnborg, S., J. Lagergren, D. Seese. 1991. Easy problems for tree-decomposable graphs. Journal of Algorithms **12**, 308–340.

[23] Arnoff, E. L., S. S. Sengupta. 1961. Mathematical programming. R. L. Ackoff, ed. *Progress in Operations Research*. John Wiley & Sons, New York, USA. 105–210.

[24] Arora, S. 1998. Polynomial-time approximation schemes for Euclidean TSP and other geometric problems. Journal of the ACM **45**, 753–782.

[25] Aurenhammer, F. 1991. Voronoi diagrams: A survey of a fundamental geometric data structure. ACM Computational Surveys **23**, 345–405

[26] Australia Telescope Compact Array. 2005. New user information. Available at www.narrabri.atnf.csiro.au/.

[27] Avner, P., T. Bruls, I. Poras, L. Eley, S. Gas, P. Ruiz, M. V. Wiles, R. Sousa-Nunes, R. Kettleborough, A. Rana, J. Morissette, L. Bentley, M. Goldsworthy, A. Haynes, E. Herbert, L. Southam, H. Lehrach, J. Weissenbach, G. Manenti, P. Rodriguez-Tome, R. Beddington, S. Dunwoodie, R. D. Cox. 2001. A radiation hybrid transcript map of the mouse genome. Nature Genetics **29**, 194–200.

[28] Bachem, A., M. Wottawa. 1992. Parallelisierung von Heuristiken für grosse Traveling-Salesman-Probleme. Report No. 92.119, Mathematisches Institut, Universität zu Köln, Cologne, Germany.

[29] Bailey, C. A., T. W. McLain, R. W. Beard. 2001. Fuel-saving strategies for dual spacecraft interferometry missions. Journal of the Astronautical Sciences **49**, 469–488.

[30] Balas, E. 1965. An additive algorithm for solving linear programs with zero-one variables. Operations Research **13**, 517–546.

[31] Balas, E. 1975. Facets of the knapsack polytope. Mathematical Programming **8**, 146–164.

[32] Balas, E., M. Fischetti. 2002. Polyhedral theory for the asymmetric traveling salesman problem. G Gutin, A. P. Punnen, eds. *The Traveling Salesman Problem and Its Variations*. Kluwer, Boston, Massachusetts, USA. 117–168.

[33] Balas, E., N. Simonetti. 2001. Linear time dynamic programming algorithms for some new classes of restricted TSPs: A computational study. INFORMS Journal on Computing **13**, 56–75.

[34] Balas, E., P. Toth. 1985. Branch and bound methods. E. L. Lawler, J. K. Lenstra, A. H. G. Rinnooy Kan, D. B. Shmoys, eds. *The Traveling Salesman Problem*. John Wiley & Sons, Chichester, UK. 361–401.

[35] Bancroft Library. 2005. The Mississippi River. University of California, Berkeley, California, USA. Material available on the World Wide Web page `bancroft.berkeley.edu/Exhibits/MTP/mississippi.html/`.

[36] Barachet, L. L. 1957. Graphic solution of the traveling-salesman problem. Operations Research **5**, 841–845.

[37] Barahona, F., R. Anbil. 2000. The volume algorithm: Producing primal solutions with a subgradient method. Mathematical Programming **87**, 385–399.

[38] Barahona, F., L. Ladányi. 2001. Branch-and-cut based on the volume algorithm: Steiner trees in graphs and max cut. IBM Research Report RC22221.

[39] Barahona, F., A. R. Mahjoub. 1986. On the cut polytope. Mathematical Programming **36**, 157–173.

[40] Barbagallo, S., M. L. Bodoni, D. Medina, F. Corno, P. Prinetto, M. S. Reorda. 1996. Scan insertion criteria for low design impact. *Proceedings of the 14th IEEE VLSI Test Symposium (VTS '96)*. 26–31.

[41] Bartholdi III, J. J., L. K. Platzman. 1982. An O(NlogN) planar travelling salesman heuristic based on spacefilling curves. Operations Research Letters **1**, 121–125.

[42] Bartholdi III, J. J., L. K. Platzman, R. L. Collins, W. H. Warden III. 1983. A minimal technology routing system for Meals on Wheels. INTERFACES **13**, 1–8.

[43] Basso, D., P. S. Bisiacchi, M. Cotelli, C. Farinello. 2001. Planning times during traveling salesman's problem: Differences between closed head injury and normal subjects. Brain and Cognition **46**, 37–42.

[44] Bazaraa, M. S., J. J. Goode. 1977. The traveling salesman problem: A duality approach. Mathematical Programming **13**, 221–237.

[45] Beardwood, J., J. H. Halton, J. M. Hammersley. 1959. The shortest path through many points. Proceedings of the Cambridge Philosophical Society **55**, 299–327.

[46] Bellman, R. 1957. *Dynamic Programming*. Princeton University Press, Princeton, New Jersey, USA.

[47] Bellman, R. 1960. Combinatorial processes and dynamic programming. R. Bellman, M. Hall, Jr., eds. *Combinatorial Analysis*. American Mathematical Society, Providence, Rhode Island, USA. 217–249.

[48] Bellman, R. 1962. Dynamic programming treatment of the travelling salesman problem. Journal of the Association for Computing Machinery **9**, 61–63.

[49] Bellmore, M., J. C. Malone. 1971. Pathology of traveling-salesman subtour-elimination algorithms. Operations Research **19**, 278–307.

[50] Bellmore, M., G. L. Nemhauser. 1968. The traveling salesman problem: A survey. Operations Research **16**, 538–558.

[51] Ben-Ameur, W., J. Neto. 2004. Acceleration of cutting planes and column generation algorithms: Application to network design. Submitted.

[52] Ben-Ameur, W., J. Neto. 2004. Multipoint separation. Submitted.

[53] Ben-Dor, A., B. Chor. 1997. On constructing radiation hybrid maps. Journal of Computational Biology **4**, 517–533.

[54] Ben-Dor, A., B. Chor, D. Pelleg. 2000. RHO-Radiation hybrid ordering. Genome Research **10**, 365–378.

[55] Bentley, J. L. 1990. Experiments on traveling salesman heuristics. *Proceedings of the First Annual ACM-SIAM Symposium on Discrete Algorithms*. ACM Press, New York, USA. 91–99.

[56] Bentley, J. L. 1992. Fast algorithms for geometric traveling salesman problems. ORSA Journal on Computing **4**, 387–411.

[57] Bentley, J. L. 1997. Faster and faster and faster yet. Unix Review **15**, 59–67.

[58] Berge, C. 1973. *Graphs and Hypergraphs*. North-Holland, Amsterdam, The Netherlands.

[59] Bern, M. W., E. L. Lawler, A. L. Wong. 1987. Linear time computation of optimal subgraphs of decomposable graphs. Journal of Algorithms **8**, 216–235.

[60] Bertsimas, D., J. N. Tsitsiklis. 1997. *Introduction to Linear Optimization*. Athena Scientific, Nashua, New Hampshire, USA.

[61] Biggs, N. L., E. K. Lloyd, R. J. Wilson. 1976. *Graph Theory 1736–1936*. Clarendon Press, Oxford, UK.

[62] Bixby, R. E. 1992. Implementing the simplex method: The initial basis. ORSA Journal on Computing **4**, 267–284.

[63] Bixby, R. E., W. Cook, A. Cox, E. K. Lee. 1999. Computational experience with parallel mixed integer programming in a distributed environment. Annals of Operations Research **90**, 19–44.

[64] Bixby, R. E., M. Fenelon, Z. Gu, E. Rothberg, R. Wunderling. 2000. MIP: Theory and practice—closing the gap. M. J. D. Powell, S. Scholtes, eds. *System Modelling and Optimization: Methods, Theory and Applications*. Kluwer Academic Publishers, Dordrecht, The Netherlands. 19–49.

[65] Bixby, R. E., M. Fenelon, Z. Gu, E. Rothberg, R. Wunderling. 2004. Mixed-integer programming: A progress report. M. Grötschel, ed. *The Sharpest Cut: The Impact of Manfred Padberg and His Work*. SIAM, Philadelphia, Pennsylvania, USA. 309–325.

[66] Bland, R. G., D. F. Shallcross. 1989. Large traveling salesman problems arising from experiments in X-ray crystallography: A preliminary report on computation. Operations Research Letters **8**, 125–128.

[67] Bleeker, H., P. Van Den Eijnden, F. De Jong. 1993. *Boundary-Scan Test: A Practical Approach*. Kluwer Academic Publishers, Dordrecht, The Netherlands.

[68] Bock, D. 2001. 350-antenna sample configurations for the Allen Telescope Array. ATA Memo 21, Radio Astronomy Laboratory, University of California at Berkeley, Berkeley, California, USA.

[69] Bock, F. 1958. An algorithm for solving "traveling-salesman" and related network optimization problems. Research Report, Armour Research Foundation. (Presented at the Operations Research Society of America Fourteenth National Meeting, St. Louis, October 24, 1958.)

[70] Bock, F. 1963. Mathematical programming solution of traveling salesman examples. R. L. Graves, P. Wolfe, eds. *Recent Advances in Mathematical Programming*. McGraw-Hill, New York, USA. 339–341.

[71] Bodlaender, H. L. 1993. A tourist guide through treewidth. Acta Cybernetica **11**, 1–21.

[72] Bodlaender, H. L. 1996. A linear time algorithm for finding tree-decompositions of small treewidth. SIAM Journal on Computing **25**, 1305–1317.

[73] Bodlaender, H. L. 1998. A partial k-arboretum of graphs with bounded treewidth. Theoretical Computer Science **209**, 1–45.

[74] Bodlaender, H. L., T. Kloks. 1996. Efficient and constructive algorithms for the pathwidth and treewidth of graphs. Journal of Algorithms **21**, 358–402.

[75] Bodlaender, H. L., D. M. Thilikos. 1997. Constructive linear time algorithms for branchwidth. P. Degano, R. Gorrieri, A. Marchetti-Spaccamela, eds. *Proceedings of the 24th International Colloquium on Automata, Languages, and Programming*. Lecture Notes in Computer Science **1256**. Springer, Berlin, Germany. 627–637.

[76] Bodlaender, H. L., D. M. Thilikos. 1999. Graphs with branchwidth at most three. Journal of Algorithms **32**, 167–194.

[77] Bonomi, E., J.-L. Lutton. 1984. The N-city travelling salesman problem: Statistical mechanics and the Metropolis algorithm. SIAM Review **26**, 551–568.

[78] Bondy, J. A., U. S. R. Murty. 1976. *Graph Theory with Applications*. Macmillan, London, UK.

[79] Booth, K. S., G. S. Lueker. 1976. Testing for the consecutive ones property, interval graphs, and graph planarity using PQ-tree algorithms. Journal of Computer and System Sciences **13**, 335–379.

[80] Borie, R. B., R. G. Parker, C. A. Tovey. 1992. Automatic generation of linear-time algorithms from predicate calculus descriptions of problems on recursively constructed graph families. Algorithmica **7**, 555–581.

[81] Borůvka, O. 1926. On a certain minimal problem (in Czech). Práce Moravské Přírodovědecké Společnosti **3**, 37–58.

[82] Bosch, R., A. Herman. 2004. Continuous line drawings via the traveling salesman problem. Operations Research Letters **32**, 302–303.

[83] Boyd, E. A. 1993. Generating Fenchel cutting planes for knapsack polyhedra. SIAM Journal of Optimization **3**, 734–750.

[84] Boyd, E. A. 1994. Fenchel cutting planes for integer programs. Operations Research **42**, 53–64.

[85] Boyd, S., S. Cockburn, D. Vella. 2001. On the domino-parity inequalities for the STSP. Computer Science Technical Report TR-2001-10. University of Ottawa, Ottawa, Ontario, Canada.

[86] Boyd, S., W. Cunningham, M. Queyranne, Y. Wang. 1995. Ladders for traveling salesmen. SIAM Journal on Optimization **5**, 408–420.

[87] Boyd, S. C. 1986. *The Subtour Polytope of the Travelling Salesman Problem.* Ph.D. Thesis. University of Waterloo, Waterloo, Ontario, Canada.

[88] Boyd, S. C., W. H. Cunningham. 1991. Small travelling salesman polytopes. Mathematics of Operations Research **16**, 259–271.

[89] Boyd, S. C., W. R. Pulleyblank. 1991. Optimizing over the subtour polytope of the travelling salesman problem. Mathematical Programming **49**, 163–187.

[90] Boyer, J. M., W. Myrvold. 2004. On the cutting edge: Simplified O(n) planarity by edge addition. Journal of Graph Algorithms and Applications **8** 241–273.

[91] Brady, R. M. 1985. Optimization strategies gleaned from biological evolution. Nature **317**, 804–806.

[92] Breen, M., C. Hitte, T. D. Lorentzen, R. Thomas, E. Cadieu, L. Sabacan, A. Scott, G. Evanno, H. G. Parker, E. F. Kirkness, R. Hudson, R. Guyon, G. G. Mahairas, B. Gelfenbeyn, C. M. Fraser, C. André , F. Galibert, E. A. Ostrande. 2004. An integrated 4249 marker FISH/RH map of the canine genome. BMC Genomics **5**.

[93] Brinkmeyer-Langford, C., T. Raudsepp, E.-J. Lee, G. Goh, A. A. Schäffer, R. Agarwala, M. L. Wagner, T. Tozaki, L. C. Skow, J. E. Womack, J. R. Mickelson, B. P. Chowdhary. 2005. A high-resolution physical map of equine homologs of HSA19 shows divergent evolution compared with other mammals. Mammalian Genome **16**, 631–649.

[94] British Association for the Advancement of Science. 1858. Report on the Twenty-Seventh Meeting held at Dublin in August and September 1857. John Murray, London. Part II, p. 3.

[95] Brockett, L. P. 1871. *The Commercial Traveller's Guide Book.* H. Dayton, New York, USA.

[96] Brüls, T., G. Gyapay, J.-L. Petit, F. Artiguenave, V. Vico, S. Qin, A. M. TinWollam, C. Da Silva, D. Muselet, D. Mavel, E. Pelletier, M. Levy, A. Fujiyama, F. Matsuda, R. Wilson, L. Rowen, L. Hood, J. Weissenbach, W. Saurin, R. Heilig. 2001. A physical map of human chromosome 14. Nature **409**, 947–948.

[97] Brusco, M., S. Stahl. 2005. *Branch-and-Bound Applications in Combinatorial Data Analysis.* Springer, Heidelberg, Germany.

[98] Bures, J., O. Buresova, L. Nerad. 1992. Can rats solve a simple version of the traveling salesman problem? Behavior Brain Research **52**, 133–142.

[99] Burkard, R. E. 1979. Travelling salesman and assignment problems: A survey. P. L. Hammer, E. L. Johnson, B. H. Korte, eds. *Discrete Optimization 1*. Annals of Discrete Mathematics **4**. North-Holland, Amsterdam, The Netherlands. 193–215.

[100] Burkard, R. E., U. Derigs. 1980. *Assignment and Matching Problems: Solution Methods with FORTRAN-Programs*. Lecture Notes in Economics and Mathematical Systems **184**. Springer, Berlin, Germany.

[101] Butler, M., P. Retzlaff, R. Vanderploeg. 1991. Neuropsychological test usage. Professional Psychology: Research and Practice **22**, 510–512.

[102] Camerini, P. M., L. Fratta, F. Maffioli. 1975. On improving relaxation methods by modified gradient techniques. Mathematical Programming Study **3**, 26–34.

[103] Caprara, A., M. Fischetti. 1996. $\{0, \frac{1}{2}\}$-Chvátal-Gomory cuts. Mathematical Programming **74**, 221–235.

[104] Caprara, A., M. Fischetti, A. N. Letchford. 2000. On the separation of maximally violated mod-k cuts. Mathematical Programming **87**, 37–56.

[105] Carbine, A., D. Feltham. 1998. Pentium pro processor design for test and debug. IEEE Design and Test of Computers **15**, 77–82.

[106] Cardiff, M., G. Hughes, R. Bosch. 2000. Maximizing fun at an amusement park. The UMAP Journal **21**, 483–498.

[107] Carlson, S. 1997. Algorithm of the gods. Scientific American **276**, 121–124.

[108] Carpaneto, G., M. Fischetti, P. Toth. 1989. New lower bounds for the symmetric travelling salesman problem. Mathematical Programming **45**, 223–254.

[109] Carr, R. 1997. Separating clique trees and bipartition inequalities having a fixed number of handles and teeth in polynomial time. Mathematics of Operations Research **22**, 257–265.

[110] Carr, R. 2000. Some results on node lifting of TSP inequalities. Journal of Combinatorial Optimization **4**, 395–414.

[111] Carr, R. 2004. Separation algorithms for classes of STSP inequalities arising from a new STSP relaxation. Mathematics of Operations Research **29**, 80–91.

[112] Cerf, N. J., J. Boutet de Monvel, O. Bohigas, O. C. Martin, A. G. Percus. 1997. The random link approximation for the Euclidean traveling salesman problem. Journal de Physique I **7**, 117–136.

[113] Chazelle, B. 2000. A minimum spanning tree algorithm with inverse-Ackermann type complexity. Journal of the ACM **47**, 1028–1047.

[114] Chekuri, C. S., A. V. Goldberg, D. R. Karger, M. S. Levine, C. Stein. 1997. Experimental study of minimum cut algorithms. *Proceedings of the Eighth Annual ACM-SIAM Symposium on Discrete Algorithms*. ACM Press, New York, USA. 324–333.

[115] Chen, X., M. L. Bushnell. 1995. *Efficient Branch and Bound Search with Application to Computer-Aided Design*. Springer, Heidelberg, Germany.

[116] Cherkassky, B. V., A. V. Goldberg. 1997. On implementing push-relabel method for the maximum flow problem. Algorithmica **19**, 390–410.

[117] Christof, T. 1997. *Low-Dimensional 0/1-Polytopes and Branch-and-Cut in Combinatorial Optimization*. Shaker Verlag, Aachen, Germany.

[118] Christof, T., M. Jünger, G. Reinelt. 1991. A complete description of the traveling salesman polytope on 8 nodes. Operations Research Letters **10**, 497–500.

[119] Christof, T., G. Reinelt. 1995. Parallel cutting plane generation for the TSP. P. Fritzson, L. Finmo, eds. *Parallel Programming and Applications*. IOS Press, Amsterdam, The Netherlands. 163–169.

[120] Christof, T., G. Reinelt. 1996. Combinatorial optimization and small polytopes. Top **4**, 1–64.

[121] Christof, T., G. Reinelt. 2001. Algorithmic aspects of using small instance relaxations in parallel branch-and-cut. Algorithmica **30**, 597–629.

[122] Christof, T., G. Reinelt. 2001. Decomposition and parallelization techniques for enumerating the facets of combinatorial polytopes. International Journal of Computational Geometry and Applications **11**, 423–437.

[123] Christofides, N. 1976. Worst-case analysis of a new heuristic for the travelling salesman problem. Report No. 388, Graduate School of Industrial Administration, Carnegie Mellon University, Pittsburgh, Pennsylvania, USA.

[124] Chrobak, M., T. Szymacha, A. Krawczyk. 1990. A data structure useful for finding Hamiltonian cycles. Theoretical Computer Science **71**, 419–424.

[125] Chvátal, V. 1973. Edmonds polytopes and weakly hamiltonian graphs. Mathematical Programming **5**, 29–40.

[126] Chvátal, V. 1983. *Linear Programming*. W. H. Freeman and Company, New York, USA.

[127] Clay Mathematics Institute. 2000. Millennium problems. www.claymath.org/millennium/ .

[128] Climer, S., W. Zhang. 2004. Take a walk and cluster genes: A TSP-based approach to optimal rearrangement clustering. *21st International Conference on Machine Learning (ICML'04)*, Banff, Alberta, Canada. 169–176.

[129] Climer, S., W. Zhang. 2005. Rearrangement clustering: Pitfalls, remedies, and applications. Journal of Machine Learning Research. To appear.

[130] Clochard, J.-M., D. Naddef. 1993. Using path inequalities in a branch and cut code for the symmetric traveling salesman problem. G. Rinaldi, L. Wolsey, eds. *Third IPCO Conference*. 291–311.

[131] Codenotti, B., G. Manzini, L. Margara, G. Resta. 1996. Perturbation: An efficient technique for the solution of very large instances of the Euclidean TSP. INFORMS Journal on Computing **8**, 125–133.

[132] ein alter Commis-Voyageur. 1832. *Der Handlungsreisende—wie er sein soll und was er zu thun hat, um Aufträge zu erhalten und eines glücklichen Erfolgs in seinen Geschäften gewiss zu sein—Von einem alten Commis-Voyageur.* B. Fr. Voigt, Ilmenau.

[133] Comstock, K. E., F. Lingaas, E. F. Kirkness, C. Hitte, R. Thomas, M. Breen, F. Galibert, E. A. Ostrande. 2004. A high-resolution comparative map of canine chromosome 5q14.3-q33 constructed utilizing the 1.5x canine genome sequence. Mammalian Genome **15**, 544–551.

[134] Cook, S. A. 1971. The complexity of theorem-proving procedures. *Proceedings of the 3rd Annual ACM Symposium on the Theory of Computing*. ACM Press, New York, USA. 151–158.

[135] Cook, W., D. Espinoza, M. Goycoolea. 2005. A study of domino-parity and k-parity constraints for the TSP. *Integer Programming and Combinatorial Optimization* (IPCO XI). Lecture Notes in Computer Science **3509**. Springer, Heidelberg, Germany. 452–467.

[136] Cook, W., D. Espinoza, M. Goycoolea. 2006. Computing with domino-parity inequalities for the TSP. To appear in INFORMS Journal on Computing.

[137] Cook, W., P. D. Seymour. 2003. Tour merging via branch decomposition. INFORMS Journal on Computing **15**, 233–248.

[138] Cook, W. J., W. H. Cunningham, W. R. Pulleyblank, A. Schrijver. 1998. *Combinatorial Optimization*. John Wiley & Sons, New York, USA.

[139] Cormen, T. H., C. E. Leiserson, R. L. Rivest. 1990. *Introduction to Algorithms*. MIT Press, Cambridge, Massachusetts, USA.

[140] Cornuéjols, G., J. Fonlupt, D. Naddef. 1985. The traveling salesman problem on a graph and some related integer polyhedra. Mathematical Programming **33**, 1–27.

[141] Courcelle, B. 1990. The monadic second-order logic of graphs I: Recognizable sets of finite graphs. Information and Computation **85**, 12–75.

[142] Cox, D. R., M. Burmeister, E. R. Price, S. Kim, R. M. Myers. 1990. Radiation hybrid mapping: A somatic cell genetic method for constructing high-resolution maps of mammalian chromosomes. Science **250**, 245–250.

[143] Cramer, A. E., C. R. Gallistel. 1997. Vervet monkeys as travelling salesmen. Nature **387**, 464.

[144] Craven, B. D. 1988. *Fractional Programming*. Heldermann, Berlin, Germany.

[145] Croes, G. A. 1958. A method for solving traveling-salesman problems. Operations Research **6**, 791–812.

[146] Crowder, H., E. L. Johnson, M. W. Padberg. 1983. Solving large-scale zero-one linear programming problems. Operations Research **31**, 803–834.

[147] Crowder, H., M. W. Padberg. 1980. Solving large-scale symmetric travelling salesman problems to optimality. Management Science **26**, 495–509.

[148] Csűrös, M., B. Li, A. Milosavljevic. 2003. Clone-array pooled shotgun mapping and sequencing: Design and analysis of experiments. Genome Informatics **14**, 186–195.

[149] Dantzig, G. 1948. Programming in a linear structure. US Air Force Comptroller, USAF, Washington, D.C., USA.

[150] Dantzig, G., R. Fulkerson, S. Johnson. 1954. Solution of a large scale traveling salesman problem. Technical Report P-510. RAND Corporation, Santa Monica, California, USA.

[151] Dantzig, G., R. Fulkerson, S. Johnson. 1954. Solution of a large-scale traveling-salesman problem. Operations Research **2**, 393–410.

[152] Dantzig, G. B. 1963. *Linear Programming and Extensions*. Princeton University Press, Princeton, New Jersey, USA.

[153] Dantzig, G. B., D. R. Fulkerson, S. M. Johnson. 1959. On a linear-programming, combinatorial approach to the traveling-salesman problem. Operations Research **7**, 58–66.

[154] Dash Optimization. 2006. Xpress-MP Suite. Documentation available at `www.dashoptimization.com`.

[155] Derigs, U., W. Meier. 1989. Implementing Goldberg's max-flow algorithm—A computational investigation. ZOR—Methods and Models of Operations Research **33**, 383–403.

[156] Diestel, R. 1997. *Graph Theory*. Springer, New York, USA.

[157] DIMACS. 2001. 8th DIMACS implementation challenge: The traveling salesman problem. `www.research.att.com/~dsj/chtsp`.

[158] Dosher, M. 1998. After 538,427 miles at the wheel: So many counties, so little time. Wall Street Journal, February 9, B1.

[159] Eastman, W. L. 1958. *Linear Programming with Pattern Constraints*. Ph.D. Thesis. Department of Economics, Harvard University, Cambridge, Massachusetts, USA.

[160] Easton, K., G. Nemhauser, M. Trick. 2003. Solving the travelling tournament problem: A combined integer programming and constraint programming approach. *Practice and Theory of Automated Timetabling IV*, Lecture Notes in Computer Science **2740**. Springer, Heidelberg, Germany. 100–109.

[161] Edberg, S. J., M. Shao, C. A. Beichman. 2005. SIM PlanetQuest: A mission for astrophysics and planet-finding. White paper, May 2. National Aeronautics and Space Administration, Jet Propulsion Laboratory, California Institute of Technology, Pasadena, California, USA.

[162] Edelsbrunner, H. 1987. *Algorithms in Combinatorial Geometry*. Springer, Heidelberg, Germany.

[163] Edmonds, J. 1965. Maximum matching and a polyhedron with 0,1-vertices. Journal of Research of the National Bureau of Standards **69B**, 125–130.

[164] Edmonds, J. 1965. Paths, trees, and flowers. Canadian Journal of Mathematics **17**, 449–467.

[165] Edmonds, J. 1965. The Chinese postman's problem. Bulletin of the Operations Research Society of America **13**, B-73.

[166] Edmonds, J., E. L. Johnson. 1970. Matching, a well-solved class of integer linear programs. R. Guy, H, Hanani, N. Sauer, J. Schönheim, eds. *Combinatorial Structures and Their Applications*. Gordon and Breach, New York, USA. 89–92.

[167] Ergun, Ö., J. B. Orlin, A. Steele-Feldman. 2006. Creating very large scale neighborhoods out of smaller ones by compounding moves. Journal of Heuristics **12**, 115–140.

[168] Espinoza, D. G. 2006. *On Linear programming, Integer Programming and Cutting Planes*. Ph.D. Thesis. School of Industrial and Systems Engineering, Georgia Institute of Technology, Atlanta, Georgia, USA.

[169] *Euler Archive*. 2005. www.eulerarchive.com.

[170] Euler, L. 1766. Solution d'une question curieuse que ne pariot soumise a aucune analyse. Mémoires de l'Academie Royale des Sciences et Belles Lettres, Année 1759, Berlin. 310–337. (Available as paper E309 in *The Euler Archive*, www.eulerarchive.com.)

[171] Extra Miler Club. 2005. www.extramilerclub.org.

[172] Fejes, L. 1940. Über einen geometrischen Satz. Mathematische Zeitschrift **46**, 83–85.

[173] Ferris, M. C., G. Pataki, S. Schmieta. 2001. Solving the *seymour* problem. Optima **66**, 2–6.

[174] Few, L. 1955. The shortest path and the shortest road through n points. Mathematika [London] **2**, 141–144.

[175] Fischetti, M., A. Lodi, P. Toth. 2002. Exact methods for the asymmetric traveling salesman problem. G. Gutin, A. P. Punnen, eds. *The Traveling Salesman Problem and Its Variations*. Kluwer, Boston, Massachusetts, USA. 169–206.

[176] Fleischer, L. 1999. Building chain and cactus representations of all minimum cuts from Hao-Orlin in the same asymptotic run time. Journal of Algorithms **33**, 51–72.

[177] Fleischer, L., É. Tardos. 1999. Separating maximally violated comb inequalities in planar graphs. Mathematics of Operations Research **24**, 130–148.

[178] Fleischer, L. K., A. N. Letchford, A. Lodi. 2004. Polynomial-time separation of a superclass of simple comb inequalities. To appear in Mathematics of Operations Research.

[179] Fleischmann, B. 1967. Computational experience with the algorithm of Balas. Operations Research **15**, 153–154.

[180] Fleischmann, B. 1985. A cutting plane procedure for the travelling salesman problem on road networks. European Journal of Operational Research **21**, 307–317.

[181] Fleischmann, B. 1988. A new class of cutting planes for the symmetric travelling salesman problem. Mathematical Programming **40**, 225–246.

[182] Flood, M. M. 1956. The traveling-salesman problem. Operations Research **4**, 61–75.

[183] Flood, M. M. 1984. Merrill Flood (with Albert Tucker), Interview of Merrill Flood in San Francisco on 14 May 1984. The Princeton Mathematics Community in the 1930s, Transcript Number 11 (PMC11). Princeton University.

[184] Fonlupt, J., D. Naddef. 1992. The traveling salesman problem in graphs with some excluded minors. Mathematical Programming **53**, 147–172.

[185] Ford, L. R., Jr., D. R. Fulkerson. 1956. Solving the transportation problem. Management Science **3**, 24–32.

[186] Ford, L. R., Jr., D. R. Fulkerson. 1957. A primal-dual algorithm for the capacitated Hitchcock problem. Naval Research Logistics Quarterly **4**, 47–54.

[187] Ford, L. R., Jr., D. R. Fulkerson. 1957. A simple algorithm for finding maximal network flows and an application to the Hitchcock problem. Canadian Journal of Mathematics **9**, 210–218.

[188] Ford, L. R., Jr., D. R. Fulkerson. 1958. A suggested computation for maximal multicommodity networks flows. Management Science **5**, 97–101.

[189] Ford, L. R., Jr., D. R. Fulkerson. 1962. *Flows in Networks*. Princeton University Press, Princeton, New Jersey, USA.

[190] Forrest, J. J., D. Goldfarb. 1992. Steepest-edge simplex algorithms for linear programming. Mathematical Programming **57**, 341–374.

[191] Forrest, J. J. H., J. A. Tomlin. 1972. Updating triangular factors of the basis to maintain sparsity in the product-form simplex method. Mathematical Programming **2**, 263–278.

[192] Forrest, J. J. H., J. A. Tomlin. 1992. Implementing the simplex method for the Optimization Subroutine Library. IBM Systems Journal **31**, No. 1, 11–25.

[193] Fortune, S. J. 1987. A sweepline algorithm for Voronoi diagrams. Algorithmica **2**, 153–174.

[194] Fraker, G. C. 2004. The real Lincoln highway: The forgotten Lincoln circuit markers. Journal of the Abraham Lincoln Association **25**, 76–97.

[195] Fredman, M. L., D. S. Johnson, L. A. McGeoch, G. Ostheimer. 1995. Data structures for traveling salesmen. Journal of Algorithms **18**, 432–479.

[196] Friedman, W. A. 2004. *Birth of a Salesman: The Transformation of Selling in America*. Harvard University Press, Cambridge, Massachusetts, USA.

[197] Frieze, A., R. Karp, B. Reed. 1995. When is the assignment bound tight for the asymmetric traveling-salesman problem. SIAM Journal on Computing **24**, 484–493.

[198] Fukasawa, R., J. Lysgaard, M. P. de Aragao, M. Reis, E. Uchoa, R. F. Werneck. 2004. Robust branch-and-cut-and-price for the capacitated vehicle routing problem. D. Bienstock, G. Nemhauser, eds. *Integer Programming and Combinatorial Optimization* (IPCO IX). Lecture Notes in Computer Science **3064**. Springer, Heidelberg, Germany. 1–15.

[199] Gamboa, D., C. Rego, F. Glover. 2005. Data structures and ejection chains for solving large-scale traveling salesman problems. European Journal of Operations Research **160**, 154–171.

[200] Garey, M. R., D. S. Johnson. 1979. *Computers and Intractability: A Guide to the Theory of NP-Completeness*. Freeman, San Francisco, California, USA.

[201] Garfinkel, R. 1985. Motivation and modeling. E. L. Lawler, J. K. Lenstra, A. H. G. Rinnooy Kan, D. B. Shmoys, eds. *The Traveling Salesman Problem*. John Wiley & Sons, Chichester, UK. 17–36.

[202] Garfinkel, R. S. 1977. Minimizing wallpaper waste, part I: A class of traveling salesman problems. Operations Research **25**, 741–751.

[203] Geographic Names Information System. 2006. /geonames.usgs.gov .

[204] Ghosh, M. N. 1949. Expected travel among random points in a region. Calcutta Statistical Association Bulletin **2**, 83–87.

[205] Gibson, B. 2005. Report in preparation. Department of Psychology, University of New Hampshire, Durham, New Hampshire, USA.

[206] Gilmore, P. C., R. E. Gomory. 1964. A solvable case of the traveling salesman problem. Proceedings of the National Academy of Sciences **51**, 178–181.

[207] Gilmore, P. C., R. E. Gomory. 1964. Sequencing a one state-variable machine: A solvable case of the traveling salesman problem. Operations Research **12**, 655–679.

[208] Goemans, M. 1995. Worst-case comparison of valid inequalities for the TSP. Mathematical Programming **69**, 335–349.

[209] Goemans, M. X., D. P. Williamson. 1995. A general approximation technique for constrained forest problems. *SIAM Journal on Computing* **24**, 296–317.

[210] Goldberg, A. V. 1985. A new max-flow algorithm. Technical Report MIT/LCS/TM 291. Laboratory for Computer Science, Massachusetts Institute of Technology, Cambridge, Massachusetts, USA.

[211] Goldberg, A. V., R. E. Tarjan. 1988. A new approach to the maximum flow problem. Journal of the Association for Computing Machinery **35**, 921–940.

[212] Goldfarb, D. 1976. Using the steepest-edge simplex algorithm to solve sparse linear programs. J. Bunch, D. Rose, eds. *Sparse Matrix Computations*. Academic Press, New York, USA. 227–240.

[213] Goldfarb, D., J. K. Reid. 1977. A practicable steepest-edge algorithm. Mathematical Programming **12**, 361–371.

[214] Gomory, R. E. 1958. Outline of an algorithm for integer solutions to linear programs. Bulletin of the American Mathematical Society **64**, 275–278.

[215] Gomory, R. E. 1960. Solving linear programs in integers. R. E. Bellman, M. Hall, Jr., eds. *Combinatorial Analysis*. Proceedings of the Symposia on Applied Mathematics **X**, 211–216.

[216] Gomory, R. E. 1963. An algorithm for integer solutions to linear programs. R. L. Graves, P. Wolfe, eds. *Recent Advances in Mathematical Programming*. McGraw-Hill, New York, USA. 269–302.

[217] Gomory, R. E. 1966. The traveling salesman problem. *Proceedings of the IBM Scientific Computing Symposium on Combinatorial Problems*. IBM, White Plains, New York, USA. 93–121

[218] Gomory, R. E. 1969. Some polyhedra related to combinatorial problems. Linear Algebra and Its Applications **2**, 451–558.

[219] Gomory, R. E., T. C. Hu. 1961. Multi-terminal network flows. Journal of the Society for Industrial and Applied Mathematics **9**, 551–570.

[220] Gonick, L. 1995. The solution. Discover Magazine. April. 36–37.

[221] Gonzales, R. H. 1962. Solution to the traveling salesman problem by dynamic programming on the hypercube. Technical Report Number 18, Operations Research Center, Massachusetts Institute of Technology, Cambridge, Massachusetts, USA.

[222] Gorges-Schleuter, M. 1997. Asparagos96 and the traveling salesman problem. *Proceedings of the IEEE International Conference on Evolutionary Computation*. 171–174.

[223] Goss, S. J., H. Harris. 1975. New method for mapping genes in human chromosomes. Nature **255**, 680–684.

[224] Graham, R. L., P. Hell. 1985. On the history of the minimum spanning tree problem. Annals of the History of Computing **7**, 43–57.

[225] Graham, S. M., A. Joshi, Z. Pizlo. 2000. The traveling salesman problem: A hierarchical model. Memory and Cognition **28**, 1191–1204.

[226] Graham-Rowe, D. 2005. SnailMail 2.0. TechnologyReview.com, April 2005.

[227] Grossman, B. 2005. Laser crystal technique. Available at www.bathsheba.com/crystal/process .

[228] Grötschel, M. 1977. *Polyedrische Charackterisierungen kombinatorischer Optimierungsprobleme*. Anton Hain Verlag, Meisenheim/Glan, Germany.

[229] Grötschel, M. 1980. On the monotone symmetric travelling salesman problem: Hypohamiltonian/hypotraceable graphs and facets. Mathematics of Operations Research **5**, 285–292.

[230] Grötschel, M. 1980. On the symmetric travelling salesman problem: Solution of a 120-city problem. Mathematical Programming Study **12**, 61–77.

[231] Grötschel, M., O. Holland. 1987. A cutting-plane algorithm for minimum perfect 2-matchings. Computing **39**, 327–344.

[232] Grötschel, M., O. Holland. 1991. Solution of large-scale symmetric travelling salesman problems. Mathematical Programming **51**, 141–202.

[233] Grötschel, M., L. Lovász, A. Schrijver. 1993. *Geometric Algorithms and Combinatorial Optimization*, 2nd edition. Springer, Berlin, Germany.

[234] Grötschel, M., M. Jünger, G. Reinelt. 1984. A cutting plane algorithm for the linear ordering problem. Operations Research **32**, 1195–1220.

[235] Grötschel, M., M. Jünger, G. Reinelt. 1991. Optimal control of plotting and drilling machines: A case study. Zeitschrift für Operations Research **35**, 61–84.

[236] Grötschel, M., M. W. Padberg. 1979. On the symmetric traveling salesman problem I: Inequalities. Mathematical Programming **16**, 265–280.

[237] Grötschel, M., M. W. Padberg. 1979. On the symmetric traveling salesman problem II: Lifting theorems and facets. Mathematical Programming **16**, 281–302.

[238] Grötschel, M., M. W. Padberg. 1985. Polyhedral theory. E. L. Lawler, J. K. Lenstra, A. H. G. Rinnooy Kan, D. B. Shmoys, eds. *The Traveling Salesman Problem*. John Wiley & Sons, Chichester, UK. 252–305.

[239] Grötschel, M., W. R. Pulleyblank. 1986. Clique tree inequalities and the symmetric travelling salesman problem. Mathematics of Operations Research **11**, 537–569.

[240] Guan, M.-g. 1960. Graphic programming using odd or even points [in Chinese]. Acta Mathematica Sinica **10**, 263–266.

[241] Gupta, P., A. B. Kahng, I. L. Mandoiu, P. Sharma. 2005. Layout-aware scan chain synthesis for improved path delay fault coverage. IEEE Transactions on Computer-Aided Design of Integrated Circuits and Systems **24**, 1104–1114.

[242] Gupta, P., A. B. Kahng, S. Mantik. 2003. Routing-aware scan chain ordering. Proceedings Asia and South Pacific Design Automation Conference. 857–862.

[243] Gutin, G., H. Jakubowicz, S. Ronen, A. Zverovitch. 2003. Seismic vessel problem. Department of Computer Science, University of London, London, UK.

[244] Guyon, R., T. D. Lorentzen, C. Hitte, L. Kim, E. Cadieu, H. G. Parker, P. Quignon, J. K. Lowe, C. Renier, Dagger, Boris Gelfenbeyn, F. Vignaux, H. B. DeFrance, S. Gloux, G. G. Mahairas, C. André, F. Galibert, E. A. Ostrander. 2003. A 1-Mb resolution radiation hybrid map of the canine genome. Proceedings of the National Academy of Sciences **100**, 5296–5301.

[245] Gyapay, G., K. Schmitt, C. Fizames, H. Jones, M. Vega-Czarny, D. Spillett, D. Muselet, J. F. Prud'homme, C. Dib, C. Auffray, J. Morissette, J. Weissenbach, P. N. Goodfellow. 1996. A radiation hybrid map of the human genome. Human Molecular Genetics **5**, 339–346.

[246] Hamilton, W. R. 1856. Letter to John T. Graves on the Icosian. H. Halberstam, R. E. Ingram, eds. *The Mathematical Papers of Sir William Rowan Hamilton*, Volume III. Cambridge University Press, Cambridge, UK. 612-624.

[247] Hammer, P. L., E. L. Johnson, U. N. Peled. 1975. Facets of regular 0-1 polytopes. Mathematical Programming **8**, 179–206.

[248] Hampson, J. 1791. *Memoirs of the late Rev. John Wesley*. J. Johnson, London, UK.

[249] Harel, D., R. E. Tarjan. 1984. Fast algorithms for finding nearest common ancestors. SIAM Journal on Computing **13**, 338–355.

[250] Harris, P. M. J. 1974. Pivot selection methods of the devex LP code. Mathematical Programming **5**, 1–28.

[251] Helbig Hansen, K., J. Krarup. 1974. Improvements of the Held-Karp algorithm for the symmetric traveling-salesman problem. Mathematical Programming **7**, 87–96.

[252] Held, M., R. M. Karp. 1962. A dynamic programming approach to sequencing problems. Journal of the Society of Industrial and Applied Mathematics **10**, 196–210.

[253] Held, M., R. M. Karp. 1970. The traveling-salesman problem and minimum spanning trees. Operations Research **18**, 1138–1162.

[254] Held, M., R. M. Karp. 1971. The traveling-salesman problem and minimum spanning trees: Part II. Mathematical Programming **1**, 6–25.

[255] Held, M., P. Wolfe, H. P. Crowder. 1974. Validation of subgradient optimization. Mathematical Programming **6**, 62–88.

[256] Heller, I. 1953. On the problem of the shortest path between points. I. Abstract 664t, Bulletin of the American Mathematical Society **59**, 551–551.

[257] Heller, I. 1953. On the problem of the shortest path between points. II. Abstract 665t, Bulletin of the American Mathematical Society **59**, 551–552.

[258] Heller, I. 1954. The traveling salesman's problem, Part 1: Basic facts. Technical Report. George Washington University, Logistics Research Project, Washington, D.C., USA.

[259] Heller, I. 1955. On the travelling salesman's problem. H. A. Antosiewicz, ed. *Proceedings of the Second Symposium in Linear Programming*. National Bureau of Standards, Washington, D.C., USA. 643–665.

[260] Helsgaun, K. 2000. An effective implementation of the Lin-Kernighan traveling salesman heuristic. European Journal of Operational Research **126**, 106–130.

[261] Henzinger, M. R., P. Klein, S. Rao, S. Subramanian. 1997. Faster shortest-path algorithms for planar graphs. Journal of Computer and System Sciences **55**, 3–23.

[262] Hibbert, C. 1969. *The Grand Tour*. G. P. Putnam's Sons, New York, USA.

[263] Hicks, I. V. 2000. *Branch Decompositions and Their Applications*. Ph.D. Thesis. Computational and Applied Mathematics, Rice University, Houston, Texas, USA.

[264] Hirech, M., J. Beausang, X. Gu. 1998. A new approach to scan chain reordering using physical design information. *Proceedings of the 1998 IEEE International Test Conference*. 348–355.

[265] Hitte, C., T. D. Lorentzen, R. Guyon, L. Kim, E. Cadieu, H. G. Parker, P. Quignon, J. K. Lowe, B. Gelfenbeyn, C. Andre, E. A. Ostrander, F. Galibert. 2003. Comparison of MultiMap and TSP/CONCORDE for constructing radiation hybrid maps. Journal of Heredity **94**, 9–13.

[266] Hoffman, A., M. Mannos, D. Sokolowsky, N. Wiegmann. 1953. Computational experience in solving linear programs. Journal of the Society for Industrial and Applied Mathematics **1**, 17–33.

[267] Hoffman, A. J., P. Wolfe. 1985. History. E. L. Lawler, J. K. Lenstra, A. H. G. Rinnooy Kan, D. B. Shmoys, eds. *The Traveling Salesman Problem*. John Wiley & Sons, Chichester, UK. 1–15.

[268] Holland, O. A. 1987. *Schnittebenenverfahren für Travelling-Salesman und verwandte Probleme*. Ph.D. Thesis. Universität Bonn, Bonn, Germany.

[269] Hong, I., A. B. Kahng, B. Moon. 1997. Improved large-step Markov chain variants for the symmetric TSP. Journal of Heuristics **3**, 63–81.

[270] Hong, S. 1972. *A Linear Programming Approach for the Traveling Salesman Problem*. Ph.D. Thesis. Johns Hopkins University, Baltimore, Maryland, USA.

[271] Hopfield, J. J., D. W. Tank. 1985. Neural computation of decisions in optimization problems. Biological Cybernetics **52**, 141–152.

[272] Hopfield, J. J., D. W. Tank. 1986. Computing with neural circuits: A model. Science **233**, 625–633.

[273] Houck Jr., D. J., J. C. Picard, M. Queyranne, R. R. Vemuganti. 1980. The travelling salesman problem as a constrained shortest path problem: Theory and computational experience. OPSEARCH **17**, 93–109.

[274] Hu, T. C. 1969. *Integer Programming and Network Flows*. Addison Wesley, Reading, Massachusetts, USA.

[275] Huffman, D. A. 1952. A method for the construction of minimum-redundancy codes. Proceedings of the IRE **40**, 1098–1101.

[276] ILOG. 2006. *User's Manual, ILOG CPLEX 10.0*. ILOG CPLEX Division, Incline Village, Nevada, USA.

[277] Isaac, A. M., E. Turban. 1969. Some comments on the traveling salesman problem. Operations Research **17**, 543–546

[278] Ivansson, L. 2000. *Computational Aspects of Radiation Hybrid Mapping*. Ph.D. Thesis. Department of Numerical Analysis and Computer Science, Royal Institute of Technology, Stockholm, Sweden.

[279] Ivansson, L., J. Lagergren. 2004. Algorithms for RH mapping: New ideas and improved analysis. SIAM Journal on Computing **34**, 89–108.

[280] Jacobsen, J. L., N. Read, H. Saleur. 2004. Traveling salesman problem, conformal invariance, and dense polymers. Physical Review Letters **93**, Number 3.

[281] Jacobson, G., D. Applegate, W. Cook. 1998. Computing minimum length nucleotide linkers via TSP. Preprint. AT&T Bell Laboratories, Murray Hill, New Jersey, USA.

[282] Jain, A. K., M. N. Murty, P. J. Flynn. 1999. Data clustering: A review. ACM Computing Surveys **31**, 264–323.

[283] Jain, V., I. E. Grossmann. 1999. Resource-constrained scheduling of tests in new product development. Industrial and Engineering Chemistry Research **38**, 3013–3026.

[284] Jellis, G. P. 2001. Knight's tour font designed by D. E. Knuth. The Games and Puzzles Journal **21**. Available at www.gpj.connectfree.co.uk.

[285] Jellis, G. P. 2004. Knight's tour notes. Available at www.ktn.freeuk.com.

[286] Jessen, R. J. 1942. Statistical investigation of a sample survey for obtaining farm facts. Research Bulletin (Iowa Agricultural Experiment Station) 304. Iowa State College of Agriculture and Mechanic Arts.

[287] Jet Propulsion Laboratory. 2005. SIM PlanetQuest: The mission. Available at planetquest.jpl.nasa.gov/SIM/sim_mission.cfm.

[288] Jewell, W. S. 1958. Optimal flow through networks. Interim Technical Report No. 8, Operations Research Center, Massachusetts Institute of Technology, Cambridge, Massachusetts, USA.

[289] Johnson, D. S. 1990. Local optimization and the traveling salesman problem. *Proceedings 17th Colloquium of Automata, Languages, and Programming*. Lecture Notes in Computer Science **443**. Springer, Berlin, Germany. 446–461.

[290] Johnson, D. S., S. Krishnan, J. Chhugani, S. Kumar, S. Venkatasubramanian. 2004. Compressing large Boolean matrices using reordering techniques. *Proceedings of the 30th International Conference on Very Large Databases (VLDB)*. 13–23.

[291] Johnson, D. S., L. A. McGeoch. 1997. The traveling salesman problem: A case study. E. Aarts, J. K. Lenstra, eds. *Local Search in Combinatorial Optimization*. John Wiley & Sons, Chichester, UK. 215–310.

[292] Johnson, D. S., L. A. McGeoch. 2002. Experimental analysis of heuristics for the STSP. G. Gutin, A. Punnen, eds. *The Traveling Salesman Problem and Its Variations*. Kluwer, Boston, Massachusetts, USA. 369–443.

[293] Johnson, D. S., L. A. McGeoch, E. E. Rothberg. 1996. Asymptotic experimental analysis for the Held-Karp traveling salesman bound. *Proceedings of the Seventh Annual ACM-SIAM Symposium on Discrete Algorithms*. ACM Press, New York, USA. 341–350.

[294] Johnson, D. S., C. H. Papadimitriou. 1985. Computational complexity. E. L. Lawler, J. K. Lenstra, A. H. G. Rinnooy Kan, D. B. Shmoys, eds. *The Traveling Salesman Problem*. John Wiley & Sons, Chichester, UK. 37–85.

[295] Jünger, M., W. R. Pulleyblank. 1993. Geometric duality and combinatorial optimization. S. D. Chatterji, B. Fuchssteiner, U. Kluish, R. Liedl, eds. *Jahrbuck Überblicke Mathematik*. Vieweg, Brunschweig/Wiesbaden, Germany. 1–24.

[296] Jünger, M., G. Reinelt, G. Rinaldi. 1995. The traveling salesman problem. M. Ball, T. Magnanti, C. L. Monma, G. Nemhauser, eds. *Handbooks on Operations Research and Management Science: Network Models*. North Holland, Amsterdam, The Netherlands. 225–330.

[297] Jünger, M., G. Reinelt, G. Rinaldi. 1997. The Traveling Salesman Problem. M. Dell'Amico, F. Maffioli, S. Martello, eds. *Annotated Bibliographies in Combinatorial Optimization*. John Wiley & Sons, Chichester, UK. 199–221.

[298] Jünger, M., G. Reinelt, S. Thienel. 1995. Practical problem solving with cutting plane algorithms. W. Cook, L. Lovász, P. Seymour, eds. *Combinatorial Optimization*. DIMACS Series in Discrete Mathematics and Theoretical Computer Science **20**. American Mathematical Society, Providence, Rhode Island, USA. 111–152.

[299] Jünger, M., G. Rinaldi, S. Thienel. 2000. Practical performance of efficient minimum cut algorithms. Algorithmica **26**, 172–195.

[300] Jünger, M., S. Thienel, G. Reinelt. 1994. Provably good solutions for the traveling salesman problem. Zeitschrift für Operations Research **40**, 183–217.

[301] Kalan, J. E. 1971. Aspects of large-scale in-core linear programming. *Proceedings of the 1971 26th Annual Conference*. ACM Press, New York, USA. 304–313.

[302] Kamil, A. C., J. E. Jones. 1997. The seed-storing corvid Clark's nutcrackers learn geometric relationships among landmarks. Nature **390**, 276–279.

[303] Kaplan, C. S., R. Bosch. 2005. TSP Art. Technical Report. School of Computer Science, University of Waterloo, Waterloo, Ontario, Canada.

[304] Karg, R. L., G. L. Thompson. 1964. A heuristic approach to solving travelling salesman problems. Management Science **10**, 225–248.

[305] Karger, D. R., C. Stein. 1996. A new approach to the minimum cut problem. Journal of the ACM **43**, 601–640.

[306] Karmarkar, N. 1984. A new polynomial-time algorithm for linear programming. Combinatorica **4**, 373–395.

[307] Karp, R., L. Ruzzo, M. Tompa. 1996. Algorithms in molecular biology (course notes). Department of Computer Science and Engineering, University of Washington, Seattle, Washington, USA.

[308] Karp, R. M. 1972. Reducibility among combinatorial problems. R. E. Miller, J. W. Thatcher, eds. *Complexity of Computer Computations*. Plenum Press, New York, USA. 85–103.

[309] Karp, R. M. 1977. Probabilistic analysis of partitioning algorithms for the traveling-salesman problem in the plane. Mathematics of Operations Research **2**, 209–224.

[310] Karp, R. M. 1986. Combinatorics, complexity, and randomness. Communications of the ACM **29**, 98–109.

[311] Karp, R. M., J. M. Steele. 1985. Probabilistic analysis of heuristics. E. L. Lawler, J. K. Lenstra, A. H. G. Rinnooy Kan, D. B. Shmoys, eds. *The Traveling Salesman Problem*. John Wiley & Sons, Chichester, UK. 181–205.

[312] Kaufer, A., L. Pasquini, M. Zoccali, H. Dekker, N. Cretton, J. Smoker. 2004. Very Large Telescope: FLAMES user manual. Document No. INS-MAN-ESO-13700-2994, European Southern Observatory.

[313] Krauth, W., M. Mézard. 1989. The cavity method and the travelling-salesman problem. Europhysics Letters **8**, 213–218.

[314] Kernighan, B. W., D. M. Ritchie. 1978. *The C Programming Language*. Prentice Hall, Englewood Cliffs, New Jersey, USA.

[315] Khachiyan, L. G. 1979. A polynomial algorithm in linear programming. Soviet Mathematics Doklady **20**, 191–194.

[316] Kirkpatrick, S., C. D. Gelatt Jr., M. P. Vecchi. 1983. Optimization by Simulated Annealing. Science **220**, 671–680.

[317] Klabjan, D., E. L. Johnson, G. L. Nemhauser, E. Gelman, S. Ramaswamy. 2001. Solving large airline crew scheduling problems: Random pairing generation and strong branching. Computational Optimization and Applications **20**, 73–91.

[318] Knuth, D. E. 1973. *Sorting and Searching, The Art of Computer Programming, Volume 3*. Addison Wesley, Reading, Massachusetts, USA.

[319] Knuth, D. E. 1994. Leaper graphs. Mathematical Gazette **78**, 274–297.

[320] Knuth, D. E. 1997. *Fundamental Algorithms, The Art of Computer Programming, Volume 1*, Third Edition. Addison Wesley, Reading, Massachusetts, USA.

[321] Knuth, D. E. 1997. *Seminumerical Algorithms, The Art of Computer Programming, Volume 2*, Third Edition. Addison Wesley, Reading, Massachusetts, USA.

[322] Kolata, G. 1994. Novel kind of computing: Calculation with DNA. New York Times. November 23. C1.

[323] Koster, A. M. C. A., H. L. Bodlaender, S. P. M. van Hoesel. 2001. Treewidth: Computational experiments. Electronic Notes in Discrete Mathematics **8**. Elsevier Science Publishers, Amsterdam, The Netherlands.

[324] Koster, A. M. C. A., S. P. M. van Hoesel, A. W. J. Kolen. 2002. Solving partial constraint satisfaction problems with tree decomposition. Networks **40**, 170–180.

[325] Krolak, P., W. Felts, G. Marble. 1971. A man-machine approach toward solving the traveling salesman problem. Communications of the ACM **14**, 327–334.

[326] Kruskal, J. B. 1956. On the shortest spanning subtree of a graph and the traveling salesman problem. Proceedings of the American Mathematical Society **7**, 48–50.

[327] Krzywinski, M., J. Wallis, C. Gösele, I. Bosdet, R. Chiu, T. Graves, O. Hummel, D. Layman, C. Mathewson, N. Wye, B. Zhu, D. Albracht, J. Asano, S. Barber, M. Brown-John, S. Chan, S. Chand, A. Cloutier, J. Davito, C. Fjell, T. Gaige, D. Ganten, N. Girn, K. Guggenheimer, H. Himmelbauer, T. Kreitler, S. Leach, D. Lee, H. Lehrach, M. Mayo, K. Mead, T. Olson, P. Pandoh, A.-L. Prabhu, H. Shin, S. Tänzer, J. Thompson, M. Tsai, J. Walker, G. Yang, M. Sekhon, L. Hillier, H. Zimdahl, A. Marziali, K. Osoegawa, S. Zhao,

A.Siddiqui, P. J. de Jong, W. Warren, E. Mardis, J. D. McPherson, R. Wilson, N. Hübner, S. Jones, M. Marra, J. Schein. 2004. Integrated and sequence-ordered BAC- and YAC-based physical maps for the rat genome. Genome Research **14**, 766–779.

[328] Kuhn, H. W. 1955. On certain convex polyhedra. Abstract 799t, Bulletin of the American Mathematical Society **61**, 557–558.

[329] Kuhn, H. W., R. E. Quandt. 1963. An experimental study of the simplex method. *Proceedings of Symposia in Applied Mathematics* **15**, 107–124.

[330] van Laarhoven, P. J., E. H. Aarts. 1987. *Simulated Annealing: Theory and Applications*. Springer, Heidelberg, Germany.

[331] Lagergren, J. 1990. Efficient parallel algorithms for tree-decomposition and related problems. *Proceedings of the 31st Annual Symposium on the Foundations of Computer Science*. IEEE Computer Society Press, Los Alamitos, California, USA. 173–182.

[332] Lambert, F. 1960. The traveling-salesman problem. Cahiers du Centre de Recherche Opérationelle **2**, 180–191.

[333] Land, A. 1979. The solution of some 100-city travelling salesman problems. Technical Report. London School of Economics, London, UK.

[334] Land, A. H., A. G. Doig. 1960. An automatic method of solving discrete programming problems. Econometrica **28**, 497–520.

[335] Land, A. H., S. Powell. 1973. *Fortran Codes for Mathematical Programming*. John Wiley & Sons, Chichester, UK.

[336] Lander, E. S., L. M. Linton, B. Birren, C. Nusbaum, M. C. Zody, J. Baldwin, K. Devon, K. Dewar, M. Doyle, W. FitzHugh, R. Funke, D. Gage, K. Harris, A. Heaford, J. Howland, L. Kann, J. Lehoczky, R. LeVine, P. McEwan, K. McKernan, J. Meldrim, J. P. Mesirov, C. Miranda, W. Morris, J. Naylor, C. Christina, M. Rosetti, R. Santos, A. Sheridan, C. Sougnez, N. Stange-Thomann, N. Stojanovic, A. Subramanian, D. Wyman, J. Rogers, J. Sulston, R. Ainscough, S. Beck, D. Bentley, J. Burton, C. Clee, N. Carter, A. Coulson, R. Deadman, P. Deloukas, A. Dunham, I. Dunham, R. Durbin, L. French, D. Grafham, S. Gregory, T. Hubbard, S. Humphray, A. Hunt, M. Jones, C. Lloyd; A. McMurray, L. Matthews, S. Mercer, S. Milne, J. Mullikin, A. Mungall, R. Plumb, M. Ross, R. Shownkeen, S. Sims, R. H. Waterston, R. K. Wilson, L. W. Hillier, J. D. McPherson, M. A. Marra, E. R. Mardis, L. A. Fulton, A. T. Chinwalla, K. H. Pepin, W. R. Gish, S. L. Chissoe, M. C. Wendl, K. D. Delehaunty, T. L. Miner, A. Delehaunty, J. B. Kramer, L. L. Cook, R. S. Fulton, D. L. Johnson, P. J. Minx, S. W. Clifton, T. Hawkins, E. Branscomb, P. Predki, P. Richardson, S. Wenning, T. Slezak, N. Doggett, J.-F. Cheng, A. Olsen, S. Lucas, C. Elkin, E. Uberbacher, M. Frazier, R. A. Gibbs, D. M. Muzny, S. E. Scherer, J. B. Bouck, E. J. Sodergren, K. C. Worley, C. M. Rives, J. H. Gorrell, M. L. Metzker, S. L.

Naylor, R. S. Kucherlapati, D. L. Nelson, G. M. Weinstock, Y. Sakaki, A. Fujiyama, M. Hattori, T. Yada, A. Toyoda, T. Itoh, C. Kawagoe, H. Watanabe, Y. Totoki, T. Taylor, J.Weissenbach, R. Heilig, W. Saurin, F. Artiguenave, P. Brottier, T. Bruls, E. Pelletier, C. Robert, P. Wincker, D. R. Smith, L. Doucette-Stamm, M. Rubenfield, K. Weinstock, H. M. Lee, J. Dubois, A. Rosenthal, M. Platzer, G. Nyakatura, S. Taudien, A. Rump, H. Yang, J. Yu, J. Wang, G. Huang, J. Gu, L. Hood, L. Rowen, A. Madan, S. Qin, R. W. Davis, N. A. Federspiel, A. P. Abola, M. J. Proctor, R. M. Myers, J. Schmutz, M. Dickson, J. Grimwood, D. R. Cox, M. V. Olson, R. Kaul, C. Raymond, N. Shimizu, K. Kawasaki, S. Minoshima, G. A. Evans, M. Athanasiou, R. Schultz, B. A. Roe, F. Chen, P. Huaqin; J. Ramser, H. Lehrach, R. Reinhardt, W. R. McCombie, M. de la Bastide, N. Dedhia, H. Blöcker, K. Hornischer, G. Nordsiek, R. Agarwala, L. Aravind, J. A. Bailey, A. Bateman, S. Batzoglou, E. Birney, P. Bork, D. G. Brown, C. B. Burge, L. Cerutti, H.-C. Chen, D. Church, M. Clamp, R. R. Copley, T. Doerks, S. R. Eddy, E. E. Eichler, T. S. Furcy, J. Galagan, J. G. Gilbert, C. Harmon, Y. Hayashizaki, D. Haussler, H. Hermjakob, K. Hokamp, W. Jang, L. S. Johnson, T. A. Jones, S. Kasif, A. Kaspryzk, S. Kennedy, W. J. Kent, P. Kitts, E. V. Koonin, I. Korf, D. Kulp, D. Lancet, T. M. Lowe, A. McLysaght, T. Mikkelsen, J. V. Moran, N. Mulder, V. J. Pollara, C. P. Ponting, G. Schuler, J. Schultz, G. Slater, A. F. A. Smit, E. Stupka, J. Szustakowki, D. Thierry-Mieg, J. Thierry-Mieg, L. Wagner, J. Wallis, R. Wheeler, A. Williams, Y. I. Wolf, K. H. Wolfe, S.-P. Yang, R.-F. Yeh, F. Collins, M. S. Guyer, J. Peterson, A. Felsenfeld, K. A. Wetterstrand, A. Patrinos, M. J. Morgan. 2001. Initial sequencing and analysis of the human genome. Nature **409**, 860–921.

[337] Laporte, G. 1992. The traveling salesman problem: An overview of exact and approximate algorithms. European Journal of Operations Research **59**, 231–247.

[338] Lawler, E. L., J. K. Lenstra, A. H. G. Rinnooy Kan, D. B. Shmoys, eds. 1985. *The Traveling Salesman Problem*. John Wiley & Sons, Chichester, UK.

[339] Lawler, E. L., D. E. Wood. 1966. Branch-and-bound methods: A survey. Operations Research **14**, 699–719.

[340] L'Ecuyer, P. 1999. Good parameters and implementations for combined multiple recursive random number generators. Operations Research **47**, 159–164.

[341] L'Ecuyer, P. 2000. RngStream. www.iro.umontreal.ca/~lecuyer.

[342] Lee, J., M. Y. Choi. 1994. Optimization by multicanonical annealing and the traveling salesman problem. Physical Review E **50**, R651–R654.

[343] Lee, M. D., D. Vickers. 2000. The importance of the convex hull for human performance on the traveling salesman problem: A comment on MacGregor and Ormerod (1996). Perception and Psychophysics **62**, 226–228.

[344] Lemke, C. E. 1954. The dual method of solving the linear programming problem. Naval Research Logistics Quarterly **1**, 36–47.

[345] Lenstra, J. K. 1974. Clustering a data array and the traveling-salesman problem. Operations Research **22**, 413–414.

[346] Lenstra, J. K., A. H. G. Rinnooy Kan. 1975. Some simple applications of the travelling salesman problem. Operations Research Quarterly **26**, 717–733.

[347] Letchford, A. N. 2000. Separating a superclass of comb inequalities in planar graphs. Mathematics of Operations Research **25**, 443–454.

[348] Letchford, A. N., A. Lodi. 2002. Polynomial-time separation of simple comb inequalities. W. J. Cook, A. S. Schulz, eds. *Integer Programming and Combinatorial Optimization*. Lecture Notes in Computer Science **2337**. Springer, Heidelberg, Germany. 93–108.

[349] Letchford, A. N., A. Lodi. 2002. Primal cutting plane algorithms revisited. Mathematical Methods of Operations Research **56**, 67–81.

[350] Letchford, A. N., A. Lodi. 2003. Primal separation algorithms. 4OR (Quarterly Journal of the Belgian, French and Italian Operations Research Societies) **1**, 209–224.

[351] Letchford, A. N., G. Reinelt, D. O. Theis. 2004. A faster exact separation algorithm for blossom inequalities. D. Bienstock, G. Nemhauser, eds. *Integer Programming and Combinatorial Optimization* (IPCO IX). Lecture Notes in Computer Science **3064**. Springer, Heidelberg, Germany. 196–205.

[352] Letchford, A. N., N. Pearson. 2005. Exploiting planarity in separation routines for the symmetric traveling salesman problem. Preprint. Department of Management Science, Lancaster University, Lancaster, UK.

[353] Levine, M. 1999. Finding the right cutting planes for the TSP. M. T. Goodrich, C. C. McGeoch, eds. *Algorithm Engineering and Experimentation, International Workshop ALEXNEX'99*. Lecture Notes in Computer Science **1619**. Springer, Heidelberg, Germany. 266–281.

[354] Levitt, M. E. 1997. Designing ultrasparc for testability. IEEE Design and Test of Computers **14**, 10–17.

[355] Lezak, M. D. 1995. *Neuropsychological Assessment*, Third Edition. Oxford University Press, New York, USA.

[356] Lin, S. 1965. Computer solutions of the traveling salesman problem. The Bell System Technical Journal **44**, 2245–2269.

[357] Lin, S., B. W. Kernighan. 1973. An effective heuristic algorithm for the traveling-salesman problem. Operations Research **21**, 498–516.

[358] Linderoth, J. T., M. W. P. Savelsbergh. 1999. A computational study of search strategies for mixed integer programming. INFORMS Journal on Computing **11**, 173–187.

[359] Lipton, R. J. 1995. DNA solution of hard computational problems. Science **268**, 542–545.

[360] Little, J. D. C., K.G. Murty, D.W. Sweeney, C. Karel. 1963. An algorithm for the traveling salesman problem. Operations Research **11**, 972–989.

[361] Lovász, L., M. Plummer. 1986. *Matching Theory.* North Holland, Amsterdam, The Netherlands.

[362] MacDonald, S. E., J. C. Pang, S. Gibeault. 1994. Marmoset (*Callithrix jacchus jacchus*) spatial memory in a foraging task: Win-stay versus win-shift strategies. Journal of Comparative Psychology **108**, 328–334.

[363] MacGregor, J. N., T. Ormerod. 1996. Human performance on the traveling salesman problem. Perception and Pyschophysics **58**, 527–539.

[364] MacGregor, J. N., T. C. Ormerod. 2000. Evaluating the importance of the convex hull in solving the Euclidean version of the traveling salesperson problem: Reply to Lee and Vickers (2000). Perception and Pyschophysics **62**, 1501–1503.

[365] MacGregor, J. N., E. P. Chronicle, T. C. Ormerod. 2004. Convex hull or crossing avoidance? Solution heuristics in the travelling salesperson problem. Memory and Cognition **32**, 260–270.

[366] Madsen, O. B. 1988. An application of travelling-salesman routines to solve pattern-allocation problems in the glass industry. Journal of the Operational Research Society **39**, 249–256.

[367] Magirou, V. F. 1986. The efficient drilling of printed circuit boards. INTER-FACES **16**, 13–23.

[368] Mahalanobis, P. C. 1940. A sample survey of the acreage under jute in Bengal. Sankhya, The Indian Journal of Statistics **4**, 511–530.

[369] Mak, K.-T., A. J. Morton. 1993. A modified Lin-Kernighan traveling-salesman heuristic. Operations Research Letters **13**, 127–132.

[370] Makar, S. 1998. A layout-based approach for ordering scan chain flip-flops. *Proceedings of the 1998 IEEE International Test Conference.* 341–347.

[371] Marchand, H., A. Martin, R. Weismantel, L. A. Wolsey. 2002. Cutting planes in integer and mixed-integer programming. Discrete Applied Mathematics **123**, 397–446.

[372] Margot, F. 1992. Quick updates for p-opt TSP heuristics. Operations Research Letters **11**, 45–46.

[373] Margot, F. 2003. Exploiting orbits in symmetric ILP. Mathematical Programming **98**, 3–21.

[374] Markowitz, H. M. 1957. The elimination form of the inverse and its application to linear programming. Management Science **3**, 255–269.

[375] Marks, E. S. 1948. A lower bound for the expected travel among m random points. Annals of Mathematical Statistics **19**, 419–422.

[376] Marsten, R. E. 1981. The design of the XMP linear programming library. ACM Transactions on Mathematical Software **7**, 481–497.

[377] Martin, A. 2001. General mixed integer programming: Computational issues for branch-and-cut algorithms. M. Jünger, D. Naddef, eds. *Computational Combinatorial Optimization: Optimal or Provably Near-Optimal Solutions*. Lecture Notes in Computer Science **2241**. Springer, Heidelberg, Germany. 1–25.

[378] Martin, G. T. 1966. Solving the traveling salesman problem by integer linear programming. Technical Report, C-E-I-R, New York, USA.

[379] Martin, O., S. W. Otto, E. W. Felten. 1991. Large-step Markov chains for the traveling salesman problem. Complex Systems **5**, 299–326.

[380] Martin, O., S. W. Otto, E. W. Felten. 1992. Large-step Markov chains for the TSP incorporating local search heuristics. Operations Research Letters **11**, 219–224.

[381] Martin, O. C., S. W. Otto. 1996. Combining simulated annealing with local search heuristics. Annals of Operations Research **63**, 57–75.

[382] Massberg, J., J. Vygen. 2005. Approximation algorithms for network design and facility location with service capacities. C. Chekuri, K. Jansen, J.D.P. Rolim, L. Trevisan, eds. *Proceedings of the 8th International Workshop on Approximation Algorithms for Combinatorial Optimization Problems (APPROX 2005)*. Lecture Notes in Computer Science **3624**. Springer, Berlin, Germany. 158–169.

[383] Matise, T. C., M. Perlin, A. Chakravarti. 1994. Automated construction of genetic linkage maps using an expert system (MultiMap): A human genome linkage map. Nature Genetics **6**, 384–390.

[384] Matoušek, J., R. Thomas. 1991. Algorithms finding tree-decompositions of graphs. Journal of Algorithms **12**, 1–22.

[385] Maurras, J. F. 1975. Some results on the convex hull of Hamiltonian cycles of symmetric complete graphs. B. Roy, ed. *Combinatorial Programming: Methods and Applications*. Reidel, Dordrecht, The Netherlands. 179–190.

[386] McCormick Jr., W. T., P. J. Schweitzer, T. W. White. 1972. Problem decomposition and data reorganization by a clustering technique. Operations Research **20**, 993–1009.

[387] Megiddo, N. 1991. On finding primal- and dual-optimal bases. ORSA Journal on Computing **3**, 63–65.

[388] Mehlhorn, K. 1984. *Multi-dimensional Searching and Computational Geometry*. Springer, Berlin, Germany.

[389] Menger, K. 1931. Bericht über ein mathematisches Kolloquium. Monatshefte für Mathematik und Physik **38**, 17–38.

[390] Menotti-Raymond, M., V. A. David, R. Agarwala, A. A. Schäffer, R. Stephens, S. J. O'Brien, W.J. Murphy. 2003. Radiation hybrid mapping of 304 novel microsatellites in the domestic cat genome. Cytogenetic and Genome Research **102**, 272–276.

[391] Menotti-Raymond, M., V. A. David, Z. Q. Chen, K. A. Menotti, S. Sun, A. A. Schäffer, R. Agarwala, J. F. Tomlin, S. J. O'Brien, W. J. Murphy. 2003. Second-generation integrated genetic linkage/radiation hybrid maps of the domestic cat (*Felis catus*). Journal of Heredity **94**, 95–106.

[392] Menzel, E. W. 1973. Chimpanzee spatial memory organization. Science **182**, 943–945.

[393] Miliotis, P. 1976. Integer programming approaches to the travelling salesman problem. Mathematical Programming **10**, 367–378.

[394] Miliotis, P. 1978. Using cutting planes to solve the symmetric travelling salesman problem. Mathematical Programming **15**, 177–188.

[395] Miller, C. E., A. W. Tucker, R.A. Zemlin. 1960. Integer programming formulation of traveling salesman problems. Journal of the Association for Computing Machinery **7**, 326–329.

[396] Miller, D. L., J. F. Pekny. 1991. Exact solution of large asymmetric traveling salesman problems. Science **251**, 754–761.

[397] Miller, D. L., J. F. Pekny. 1995. A staged primal-dual algorithm for perfect *b*-matching with edge capacities. ORSA Journal on Computing **7**, 298–320.

[398] Minkowski, H. 1896. *Geometrie der Zahlen (Erste Lieferung)*. Teubner, Leipzig, Germany.

[399] Mitchell, J. E. 2000. Computational experience with an interior point cutting plane algorithm. SIAM Journal on Optimization **10**, 1212–1227.

[400] Mitchell, J. E., B. Borchers. 2000. Solving linear ordering problems with a combined interior point/simplex cutting plane algorithm. H. Frenk, K. Roos, T. Terlaky, S. Zhang, eds. *High Performance Optimization*. Kluwer Academic Publishers, Dordrecht, The Netherlands. 349–366.

[401] Monet, D. G. 1998. The 526,280,881 objects in the USNO-A2.0 catalog. Bulletin of the American Astronomical Society **30**, 1427.

[402] Moret, B. M. E., S. Wyman, D. A. Bader, T. Warnow, M. Yan. 2001. A new implementation and detailed study of breakpoint analysis. R. B. Altman, A. K. Dunker, L. Hunker, K. Lauderdale, and T. E. Klein, eds. *Proceedings of the 6th Pacific Symposium on Biocomputing, Hawaii.* World Scientific Publication. 583–594.

[403] Morton, G., A. H. Land. 1955. A contribution to the 'travelling-salesman' problem. Journal of the Royal Statistical Society, Series B, **17**, 185–194.

[404] Mudrov, V. I. 1963. A method of solution of the traveling salesman problem by means of integer linear programming (the problem of finding the Hamiltonian paths of shortest length in a complete graph) (in Russian). Zhurnal Vychislennoĭ Fiziki (USSR) **3**, 1137–1139. (Abstract in: International Abstracts in Operations Research **5** (1965), Abstract Number 3330.)

[405] Mühlenbein, H., M. Gorges-Schleuter, O. Krämer. 1988. Evolution algorithms in combinatorial optimization. Parallel Computing **7**, 65–85.

[406] Müller-Merbach, H. 1961. Die Ermittlung des kürzesten Rundreiseweges mittels linearer Programmierung. Ablauf und Planungsforschung **2**, 70–83.

[407] Müller-Merbach, H. 1966. Drei neue Methoden zur Lösung des Travelling Salesman Problems, teil 1. Ablauf und Planungsforschung **7**, 32–46.

[408] Müller-Merbach, H. 1966. Drei neue Methoden zur Lösung des Travelling Salesman Problems, teil 2. Ablauf und Planungsforschung **7**, 78–91.

[409] Müller-Merbach, H. 1970. *Optimale Reihenfolgen.* Springer, Berlin, Germany.

[410] Müller-Merbach, H. 1983. Zweimal travelling salesman. DGOR Bulletin **25**, 12–13.

[411] Murphy, W. J., R. Agarwala, A. A. Schäffer, R. Stephens, C. Smith, Jr., N. J. Crumpler, V. A. David, S. J. O'Brien. 2005. A rhesus macaque radiation hybrid map and comparative analysis with the human genome. Genomics **86**, 383–395.

[412] Murphy, W. J., A. J. Pearks Wilkerson, T. Raudsepp, R. Agarwala, A. A. Schäffer, R. Stanyon, B. P. Chowdhary. 2006. Novel gene acquisition on carnivore Y chromosomes. Public Library of Science Genetics.

[413] Murray, H. J. R. 1913. *A History of Chess.* Oxford University Press, Oxford, UK.

[414] Murray, H. J. R. 1930. *The Early History of the Knight's Tour.* Unpublished manuscript.

[415] Murtagh, B. A. 1981. *Advanced Linear Programming: Computation and Practice.* McGraw-Hill, New York, USA.

[416] Naddef, D. 1992. The binested inequalities for the symmetric traveling salesman problem. Mathematics of Operations Research **17**, 882–900.

[417] Naddef, D. 2002. Polyhedral theory and branch-and-cut algorithms for the symmetric TSP. G. Gutin, A. Punnen, eds. *The Traveling Salesman Problem and Its Variations*. Kluwer, Boston, Massachusetts, USA. 29–116.

[418] Naddef, D., Y. Pochet. 2001. The symmetric travelling salesman polytope revisited. Mathematics of Operations Research **26**, 700–722.

[419] Naddef, D., G. Rinaldi. 1991. The symmetric traveling salesman polytope and its graphical relaxation: Composition of valid inequalities. Mathematical Programming **51**, 359–400.

[420] Naddef, D., G. Rinaldi. 1992. The crown inequalities for the traveling salesman polytope. Mathematics of Operations Research **17**, 308–326.

[421] Naddef, D., G. Rinaldi. 1993. The graphical relaxation: A new framework for the symmetric traveling salesman polytope. Mathematical Programming **58**, 53–88.

[422] Naddef, D., G. Rinaldi. 2006. The symmetric traveling salesman polytope: New facets from the graphical relaxation. Preprint.

[423] Naddef, D., S. Thienel. 2002. Efficient separation routines for the symmetric traveling salesman problem I: General tools and comb separation. Mathematical Programming **92**, 237–255.

[424] Naddef, D., S. Thienel. 2002. Efficient separation routines for the symmetric traveling salesman problem II: Separating multi handle inequalities. Mathematical Programming **92**, 257–283.

[425] Naddef, D., E. Wild. 2003. The domino inequalities: Facets for the symmetric traveling salesman polytope. Mathematical Programming **98**, 223–251.

[426] Narayanan, S., R. Gupta, M. A. Breuer. 1993. Optimal configuring of multiple scan chains. IEE Transactions on Computers **42**, 1121–1131.

[427] National Imagery and Mapping Agency. 2006. `erg.usgs.gov/nimamaps` .

[428] Nemhauser, G. L., L. A. Wolsey. 1988. *Integer and Combinatorial Optimization*. John Wiley & Sons, New York, USA.

[429] Nešetřil, J. 1997. A few remarks on the history of MST-problem. Archivum Mathematicum (Brno) **33**, 15–22.

[430] Neto, D. M. 1999. *Efficient Cluster Compensation for Lin-Kernighan Heuristics*. Ph.D. Thesis. Department of Computer Science, University of Toronto, Ontario, Canada.

[431] Neumaier, A., O. Shcherbina. 2004. Safe bounds in linear and mixed-integer linear programming. Mathematical Programming **99**, 283–296.

[432] Nguyen, D. H. 2004. *Hybrid Genetic Algorithms for Combinatorial Optimization Problems*. Ph.D. Thesis. Department of Systems Engineering, University of Miyazaki, Japan.

[433] Nguyen, D. H., I. Yoshihara, K. Yamamori, M. Yasunaga. 2004. A parallel hybrid genetic algorithm for large-scale TSPs. *Proceedings of the 5th International Conference on Simulated Evolution and Learning (SEAL04)*.

[434] Nicolici, N., B. Al-Hashimi. 2002. Multiple scan chains for power minimization during test application in sequential circuits. IEEE Transactions on Computers **51**, 721–734.

[435] Norman, R. Z. 1955. On the convex polyhedra of the symmetric traveling salesman polytope. Bulletin of the American Mathematical Society **61**, 559.

[436] Oliver, I. M., D. J. Smith, J. R. C. Holland. 1987. A study of permutation crossover operators on the traveling salesman problem. J. J. Grefenstette, ed. *Genetic Algorithms and Their Applications: Proceedings of the Second International Conference on Genetic Algorithms*. Lawrence Erlbaum Associates, Hillsdale, New Jersey, USA. 224–230.

[437] Ong, H. L., H. C. Huang. 1989. Asymptotic expected performance of some TSP heuristics: An empirical evaluation. European Journal of Operational Research **43**, 231–238.

[438] Orchard-Hays, W. 1990. History of the development of LP solvers. Interfaces **20:4**, 61–73.

[439] Orden, A. 1952. Solution of systems of linear inequalities on a digital computer. *Proceedings of the 1952 ACM National Meeting (Pittsburgh)*. ACM Press, New York, USA. 91–95.

[440] Ormerod, T. C., E. P. Chronicle. 1999. Global perceptual processing in problem solving: The case of the traveling salesperson. Perception and Psychophysics **61**, 1227–1238.

[441] Oswald, M., G. Reinelt, D. O. Theis. 2005. Not every GTSP induces an STSP facet. M. Jünger, V. Kaibel, eds. *Integer Programming and Combinatorial Optimization* (IPCO XI). Lecture Notes in Computer Science **3509**. Springer, Heidelberg, Germany. 468–482.

[442] Oswald, M., G. Reinelt, D. O. Theis. 2006. On the graphical relaxation of the symmetric traveling salesman polytope. Mathematical Programming. To appear.

[443] *Oxford English Dictionary*, Second Edition. 1989. Clarendon Press, Oxford.

[444] Padberg, M., G. Rinaldi. 1987. Optimization of a 532-city symmetric traveling salesman problem by branch and cut. Operations Research Letters **6**, 1–7.

[445] Padberg, M., G. Rinaldi. 1990. An efficient algorithm for the minimum capacity cut problem. Mathematical Programming **47**, 19–36.

[446] Padberg, M., G. Rinaldi. 1990. Facet identification for the symmetric traveling salesman polytope. Mathematical Programming **47**, 219–257.

[447] Padberg, M., G. Rinaldi. 1991. A branch-and-cut algorithm for the resolution of large-scale symmetric traveling salesman problems. SIAM Review **33**, 60–100.

[448] Padberg, M. W. 1973. On the facial structure of set packing polyhedra. Mathematical Programming **5**, 199–215.

[449] Padberg, M. W. 1975. A note on zero-one programming. Operations Research **23**, 833–837.

[450] Padberg, M. W., M. Grötschel. 1985. Polyhedral computations. E. L. Lawler, J. K. Lenstra, A. H. G. Rinnooy Kan, D. B. Shmoys, eds. *The Traveling Salesman Problem*. John Wiley & Sons, Chichester, UK. 307–360.

[451] Padberg, M. W., S. Hong. 1980. On the symmetric travelling salesman problem: A computational study. Mathematical Programming Study **12**, 78–107.

[452] Padberg, M. W., M. R. Rao. 1980. The Russian method and integer programming. GBA Working Paper. New York University, New York, USA.

[453] Padberg, M. W., M. R. Rao. 1982. Odd minimum cut-sets and b-matchings. Mathematics of Operations Research **7**, 67–80.

[454] Pardalos, P. M., L. S. Pitsoulis, M. G. C. Resende. 1997. Algorithm 769: Fortran subroutines for approximate solution of sparse quadratic assignment problems using GRASP. ACM Transactions on Mathematical Software **23**, 196–208.

[455] Parker, K. P. 2003. *The Boundary-Scan Handbook*, Third Edition. Springer, Berlin, Germany.

[456] Parker, R. G., R. L. Rardin. 1983. The traveling salesman problem: An update of research. Naval Research Logistics Quarterly **30**, 69–96.

[457] Paun, G., G. Rozenberg, A. Salomaa. 1998. *DNA Computing: New Computing Paradigms*. Springer, Berlin, Germany.

[458] Pekny, J. F., D. L. Miller. 1992. A parallel branch and bound algorithm for solving large asymmetric traveling salesman problems. Mathematical Programming **55**, 17–33.

[459] Percus, A. G., O. C. Martin. 1996. Finite size and dimensional dependence in the Euclidean traveling salesman problem. Physical Review Letters **76**, 1188–1191.

[460] Perttunen, J. 1994. On the significance of the initial solution in travelling salesman heuristics. Journal of the Operational Research Society **45**, 1131–1140.

[461] Pfluger, P. 1968. Diskussion der Modellwahl am Beispiel des Traveling-salesman Problems. M. Bechmann, H. P. Künzi, eds. *Einführung in die Methode Branch and Bound.* Lecture Notes in Operations Research and Mathematical Economics **4**. Springer, Berlin, Germany. 88–106.

[462] Pohle, T., E. Pampalk, G. Widmer. 2005. Generating similarity-based playlists using traveling salesman algorithms. *Proceedings of the 8th International Conference on Digital Audio Effects.*

[463] Polivanova, N. I. 1974. On some functional and structural features of the visual-intuitive components of a problem-solving process. Voprosy Psikhologii **4**, 41–51.

[464] Pudney, J. 1953. *The Thomas Cook Story.* Michael Joseph, London.

[465] Pulleyblank, W. R. 1973. *Faces of Matching Polyhedra.* Ph.D. Thesis. Department of Combinatorics and Optimization, University of Waterloo, Waterloo, Ontario, Canada.

[466] Punnen, A. P. 2002. The traveling salesman problem: Applications, formulations, and variations. G. Gutin, A. Punnen, eds. *The Traveling Salesman Problem and Its Variations.* Kluwer, Boston, Massachusetts, USA.

[467] Queyranne, M., Y. Wang. 1993. Hamiltonian path and symmetric travelling salesman polytopes. Mathematical Programming **58**, 89–110.

[468] Rahimi, K., M. Soma. 2003. Layout driven synthesis of multiple scan chains. IEE Transactions on Computer-Aided Design of Integrated Circuits and Systems **22** 317–326.

[469] Ralphs, T. K., L. Kopman, W. R. Pulleyblank, L. E. Trotter. 2003. On the capacitated vehicle routing problem. Mathematical Programming **94**, 343–359.

[470] Ratliff, H. D., A. S. Rosenthal. 1983. Order-picking in a rectangular warehouse: A solvable case of the traveling salesman problem. Operations Research **31**, 507–521.

[471] Reed, B. 1992. Finding approximate separators and computing tree-width quickly. *Proceedings of the 24th Annual ACM Symposium on the Theory of Computing.* ACM Press, New York, USA. 221–228.

[472] Reinelt, G. 1991. TSPLIB—A traveling salesman problem library. ORSA Journal on Computing **3**, 376–384.

[473] Reinelt, G. 1991. TSPLIB—Version 1.2. Report No. 330, Schwerpunkt-programm der Deutschen Forschungsgemeinschaft, Universität Augsburg, Augsburg, Germany.

[474] Reinelt, G. 1992. Fast heuristics for large geometric traveling salesman problems. ORSA Journal on Computing **4**, 206–217.

[475] Reinelt, G. 1994. *The Traveling Salesman: Computational Solutions for TSP Applications*. Lecture Notes in Computer Science **840**. Springer, Berlin, Germany.

[476] Reinelt, G. 1995. TSPLIB95. Interdisziplinäres Zentrum für Wissenschaftliches Rechnen (IWR), Heidelberg.

[477] Reitan, R. M., D. Wolfson. 1993. *The Halstead-Reitan Neuropsychological Test Battery: Theory and Clinical Interpretation*. Neuropsychology Press, Tucson, Arizona, USA.

[478] Reiter, S., G. Sherman. 1965. Discrete optimizing. SIAM Journal on Applied Mathematics **13**, 864–889.

[479] Riley, W., III. 1959. A new approach to the traveling-salesman problem. Operations Research (Supplement 1) **7**, Abstract B4.

[480] Riley, W., III. 1960. Micro-analysis applied to the travelling salesman problem. Operations Research (Supplement 1) **8**, Abstract D4.

[481] Robacker, J. T. 1955. Some experiments on the traveling-salesman problem. Research Memorandom RM-1521. RAND Corporation, Santa Monica, California, USA.

[482] Robertson, N., P. D. Seymour. 1991. Graph minors. X. Obstructions to tree-decomposition. Journal of Combinatorial Theory, Series B, **52**, 153–190.

[483] Robinson, J. 1949. On the Hamiltonian game (a traveling salesman problem). RAND Research Memorandum RM-303. RAND Corporation, Santa Monica, California, USA.

[484] Rohe, A. 1997. *Parallele Heuristiken für sehr grosse Traveling Salesman Probleme*. Diplomarbeit, Research Institute for Discrete Mathematics, Universität Bonn, Bonn Germany.

[485] van Rooij, I., A. Schactman, H. Kadlec, U. Stege. 2003. Children's performance on the Euclidean traveling salesperson problem. Technical Report. University of Victoria, Victoria, British Columbia, Canada.

[486] van Rooij, I., U. Stege, A. Schactman. 2003. Convex hull and tour crossings in the Euclidean traveling salesperson problem: Implications for human performance studies. Memory and Cognition **31**, 215–220.

[487] Rosenkrantz, J. 1986. Interns develop 'LaserLogic'. Bell Labs News, March 3.

[488] Rossman, M. J., R.J. Twery. 1958, A solution to the travelling salesman problem. Bulletin of the Operations Research Society of America Fourteenth National Meeting (St. Louis, October 23–24). Abstract E3.1.3.

[489] Sault, R. J., P. J. Teuben, M. C. H. Wright. 1995. A retrospective view of MIRIAD. R. A. Shaw, H. E. Payne, J. J. E. Hayes, eds. *Astronomical Data Analysis Software and Systems IV*. ASP Conference Series **77**, 433–436.

[490] Schäfer, M. 1994. *Effiziente Algorithmen für sehr grosse Traveling Salesman Probleme*. Diplomarbeit, Research Institute for Discrete Mathematics, Universität Bonn, Bonn, Germany.

[491] Schäffer, A. A. 2005. Human genetic linkage analysis. S. Aluru, ed. *Handbook of Computational Molecular Biology*. Chapman and Hall/CRC, Boca Raton, Florida, USA. Chapter 17.

[492] Schieber, B., U. Vishkin. 1988. On finding lowest common ancestors: Simplification and parallelization. SIAM Journal on Computing **17**, 1253–1262.

[493] Schilham, R. M. F. 2001. *Commonalities in Local Search*. Ph.D. Thesis. Technische Universiteit Eindhoven, Eindhoven, The Netherlands.

[494] Schrijver, A. 1986. *Theory of Linear and Integer Programming*. John Wiley & Sons, Chichester, UK.

[495] Schrijver, A. 2003. *Combinatorial Optimization*. Springer, Berlin, Germany.

[496] Secord, A. 2002. Weighted Voronoi stippling. Proceedings of the 2nd International Symposium on Non-Photorealistic Animation and Rendering. ACM Press, New York, USA. 37–43.

[497] Seymour, P. D., R. Thomas. 1994. Call routing and the ratcatcher. Combinatorica **14**, 217–241.

[498] Shapiro, D. M. 1966. *Algorithms for the Solution of the Optimal Cost and Bottle-neck Traveling Salesman Problems*. Sc.D. Thesis. Washington University, St. Louis, Missouri, USA.

[499] Shor, N. Z. 1970. Utilization of the operation of space dilation in the minimization of convex functions. Cybernetics **6**, 7–15.

[500] Shortridge, K., M. Zoccali, F. Primas, A. Kaufer. 2004. Very Large Telescope: EPOSS user manual. Document Number INS-MAN-AUS-13271-0079, European Southern Observatory.

[501] Skidelsky, S. S. 1916. *The Tales of a Traveller: Reminiscences from Twenty-Eight Years on the Road*. A. T. De La Mare, New York, USA.

[502] Sleator, D. D., R. E. Tarjan. 1983. Self-adjusting binary trees. *Proceedings of the Fifteenth Annual SCM Symposium on Theory of Computing.* ACM Press, New York, USA. 235–245.

[503] SMAPO. 2006. Symmetric and graphical traveling salesman polyhedra. Available at www.iwr.uni-heidelberg.de/groups/comopt /software/SMAPO/tsp/tsp.html .

[504] Smith, T. H. C., T.W.S. Meyer, G. L. Thompson. 1990. Lower bounds for the symmetric travelling salesman problem from Lagrangean relaxations. Discrete Applied Mathematics **26**, 209–217.

[505] Smith, T. H. C., G. L. Thompson. 1977. A LIFO implicit enumeration search algorithm for the symmetric traveling salesman problem using Held and Karp's 1-tree relaxation. P. L. Hammer, E. L. Johnson, B. H. Korte, G. L. Nemhauser, eds. *Studies in Integer Programming.* Annals of Discrete Mathematics **1**. North-Holland, Amsterdam, The Netherlands. 479–493.

[506] Spears, T. B. 1994. *100 Years on the Road: The Traveling Salesman in American Culture.* Yale University Press, New Haven, Connecticut, USA.

[507] Stancu-Minasian, I. M. 1997. *Fractional Programming: Theory Methods and Applications.* Kluwer, Dordrecht, The Netherlands.

[508] Steckhan, H., R. Thome. 1972. Vereinfachungen der Eastmanischeen Branch-Bound-Lösung für symmetrische Traveling Salesman Probleme. Methods of Operations Research **14**, 360–389.

[509] Steele, J. M. 1990. Probabilistic and worst case analyses of classical problems of combinatorial optimization in Euclidean space. Mathematics of Operations Research **15**, 749–770.

[510] Steele, J. M. 1997. *Probability Theory and Combinatorial Optimization.* SIAM, Philadelphia, Pennsylvania, USA.

[511] Stein, D. M. 1977. *Scheduling Dial-a-Ride Transportation Systems: An Asymptotic Approach.* Ph.D. Thesis. Harvard University, Cambridge, Massachusetts, USA.

[512] Stewart, E. A., K. B. McKusick, A. Aggarwal, E. Bajorek, S. Brady, A. Chu, N. Fang, D. Hadley, M. Harris, S. Hussain, R. Lee, A. Maratukulam, K. O'Connor, S. Perkins, M. Piercy, F. Qin, T. Reif, C. Sanders, X. She, W.-L. Sun, P. Tabar, S. Voyticky, S. Cowles, J.-B. Fan, C. Mader, J. Quackenbush, R. M. Myers, D. R. Cox. 1997. An STS-based radiation hybrid map of the human genome. Genome Research **7**, 422–433.

[513] Suh, J. Y., D. Van Gucht. 1987. *Proceedings of the Second International Conference on Genetic Algorithms on Genetic Algorithms and their Application.* Lawrence Erlbaum Associates, Mahwah, New Jersey, USA. 100–107

[514] Suhl, U. H., L. M. Suhl. 1990. Computing sparse LU factorizations for large-scale linear programming bases. ORSA Journal on Computing **2**, 325–335.

[515] Suri, M. 2001. *The Death of Vishnu*. W. W. Norton and Company, New York, USA.

[516] Suri, M. 2001. A fundamental decomposition theorem for fiction? SIAM News **34**, p. 1.

[517] Suurballe, W. 1964. Network algorithm for the traveling salesman problem. Operations Research (Supplement 1) **12**, Abstract 3Tc.1.

[518] Tamaki, H. 2003. Alternating cycle contribution: A tour-merging strategy for the travelling salesman problem. Max-Planck Institute Research Report, MPI-I-2003-1-007. Saarbrücken, Germany.

[519] Tarjan, R. E. 1975. Efficiency of a good but not linear set union algorithm. Journal of the Association of Computing Machinery **22**, 215–225.

[520] Tarjan, R. E. 1983. *Data Structures and Network Algorithms*. SIAM, Philadelphia, Pennsylvania, USA.

[521] Tarjan, R. E., J. van Leeuwen. 1984. Worst-case analysis of set union algorithms. Journal of the Association of Computing Machinery. **31**, 245–281.

[522] Thüring, B. 1961. Zum Problem der exakten Ermittlung des kürzesten Rundreiseweges. Elektronische Datenverarbeitung **3**, 147–156.

[523] Tomlin, J. A. 1989. The influences of algorithmic and hardware developments on computational mathematical programming. M. Iri, K. Tanabe, eds. *Mathematical Programming: Recent Developments and Applications*. Kluwer, Boston, Massachusetts, USA. 159–175.

[524] Trick, M. A. 1987. *Networks with Additional Structured Constraints*. Ph.D. Thesis. Georgia Institute of Technology, Atlanta, Georgia, USA. 1986.

[525] Troitski, I. 1999. Laser-induced damage creates interior images. Optical Engineering Reports, November.

[526] Tucker, A. 1972. A structure theorem for the consecutive 1's property. Journal of Combinatorial Theory Series B **12**, 153–162.

[527] Tucker, A. 1983. Letter from Albert Tucker to David Shmoys, February 18, 1983.

[528] Twain, M. 1869. The Innocents Abroad. American Publishing Company, Hartford, Connecticut, USA.

[529] Ulder, N. L. J., E. H. L. Aarts, H.-J. Bandelt, P. J. M. Laarhoven, E. Pesch. 1991. Genetic local search algorithms for the traveling salesman problem. H.-P. Schwefel, R. Männer, eds. *Parallel Problem Solving from Nature*. Elsevier Science Publishers, Amsterdam, The Netherlands. 109–116.

[530] Vandenbussche, D., G. L. Nemhauser. 2005. A branch-and-cut algorithm for nonconvex quadratic programs with box constraints. Mathematical Programming **102**, 559–575.

[531] Vanderbei, R. J. 1992. LOQO User's Manual. Technical Report 92-5, Princeton University.

[532] Vanderbei, R. J. 2001. *Linear Programming: Foundations and Extensions.* Kluwer, Boston, Massachusetts, USA.

[533] Vella, D. 2001. *Using DP-Constraints to Obtain Improved TSP Solutions.* M. S. Thesis. University of Ottawa, Ottawa, Ontario, Canada.

[534] Venter, J. C., M. D. Adams, E. W. Myers, P. W. Li, R. J. Mural, G. G. Sutton, H. O. Smith, M. Yandell, C. A. Evans, R. A. Holt, J. D. Gocayne, P. Amanatides, R. M. Ballew, D. H. Huson, J. R. Wortman, Q. Zhang, C. D. Kodira, X. H. Zheng, L. Chen, M. Skupski, G. Subramanian, P. D. Thomas, J. Zhang, G. L. G. Miklos, C. Nelson, S. Broder, A. G. Clark, J. Nadeau, V. A. McKusick, N. Zinder, A. J. Levine, R. J. Roberts, M. Simon, C. Slayman, M. Hunkapiller, et al. 2001. The sequence of the human genome. Science **291**, 1304–1351.

[535] Verblunsky, S. 1951. On the shortest path through a number of points. Proceedings of the American Mathematical Society **2**, 904-913.

[536] Verhoeven, M. G. A., P. C. J. Swinkels, E. H. L. Aarts. 1995. Parallel local search for the traveling salesman. Working Paper, Philips Research Laboratories, Eindhoven.

[537] Verhoeven, M. G. A., E. H. L. Aarts, E. van de Sluis, R. J. M. Vassens. 1992. Parallel local search and the travelling salesman problem. R. Männer, B. Manderick, eds. *Parallel Problem Solving from Nature 2.* North-Holland, Amsterdam, The Netherlands. 543–552.

[538] Vermeulen, B., S. K. Goel. 2002. Design for debug: Catching design errors in digital chips. IEEE Design and Test of Computers **19**, 37–45.

[539] Vickers, D., M. Butavicius, M. Lee, A. Medvedev. 2001. Human performance on visually presented traveling salesman problems. Pyschological Research **65**, 34–45.

[540] Vickers, D., M. D. Lee. 1998. Never cross the path of a traveling salesman: The neural network generation of Halstead-Reitan Trail Making Tests. Behavior Research Methods, Instruments, and Computers **30**, 423–431.

[541] Vickers, D., M. D. Lee, M. Dry, P. Hughes. 2003. The roles of the convex hull and the number of potential intersections in performance on visually presented traveling salesperson problems. Memory and Cognition **31**, 1094–1104.

[542] Vickers, D., M. D. Lee, P. Hughes, M. Dry, J. A. McMahon. 2006. The aesthetic appeal of minimal structures: Judging the attractiveness of solutions to traveling salesperson problems. Perception and Psychophysics **68**, 32–42

[543] Vickers, D., T. Mayo, M. Heitmann, M. D. Lee, P. Hughes. 2004. Intelligence and individual differences in performance on three types of visually presented optimization problems. Personality and Individual Differences **36**, 1059–1071.

[544] Volgenant, T., R. Jonker. 1982. A branch and bound algorithm for the symmetric traveling salesman problem based on the 1-tree relaxation. European Journal of Operational Research **9**, 83–89

[545] Volgenant, T., R. Jonker. 1983. The symmetric traveling salesman problem and edge exchanges in minimal 1-trees. European Journal of Operational Research **12**, 394–403.

[546] Volgenant, T., R. Jonker. 1990. Fictitious upper bounds in an algorithm for the symmetric traveling salesman problem. Computers and Operations Research **17**, 113–117.

[547] Wagner, M. L., T. Raudsepp, G. Goh, R. Agarwala, A. A. Schäffer, P. K. Dranchak, C. Brinkmeyer-Langford, L. C. Skow, B. P. Chowdhary, J. R. Mickelson. 2006. A 1.3-Mb interval map of equine homologs of HSA2. Cytogenetic and Genome Research **112**, 227–234.

[548] Walshaw, C. 2002. A multilevel approach to the travelling salesman problem. Operations Research **50**, 862–877.

[549] Warme, D. M., P. Winter, M. Zachariasen. 2000. Exact algorithms for plane Steiner tree problems: A computational study. Combinatorial Optimization **6**, 81–116.

[550] Waud, A. R. 1867. Harper's Weekly **11**, p. 641.

[551] Wenger, K. 2004. *Generic Cut Generation Methods for Routing Problems*. Shaker Verlag, Aachen, Germany.

[552] Wenger, K. M. 2002. A new approach to cactus construction applied to TSP support graphs. W. J. Cook, A. S. Schulz, eds. *Integer Programming and Combinatorial Optimization*. Lecture Notes in Computer Science **2337**. Springer, Heidelberg, Germany. 109–126.

[553] West, D. B. 1996. *Introduction to Graph Theory*. Prentice Hall, Englewood Cliffs, New Jersey, USA.

[554] Wilkens, D. R. 2005. Sir William Rowan Hamilton (1805–1865). www.maths.tcd.ie/pub/HistMath/People/Hamilton/.

[555] Everts-van der Wind, A., D. M. Larkin, C. A. Green, J. S. Elliott, C. A. Olm-stead, R. Chiu, J. E. Schein, M. A. Marra, J. E. Womack, H. A. Lewin. 2005. A high-resolution whole-genome cattle-human comparative map reveals details of mammalian chromosome evolution. Proceedings of the National Academy of Sciences **102**, 18526–18531.

[556] Woeginger, G. J. 2003. Exact algorithms for NP-hard problems: A survey. M. Jünger, G. Reinelt, G. Rinadli, eds. *Combinatorial Optimization—Eureka, You Shrink!* Lecture Notes in Computer Science **2570**. Springer, Heidelberg, Germany. 185–207.

[557] Wolsey, L. A. 1975. Faces for a linear inequality in 0-1 variables. Mathematical Programming **8**, 165–178.

[558] Wolsey, L. A. 1976. Facets and strong valid inequalities for integer programs. Operations Research **24**, 367–372.

[559] Wolsey, L. A. 1998. *Integer Programming*. John Wiley & Sons, New York, USA.

[560] Wunderling, R. 1996. *Paralleler und Objektorientierter Simplex-Algorithmus*. Ph.D. Thesis. TR-96-09. Konrad-Zuse-Zentrum für Informationstechnik Berlin, Berlin, Germany.

[561] Yudin, D. B., A. S. Nemirovskiĭ. 1976. Informational complexity and efficient methods for the solution of convex extremal problems (in Russian). Èkonomika i Matematicheskie Metody **12**, 357–369.

Index